# Physical Adsorption:
# Experiment, Theory and Applications

# NATO ASI Series

## Advanced Science Institutes Series

*A Series presenting the results of activities sponsored by the NATO Science Committee, which aims at the dissemination of advanced scientific and technological knowledge, with a view to strengthening links between scientific communities.*

The Series is published by an international board of publishers in conjunction with the NATO Scientific Affairs Division

| | |
|---|---|
| **A  Life Sciences** | Plenum Publishing Corporation |
| **B  Physics** | London and New York |
| | |
| **C  Mathematical and Physical Sciences** | Kluwer Academic Publishers |
| **D  Behavioural and Social Sciences** | Dordrecht, Boston and London |
| **E  Applied Sciences** | |
| | |
| **F  Computer and Systems Sciences** | Springer-Verlag |
| **G  Ecological Sciences** | Berlin, Heidelberg, New York, London, |
| **H  Cell Biology** | Paris and Tokyo |
| **I   Global Environmental Change** | |

### PARTNERSHIP SUB-SERIES

| | |
|---|---|
| 1. **Disarmament Technologies** | Kluwer Academic Publishers |
| 2. **Environment** | Springer-Verlag / Kluwer Academic Publishers |
| 3. **High Technology** | Kluwer Academic Publishers |
| 4. **Science and Technology Policy** | Kluwer Academic Publishers |
| 5. **Computer Networking** | Kluwer Academic Publishers |

*The Partnership Sub-Series incorporates activities undertaken in collaboration with NATO's Cooperation Partners, the countries of the CIS and Central and Eastern Europe, in Priority Areas of concern to those countries.*

### NATO-PCO-DATA BASE

The electronic index to the NATO ASI Series provides full bibliographical references (with keywords and/or abstracts) to more than 50000 contributions from international scientists published in all sections of the NATO ASI Series.
Access to the NATO-PCO-DATA BASE is possible in two ways:

– via online FILE 128 (NATO-PCO-DATA BASE) hosted by ESRIN,
Via Galileo Galilei, I-00044 Frascati, Italy.

– via CD-ROM "NATO-PCO-DATA BASE" with user-friendly retrieval software in English, French and German (© WTV GmbH and DATAWARE Technologies Inc. 1989).

The CD-ROM can be ordered through any member of the Board of Publishers or through NATO-PCO, Overijse, Belgium.

Series C: Mathematical and Physical Sciences – Vol. 491

# Physical Adsorption:
# Experiment, Theory
# and Applications

edited by

## Jacques Fraissard

Université Pierre et Marie Curie,
Laboratoire de Chimie des Surfaces,
Paris, France

SPRINGER SCIENCE+BUSINESS MEDIA, B.V.

Proceedings of the NATO Advanced Study Institute on
Physical Adsorption: Experiment, Theory and Applications
La Colle sur Loup, France
May 19–June 1, 1996


A C.I.P. Catalogue record for this book is available from the Library of Congress


ISBN 978-94-010-6392-0      ISBN 978-94-011-5672-1 (eBook)
DOI 10.1007/978-94-011-5672-1

*Printed on acid-free paper*

# TABLE OF CONTENTS

*Poster Communications*

# PREFACE

The subject of Physical Adsorption has enormous economic and technological value while it continues to present significant scientific challenges with prospects for further important technological developments. The literature on the subject is truly enormous. Particularly during the last few years there have been three developments that led us to organize the Advanced Study Institute on which this volume is based:

- Significant development in the theory of physical adsorption;

- Developments in instrumentation that allow the detailed characterization of materials including microporous solids;

- The realization that closer coupling of scientific and technological pursuits can lead to greater scientific understanding and better technology.

The structure of the ASI reflected the coming-together of these three factors. Following an incisive historical review of the subject by K.S.W. Sing, the ASI and the book focus essentially equally on theory, assessment and applications. Topics covered include:

- The dramatic progress in theoretical analysis (statistical thermodynamic and quantum mechanics), fuelled by access to exponential increases in computational capabilities, and reflected in the lectures of K. Gubbins, G. Horvath, D. Nicholson and W. Rudzinsky.

- The breakthroughs in experimentation, among others: high resolution adsorption, NMR of solids and of adsorbed inert gases such as $^{129}$Xe or $^{15}$N, which are described in the lectures by W.C. Conner, I. Dekany, G. Findcnegg, J. Rouquerol and M.A. Springuel-Huet.

- The unique aspects of transport in micropores, analysed by J. Karger, D. Ruthven and K.K. Unger.

- Finally, the application of physical adsorption to industrial processes reviewed by N. Kanellopoulos, C. Mellot and C. Monereau.

This book is addressed to a wide readership. Specialized workers in the field should find the updated material on several areas of this topic very useful. University teachers could use the material in the book for introductory or graduate courses, and those who have a general interest in the subject should find the overviews offered particularly interesting. There are extensive literature references for further detailed studies.

Many people have contributed to the success of the ASI on which this volume is based. We thank of course all the participants for contributing to the intellectual dialogue. It is also a great pleasure to acknowledge the financial support provided by the Scientific Affairs Division of the North Atlantic Treaty Organization; also the National Science Foundation for the fellowships to American Students and post-doctorate fellows; and the International Science Foundation for the support it provided to Russian scientists. Many individuals whose great help with the organization of the ASI we acknowledge with gratitude include Drs. M.A. Springuel-Huet, J.L. Bonardet, A. Gédéon, Mrs C. Bonmkratz and Mrs F. Sarrazin.

Jacques Fraissard
Wm Curtis Conner

# LECTURES

# HISTORICAL PERSPECTIVES OF PHYSICAL ADSORPTION

K.S.W.SING
*School of Chemistry, University of Bristol,*
*BS8 1TS, U.K.*

## 1. Early History

Solid adsorbents such as charcoal and clays were used in ancient times, but the earliest quantitative measurements of the uptake of gases by such materials appear to have been made by Scheele in 1773. Other studies around this time were undertaken independently by Priestley and the Abbe Fontana, but it was not until 1881 that the first attempts were made by Chappuis and Kayser to relate the amount of gas taken up by charcoal to the gas pressure. In that year, Kayser introduced the term *adsorption* and within the next few years the terms *isotherm* and *isothermal curve* were applied to the results of adsorption measurements at constant temperature.

In 1909 McBain reported [1] that the uptake of hydrogen by carbon appeared to occur in two stages: a rapid process of adsorption was followed by a slow process of absorption into the interior of the solid. McBain proposed the use of the term *sorption* to cover both phenomena. Even now, with some systems it is difficult to distinguish clearly between adsorption and absorption and in that case it is convenient to employ the term sorption - together with *sorbent, sorbate* and *sorptive* [2].

Many adsorption isotherms were determined during the early years of this century by Coolidge, Homfray, Lamb, Polanyi and others, the most popular adsorbents being activated charcoal and silica gel. In 1932, Lennard-Jones commented [3] "The literature pertaining to the sorption of gases is now so vast that it is impossible for any, except those who are specialists in the experimental technique, rightly to appraise the work---". It was already apparent that different physical and chemical interactions may bring about adsorption, depending on the nature of the gas-solid system and the conditions of temperature and pressure.

It is sometimes forgotten that London's first work on dispersion forces was published as long ago as 1930 and that in 1937 the Lennard-Jones attraction-repulsion (6-12) potential was proposed. It is remarkable that this form of the pair potential is used to-day in the application of density functional theory and other computational procedures.

The early work on physical adsorption (physisorption) has been well reviewed by Brunauer [4], Deitz [5] and Forrester and Giles [6].

*J. Fraissard (ed.), Physical Adsorption: Experiment, Theory and Applications, 3–8.*
© *1997 Kluwer Academic Publishers.*

## 2. The Monolayer Concept

The publication of Langmuir's monumental papers over the period 1915 - 1918 led to a radical change in the approach to surface science. Earlier theories, such as that of Polanyi (1914) had pictured the adsorbed layer as a thick film of decreasing density at increasing distance from the solid surface. Langmuir's great contribution was to bring together all the available experimental and theoretical evidence in support of the unifying concept of the monomolecular layer (the *monolayer*).

Langmuir argued [7] that " If any molecules impinging on the surface are condensed, a certain time interval must elapse before they can evaporate. This time lag will bring about the accumulation of molecules in the surface layer, and may thus be looked upon as the cause of adsorption". He considered, however, that "The forces acting between two layers of (adsorbed) molecules will usually be very much less than those between the crystal surface and the first layer of molecules. The rate of evaporation in the second layer will, therefore, generally be so much more rapid than in the first, that the number of molecules in the second layer will be negligible".

By equating the rates of condensation and evaporation, Langmuir obtained [7] his isotherm equation in the form

$$\theta_1 = \alpha\mu/(v_1 + \alpha\mu) \tag{1}$$

where $\theta_1$ is the fraction of surface covered by the adsorbed molecules, $\mu$ is the number of molecules striking 1 $cm^2$ of surface per second, $\alpha$ is a condensation coefficient ($\alpha<1$) and $v_1$ is the rate of evaporation from a completely covered surface. According to the collision theory, $\mu$ is proportional to the gas pressure, $p$, and therefore we may transform Equation (1) into the well-known form of the Langmuir equation [8]

$$n \ n_s = bp \ (1 + bp) \tag{2}$$

where $n_s$ and $b$ are empirical constants. Of course, if the Langmuir model is obeyed, $n_s$ must correspond to the limiting coverage of $\theta = 1$ (i.e. the monolayer capacity).

In his early paper [7], Langmuir expressed the view that as the pressure approached the saturation vapour pressure, "there will be a tendency for the adsorbed film to become several molecules deep" and went on to state "According to this theory it is very improbable that films more than one or two molecules deep would ever be held on a surface by adsorption, except with nearly saturated vapors".

However, Langmuir did recognise that with a porous adsorbent such as charcoal "it is impossible to know definitely the area on which the adsorption takes place" and a little later in the same paper we find the very clear statement that the equations "which apply to adsorption by plane surfaces, could not apply to adsorption by charcoal". In the light of these comments, it may seem surprising that Equation (2) has continued to be one of the most popular equations for the analysis of isotherms on activated carbons!

In his 1918 paper [8], Langmuir discussed the extension of his theory to more complex systems by taking account of heterogeneous surfaces and multilayer adsorption. Many other attempts were made to refine the Langmuir model to allow for the mobility of adsorbed molecules and lateral interactions. The Langmuir equation itself may be regarded as a useful empirical equation, whereas the model on which it is based is essentially for an ideal localized monolayer.

## 3. Multilayer Adsorption

Another important stage in the history of physisorption was the work of Brunauer and Emmett [9], which preceded the appearance of the Brunauer-Emmett-Teller (BET) theory in 1938 [10]. In 1935 Brunauer and Emmett [11] set out to determine the surface area of an iron synthetic ammonia catalyst. This was probably the first systematic attempt to use gas adsorption for surface area determination and involved the measurement of the isotherms of six different gases ($N_2$, Ar, $O_2$, CO, $CO_2$ and $C_4H_{10}$) at, or near, their respective boiling points.

The adsorption isotherms were all found to be S-shaped with three distinctive regions: the low pressure region was concave to the pressure axis, the high pressure region was convex and the middle region was approximately linear. At first, it was believed that extrapolation of the middle linear portion to zero pressure gave the amount of gas required to cover the adsorbent surface with a monomolecular layer. Later [9], convincing evidence was obtained in favour of the beginning of the middle linear section (i.e. Point B). The surface area, $A$, was then evaluated from the amount adsorbed at Point B, the assumption being made that the completed monolayer was in a close-packed state.

The BET theory appeared to provide a theoretical basis for the selection of Point B as representing the stage of monolayer completion and the beginning of multilayer development. By making a number of simpifying assumptions, Brunauer, Emmett and Teller [10] were able to extend the Langmuir monolayer model to multilayer adsorption. Thus, adsorption of the first layer was assumed to take place on an array of surface sites of uniform energy. Molecules in the first layer would then act as sites for second and higher layer adsorption and in the simplest case this would lead to the adsorption of an infinite number of layers at $p^o$, the saturation pressure.

Under equilibrium conditions, the rates of condensation and evaporation were equated for each adsorbed layer and it was assumed that the rate constants were the same for all layers after the first and that the adsorption energy for the higher layers was equal to the energy of condensation.

Summation of the amounts adsorbed in all layers then gave the BET equation

$$n \, n_{\mathrm{m}} = C(p \, p^o)/[(1 - p \, p^o)(1 + |C - 1|p \, p^o)] \qquad (3)$$

where $n_{\mathrm{m}}$ is the monolayer capacity and $C$ is a constant. According to the BET theory, $C$ is related exponentially to the energy of adsorption for the first layer.

Equation (3) requires a linear relation between $p \, n(p^o-p)$ and $p \, p^o$, which is known as the BET plot. In practice, the range of linearity of the BET plot is always restricted to a fairly small part of the isotherm and with some systems the linear range does not extend above $p \, p^o$ ~0.1. Brunauer, Emmett and Teller [10] were able to obtain some improvement in the range of fit by restricting the amount adsorbed at the saturation pressure to a finite number of layers.

A few years later, Brunauer, Deming, Deming and Teller put forward [12] a much more elaborate equation (the BDDT equation) in an attempt to allow for an additional contribution to the energy of adsorption as a result of capillary condensation. The BDDT equation contains four adjustable constants, which make it possible to fit some isotherms over a wide range of $p \, p^o$ However, it is not possible to evaluate independently the BDDT parameters and very few attempts have been made to apply the equation.

An important contribution by Brunauer and his co-workers [12] was the identification of five different types of physisorption isotherms, now generally known as the BDDT classification. Type I was referred to as the Langmuir isotherm since it appeared to represent monomolecular adsorption in accordance with Equation (2). Type II was the S-shaped, or sigmoid, isotherm, which had been characterized by Brunauer and Emmett [11] as the normal form of multilayer isotherm. The remaining types (III, IV and V) represented more complex systems in which weaker adsorbent-adsorbate interactions were involved (i.e. III and V) or pore filling by capillary condensation (i.e. IV and V). The BDDT classification has served as the basis for a more complete classification proposed by the IUPAC [2].

The BET theory attracted an enormous amount of interest in the late 1940's. The artificiality of the model was pointed out by Hill, Halsey, Gregg and others and various alternative monolayer-multilayer isotherm equations were proposed (e.g. by Anderson, Huttig and Harkins and Jura). These developments are reviewed in some detail by Young and Crowell [13].

Although it is now generally agreed that the BET theory was based on an over-simplified model of physisorption, the BET-nitrogen method continues to be used as a standard prcedure for the determination of the surface area of fine powders and porous materials. There are probably two main reasons for its continued popularity: first, under favourable conditions, the BET plot does appear to provide a fairly reliable estimate of the monolayer capacity - especially for nitrogen adsorption at 77K; secondly, the method is not difficult to apply or to comprehend!

## 4. Capillary Condensation

In 1911, Zsigmondy pointed out [14] that the condensation of a vapour can occur in narrow pores at pressures well below the normal saturation vapour pressure. At first, there appeared to be a simple exponential relationship between the pore radius, $r_k$, and the relative pressure, $p\,p^o$, at which condensation takes place. Thus, for a cylindrical pore

$$p\,p^o = \exp[-2\sigma v^l/RTr_k] \qquad (4)$$

where $\sigma$ is the surface tension and $v^l$ the molar volume of the liquid adsorptive. Equation (4), which is a modified form of a relation originally proposed by Lord Kelvin (W. Thomson) in 1871, is now universally known as the Kelvin equation.The range of reliability of the Kelvin equation has been under discussion for over 75 years. It is still widely used for pore size analysis, but its limitations remain unresolved.

It is now evident that capillary condensation plays an important secondary role in the physisorption of vapours by mesoporous solids (with pore widths in the range 2-50nm ). Capillary condensation is always preceded by adsorption on the pore walls and therefore the true pore width can be evaluated only if the adsorbed layer thickness is known. Following the work of Foster [15], various attempts have been made to derive the mesopore size distribution from the capillary condensation region of physisorption isotherms. However, none of these procedures is likely to be entirely reliable unless certain conditions are satisfied [2].

## 5. Micropore Filling

As we have already seen, Langmuir was already aware in 1916 of the complexity of the adsorptive properties of porous charcoal. In 1932, McBain [16] drew attention to the fact that many adsorption studies had been undertaken on solids "whose porosity is so extreme as to approach molecular or atomic dimensions". Certain materials of this type were already known as *molecular sieves*, but the full significance of their adsorptive behaviour became evident only as a result of Barrer's systematic work on zeolites [17].

Extensive studies of physisorption by activated carbons, synthetic zeolites and other microporous adsorbents (having pore widths<2nm) were carried out by Dubinin [18] and Kiselev [19] and their co-workers.

The period 1950 - 70 was a time of fierce discussion on the interpretation of Type I isotherms. Brunauer and de Boer and their co-workers maintained that monolayer adsorption could occur on the micropore walls before the onset of pore filling, whereas Dubinin and others [20] believed that the monolayer concept could not be applied to the Type I isotherm - as Langmuir had originally warned.

The Russian work and the more recent experimental measurements on well characterized microporous adsorbents have confirmed the theoretical predictions [21] that the mechanism of physisorption in pores of molecular dimensions is not the same as on the open surface or in mesopores. However, it now seems likely that monolayer adsorption can occur on the walls of the wider micropores at low $p$ $p^o$ before the onset of pore filling by a cooperative process. The mechanisms of micropore filling are thus dependent on the ratio of pore width/molecular diameter as well as the nature of the adsorption system. Further progress in this topical area of research will depend on the development of improved model adsorbents and computational procedures.

## 6. References

1. McBain, J.W. (1909) *Phil. Mag.* **18**, 916.
2. Sing, K.S.W., Everett, D.H., Haul, R.A.W., Moscou, L., Pierotti, R.A., Rouquerol, J.and Siemieniewska, T. (1985) *Pure Appl. Chem.* **57**, 603-619.
3. Lennard-Jones, J.E. (1932) *The Adsorption of Gases by Solids*, Farad. Soc. 333-359.
4. Brunauer, S. (1945) *The Adsorption of Gases and Vapours*, Oxford University Press.
5. Deitz, V.R. (1944) *Bibliography of Solid Adsorbents*, National Bureau of Standards Washington.
6. Forrester, S.D. and Giles, C.H. (1971) *Chem. & Ind.* 831-839.
7. Langmuir, I. (1916) *J. Amer. Chem. Soc.* **38**, 2221-2295.
8. Langmuir, I. (1918) *J. Amer. Chem. Soc.* **40**, 1361-1403.
9. Brunauer, S. and Emmett, P.H. (1937) *J. Amer. Chem. Soc.* **59**, 2682.
10. Brunauer, S., Emmett, P.H. and Teller, E.(1938) *J. Amer. Chem. Soc.* **60**, 309-319.
11. Brunauer, S. and Emmett, P.H. (1935) *J. Amer. Chem. Soc.* **57**, 1754.
12. Brunauer, S., Deming, L.S., Deming, W.E. and Teller,E. (1940) *J. Amer. Chem. Soc.* **57**, 1754.
13. Young, D.M. and Crowell, A.D. (1962) *Physical Adsorption of Gases*, Butterworths, London.
14. Zsigmondy, R. (1911) *Z. anorg. Chem.* **71**, 356.
15. Foster, A.G. (1932) *Trans. Farad. Soc.* **28**, 645.
16. McBain, J.W. (1932) *The Adsorption of Gases by Solids*, Farad. Soc. 408-409.
17. Barrer, R.M. (1978) *Zeolites and Clay Minerals as Sorbents and Molecular Sieves*, Academic Press, London.
18. Dubinin, M.M. (1967) *J. Colloid Interface Sci.* **23**, 487-499.
19. Kiselev, A.V. (1965) *Discuss. Farad. Soc.* **40**, 205-218.
20. Gregg, S.J. and Sing, K.S.W. (1982) *Adsorption, Surface Area and Porosity*, Academic Press, London.
21. Everett, D.H. and Powl, J.C. (1976) *J. Chem. Soc., Farad. Trans.* **I**, **72**, 619-636.

# ANALYSIS OF PHYSISORPTION ISOTHERMS
*Determination of Surface Area and Porosity*

K.S.W.SING
*School of Chemistry, University of Bristol*
*BS8 1TS, UK*

## 1. Introduction

An adsorption isotherm is the quantitative relationship at constant temperature between the amount of gas adsorbed, $n$, by unit mass of solid (the adsorbent) and the equilibrium pressure, $p$, or relative pressure, $p\,p^o$, of the gas (the adsorptive). The material in the adsorbed state is known as the adsorbate. Physical adsorption (physisorption) isotherms are usually displayed in graphical form, as in Figure 1.

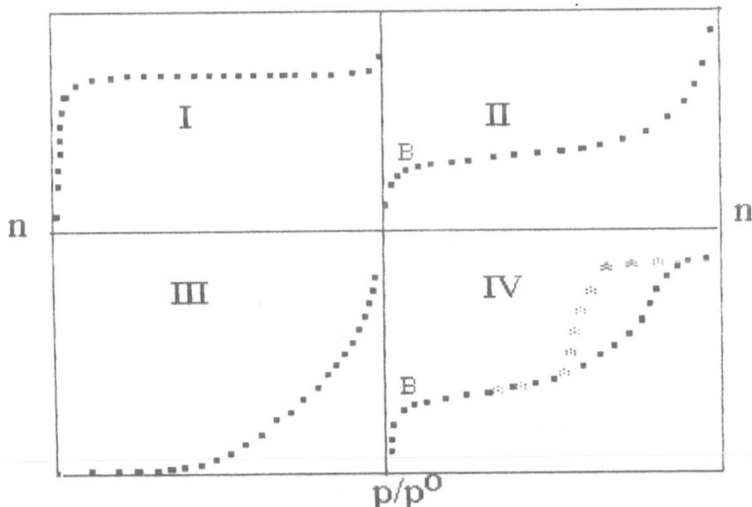

*Figure 1.* Types of physisorption isotherms

Non-porous adsorbents give Types II or III isotherms, the individual shape being mainly dependent on the nature of the adsorbent-adsorbate interactions and the operational

9

*J. Fraissard (ed.), Physical Adsorption: Experiment, Theory and Applications, 9–16.*
© *1997 Kluwer Academic Publishers.*

temperature. Types I and IV are the chararacteristic shapes exhibited by isotherms obtained with microporous and mesoporous adsorbents, respectively. The former have pore widths < 2 nm, whereas the latter have pore widths in the range 2 - 50 nm.

Two other isotherm types, which are less common than those depicted in Figure 1, are the stepwise isotherms (Type VI) given by the adsorption of non-polar molecules on highly uniform surfaces and isotherms (Type V) with convex curvature with respect to the $p\,p^o$ axis at low pressure and limiting adsorption (similar to Type IV) at high $p\,p^o$. The first stage in the analysis of an experimental isotherm is the identification of its type, since this allows a tentative interpretation to be made of the mechanism of physisorption. This approach is consistent with the IUPAC recommendations on reporting physisorption data [1] and the characterization of porous solids [2].

## 2. Type II isotherms

A Type II isotherm is the result of monolayer-multilayer adsorption on an open surface. The isotherm is completely reversible provided that the adsorbent is rigid and that it does not interact chemically with the adsorptive. If the knee of the isotherm is fairly sharp, an initial assessment of the monolayer capacity, $n_m$, can be obtained from the amount of gas adsorbed at Point B - as originally proposed by Brunauer and Emmett [3]. This feature is typical of many nitrogen isotherms at 77 K and is one reason why nitrogen is the generally favoured adsorptive for the determination of surface area [4].

The monolayer capacity is related to the surface area, $A$, by the simple equation

$$A = n_m\, a_m\, L \tag{1}$$

where $a_m$ is the average area occupied by a molecule of the adsorbate in the completed monolayer and $L$ is the Avogadro constant. It is evident that to evaluate $A$ it is necessary to know both $n_m$ and $a_m$ and therefore an error in either one must affect the level of accuracy of $A$.

The most widely used method for the determination of $n_m$ is that proposed by Brunauer, Emmett and Teller [4], universally known as the BET method. The BET theory was based on an extension of the Langmuir model to multilayer adsorption, the assumption being made that localized monolayer adsorption occurs on a uniform surface and that the adsorbed molecules can then act as sites for multilayer adsorption. A further assumption was that all layers after the first have liquid-like properties.

For ease of application, the BET equation is generally expressed in the linear form

$$\frac{p}{n(p^o - p)} = \frac{1}{n_m\,C} + \frac{(C-1)\,p}{n_m\,C\,p^o} \tag{2}$$

In Equation (2), $C$ is an empirical constant which, according to the BET theory, is related exponentially to the energy of adsorption for the first layer. However, although the value of $C$ does provide a useful indication of the isotherm shape, it is no longer thought to be quantitatively related to the energy of adsorption [5].

The value of $n_m$ is derived from the slope and intercept of the BET plot of $p/n(p^o\text{-}p)$ against $p\,p^o$. All known BET plots have very limited ranges of linearity, which rarely extend above $p\,p^o \sim 0.3$; with some systems the linear range is confined to $p\,p^o < 0.1$. A high value of $C$ is consistent with a sharp knee in the isotherm and a well-defined Point B. On the other hand, a low value of $C$ (i.e. $C < 50$) is the result of a more gradual curvature and there is then uncertainty on the exact location of Point B. In the most favourable case, the agreement between the two vlues of $n_m$ is within a few per cent.

At present, nitrogen is generally considered to be the most suitable adsorptive (at 77K) for surface area determination [1]. It is usually assumed that the BET-nitrogen monolayer is close-packed, giving $a_m(N_2) = 0.162$ nm$^2$ at 77K. With some non-porous adsorbents, the use of this value leads to BET-areas, which agree with independently determined surface areas to within $\sim$ 10%. However, if the monolayer is localized, a constant value of $a_m$ would not be expected and therefore it is not surprising to find that much larger differences have been reported [5].

When another adsorptive is used an arbitrary adjustment in the effective $a_m$ value is usually required to bring the BET area into agreement with the nitrogen value. The adjusted values of $a_m$ for a particular adsorptive are dependent on the temperature and the nature of the adsorbent surface.

Krypton adsorption at 77K is often employed for the determination of small surface areas ($<5$ m$^2$g$^{-1}$). Its low vapour pressure at the operational temperature allows the dead space corrections to be minimised in the volumetric determination of the amount adsorbed. Since 77K is below the triple point of krypton, the decision has to be taken: should the effective $p^o$ be that of the solid phase or that of the supercooled liquid? There are arguments in favour of each decision and the choice is not easy, but it should be clearly stated along with an indication of the adopted value of $a_m$ (e.g. 0.195 nm$^2$, coresponding to $p^o$ for the supercooled liquid).

If the fine particles of a powder are porous, the experimenal isotherm will be made up of two or more parts: it may be regarded as a composite isotherm. In this case, if the pore size is confined to the micropore range, the composite isotherm may have the overall appearance of a normal Type II isotherm. Any mesoporosity is likely to lead to some Type IV character, but this is not always distinctive.

Resolution of the isotherm into its component parts can be achieved by application of the $\alpha_s$-method [6]. This involves plotting the amount adsorbed against the reduced adsorption, $\alpha_s$, determined on a non-porous reference material having a similar surface structure to the adsorbent under investigation. A linear $\alpha_s$-plot with a zero intercept is obtained, of course, when the two isotherms are of identical shape and the only difference between the adsorbents is in the magnitude of their surface areas. We can then conclude that the powder particles have no detectable micropore or mesopore structure.

The effects of porosity are revealed by a deviation of the $\alpha_s$-plot from linearity and/or a positive intercept on the $n$ axis. If the $\alpha_s$-plot is sufficiently linear in the multilayer range, it is possible to evaluate the external area  from the slope and the micropore capacity by back-extrapolation to $\alpha_s = 0$.

Pseudo-Type II isotherms are given by some materials containing slit-shaped pores or aggregates of platy particles. Such systems tend to give isotherms exhibiting adsorption hysteresis; but, unlike the Type IV isotherm, there is no detectable plateau in the region of high $p\,p^o$.

## 3. Type III isotherms

A true Type III isotherm is convex to the pressure axis over its complete range. An extreme example of this behaviour is given by the adsorption of water vapour on the clean basal plane of graphite (or grapitized carbon black). In this case, the adsorption is very low until $p\,p^o \sim 0.8$: the slope of the isotherm then continues to increase until $p\,p^o \cong 1$. Some adsorption systems give isotherms of intermediate character, with low values of $C$ ($<10$) and no identifiable Point B.

The absence of a well-defined Point B is a clear indication that the BET method cannot be expected to give a reliable assessment of the overall surface area. However, the derived value of $n_m$ may provide a useful indication of the extent of the area of an "active" part of the surface. The specificity of water adsorption is especially high [7,8] and therefore the adsorption of water vapour can be used to obtain the effective area of the hydrophilic part of the surface of a carbon black.

## 4. Type IV isotherms

In the monolayer region, a Type IV isotherm on a mesoporous adsorbent follows exactly the same path as the corresponding Type II isotherm on a non-porous solid having the same surface area and surface composition. The completion of  monolayer coverage is therefore indicated by the appearance of a sharp Point B and the application of the BET method can be expected to give a reliable assessment of $n_m$ and $A$.

The upward deviation, which occurs in the multilayer region, is the direct result of capillary condensation within the mesopore structure. The isotherm reaches the plateau (see Figure 1) when the mesopore filling is complete. The mesopore capacity, $n_p$, is given by the amount adsorbed at the plateau and it is generally assumed that this can be converted into the mesopore volume, $V_p$, by taking the normal liquid density, $\rho_1$, as the density of the capillary condensate. This assumption appears to be justified, provided that the pore size distribution does not extend into the micropore range.

In the classical approach [1], the mesopore size is obtained by the application of the Kelvin equation equation in the form

$$\frac{1}{r_1} + \frac{1}{r_2} = - \frac{RT \ln (p\, p^o)}{\sigma_1 v_1} \tag{3}$$

which relates the principal radii, $r_1$ and $r_2$, of curvature of the liquid meniscus in the pore to the relative pressure, $p\, p^o$, at which the capillary condensation occurs. Here, $\sigma_1$ is the surface tension of the liquid condensate and $v_1$ is its molar volume.

When we use Equation (3) to obtain the pore radius or pore width, we necessarily make the following assumptions: a that the curvature of the meniscus is controlled by the pore size; b that all the pores have the same uniform shape; c that the condensate has the same physical properties as the bulk liquid. The pores are usually assumed to be either cylindical or slit - shaped : in the former case, the meniscus is hemispherical and $r_1 = r_2$; in the latter case, the meniscus is hemicylindrical and $r_2$ is infinite.

Rearrangement of Equation (3) and repacement of the left-hand side by $2/r_k$ gives

$$r_k = 2\sigma_1 v_1/[R\, T \ln(p\, p^o)] \tag{4}$$

which is the most frequently used form of the Kelvin equation, with $r_k$ usually referred to as the Kelvin radius.

Since capillary condensation can occur only after some adsorption has already taken place on the mesopore walls, it is necessary to correct for the adsorbed layer thickness, $t$, in order to calculate the pore radius, $r_p$, or the pore width, $d_p$. For a cylindrical pore we then have :

$$r_p = r_k + t \tag{5}$$

and for a parallel-sided slit we have

$$d_p = r_k + 2t \tag{6}$$

Values of $t$ are obtained from the isotherm data for the adsorption of the same adsorptive on a non-porous reference material having similar surface structure to the mesoporous solid.

Various computational procedures have been proposed for the derivation of the mesopore size distribution from an experimental isotherm. It is necessary to assume that the pores are rigid and that the distribution does not extend continuously into either the micropore or the macropore range. The pore size distribution is expressed in the graphical form of $\delta v_p / \delta r_p$ vs. $r_p$ and is usually arrived at by the notional removal of the condensate step-by-step from the mesopores whilst also allowing for the thinning of the multilayer on the pore walls.

The location and shape of the distribution curve is, of course, dependent on which branch of the hysteresis loop is used to compute the pore size. Although there is still no definitive answer to this problem, it is now apparent that both pore shape and connectivity should be taken into account. Reversible capillary condensation-evaporation would occur if the meniscus could develop from the apex of a conical pore. In an open-ended cylindrical pore there is likely to be some delay in the condensation and in a slit-shaped pore the delay is even greater [9]. On the other hand, pore blocking effects must be more likely to affect the desorption branch [10]. When we consider that most porous adsorbents contain a complex network of pores of different size and shape, we arrive at the tentative conclusion that the adsorption branch is likely to provide a more reliable basis for the computation of the mesopore size distribution. However, we should keep in mind that with some systems, such as an assemblage of slit-shaped pores, this may also give a misleading picture of the pore structure.

Recent work [11] on MCM-41, a model mesoporous adsorbent, has revealed that it is possible to obtain a well-defined *reversible* Type IV isotherm. The pore structure in this material is in the form of hexagonal arrays of uniform tubular channels [12]. Pore filling by nitrogen at 77K has been found [11] to occur reversibly at $p\,p^o = 0.41$-$0.46$: according to the corrected Kelvin equation, this corresponds to $r_p \cong 4$ nm. It is significant that the isotherms of oxygen and argon adsorption at 77K, although having the same general shape as the nitrogen isotherm, have stable and reproducible hysteresis loops. These results are consistent with other evidence [5] that capillary condensation hysteresis does not appear below a critical $p\,p^o$, which is dependent on the adsorptive and temperature. In the case of nitrogen at 77K, this lower limit is at $p\,p^o \cong 0.42$.

## 5. Type I isotherms

The plateau of a well-defined Type I isotherm extends over a wide range of $p\,p^o$ and its low slope is a consequence of relatively little multilayer adsorption on a small external surface. Molecular sieve carbons and many zeolites give isotherms of Type Ia, having a high degree of rectangularity with the plateau approached at very low $p\,p^o$. Type Ib isotherms, which are given by adsorbents with wider micropores, have a more rounded appearance and a gradual approach to a shorter plateau.

The difference in shape between the Type Ia and Ib isotherms is mainly dependent on the micropore width and geometry in relation to the molecular size. The micropores in activated carbons tend to be slit-shaped; in a narrow pore of this shape the overlap of the adsorption forces is significant provided that the pore width is not much larger than two molecular diameters [13]. This enhancement of adsorption energy is responsible for the high adsorption affinity, which is associated with the "primary micropore filling" process at very low $p\,p^o$ [9]. Wider micropores in the range 2 - 5 molecular diameters are filled by a cooperative process, which occurs over a range of higher $p\,p^o$ (to ~ 0.2). This "secondary micropore filling" is preceded by monolayer coverage of the pore walls, which allows the internal surface area to be evaluated from the slope of the initial section of the $\alpha_s$-plot.

A popular equation for the analysis of Type I isotherms is that proposed by Dubinin and Radushkevich (the DR equation) [14]. In its linear form the DR equation is essentially

$$\log (n\, n_{dr}) = -D \log^2 (p\,p^o) \tag{7}$$

where, according to the theory, $n/n_{dr}$ is the fractional micropore filling (i.e. $n_{dr}$ is the micropore capacity) and $D$ is a constant related to the pore structure.

Although many Type Ia isotherms on activated carbons (those of low burn-off) do give linear DR plots over wide ranges of $p\,p^o$, the theoretical significance of the derived values of $n_{dr}$ and $D$ is not clear. Generalized treatments have been introduced by Stoeckli [15] and others to allow for heterogeneity in the micropore structure. In practice, these elegant procedures are of questionable value because of the difficulty in arriving at an unambiguous distinction between the effects of micropore size distribution and surface heterogeneity. For this reason, many workers now prefer to rely on empirical methods of isotherm analysis - such as the $\alpha_s$-method - which are capable of providing effective values of surface area and micropore capacity [16].

It must be kept in mind that the internal surface area and micropore volume of molecular sieve adsorbents are difficult to define and to determine. In the context of gas adsorption, it is not advisable to attempt any assessment of the internal surface area of such materials (by, say, the application of the Langmuir equation) [1]. An effective value of the micropore volume can be obtained, but only in relation to a particular adsorptive at a given temperature.

The use of nitrogen alone cannot be expected to provide a reliable assessment of the micropore size distribution. It is therefore advisable to employ a number of probe molecules of different size. This number can be minimised by taking account of the different stages of micropore filling [16,17].

## 6. References

1. Sing, K.S.W., Everett, D.H., Haul, R.A.W., Moscou, L., Pierotti, R.A., Rouquerol, J. and Siemieniewska, T. (1985) *Pure Appl. Chem.* **57**, 603-619.
2. Rouquerol, J., Avnir, D., Fairbridge, C.W., Everett, D.H., Haynes, J.M., Pernicone, N., Ramsay, J.D., Sing, K.S.W. and Unger, K.K. (1994) *Pure Appl. Chem.* **66**, 1739-1758.
3. Brunauer, S. and Emmett, P.H. (1937) *J. Amer. Chem. Soc.* **59**, 2682.
4. Brunauer, S., Emmett, P.H. and Teller, E. (1938) *J. Amer. Chem. Soc.* **60**, 309-319.
5. Gregg, S.J. and Sing, K.S.W. (1982) *Adsorption, Surface Area and Porosity*, Academic Press, London.
6. Sing, K.S.W. (1970) in *Surface Area Determination*, Butterworths, London, eds. Everett, D.H. and Ottewill, R.H., 25-41.
7. Barrer, R.M. (1966) *J. Colloid Interface Sci.* **21**, 415-434.
8. Carruthers, J.D., Payne, D.A., Sing, K.S.W. and Stryker, L.J. (1971) *J. Colloid Interface Sci.* **36**, 205-216.
9. Sing, K.S.W. (1989) *Colloids & Surfaces* **38**, 113-124.
10. Seaton, N.A. (1991) *Chem. Eng. Sci.*. **46**, 1895-1909.
11. Branton, P.J., Hall, P.G., Sing, K.S.W., Reichert, H., Schüth, F. and Unger, K.K. (1994) *J. Chem. Soc. Faraday Trans.* **90**, 2965-2967.
12. Kresge, C.T., Leonowicz, M.E., Roth, W.J., Vartuli, J.C. and Beck, J.S. (1992) *Nature* **359**, 710.
13. Everett, D.H. and Powl, J.C. (1976) *J. Chem. Soc. Faraday Trans.* 1, **72**, 619-636.
14. Dubinin, M.M. and Radushkevich, L.V. (1947) *Proc. Acad. Sci. USSR* **55**, 331.
15. Stoeckli, H.F. (1995) in *Porosity in Carbons*, Edward Arnold, London, ed. Patrick, J. 67-92.
16. Carrott, P.J.M., Roberts, R.A. and Sing, K.S.W. (1988) in *Characterization of Porous Solids*, Elsevier, Amsterdam, eds. Unger, K.K., Rouquerol, J., Sing, K.S.W. and Kral, H., 89-100.
17. Kaneko, K. (1996) in *Adsorption on New and Modified Inorganic Sorbents*, Elsevier, Amsterdam, eds. Dabrowski, A. and Tertykh, V.A., 573-598.

# METHODOLOGICAL PROBLEMS RELATED TO THE PREPARATION OF THE SURFACE (OUTGASSING) AND TO THE DETERMINATION OF THE ADSORPTION ISOTHERMS

**J. ROUQUEROL**

*CNRS*

*Centre de Thermodynamique et de microcalorimètrie*

*26 rue du 141éme R.I.A., France*

## 1. Introduction

Whatever the aim of a gas adsorption experiment it must, of course, be prepared and carried out with a care depending on the objective : a routine surface area determination on a well-known material may simply need a reproducible procedure, whatever it is, whereas the understanding of the adsorption mechanism or a minute and safe characterization of the adsorbing surface are of course much more demanding. What are, commonly to-day, the *limits to the quality of the experiments* and are there *promissing ways to move them back* ? These are the questions we shall try to address here, in two steps : we shall first examine the *preparation* of the sample and then *the assessment of the adsorption isotherm.*

## 2. Preparation of the adsorbent

### 2.1. Problems encountered

After selecting the appropriate sample mass (which depends both on the sample heterogeneity and on the performance of the gas adsorption equipment to be used) one is faced with the much more delicate problem of *selecting the procedure of thermal outgassing.* This is not at all trivial and can influence deeply the subsequent adsorption

*J. Fraissard (ed.), Physical Adsorption: Experiment, Theory and Applications,* 17–32.

data (specially when the adsorbent is microporous). This requires a clear goal and a safe information about the behaviour of this particular sample on outgassing.

The *clear goal* involves the exact state at which we want to study the sample. For instance, in the case of a fundamental study on a crystalline surface which is expected to be as perfect as possible, all physisorbed and chemisorbed species must be eliminated. This may require, for instance, for graphite, a heating up to 900° C. More often, the physical chemist or the chemical engineer will prefer an outgassing which will keep untouched the surface chemistry, surface area and porous structure of the sample. In this case the outgassing mainly aims to eliminate physisorbed water (and also, but usually in much lower proportion, some organic vapours caught by the sample from its environment).

For this purpose, *a safe information* about the behaviour of the sample *on outgassing* is required in the form of a thermal analysis curve. This can be a thermogravimetric curve which could allow for instance to detect successive steps separated by plateaus or inflexions and to decide on the optimum outgassing temperature. Unfortunately, one is then faced with one or several of the following limitations, which considerably weaken the meaning of the experiment :

(a) The sample prepared for adsorption is not prepared in a thermobalance but in an adsorption bulb : the quality of the vacuum is not the same, neither, quite often, the sample mass (and therefore the temperature gradients and final homogeneity of the surface)

(b) in order to increase the homogeneity and final extent of outgassing, one tends to keep the sample, over a number of hours, at the final, selected, temperature : *the state of this sample can be relatively different from the one expected* from a dynamic TG curve (which involves significant temperature lags)

(c) sometimes, one may prefer to outgass under *a flow of carrier gas* : the shapes of the TG crucible and of the adsorption bulb may strongly influence he rate of outgassing

(d) sometimes, also, one may need to used *a frit, or a glass-wool pad*, to prevent a sample of micrometer grain-size from spurting out of its bulb during outgassing : nothing is known any more about the quality of the vacuum surrounding the sample.

*2.2. An answer to the above problems : sample-controlled outgassing*

The idea is here to control the sample outgassing in such a way that
(a) it is closely reproducible, even when the sample mass and crucible (or bulb) shape change significantly
(b) it allows a simultaneous determination of a thermal analysis curve, in the actual conditions selected for the preparation of the sample
(c) it provides a highly homogeneous sample
(d) it eliminates the "spurting out" problem, whatever the quality of the vacuum.

The solution is provided by one of the many forms of "Controlled transformation Rate Thermal Analysis" (CRTA). In this general method of heat treatment or thermal analysis [1] the temperature of the sample does not follow any pre-determined programme. Instead, it is the extent of the thermal transformation of the sample which is forced to follow a pre-determined programme, by appropriate heating. For this purpose, a physical quantity related to the extent of the thermal transformation must be selected. This can be a gas flow, a mass, a dimension of the sample, a heat flow etc... In most cases it is convenient to keep a constant rate of transformation.

A simple set-up for Sample Controlled Outgassing is represented in Figure 1. The sample, which is located in a standard adsorption bulb, is continuously evacuated through diaphragm D. Here, the physical quantity used to follow the thermal transformation of the sample (i.e. its outgassing) is directly the gas flow rate originating from the sample. This rate is assessed from the pressure drop occurring through diaphragm D. Since the vacuum on the right hand side is constant and good, the pressure drop can be measured by a pressure gauge G (conveniently a Pirani, or

Penning or capacitance gauge) located in the left hand side. The signal from the gauge is sent to the PID controller which heats the sample in such a way as to keep this stationary pressure constant. Depending on the vacuum, gauge and diaphragm available, this residual pressure over the sample can be controlled, with this equipment, in the range between $10^{-5}$ and 100mbar. The recording of the temperature *vs* time is comparable to a thermogravimetric recording, since -provided the gas evolved on outgassing is of constant composition- time is a linear function of the mass lost by the sample. This means that each sample outgassed by this technique is provided with its own thermal analysis curve and with its precise thermal history.

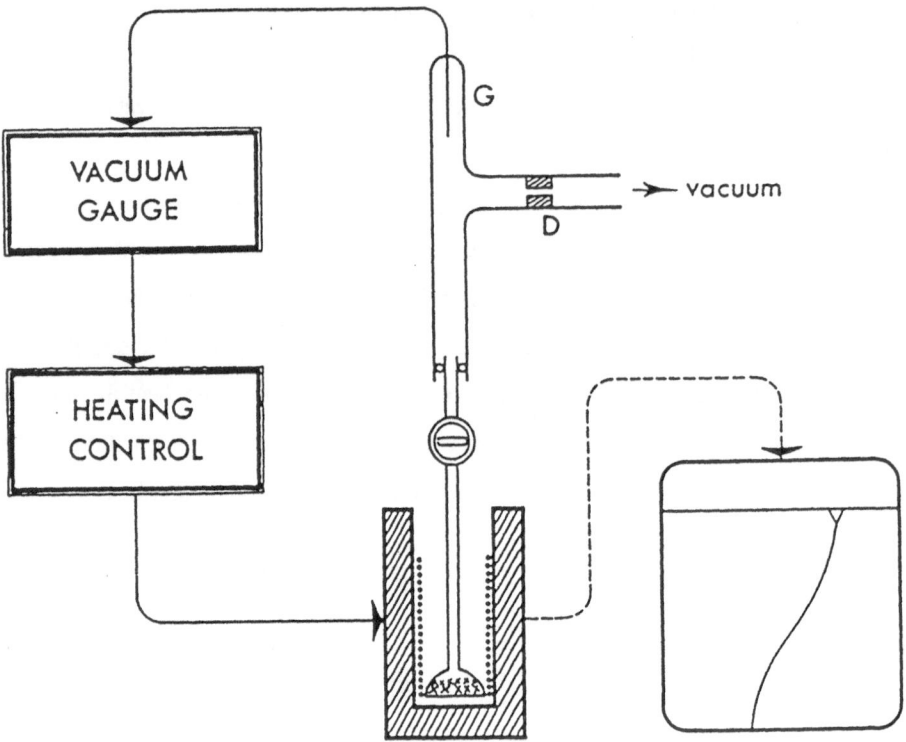

Figure 1 : Simple set-up for Sample-Controlled Outgassing

In case the adsorption is to be carried out by gravimetry, it is then better of course to get the Sample-Controlled Outgassing directly *in situ*. This can be done with the help of one of the two control loops shown in Figure 2 - Loop I uses the derivative, *vs* time, of the thermogravimetric signal, whereas loop II makes use of the same control as in Figure 1.

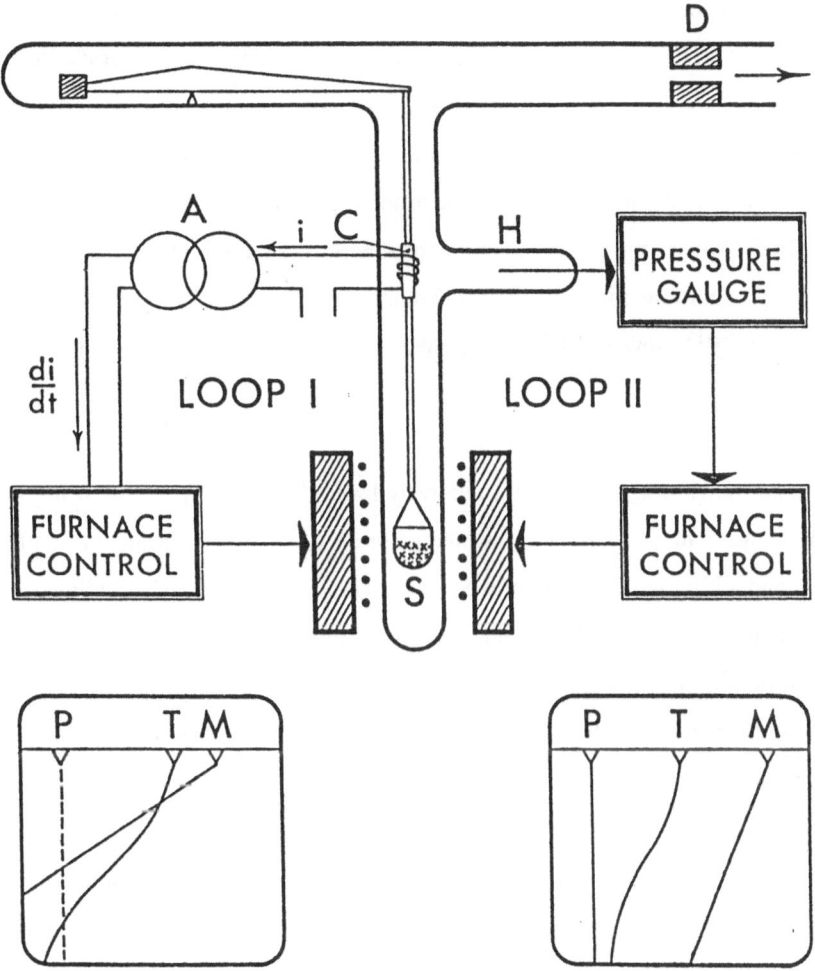

Figure 2 : Two ways to get a Sample-Controlled Outgassing in a thermobalance

Loop I : from the derivative, vs time, of the thermogravimetric signal

Loop II : same control as in figure 1.

In the case of an unknown sample, it is wise to carry out the Sample-Controlled Outgassing over a broad temperature range and to then examine the thermal analysis curve (the CRTA curve) to decide on the most appropriate point to end the outgassing. In the frequent case when one wishes to get ride of physisorbed water, it is likely that the best choice will be -when it does exist- an inflexion point located between 100 and 200°C, which shows that a first outgassing mechanism (dehydration) is arriving to an end, whereas a second mechanism, probably corresponding to a degradation of the sample, has not yet taken place. In case the latter mechanism is not expected to provide only water vapour, a confirmation of the meaning of this inflexion point can be obtained by connecting directly to a quadrupole gas analyzer the set up for Sample-Controlled Outgassing [2]. The above procedure was used to determine the best outgassing conditions for a set of six reference adsorbents (surface areas ranging from 0.1 to 8 $m^2g^{-1}$) now available from the Bureau Communautaire de Référence in Brussels [3].

The rate of outgassing is fixed at a relatively low value (Usually : 1 to 10 mg lost per hour) in order to get (at will) low temperature and pressure gradients within the sample and a high homogeneity of the final product. This is also which allows to be, at any time, in quasi-equilibrium conditions, so that the heat treatment can be stopped at the pre-determined temperature : the sample is then quenched, without any temperature plateau, so that its representative point really remains on the recorded thermal analysis curve.

This type of pretreatment of the adsorbent is well-suited either for a simple outgassing, as we saw, or *to study the influence of the heat treatment on the adsorbing properties* of the sample (including changes in the surface area and pore-size) -Figure 3 shows the CRTA curve of a $Be(OH)_2$ sample. The heat treatment was stopped at each point indicated on the curve. The adsorption sample bulb was then attached to a gas adsorption equipment and the full nitrogen adsorption-desorption isotherm was then determined.

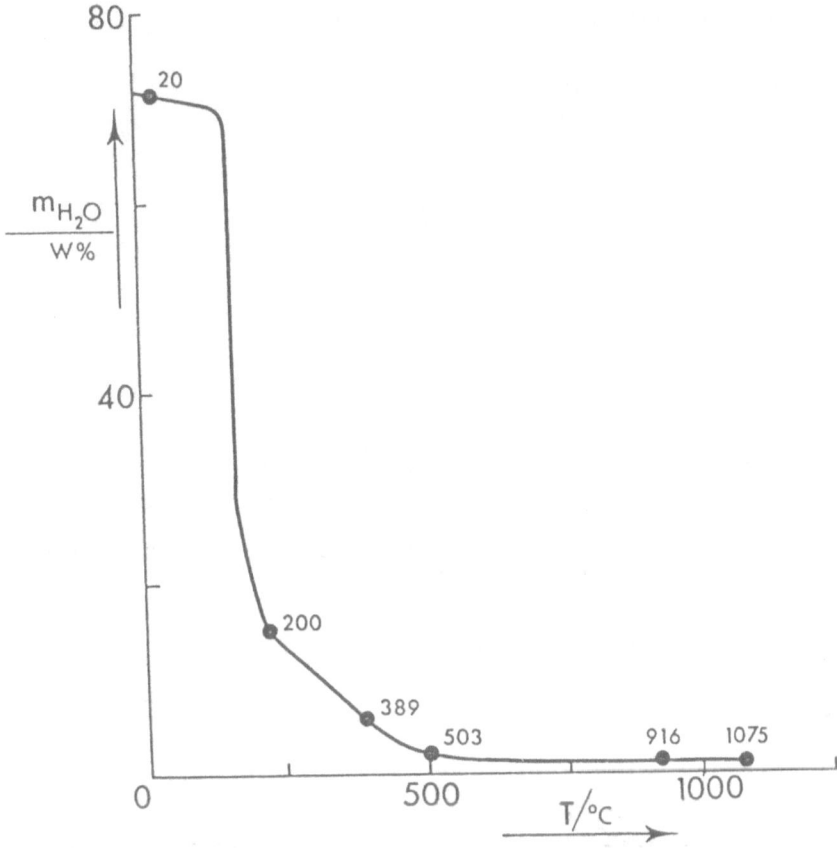

Figure 3 : CRTA curve of a Be(OH)$_2$ sample.

The sample bulb was then connected again to the CRTA set up of Figure 1 and the next portion of the CRTA treatment was then carried out. The full set of adsorption-desorption isotherms is given in figure 4. One sees that as soon as the starting beryllium hydroxide is decomposed, the resulting beryllia samples (obtained at 200° C or above) show, on their adsorption isotherms, a typical kink (indicated by an arrow) corresponding to the filling of pores with a very narrow pore size distribution. As the final temperature of the heat-treatment increases, this kink is shifted towards higher and higher pressures : this is simply because of the pore-size increase. Gas adsorption

24

and transmission electron microscopy show an increase from 0.5nm (200°C) to about 4nm (1075° C) [4]. This illustrates the high homogeneity of the CRTA treatment, and the way it can be used to prepare taylor-made adsorbents.

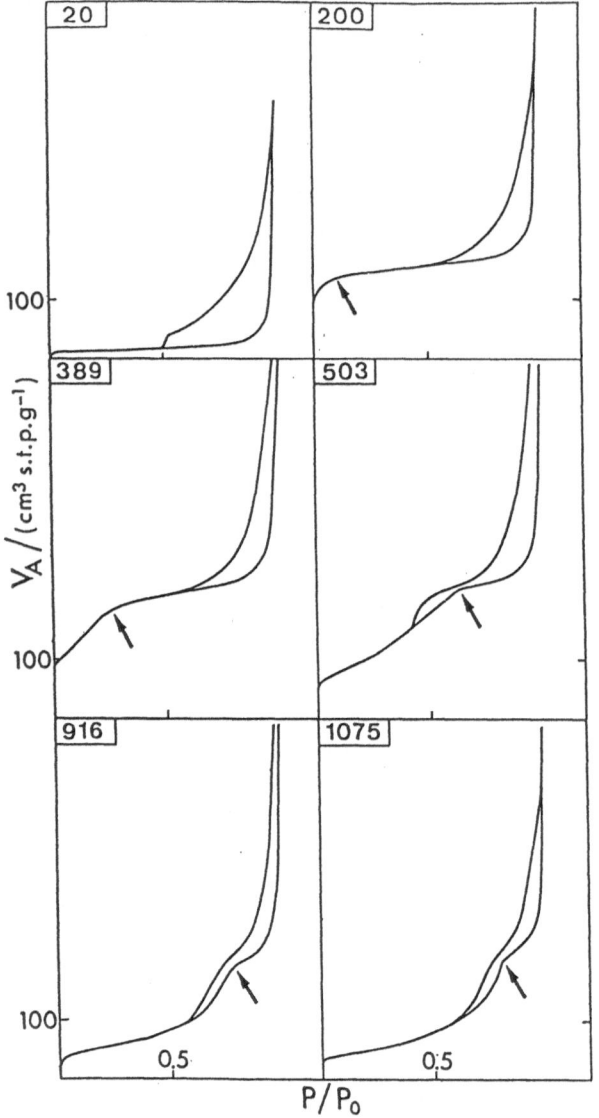

Figure 4 : Nitrogen adsorption-desorption isotherms obtained at 77 K on a set of beryllia samples. Figure on top left of each frame indicates final temperature of CRTA treatment.

# 3. Determination of the gas adsorption isotherms

## 3.1. A few problems encountered

The reasons for *discrepancy*, in adsorption measurements, from one laboratory to the other, are *extremely numerous*. They include any type of uncertainty : in pressure or volume calibration and measurement, in temperature homogeneity and measurement, in control of the liquid nitrogen level and temperature (when it is used), in control of the adsorption equilibrium ... *A good understanding* of the adsorption mechanism *requires a large number of points on the adsorption* isotherm : with the usual, point by point, procedure, this is not only time-consuming but it also leads to cumulated errors in the case of volumetric (or, better said to-day, manometric) experiments.

## 3.2. A rewarding procedure : continuous quasi-equilibrium adsorption

The idea of a continuous introduction of the adsorptive was proposed in the fifties [5], but under conditions which were not found satisfactory : either equilibrium was never reached, or only a few gases, with relatively low vapour pressure (like argon 77 K) could be used, but only to determine the first portion of the adsorption isotherm (monolayer region). [6]

It is only in the seventies that conditions for a real quasi-equilibrium experiment were proposed and experimented [7] [8]. Systematic comparison with the conventional, point by point, procedure was carried out and the interest of the new approach was then shown [9] :

(a) possibility to reach and check satisfactory quasi-equilibrium conditions (simple check : two experiments are carried out at two different rates of adsorptive introduction; if the adsorption curves recorded are superimposed, they can be considered as real adsorption isotherms ; if they are not superimposed, a lower rate must be chosen).

(b) infinite number of meaningful points of the adsorption isotherm (sometimes expressed by the term "high-resolution procedure")

(c) possibility (even before the use of microcomputers) to get the adsorption isotherm in a real time

(d) time-saving

(e) in some cases, large experimental simplification.

Finally, just a warning before presenting a few set-ups : this *continuous adsorption* procedure must not be confused with the procedures which, like that proposed by Nelson and Eggertsen [10], use a continuous flow of a carrier gas but give actually rise to a discontinuous, point by point, adsorption.

*3.3. A few set-ups for continuous, quasi-equilibrium*

A number of set-ups were proposed to carry out continuous, quasi-equilibrium adsorption experiments.

The simplest and most rewarding set-up is probably the one using a *leak-value and an adsorption balance* [11]. The arrangement is presented in Figure 5. The needle-value is used for the continuous introduction of the adsorptive, either from a compressed gas cylinder or from a liquid source in equilibrium with its vapour. This leak valve can be manual (which is usually satisfactory) or automated in case one wishes to save time and to adjust the opening in order to get a relatively constant rate of either pressure change or gas uptake by the sample. One can easily record the signal from the balance *vs* the signal from a pressure gaugeand then get a direct, on-line, recording of the adsorption-desorption isotherm : this was the first set-up which ever allowed such a recording (with a simple X,Y recorder, when microcomputers were not yet available.)

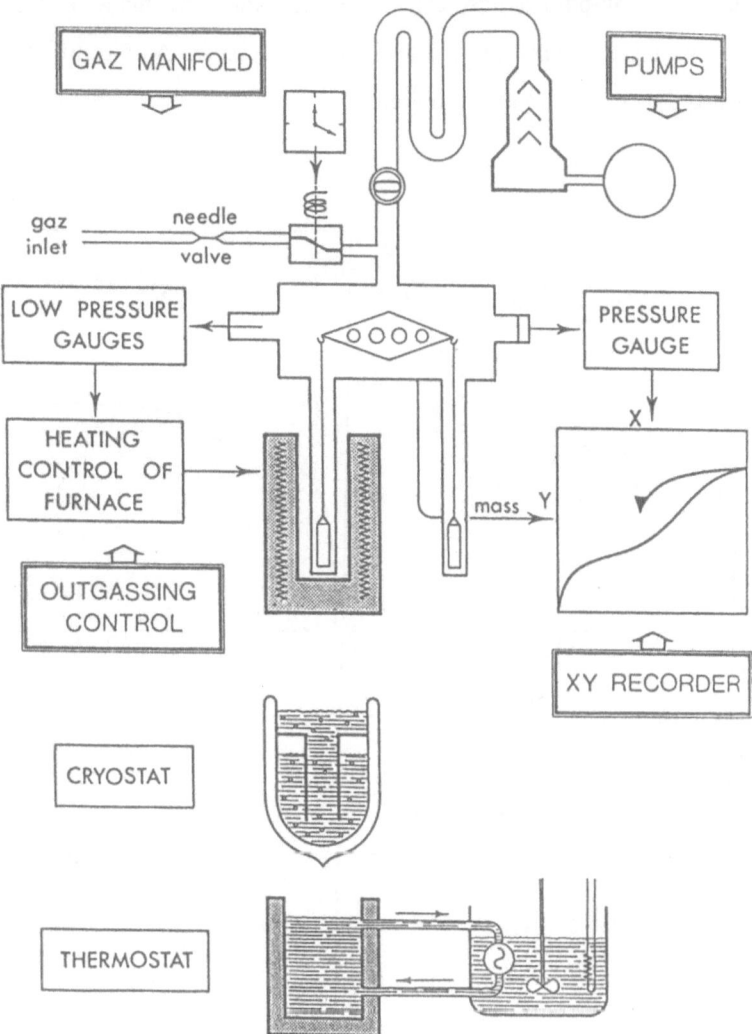

Figure 5 : Principle of continuous adsorption gravimetry.

In the case of non-porous adsorbents provision must be made for a buoyancy correction, which is simply proportional to the pressure. A few direct recordings obtained with this set-up are given in Figure 6.

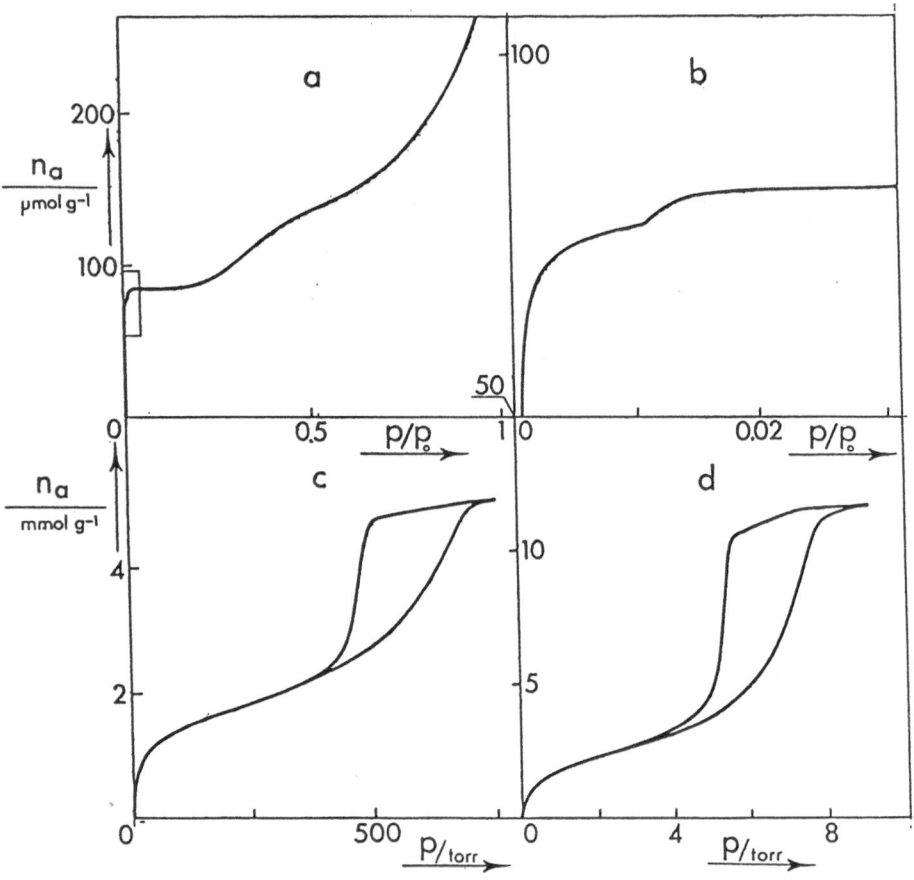

Figure 6 : Adsorption isotherms directly recorded by continuous quasi-equilibrium gravimetry :

Nitrogen-graphite at 77 K (a and b)

Nitrogen-Vycor glass at 77 K (c)

Water-Vycor glass at 282.5 K (d)

Another, relatively simple set-up, which was the first to provide at will quasi-equilibrium conditions [7] [8] makes use of one or several *sonic nozzles*. It is represented in Figure 7. The adsorptive, from a compressed gas cylinder fitted with a high stability manometer is fed to the adsorption volume through one of the three sonic nozzles labelled S. The major interest of these sonic nozzles is to provide a constant flow-rate, irrespective of the downstream pressure, if the upstream pressure is high enough and constant. The three nozzles are selected in such a way as to cover the range from 0.01 up to 100 cm$^3$ STP per hour. The most standard rates needed to achieve good quasi-equilibrium conditions range between 1 and 10 cm$^3$ STP per hour. The quasi-equilibrium pressure is continuously recorded *vs* time, i.e. *vs* the amount of adsorptive introduced into the system. The recorded curve must simply be corrected for the amount of adsorptive remaining in the dead space : this results (like the buoyancy correction in adsorption gravimetry) into a correction proportional to the pressure reached.

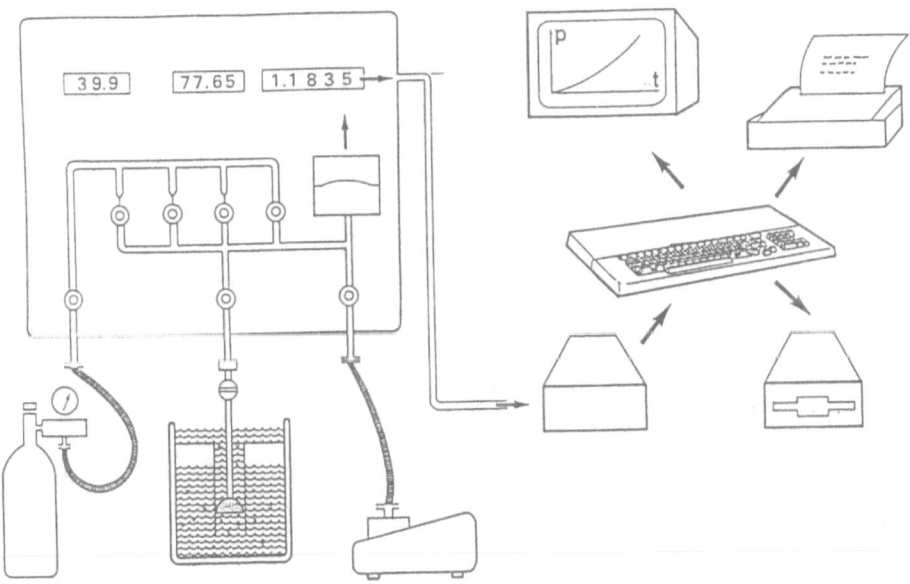

Figure 7 : Principle of continuous adsorption flowmetry making use of sonic nozzles.

A more sophisticated set-up associates a *mass-flowmeter* (i.e. a gas flowmeter based upon a thermal principle) with an *automated leak-valve* [12]. This is a flexible way to control a constant rate of introduction or extraction of the adsorptive. It therefore allows to determine a full adsorption-desorption isotherm (whereas the sonic nozzle set-up is limited to the adsorption branch). In counterpart, it does not seem that, to-day stable, flow rates lower than 5 $cm^3$ STP per hour can be achieved in this way : this may still be found too high to always ensure satisfactory quasi-equilibrium conditions.

A reservoir of adsorptive, an automated leak-valve and a pressure transducer can also be used to ensure a constant consumption of adsorptive from the reservoir [13] : the possibilities of this set-up are comparable to those of the former.

Finally, the differential introduction of adsorptive (along the lines of the set-up already commented on [6] can be using two independent reservoirs and automated leak-valves to feed the sample-side and balance-side, respectively, with the same pressure of adsorptive : the amount adsorbed by the sample is simply derived from the pressure difference between the two reservoirs.

## 4. Conclusion

As can be seen, both approaches presented here (sample controlled outgassing and continuous quasi-equilibrium gas adsorption) are promising but, still in development. A few studies (specially involving 2D phase changes or minute changes in pore-size) could not have been possible without them. Nevertheless, in many cases, (depending on the goal) more conventional procedures are still quite adequate or, at least, complementary.

# 5. References

[1] Rouquerol, J (1989) Controlled transformation rate thermal analysis : the hidden face of thermal analysis, *Thermochimica Acta* **144**, 209-224.

[2] Rouquerol, J., Bordère, S., and Rouquerol, F. (1992) Controlled rate evolved gas analysis : recent experimental set-up and typical results, *Thermochimica Acta* **203**, 193-202.

[3] Rouquerol, J., Rouquerol, F., Triaca, M., and Cerclier, O. (1985) The use of thermal analysis to select the standard outgassing conditions of a set of reference assorbents, *Thermochimica Acta* **85**, 305-310.

[4] Rouquerol, F., Rouquerol, J., and Imelik, B. (1985) Problems Raised by the Complex Adsorption Isotherms of the Nitrogen/Beryllium Oxide System, in "Principles and Applications of pore Structural Characterization", J.M. Haynes and P. Rossi-Doria editors., J.W. Arrowsmith, Bristol, 213-226.

[5] Innes, W.B., (1951) Apparatus and Procedure for Rapid Automatic Adsorption, Surface Area, and Pore volume Measurement, *Analytical Chemistry* **23**, N° 5, 759-763.

[6] Bültemann, H.J., (1962) Apparcil pour la Determination de l'Aire Superficielle selon BET, *Le Vide* **99**, 201-207.

[7] Rouquerol, J., Calorimétrie d'adsorption aux basses températures I.- Les calorimètres d'adsorption, *Thermochimie* **201**, 537-545.

[8] Grillet, Y., Rouquerol, F., et Rouquerol, J., (1977) Etude de l'adsorption physique des gaz par une procédure continue. I.- Application à la détermination des aires spécifiques d'adsorbants mésoporeux, *Journal de Chimie Physique* **2**, 179-182.

32

[9] Rouquerol, J., Rouquerol, F., Grillet, Y. and Ward, R.J. (1988) A critical assessment of quasi-equilibrium gas adsorption techniques in "volumetry, gravimetry or calorimetry in Characterization of porous solids", K.K. Unger, J. Rouquerol, K.S.W. Sing and H. Kral editors, Elsevier Science Publishers, 67-76.

[10] Nelson, F.M., and Eggertsen, F.T., (1958) Determination of Surface Area Adsorption Measurements by a Continuous Flow Method, *Anal. Chem.* **30**, 1387-1390.

[11] Rouquerol, J., and Davy, L., (1978) Automatic gravimetric apparatus for recording the adsorption isotherms of gases or vapours onto solids, *Thermochimica Acta* **24**, 391-224.

[12] Venero, A.F., and Chiou, J.N., (1988) in "Procceding of the materials research society, micostructure, properties and catalysis", **111**, 235.

[13] Ajot, H., Joly, J.F., Raatz, F., and Russmann, C., (1991) A new apparatus for continuous adsorption. Application to the characterization of micoporous solids, *Studies in Surface Science and Catalysis* **62**, 161-167.

# PHYSICAL ADSORPTION IN MICROPOROUS SOLIDS

*Measurements and Insight into the Interactions*

**Wm. Curtis Conner**
Dept. Chem. Engineering
University of Massachusetts
Amherst, MA, 01003, USA

## 1. Introduction

The adsorption of gases onto porous solids is a primary method by which the morphology of solids is characterized. It is conventionally employed for high surface area solids comprising pores greater than 1nm in radius. Specifically, the surface area, pore volume and pore size distribution can be inferred and calculated from analyses of the relationship between the volume adsorbed and the pressure of a physically adsorbing gas. These morphological characteristics (surface area, pore volume and pore size distribution) are employed in the design and analyses of solids employed in catalysis, as adsorbents or for separations.

Unfortunately, conventional automated approaches for the measurement of physical adsorption do not work for adsorption in microporous materials. Further, conventional analyses cannot be employed to calculate or to infer any of the morphological characteristics: surface area, pore volume or pore size distributions. For this reason, considerable recent theoretical studies have focused on interpreting physical adsorption in microporous solids and equipment is being modified that improves the experimental techniques for the measurement of adsorption in microporous solids. The state-of-the-art in the interpretation of physical adsorption in micropores is presented at this NATO school as are the state-of-the-art in the measurements being demonstrated. For good reason, there is still much work to do.

These two presentations will attempt to review the experimental evidence related to adsorption in microporous materials. First, the experimental parameters required for the measurement of physical adsorption in micropores are analyzed. This adsorption occurs at relative pressures substantially lower than that for adsorption in mesoporous materials and spans several orders of magnitude; thus, unique techniques are required for these measurements. Second, the experimental isotherms for adsorption of $N_2$ and Ar in micropores are discussed as they relate to differences between adsorbates and conditions of the measurements. The pressures at which the pores fill vary by three orders of magnitude over the range of known zeolites and differ substantially with adsorbent and temperature. Conventional surface area and pore volume analyses cannot be employed since it is apparent that this is not a "conventional" condensation process. However, these

33

*J. Fraissard (ed.), Physical Adsorption: Experiment, Theory and Applications, 33–63.*

results give some insight into the unique nature of "physisorption" in micropores, and more detailed studies of the adsorbed species are warranted. The final part of this presentation will focus on the spectroscopic *in situ* studies of the species physisorbed in micropores. These preliminary studies document that considerable additional *in situ* studies are required to determine the nature of the species adsorbed within micropores and modified theories based on these determinations are needed.

## 2. Measurement of Physical Adsorption in Samples with Micropores

The criteria required to measure physical adsorption for adsorption in microporous materials will be analyzed. Further, the fact that these samples also can posses mesopores between microporous domains (agglomerates of microporous subparticles) needs to be realized in the physical adsorption measurements and their analyses. Finally, the optimization of these processes will be discussed.

It is first necessary to realize that the conditions under which micropores are filled differ substantially from those normally encountered in physical adsorption measurements (for meso-porous solids) and span several orders of magnitude in pressure. The adsorption in pores of dimension 3-20Å occurs at relative pressures of $10^{-7}$ to $10^{-3}$ (bar, if at the adsorbent boiling point). Most importantly, these measurements must reflect adsorption equilibrium: the amount adsorbed at a fixed constant temperature over this variation of pressures. The amount of adsorbent gas exposed to the sample needs to span several orders of relative pressure, the heat released during adsorption needs to be equilibrated with the thermal bath, and time required for the measurements must be practical. These constraints dictate that the system employed for these analyses differs substantially from that employed in the analysis of physical adsorption in meso-porous solids and that the methods by which the adsorbent is exposed to the adsorbate must differ from those normally employed for physical adsorption measurements.

In addition, the total range of pores within a sample which contain micropores can include the pores between the individual microcrystalline crystallites. This extends the range of data (pressure versus volume adsorbed) that may need to be collected all the way up to saturation.

This section will discuss the unique experimental requirements to measure equilibrium adsorption isotherms for adsorption in micropores (Conditions in section 2.1). The systems to be employed in these adsorption measurements differ from systems conventionally used to measure adsorption equilibrium for mesoporous solids. These System requirements are discussed in section 2.2. The methods by which these measurements should be made differ from more conventional sorption measurements and these Methods are discussed in 2.3. Samples containing micropores most often also have larger pores that exist between the microporous domains. The ability to measure and also to analyze these solids which comprise complex porous morphologies are then discussed in 2.4.

## 2. 1. Conditions for HRADS Measurements

In general, the major difficulties in the measurement of adsorption isotherms at low pressures (in the order of $10^{-5}$ Torr in the case of micropores) is the addition of small amounts of adsorbate gas and the accurate measurement of the resulting equilibrium pressures. Unfortunately, heat transfer is not good in the range of pressures where micropores are filled. As a result, considerable time is required to maintain thermal equilibrium transfering the high heat of adsorption as the pores fill. At pressures below $10^{-3}$ torr, we have measured that it can take to over an hour for equilibrium to be achieved (i.e., a stable pressure) for even small samples (< 0.1g). If adsorption is not occuring or the pressure is higher, less time is required. The time required between individual measurements changes. Moreover, constant flow measurements cannot be employed for the lowest pressures... to measure the smaller pores. We have dosed the adsorbate gas over the sample through a dosing valve such as employed in chromatorgaphy (or a combination of vacuum valves). In this way, the amount of adsorbate gas admitted into the system is measured and is controlled. Each dose corresponds to increases in relative pressures ($P/P_0$) of less than $10^{-6}$ per dose for Ar at 87K or $N_2$ at 77K for the initial doses of gas over the sample. The system is allowed to equilibrate for a variable time before the next aliquot of adsorbate is introduced over the sample.

It is apparent that small doses of adsorbate must be added to study adsorption which can significantly occur between increments of pressure as low as $10^{-4}$ torr (see the data in section 2). Yet the range of pressures that need to be studied, varies up to 1 torr. This would be variations in relative pressure, $P/P_0$, from $10^{-7}$ to $10^{-3}$ in increments of $10^{-7}$ for adsorption at the boiling point of the adsorbent. The size of the dose is varied throughout the measurement of the isotherm. Initially doses in increments less than $10^{-3}$ torr ($P/Po < 10^{-6}$) are added. If equal size doses were added, over a thousand doses in increments of $P/P_0 < 10^{-6}$ would be required to span the pressures to $P/P_0 = 10^{-3}$. If equal times were allowed for equilibrium between doses of adsorbent, one thousand increments of equilibrium time would be needed. As will be documented in section 2, the adsorption for uniform sized micropores, as in zeolites, span almost an order of magnitude in relative pressure for any of the samples. Thus, measurements need to be made over each decade of relative pressure. Yet, the precision of measurements for any sample should be similar and, thus, not increase or decrease as the range of pressures increase. This is particularly crucial for samples with unknown pore dimensions or which comprise a variety of pore dimensions over the range of 0.3-2nm.

In summary, the amount of the gas dosed needs to be varied by up to an order of magnitude (increasing with pressure) and the time for equilibrium (decreasing with pressure) to assure accuracy in the equilibrium pressure and to minimize the total time of the measurement.

Helium may be first added into the system to determine the "dead volume" of the sample and sample cell from the relationship between pressure and volume added. The dead volume refers to the volume of the sample holder excluding the sample itself. After the evacuation of the helium, the adsorbate gas addition follows. Note that helium may be run into the system continuously through a flow controller for the determination of the dead volume in order to reduce the measurement time since helium does not adsorb at the same pressures where studies of physical adsorption of other gases are studied. An added advantage of the exposure to helium prior to the adsorption measurement is that the large heat transfer coefficient of helium assures rapid initial thermal equilibration between the sample and the liquid nitrogen or argon temperature baths. It should be noted that Kaneko et al. (1) have recently pioneered in studies of He physisorption. We anxiously await further studies of He physical adsorption in microporous solids; although, the experimental requirements for these studies will also need to be optimized as it is inherently more complex than adsorption of $N_2$ or Ar at 77 or 87K.

## 2.2. Requirements

Another practical constraint on the measurement of physical adsorption in micropores is that the pressures at which adsorption occurs span a broad spectrum of presures... up to seven orders of magnitude in pressure. This requires that more than one pressure transducer be employed to measure the adsorbent pressure at equilibrium. Since the state-of-the-art in pressure measurements dictate that any single pressure transducer is, at best, only acurate over three orders of magnitude in pressure, i.e., 0.1% of the measurement range. At least three pressuring divices are required to span these total of nine orders of pressure measurements. Conventional automated systems employ 10-Torr and 1000-Torr (maximum range) pressure transducers for the measurement of pressures at equilibrium. The transducer manufacturers claim accuracy down to <1% of their full range. So, the combination of these transducers will accurately measure pressures from 0.1 up to 1000 Torr. For adsorption of a gas at its saturation pressure this means accurate measurement in relative pressure, $P/P_0$, from $10^{-4}$ to 1. So, in order to study any adsorption that may occur over ranges in relative pressures from $10^{-6}$ to $10^{-3}$, where micropores (< 2nm in diameter) fill, it is necessary to employ a more accurate measurement. We have used a 1-Torr, high accuracy capacitance transducer to obtain measurements of the pressure at equilibrium in the range of $10^{-6}<P/P_0<10^{-2}$ and, more recently, we are using transducers to measure the lowest range of pressures to 10 millitorr full scale.

The calculation of the volume adsorbed at a given relative pressure in a fixed volume system involves calculating the amount of adsorbent still in the gas phase. This is subtracted from the total amount added to the system at this point. The difference is the amount adsorbed. The amount of adsorbate in the gas phase, a "dead volume correction", is calculated from the product of the dead volume of the adsorption manifold (including that over the sample and accounting for the possible differences in temperature) multiplied by the measured pressure. Inaccurate calculations of the amount adsorbed are a result of inaccuracies in the dead volume, in the pressure measurements, or in the corrections for differences in temperatures of the volumes which comprise the dead volume. Prior adsorption systems developed primarily for measurements at relative pressures above 0.01 decrease the inaccuracies by minimizing the dead volume. This is accomplished by employing small volume valves and small diameter tubing to interconnect the parts of the manifold.

Adsorption in micropores occurs at substantially lower pressures than for samples which contain larger pores. Pores less than 2nm in dimension (diameter or width) are filled at relative pressures less than 0.001. Under these conditions, the dead volume correction is small because the relative pressure is small. It is less necessary to minimize the dead volume in order to achieve accuracy for measurements of adsorption which occurs at low relative pressures (i.e., less than 0.01).

For accurate measurements at low relative pressures, it is necessary to evacuate the sample and manifold to even lower pressures than the pressure of the initial measurement. Smaller tubing and valves in the system make the evacuation more difficult to achieve and require longer times to achieve these low pressures. Further, smaller tubing and valves mean that there is an increased possibility that there are differences in the pressure at different volumes of the system being evacuated; thus, pressure measurements are less accurate.

It is best to avoid any small diameter tubing and small valves between the vacuum pump and the sample employed in prior adsorption systems. Preferably this tubing is 0.5 inch OD or larger. The only tubing that most directly connects the sample and the high vacuum pump that is less than 0.5 inches OD is that volume which connects the sample to the adsorption manifold. This is because the tubing which connects immediately to the sample is less than 0.5 inch OD in order to reduce any inaccuracies which would occur if there were any variation in the height of a thermal bath in which the sample is immersed.

To achieve the low initial pressures required for the measurement of adsorption in small pores, it is necessary to employ an efficient pumping system capable of achieving a high vacuum (low pressure) over the sample. This should be at least an order of magnitude (and preferably at least two orders of magnitude) lower in pressure than the initial pressure at which adsorption may occur. Adsorption occurs below a relative pressure of $10^{-7}$ to $10^{-6}$ for adsorption in pores of a diameter less than 0.4 nm. Preferably, the pumping system should be able to efficiently evacuate the manifold and sample to less than a relative pressure of $10^{-9}$. The vacuum systems employed for most of the current automated adsorption systems employ oil diffusion pumps with a rough, backing pump. This is not adequate to reach these required vacuums efficiently. A turbomolecular pump is preferred; although, other vacuum systems could also fulfill these requirements.

**Summary of requirements for the efficient, accurate measurement of adsorption at low relative.**

    a) The pressure of adsorption is measured by more than one pressure transducer and/or employing transducers that have more than a single maximum range. The range of the pressure measuring device is such that pressures of 0.001 torr (and preferably 0.0001 torr, i.e., ca. $10^{-7}$ atm.) are measured accurately.

    b) Gas is dosed over the sample and the volume of the gas dosed between measurements is changed as the measurement pressure changes.

    c) The time between doses varies as the time required to achieve equilibrium.

    d) The tubing connecting the pumping system and the sample and the sample and the pressure transducers is of diameter greater than 0.25 inch in outer diameter, OD (and preferably equal to or greater than 0.5 inches OD), however, the tubing immediately connected to the sample may be of OD equal to or less than 0.25 inch. The volume of this smaller diameter tubing shall be less than 10% of the total volume of the closed system into which the adsorbing gas is dosed.

    e) The pumping system employed to evacuate the closed system into which the adsorbing gas is dosed must be capable of reaching pressures less than $10^{-8}$ atmospheres in this system (and preferably less than $10^{-9}$ atmospheres pressure).

    The adorption of other gases at higher temperature or even the same gases at higher temperatures may seem attractive alternatives. In this manner, the requirements of reaching and measuring the lower pressures would be less. However, the volume of gas which adsorbes to fills the pores decreases at higher temperatures as it is apparent that the density of the adsorbing gas within the micrpores decreases significantly as temperature increases. As as a result, the steepness of the isotherm ($dV_{ads}/dP$) decreases and sensitivity is reduced.

In order to overcome these difficulties, we developed a modified static technique and apparatus to employ this technique(2). When the adsorbate gas is dosed into the sample cell, the pressure goes up. As adsorption takes place, the pressure decreases. The system is allowed to equilibrate before the next aliquot of adsorbate is dosed. Equilibrium is reached and the pressure no longer changes with time. The time required for equilibrium to be achieved varies from more than 30 minutes for the initial doses at residual pressures below $10^{-4}$ torr to less than 3 minutes for pressures approaching one torr (P/Po ~$10^{-3}$). There is no known method by which a system that employs continuous flow of gas over the sample can measure equilibrium pressures during the flow process for the measurement of adsorption at the low pressures required for the adsorption and filling of pores less than 2 nm in dimension.

### 2.3. Microporous- Mesoporous Pore Networks

In general, when a sample comprises both micro- and meso-porosity, the regions of pressure where the pores fill will be separate and can be analyzed independantly. Thus, any region where hysteresis is evident above a relative pressure of 0.5 might be analyzed to calculate a pore size distribution of the mesopores. This would be found, for example, for an agglomeration of small subparticles which contain microporosity. However, there are several potential problems in that the surface areas of both regions can be mostly within the micropoous regions of the samples. During the analyses of the mesopores by adsorption or by desorption, the surface area may be needed, depending on the type of analysis being performed. As discussed below, the surface area of samples which contain micropores cannot be estimated by conventional BET analysis. In this case, desorption analysis over the region where hysteresis occurs will yield an estimate of the surface area that could then be employed for adsorption analyses; however, this surface area is artifically large since the radius employed in the calculation of area from volume may be smaller than the actual pore dimension(3, 4).

### 3. Adsorption Isotherms in Zeolites

We have studies a broad spectrum of zeolites from rho- to VPI-5 zeolites(5). Nitrogen and argon were adsorptions were studied at both 77k and 87K. Initial studies of adsorption in zeolites had suggested that Ar adsorption at 87K was to be preferred as it was more sensitive to the pore dimensions, i.e., it exhibited greater differences in relative pressures as the pore dimensions changed than did $N_2$ adsorption at 77K. This was initially suggested as due to quadrupolar interactions between Al and $N_2$. The adsorption isotherms for adsorption of Ar and of $N_2$ in two zeolites, Linde 5A and VPI-5, are shown below in Figures 1 and 2.

40

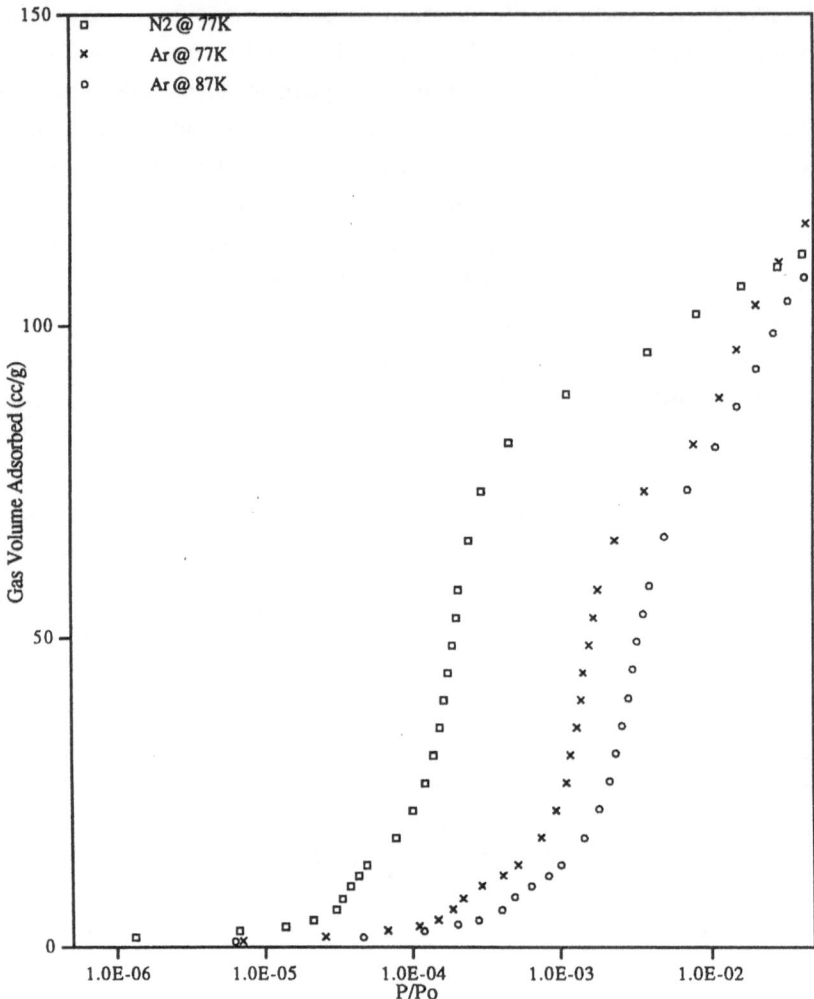

**Figure 1: VPI-5 Adsorption Isotherms for $N_2$ at 77K, and Ar at 77 and 87K.**

Comparisons of the isotherms in these figures shows that the for any specific zeolite the adsorption occurs at slightly over one order of magnitude in relative pressure for $N_2$ adsorption at 77K than for Ar adsorption at 87K, i.e., both at their boiling points. Adsorption of Ar at 87K and at 77K are essentially identical, although the lower temperature adsorption occurs at about half the pressure. The shifts of the isotherms between the two samples is relatively the same. VPI-5 possesses pores that are nominally three times the dimensions of the pores in the Linde 5A zeolite. This difference in pore dimension by roughly a factor of three (and in pore dimensions by roughly 1nm) results in a shift in the adsorption isotherms for a specific adsorbent and temperature by two orders of magnitude. Indeed, this high resolution adsorption, HRADS, is very sensitive to the dimensions of the pores and smaller pores fill with adsorbent at lower pressures.

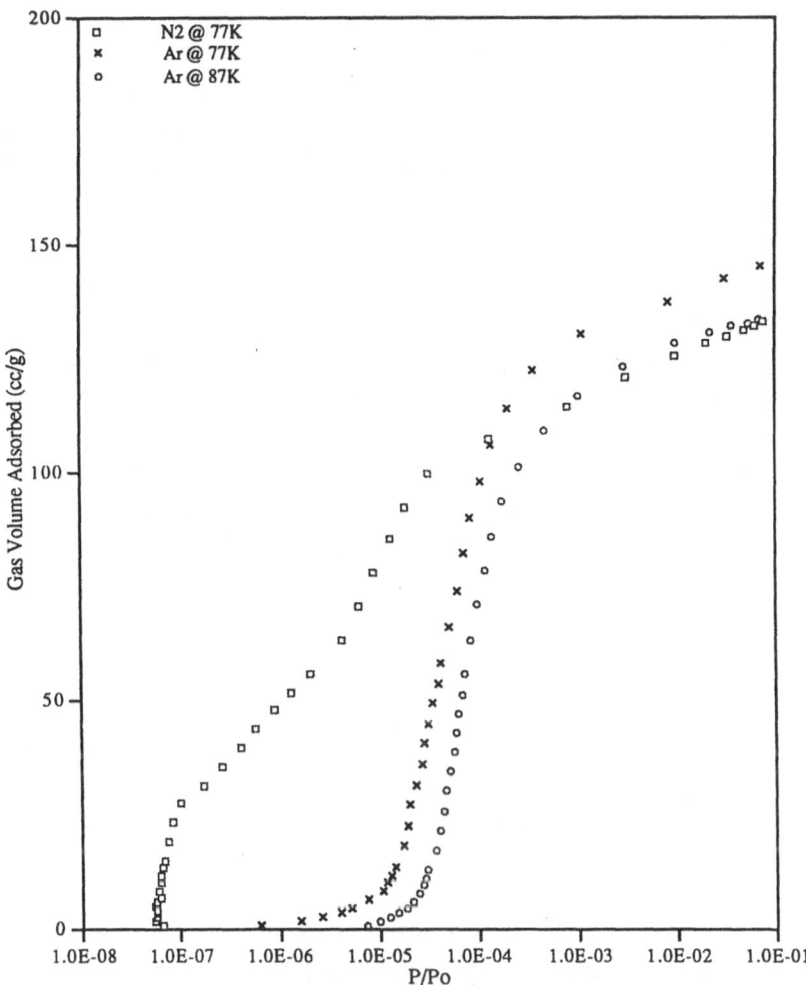

**Figure 2: Linde Type 5A adsorption isotherms for $N_2$ at 77K, and Ar at 77 and 87K.**

The adsorption isotherms for adsorption in zeolites document the behavior of different adsorbents/adsorbates and the temperatures of adsorption. They suggest that $N_2$ and Ar can each be employed in the characterization of pores from 0.4-2nm in diameter. Further, the adsorption isotherms give evidence that the interactions of the adsorbing molecules and the microporous environment differ notably for pores of similar dimensions. These data give indirect evidence of these complex interactions.

### 3.1. Insight and Choices: Adsorbed Species

The adsorption of nitrogen occurs lower in relative pressure than argon at the same temperature for the same samples containing micropores. This contrasts with adsorption in mesopores where adsorption for argon take place at lower relative pressures than for nitrogen gas(6). Thus, microporosity is not simply a three dimensional extension of adsorption onto a surface. The fact that the interaction potential with the surface can overlap for micropores does not mean that the effects are simpy multiplied, otherwise argon would still fill the micropores at lower relative pressures than does nitrogen. Nitrogen does have a quadrupole moment and many of the solids known to posses micropores contain aluminum (e.g., zeolites) which also has a quadrupole moment. It might seem that nitrogen quadrupolar interactions with the solid will increase the interaction potential and induce adsorption at lower pressures. The data does not support this as adsorption for nitrogen occurs at lower pressures for samples which are aluminum free, e.g., for silicalite.

Figure 3 contrasts the adsorption of Nitrogen at 77K in ZSM-5, Silicalite and ZSM-11. Adsorption in Silicalite does not occur at substantially lower pressures than does adsorption in ZSM-5 although there is essentially no aluminum in the silicalite. Also, ZSM-5 and ZSM-11 are similar structurally however the "side-channels" are "zig-zag" in the ZSM-5 while they are straight in the ZSM-11, all else being the same. There is no evidence that this difference in pore configuration has any substantial effect on the adsorption isotherms.

Thus, we are unaware of any explanation for the e nhanced adsorption of nitrogen at an order of magnitude lower in pressure than for argon. This does seem to be a significant clue to the nature of the adsorption in micropores. Possibly, the explanation is even simpler than involking quadrupolar interactions or other higher order interations. A eliptical or "sausage-like" molecule can interact with multiple surfaces within a pore more than can a spherical molecule. Several configurations of interaction are available to span or to orient itself to minimize the interaction potential and, thus, to maximize the depth of the free energy well. In any case, no accepted quantitative explanation is available.

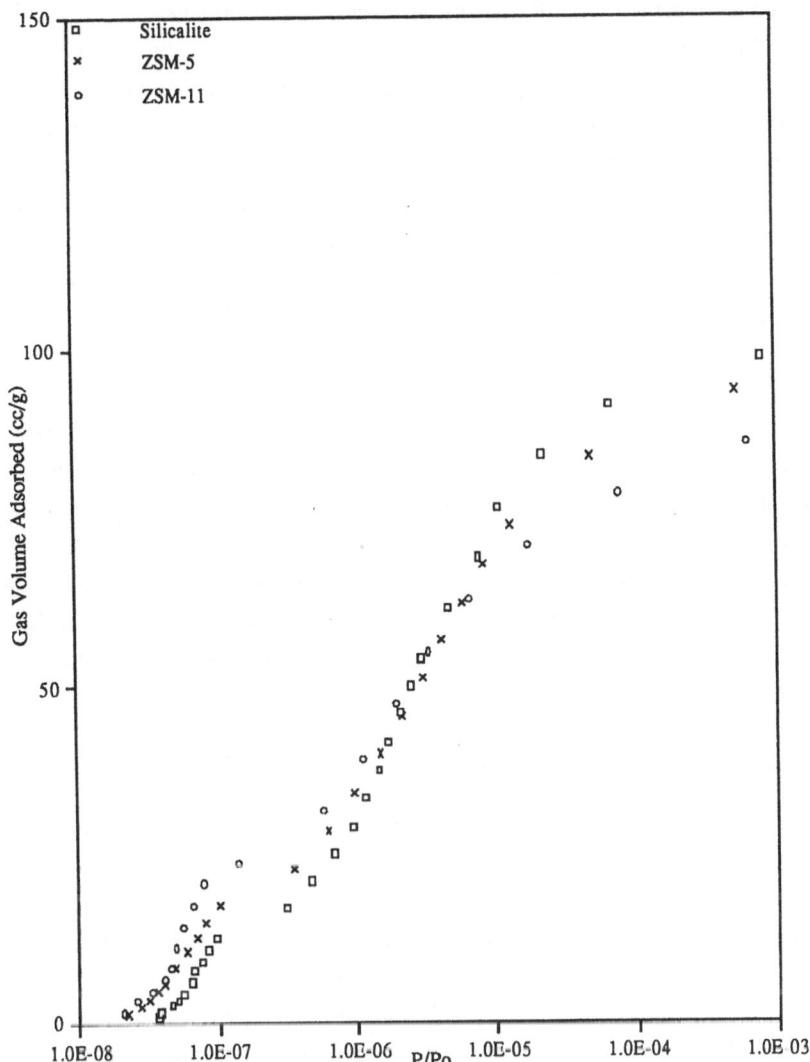

**Figure 3:** Adsorption Isotherms for nitrogen at liquid nitrogen temperature on Silicalite, ZSM-5 and ZSM-11. The apparent break at P/Po ~ 1-2 $\times 10^{-7}$ is an artefact.

The choice of adsorbent and the temperature at which the measurements should be performed is not unique. Nitrogen adsorption at liquid nitrogen temperature will yield data similar to Ar adsorption but at lower presures. This taxes the conventional pressure transducers as adsorption in samples which contain micropores and which are to be studied up to saturation (P/Po ->1) can span more than seven orders of magnitude in pressure. Since the most accurate diaphram transducers have workable ranges of three orders of magnitude, this would require at least three transducers, see section 2. Thus, Ar adsorption would seem to be better as pore filling occurs at higher relative pressures. The next question is whether it is necessary to conduct the adsorption at 87K (liquid Ar) or can the meaurements be made at 77K (liquid $N_2$) since liquid Ar is less easily available and is more expensive. One might worry that the Ar would solidify within the porous network at 77K. There is no experimental evidence that Ar solidifies within micropores (<2nm) at 77K and thus, all other things being equal, Ar adsorption at liquid nitrogen temperature would appear to be the best choice.

However, we do have evidence that the smaller kinetic diameter of $N_2$ allows it to access smaller pores than does Ar and, thus, $N_2$ adsorption might yield additional data for samples containing smaller pores, in rho zeolite for example (5). .Indeed, we strongly believe that both Ar and $N_2$ adsorption isotherms give more data than either does alone. As an example, $N_2$ adsorption can show adsorption hysteresis while Ar adsorption does not. Further, if samples contain large mesoporous volumes, Ar at 77K could solidify in the larger pores, and this would make the analyses slightly more complex. Thus, there is no clear choice for adsorbent and temperature; although, Ar adsorption at 77K appears to be the best place to start.

### 3.2. Interpretation: Pore Volumes and Surface Areas

Several papers in this school and in the literature have realized that the conditions associated with physical adsorption in micropores will differ substantially from the condensation found for adsorption and the filling of larger pores. Many recent studies have advanced the theoretical basis by which these unique interactions have been analyzed. In these studies we will primarily first document the range and nature of the adsorption isotherms for adsorption of Ar and $N_2$ in zeolites. We will then discuss two of the most commonly employed morphological characteristics that are inferred in the analysis of physisorption in microporous solids: surface area and pore volume. Interpretation of micropore size distributions will only be discussed as they relate to the actual adsorption data.

Surface areas and pore size distributions, most often determined by $N_2$ or Ar adsorption, are underlineunderlinelyunderline employed to describe the morphology of zeolites and other microporous solids such as activated carbons and pillared clays. BET theory and T plots (etc.) are employed in these analyses. As examples, pore for $N_2$ at 77K or Ar at 77K or 87K will be discussed. Figure 4 shows a comparison of different adsorbents in Y zeolite. The pore volumes have been related to amount of gas required to fill the pores when converted to the volume of the corresponding liquid using the liquid density which is assumed constant. It is further understood that all the porous space is filled below $P/Po = 10^{-3}$. It is apparent from this data that different volumes of Ar gas fill the pores of Y zeolite at 77 and 87K while the pore volume is not changing by 20+%. Likewise, the similarities in the amount of Ar at 87K and $N_2$ at 77K required to fill the pores does not reflect the > 20% difference in liquid densities for argon and nitrogen. Indeed, the decrease in volume required to fill the pores of a microporous solid continues to decrease as temperature increases as will be discussed below.

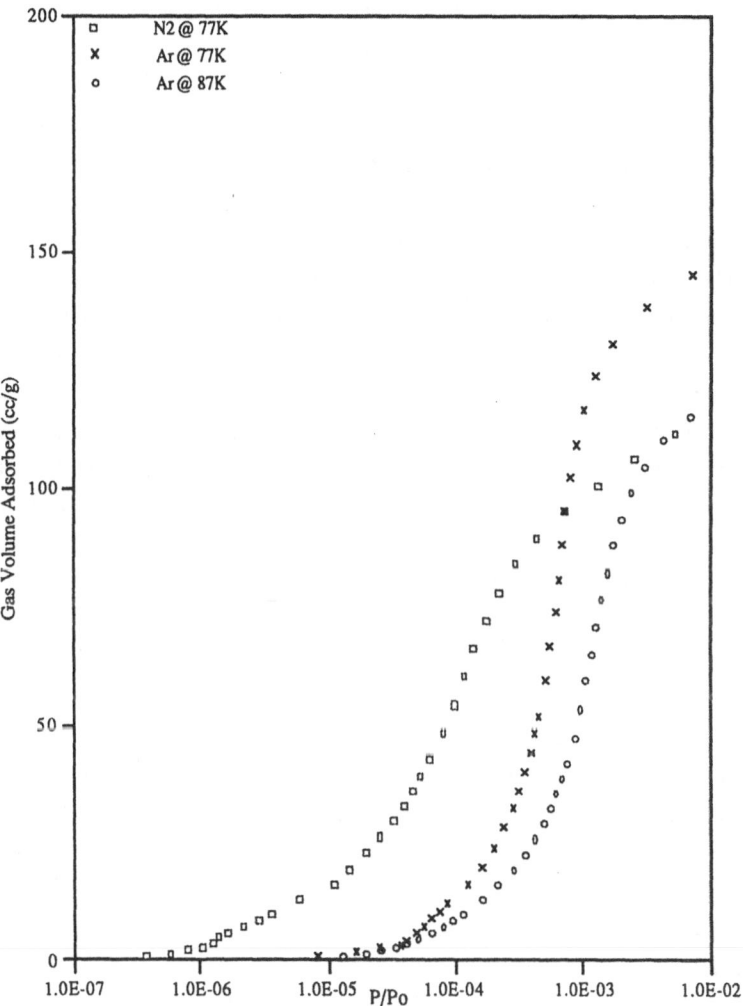

Figure 4: Adsorption Isotherms for nitrogen and argon in Y zeolite.

It is found that there are substantial differences (to greater than 20%) in the estimates of pore volumes depending on the adsorbate and on the temperature for a specific adsorbate. Even less precise are the estimations of the total surface areas using BET theory for any solid which comprises a significant fraction of micropores. Pores from 0.3->~2nm are all filled below relative pressures (P/Po) of 0.05. The amount of $N_2$ or Ar adsorbed at 0.05 = P/Po represents several multilayers for the larger pores and less than a monolayer for the smaller pores... the fraction of a monolayer adsorbed below 0.05 varies by over an order of magnitude depending on the pore dimensions. BET theory is based on estimating the monolayer volume for data on physical adsorption at relative pressures from 0.05 to ~0.3. BET theory will vastly over estimate the surface area for solids which contain the larger pores and will under estimate the effective surface areas for samples which comprise the smaller pores. This variation of in accuracy in the calculated surface areas by up to over an order of magnitude needs to be realized.

## 4. Spectroscopic Insight into Physisorption in Micropores

The state of molecules "physically" adsorbed into micropores is obviously different from the states of species adsorbed in solids which do not contain micropores. Theories which describe the species adsorbed by high surface area solids which comprise pores larger than 2nm in diameter are universally accepted as reasonable. These theories span the sub-monolayer to multi-layer to condensation phenomena involved in these adsorption isotherms. Adsorption in micropores is different in that the interactions between the adsorbing species and the microporous surface is larger for surfaces that are not atomically flat. Heats of adsorption are substantially greater and the amounts of steric interactions are vastly enhanced.

It is first necessary to understand the nature of the species adsorbed. Unfortunately, there are only a limited number of *in situ* studies of adsorbed species. These studies will be reviewed and the resulting insight into the nature of the state of the physisorbed species will be summarized. These include [15]N, [13]C, H and [129]Xe NMR studies of adsorbing species. It is found that the nuclear motion of the adsorbed species is modified compared to the gas phase but that the mobility is still larger than that for solid adsorbed phases. The rotational mobility for adsorbed molecules is significantly restricted. Infrared *in situ* studies document that the vibrational motion of adsorbed species is also modified. ESR studies document that the valence electrons of physisorbed species are perturbed.

In addition, spectroscopic studies of the solid during physisorption have documented that the solid phase can be perturbed due to adsorption. These studies include solids NMR (e.g., [29]Si and [27]Al MAS-NMR) as well as in situ X-ray diffraction and far infrared studies during adsorption in crystalline solids (e.g., zeolites). The solids can flex to accommodate the adsorbing species.

It should be apparent that the state of the adsorbing species as well as the state of the adsorbent are modified during the adsorption process. This presentation will document the experimental evidence (primarily *in situ*) that the nature of the species adsorbed differs from gaseous, liquid, solid phases or from conventional adsorption phenomena for adsorption in high surface areas solids whose pores are larger than 2nm.

Prior studies have employed a variety of spectroscopies to study the species adsorbed and the adsorbents during adsorption in microporous solids. These changes in the adsorbed species and the adsorbate are discussed in the two sections of this chapter: The state of the adsorbed species in 4.1 and the influence of the solid structure in 4.2. Further, the insight into nature of the adsorbed species will focus on static in situ studies of physisorbed species and not on studies of diffusion in microporous materials which are discussed in other sections of the Advanced Study Institute, ASI.

### 4.1 Physically Adsorbed Species

### 4. 1.1. Adsorption of Homonuclear Molecules

This section will first review what is known about the nitrogen and hydrogen during physisorption. Infrared, NMR and microwave studies have given insight into the state of the adsorbed nitrogen as it covers the surface and fills the micropores of a solid. As discussed elsewhere in this chapter (see sections 3) and throughout this ASI, the interactions between an adsorbing species and micropores differ substantially from conventional physisorption phenomena. So, what is the state of nitrogen physisorbed in micropores ? The calorimetric studies recently conducted by Roquerol et al. and the Xe studies by Fraissard et al. will not, however, be discussed at length in this chapter as they are also presented and analyzed in another section of this ASI.

### 4.1.1.1. Infrared Studies

The infrared of adsorbed molecular nitrogen seems at first to be a difficult study as there is no infrared spectra for $N_2$ in the gas phase, a diatomic. Chang and Kokes were the first to detect the infrared spectra for $N_2$ adsorbed on ZnO(7). A narrow band appears at 2337cm-1 and displaces the band for molecular hydrogen which was present at 4019cm$^{-1}$ (see brief discussion below). The nitrogen vibrational band is slightly shifted from the gas phase Raman band (2335 cm-1). At the same time the ZnH (~1700 cm-1) and OH (~3500cm-1) bands for hydrogen dissociated on ZnO are perturbed by about 20cm-1 (up for ZnH and down for OH). The heat of adsorption of nitrogen is ~ 10 to 15% of the surface and the heat of adsorption is ~4.6 kcal/mole. The nitrogen is believed to adsorb due to the interaction between the $N_2$ quadrupole and the field gradient at the surface approaching the active sites on ZnO; however, somewhat surprisingly, the change in vibrational frequency from the gas phase was not large.

48

Zubkov et al.(8) also used infrared is studies of $N_2$ adsorption on η- alumina and detected a substantially larger shift in the vibrational band, i.e., ~30 cm$^{-1}$ to 2360 cm$^{-1}$. The authors conclude that the shift is due to interactions between $N_2$ and the Lewis-acid sites on the surface (as Al$^+$).

Sasse and Förster(9) studied the infrared spectra of $D_2$, $N_2$ and $O_2$ adsorpbed in NaY zeolite at 90K as seen in Figure 5. They found that the infrared signal for nitrogen exhibited two lines in the infrared at 2337 and 2334 cm$^{-1}$ are (slightly) blue shifted in comparison with the gas phase free molecule at 2330 cm$^{-1}$. Interaction potential calculations were employed to determine that negative quadrupolar interactions dominated the resultant vibrational spectra. The authors conclude that the two bands are due to $N_2$ adsorbed near the cation (Na$^+$) in the zeolite and a second band (at 2334 cm-1) is due to the formation of a dimer complex on the surface [Na$^+$(N$_2$)$_2$]. In contrast the spectrum of $D_2$ exhibited three bands at 2987 2963 and 2946cm-1 which were red shifted from the gaseous Raman spectra (2993 cm$^{-1}$). The infrared spectra for adsorbed $O_2$ was not shifted from its gas phase spectra when adsorbed.

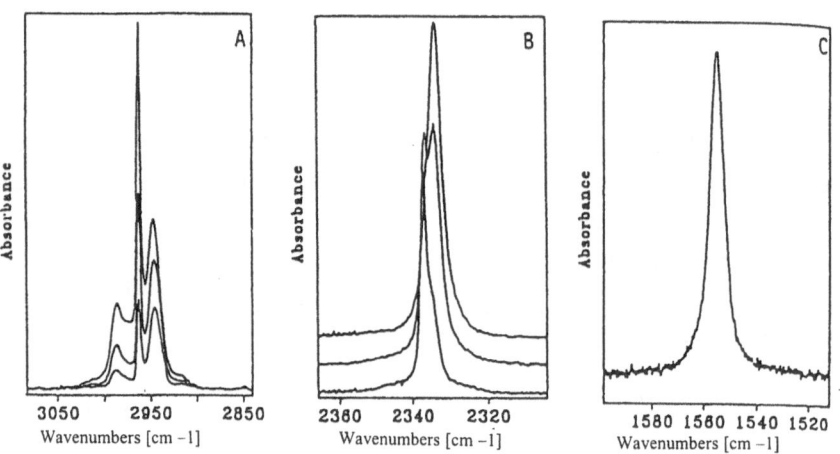

**Figure 5: Infrared adsorption spectra of $D_2$ (left), $N_2$ (center) and $O_2$ (right) physically adsorbed on NaY zeolite at 90K.** *(9)*

The infrared spectra of adsorbed hydrogen has also been used as a probe of oxide surfaces. Again, Kokes at al.(10) were one of the first to detect the infrared spectra for molecular hydrogen. They detected molecular bands for $H_2$(at 4019cm-1), HD(at 3507cm-1) and $D_2$ (at 2887cm-1) adsorption on ZnO. These do represent substantial red shifts from the gaseous Raman bands at 4161, 3627 and 2990cm$^{-1}$ respectively. Interestingly these shifts are in the same direction as noted above for the interaction between $D_2$ and zeolites, whereas the sites on ZnO are certainly a basic. Kazanski et al.(8,

*11, 12*) have studied the adsorption of molecular $H_2$ at 77K on a series of zeolites. All of the infrared bands that were detected were red shifted by 10 to 90cm-1 from the gas phase Raman band. As many as three bands were evident for hydrogen adsorption (at 4096, 4125 and 4150cm$^{-1}$ on NaY zeolite for example). Their quantum-chemical calculations were employed to describe the interactions between molecular $H_2$ and the Lewis-acid sites (and the basicity of neighboring oxygen atoms) modelled on the zeolites. However, a blue shifted band (4229-4239cm$^{-1}$) is also detected at high pressure. Even more blue shifted bands are evident for adsorption of molecular $H_2$ on the protonic (acidic) forms of zeolites. Bands at 4105 and 4125cm-1 are attributed to $H_2$ interacting with the Brønsted acidic bridging protons while bands at 4010, 4035 and 4060cm-1 are attributed to interacting with the Lewis-acid sites formed by degassing the zeolites at higher temperatures.

**Infrared Frequencies (in cm$^{-1}$) of Molecular $H_2$ and $N_2$ on Oxide Surfaces at ~77K**

| Species adsorbed | | $N_2$ | $D_2$ | $H_2$ |
|---|---|---|---|---|
| **Gas (Raman Bands)** | | 2330-5 | 2990-3 | 4161 |
| **Surface** | ZnO | 2337 | 2887 | 4019 |
| | η-Al$_2$O$_3$ | 2360 | 2785 to 2600 | 4030 4125 |
| | NaY | 2334 2337 | 2987 to 2946 | 4096 to 4150 |
| | H-Y | | | 4035 4105 |
| | NaZSM-5 | | | 4110 |
| | HZSM-5 | | | 4125$_b$ 4105$_b$ 4060$_L$ to 4010$_L$ |
| | SiO$_2$ | | | 4125 |
| | MgO | | | 4125 4030 |

Although the infrared studies of the adsorption of molecular $N_2$ or $H_2$ have primarily been interpreted as if the molecules were probes to the acidity of the surface of oxides, the infrared spectra of adsorbing diatomic molecules are also probes to the interaction between these molecules and the surfaces. It seems clear that neutral surfaces ($SiO_2$) as well as basic surfaces (ZnO and MgO) and the "acidic" zeolites (in Na or protonic forms) induce quite similar changes in the vibrational frequencies for $N_2$, $H_2$ or $D_2$. All of the infrared spectra of these adsorbed diatomics are shifted in the same direction (decreasing wave number) compared to gaseous Raman resonances. The notable exception is the increase in resonance wave number for adsorption of $N_2$ on η−alumina. We infer from these comparisons that adsorption on surfaces induces a polarizability in the $N_2$ or $H_2$ (or $D_2$) which gives rise to a detectable resonance in the infrared. These interactions result in an increase in the vibrational frequency of the adsorbed species as they increase the interactions, and thus, the perturbation of the polarizability of the adsorbed species.

### 4.1.1.2. Microwave Studies

The adsorption of microwave radiation during nitrogen adsorption was employed in a series of articles by Guermeur et al.(*13, 14*). The silica was placed in a controlled environment within a microwave cavity approximately resonant at 9600MHz. As nitrogen is adsorbed on the surface and within the pores of silica, the resonant frequancy of the cavity changes. First, the resonant frequency decreases and then it increases as seen in Figure 6. The maximum shifts in the resonant frequency can exceed 2500kHz and depends on the temperature at which the silica had been evacuated up to 800°C. The changes in resonant frequancies seem to depend on the concentration of surface hydroxyls. The polarizability of the hydroxyls is changed by adsorption and interaction with physisorbed molecular $N_2$. The authors explain the resonant frequancy variation by this interaction between $N_2$ and the hydroxyls. Several states of physisorbed nitrogen are inferred. One type of "adsorption sites" involve interactions with isolated hydroxyls while another is due to physisorption without interaction with hydroxyls. Another type of interaction with associated hydroxyls (geminal and viscinal) is also inferred. The authors illustrate how the surface area can be calculated from the changes in resonant frequancy with the amount adsorbed. Again, these studies document that there are several forms of nitrogen physisorption on porous solids. This adsorption changes the dielectric susceptability of the surface but not in a monotonic fashion as the physisorption can both increase and decrease the dielectric susceptability. During these studies the authors were able to measure the changes in surface polarizability, p, with temperatures from 63 to 100K by the following relationship: $p \sim (T_0 - K)/T$, where $T_0$ is 100K. Although these preliminary studies were conducted for adsorption on Aerosil (from SERVA®) and controlled pore glasses (from Electronucleonics®), the technique should yield additional insight for adsorption in more microporous solids.

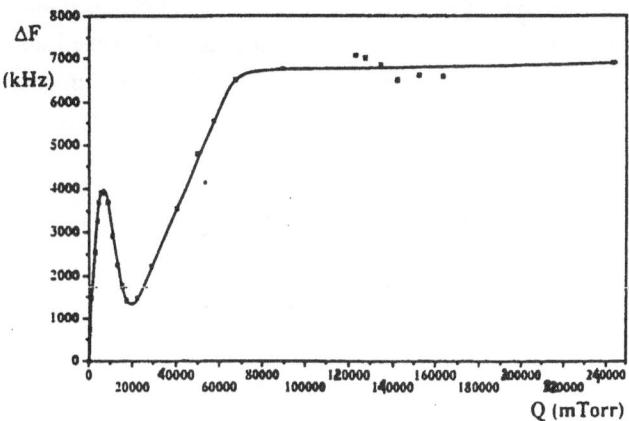

**Figure 6: The Change in resonant microwave frequence due to the adsorption of nitrogen at 77K. (*13*)**

**4.1.1.3. NMR Studies (*15*)**

We have studied the state of physisorbed nitrogen above and below liquid nitrogen temperature with the use to $^{15}N$ NMR to gain insight into the state of physically adsorbed nitrogen within zeolites, e.g., ZSM-5. The samples were evacuated and exposed to 96+% $^{15}N_2$ at 77K at pressures of 0.08 and 0.25 Atm. for several days prior to measuring the NMR spectra. At higher and lower temperatures the pressures will vary accordingly in the sealed sample ampoules. However, the gas volume in the sample ampoules was small (est. $\ll 1$ cm$^3$ STP), the pressure was reduced ($<0.25$ bar), and the adsorption was done in the region of the isotherm where the changes in volume adsorbed with pressure are insignificant (a plateau in the isotherm). Therefore, even these changes in temperature (and pressure) did not substantially change the volume adsorbed within the micropores of the ZSM-5. The temperature in the NMR was controlled to within 0.1K by a liquid He cryostat, i.e., ATC4 controller from Oxford Instruments. We confirmed that the $^{15}N$ NMR spectra for adsorbed nitrogen were completely reversible by retaking the spectra after the temperature excursions and even after the samples were removed from the cavity for over 24 hours and then were reintroduced.

Our first studies (*16*) involved adsorption at a relative pressure of 0.25. As the temperature is decreased from 77K, a phase transition from a liquid-like condensate to a restricted/confined liquid with reduced molecular motion is inferred from changes in the NMR signal width. At least two superimposed NMR signals are evident and the broader signal dominates the spectra as temperatures below 65K. This is seen in Figure 7 which shows the decomposition of the signal at 63K and P/Po =0.25. However, the transition from this state (phase) to a solid phase is not evident as the temperature is lowered to below 11K. The μ-pores restrict the molecular motion and, at the same time, change (depress the transition temperature for) the Liquid-Solid equilibrium of the condensing nitrogen.

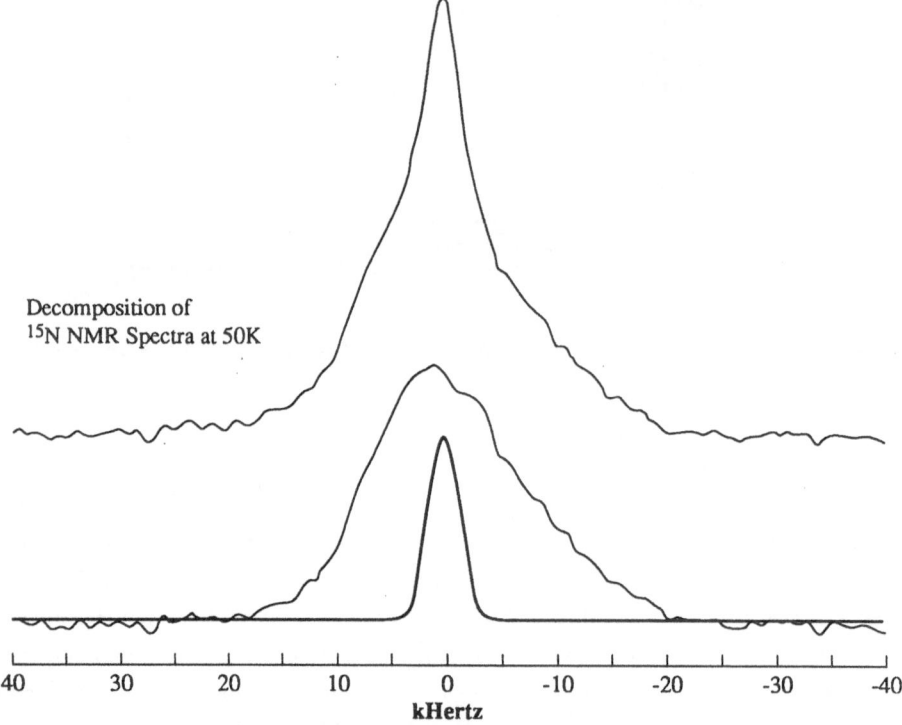

Figure 7: Decomposition of the $^{15}$N NMR spectra at 50K into two component lines for enriched Nitrogen initially adsorbed in ZSM-5 zeolite at P/P$_0$ = 0.25 at 77K.

The $^{15}$N NMR for nitrogen adsorbed at a relative pressures of 0.08 is shown in Figure 8. There is a difference in the breadth of the NMR signal with temperature. The narrow line seen for the sample at 0.08 relative pressure and 77K is characteristic of nitrogen solely in a liquid state. Note that >>95% of the nitrogen in each of the samples is adsorbed within the zeolite pores, i.e., there is little nitrogen present in the gas phase over the samples. It is apparent that the intensity of the broader signal is increasing as the temperature is lowered and that this increase in intensity is at the expense of the narrower signal.

The broader signal that is evident in the spectra at a relative pressure of 0.25 is in a more restrictive environment than the sample at a relative pressure of 0.08. The rotational motion of nitrogen is substantially more limited at the higher pressure; although, the difference in the volume adsorbed is only ca. 10%.  This is seen in Figure 8 for samples at the two pressures at 63K.

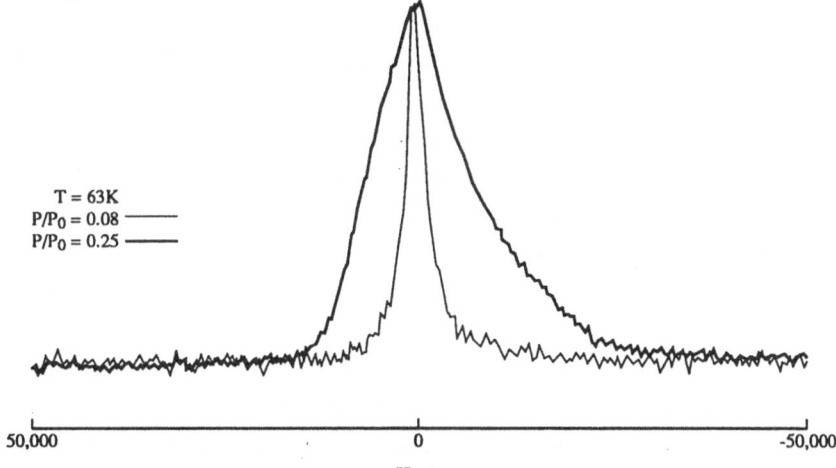

Figure 8: Comparison of the $^{15}$N NMR for nitrogen adsorbed in ZSM-5 at relative pressures of 0.08 and 0.25 at 63K.

The change from a liquid-like to a broader signal which reflects a more restricted environment is progressive as contrasted with a standard temperature dependent phase transition.

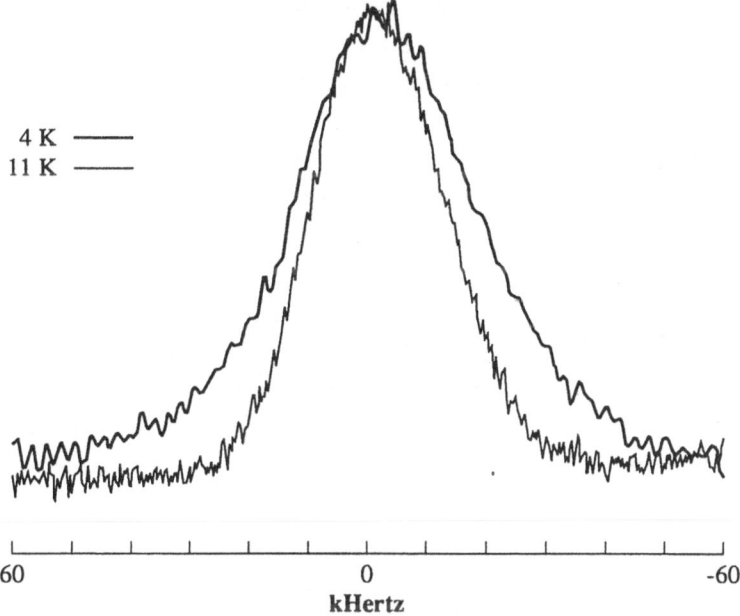

Figure 9: Comparison of the $^{15}$N NMR spectra at 11K and 4K (upper spectra) for enriched Nitrogen initially adsorbed in ZSM-5 zeolite at P/P$_0$ = 0.25 at 77K.

The spectra at 11 and 4K are overlayed in Figure 9. It is apparent that the signal is continuing to broaden between these two temperatures. The NMR signal for a solid would not continue to become broad as the temperature decreases, i.e., the solid state does not substantially change molecular mobility with temperature. This continued decrease in molecular motion and the resultant broadening of the NMR spectra suggest that the nitrogen is not yet a solid to below 11K. This is the first evidence for this freezing point depression of nitrogen due to adsorption within micropores. It is well known that the freezing point of water and, more generally, any physisorbed species is depressed when adsorbed in microporous solids.

The transition between a narrow and a broad NMR spectra is not abrupt either as the temperature is lowered or the pressure is increased. It can be inferred that the more restricted phase that is evident in the broader signal is associated with a denser phase than the narrow liquid-like signal evident at a relative pressure of 0.08 and 77K. Thus, a slightly greater volume of nitrogen is required to fill the pores. This more restricted phase is progressively formed with increased gas phase pressure and/or decreased temperature.

The two phases that are simultaneously present within the pores of the zeolite do not readily exchange on an NMR time scale and, thus, two superimposed spectra are evident and not an average of a broad and a narrow line that is averaged by exchange to yield a single spectra with a single intermediate line width. We propose that this is because the two phases are spatially separated, i.e., the two phases are formed within different portions of the void network within the zeolite. Since the new phase that is formed is more restricted in its rotational mobility than the liquid-like phase that is also evident, we propose that the broader spectra is due to a phase forming within the smaller dimensions of the void network, possibly first in the zig-zag channels. Next (with increasing pressure or decreasing temperature), it might be forming in the straight channels and finally at the channel intersections.

The progressive nature of the transformation may reflect limited exchange at the interface between these various spatial regions of the void network. Alternately, there might be a matching between the spacing between the nitrogen molecules in the more restricted configuration and the pore dimensions. Thus, for example, if the channel intersections were exactly an integral $n$ times the distance between nitrogen molecules in the more restricted phase, the more restricted phase might first form at the channel intersections.

Without further $^{15}N$ NMR data for nitrogen adsorption in other zeolites (or with variation in the Si/Al ratio's or within ZSM-11, as an example), we suggest that formation of the more restricted sorbed phase would occur within the smaller dimensions of the void network first. This would be followed, progressively, by formation of this phase within void spaces of increasing dimensions.

### 4.1.2 Adsorption of other Molecules

Infrared and, to some extent, NMR spectroscopies have been employed extensively to study adsorption in zeolites. However, the vast majority of these studies have involved probe molecules to determine the interactions involved in chemisorption by the acidic sites, associated with the exchangeable cations or by metals supported within the zeolite. The purpose of this chapter is to understand physisorption and there are far fewer studies of the spectra of physisorbed species. The exceptions are a number of studies of water, benzene and methane. The studies of benzene adsorption will be briefly reviewed as they give considerable insithe into the similarities and differences between interactions with small and large physically adsorbed molecules.

Benzene adsorption in the micropores of zeolites has been a popular area for study primarily as it relates to transport and selectivity for many petrochemical reactions(17-26). Our focus is on the state of the physisorbed benzene and not on the transport which is extensively reviewed by others in this ASI and is described in other sections of this book.

Wang et al.(25) employed diffuse reflectance FTIR to study the adsorption of benzene at 300K in HZSM-5 zeolite. The authors cite the several studies (including NMR, neutron diffraction and IR) for adsorption of benzene in faujasite zeolites which conclude that benzent can interact with the cations, the oxygen atoms or the 12-membered oxygen rings. Perturbation of the various hydroxyl groups (at 3611cm-1 and 3739 cm-1) show that the first benzene molecules adsorbed on the more acidic protons, i.e., at 3611cm-1 for SiOHAl surface species. The difference spectra shows that these bands are surpressed while the benzene only supresses the isolated hydroxyls, i.e., at 3739 cm-1 for the SiOH surface species. The changes in the spectra of adsorbed benzene is minimal in these studies compared to liquid benzene except a weakening of the ring stretching vibration at ~1480cm-1 is detected. This contrasts with infrared studies of benzene adsorption for adsorption on NaZSM-5 where Jentz and Lercher found that two pairs of out-of-plane CH vibrations. One was attributed to interactions between the benzene and the cations and the second attributed to the interactions with the Brønsted acidic sites. Coughlan et al(20), compared hydrogen and cation forms of Y-zeolitre and also concluded that the benzene interacted strongly with the protons in the lattice. They also detected out-of-plane C-H stretches for benzene strongly adsorbed in Y-zeolite.

### 4.2 The Adsorbent During Physical Adsorption

Our earliest perceptions as to the pore structure within a zeolite has depended upon the visualization of the Si(Al)-oxygen crystalline bond network. This representation and analyses depends upon a negative image of the pores based primarily upon X-ray diffraction, XRD, studies of the solid structure. Until recently the inorganic lattice has been considered as an inflexible network structure. They are, however, effectively three dimensional, inorganic polymers. Within in the last few years, several studies of zeolites have demonstrated that the framework structure of zeolitic materials is flexible. This flexibility was seen during sorption of organic molecules. Recent studies employing solids nuclear magnetic resonance, NMR and in situ XRD have documented that the shape of the adsorbing pores can change on adsorption. More recently, detailed spectroscopic studies of adsorption and of adsorbing molecules have begun to provide a picture of the pore structure and the sorbing species during sorption. In situ infrared (specifically FTIR) can also be employed to understand the dynamic configurational changes in the sorbing species and the energetics of these interactions as documented elsewhere in this chapter.

High silica zeolites are crystalline solids primarily consisting of three dimensional linked silica tetrahedra. For inorganic solids the term crystalline connotes a rigid framework structure. One readily envisions dense solid particles or powders. For zeolites the term crystalline denotes a regular geometric arrangement of the atoms which can involve void fractions well in excess of 50%. The atomic coordination is explicitly four for the silicon and two for the oxygen within most zeolites, this contrasts with atomic coordination for close packed crystalline solids of up to twelve. This "looser", but still regular, arrangement can have profound consequences. Fewer bonds or bond angles need to change for long range fundamental changes in the framework to occur.

### 4.2.1. Flexibility of Zeolites: Changes in Pore Shapes

In 1984 Hay and Jaeger(27) used powder X-ray diffraction to demonstrate that a transformation between the orthorhombic and monoclinic symmetric arrangement of the lattice occurs with heating ZSM-5 zeolites to modest temperatures (<350K). The transformation was affected by the Si/Al ratio or the adsorption of hydrocarbons within the zeolite. They found that the transition occurred at 317-325K and for the high silica ZSM-5 (as silicalite), at 295K for an intermediate Si/Al ratio of 198/1 and at <272°K for ratios <110/1. This facile transformation between crystal symmetries due to different pretreatments of ZSM-5 has been known for the last several years(28).

The use of MAS-NMR, and XRD on a variety of high silica zeolites was studied by Kokotailo et al.(29). For ZSM-5 they found a reversible discrete transition from monoclinic to orthorhombic symmetry between 343 and 362K. The change was evident in both the NMR and XRD spectra. More recently, a similar transition was found at 318-323K for ZSM-11, a closely related pentasil zeolite(30).

The temperature at which this transition occurs depends on the Si/Al ratio within ZSM-5 and the discrete, reversible transition can be measured by [29]Si NMR, confirming prior x-ray studies. A similar dependence is found for ZSM-11. Phosphorous modification of the ZSM-5 does not change the transition temperautre; however, steam treating of the zeolite does. These trends are seen in the data below:

**Monoclinic to Orthorhombic Transition Zeolites from [29]Silicon MAS NMR**

| Zeolite | Si / Al(Al-NMR* ) | Additives | Transition Temperature |
|---|---|---|---|
| **From Union Carbide, Mobil & P. Jacobs** | | | |
| Silicalite | > 1000 est. | | 340 K |
| H-ZSM-5 | 263* | | 322 K |
| H-ZSM-5 | 24.6* | | 313 K |
| ZSM-5 | 11.6* | | no transition |
| **From Haldor Topsøe** | | | |
| ZSM-5 | 500 est. | | 343 K |
| ZSM-5 | 43 est. | Steamed | 332 K |
| **From J. Lercher** | | | |
| H-ZSM-5 | 82* | | 327 K |
| H-ZSM-5 | 86.5* | 1% P | 328 K |
| H-ZSM-5 | " | 5% P | 326K |
| **ZSM-11 from Mobil & M. Davis** | | | |
| ZSM-11 | Davis >1000 est. | | 320 K |
| ZSM-11 | Mobil 160 est. | | 313 K |

**est.= estimated by those providing the samples but not independently measured.**

### 4.2.2. Pore Shape Changes due to Adsorption

Early in 1984, Fyfe et al.(31) used [29]Si MAS-NMR to demonstrate that the adsorption of xylene, benzene and pyridine induced a change in the Si environment in the zeolitic framework of ZSM-5. In other words, the lattice flexed on adsorption. Concurrently West used [29]Si MAS-NMR to document that water, which can enter the channels, and trimethylbenzene, which cannot, do not change the silicon spectra; however, other hydrocarbons, particularly benzene, induce significant shifts to higher field values for up to two-thirds of the lattice silicons(32). Parise et al.(33) used neutron diffraction to demonstrate a change in the "degree of ellipticity" of zeolite RHO with temperature (from 11 to 573K). These studies document both the flexibility and the sensitivity of the lattice force constants to modest temperatures. Adsorption gives rise to channel shape changes(33). Mägi et al.(34) showed, again by [29]Si MAS-NMR, that the nature of the silica tetrahedra, specifically the Si-O-Si bond angles, change with the nature and concentration of both anions and cations in silicate lattices.

The infrared studies of Primet et al. were used to infer a loss of ring symmetry and an out-of-plane bending for non-dissociative benzene sorption on Y-zeolite(24). Jacobs et al. showed a correlation between the infrared vibration frequencies of the T-O bonds (T= Al or Si) and the calculated average electronegativities of the solid molecular bonds(35). Increases in the IR stretching frequencies with increasing silica content were found The calculations were used to estimate the force constants and equilibrium distances of the T-O bonds from the lattice electronegativities. These correlated very well with the measured IR frequencies attributed to the bonds. Each of the modes (symmetric: 990-1100cm$^{-1}$, asymmetric: 1150cm$^{-1}$, bending: 420-480cm$^{-1}$, double ring: 580-610cm$^{-1}$ and external linkage: 550-620 & 750-840cm$^{-1}$) shifts with the lattice modifications. By implication, these frequencies can be used to follow the changes in the force constants of the zeolitic framework.

To understand the flexibility of the zeolitic lattice, it is necessary to be able to study the nature of the Si-O bonds which comprise the solid network. It is known that the amount of hydrocarbon (e.g., isobutane) adsorption can increase dramatically as the temperature is increased. It then decreases as temperatures are further increased (as expected). The explanation may be that in this temperature range the lattice becomes flexible.

Cyclohexane adsorption on ZSM-5 (and other ZSM-like zeolites) was recently studied by Chester et al(36) They concluded that, "the effective pore dimensions are larger than the crystallographically calculated pore sizes... thus siliceous zeolites appear to behave as if the pores are elliptical under hydrocarbon adsorption... " Zeolite flexibility was not envisioned. The amount and rate of cyclohexane adsorption was half that for n-hexane adsorption. Wu et al.(37) have shown that the diffusion coefficients do not change as ZSM-5 approaches saturation with cyclohexane but it increases as benzene or xylene adsorb. They also note that steaming and the addition of K or P increases the diffusion coefficient for each of the species while it decreases the pre-exponential and activation energies for diffusion. The calculated "activation energy" for diffusion of cyclohexane was 53KJ/mole, much larger than found for fluid phase diffusion. In general the diffusion coefficient decreased an order of magnitude with phosphorus or magnesium impregnation, and they attributed the change to pore blockage. They did conclude, however, that geometric factors play "an important role in the diffusion process."

We have investigated the changes in the lattice of ZSM-5 due to the adsorption of cyclohexane. Cyclohexane was chosen as it is slightly larger than benzene (6.0Å for cyclohexane, and 5.85Å for benzene and p-xylene)(8) whose sorption into ZSM-5 has already been studied in considerable detail (17, 18, 25, 31, 38). The increase in kinetic diameter from benzene to cyclohexane (and the increase in molecular flexibility) implies that there will increased interactions between the sorbent and sorbate. These results will be compared to the adsorption of p-xylene, a molecule of nominally the same kinetic diameter as benzene and close to the same overall dimension as cyclohexane but with less flexibility.

Figure 10 shows a comparison of p-xylene and cyclohexane adsorption in H-ZSM-5. It is readily seen that almost exactly half the amount of cyclohexane is adsorbed than p-xylene. It had been suggested that the transition in the p-xylene adsorption at half filling was due to initial adsorption into the straight channels followed by adsorption into the zig-zag channels of ZSM-5.

**Adsorption of p-xylene and cyclohexane on ZSM-5**

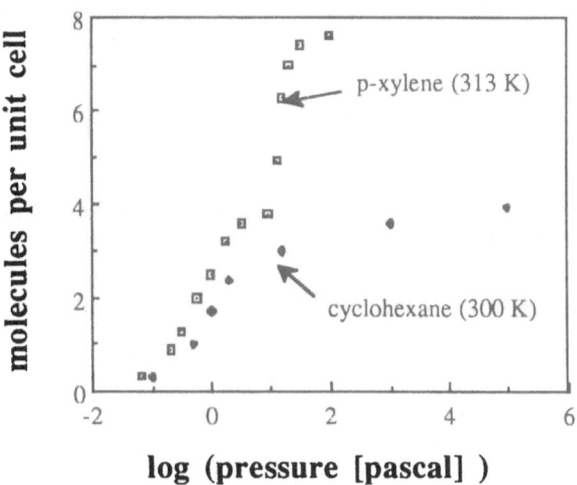

**Figure 10: Isotherm of cyclohexane adsorption in ZSM-5 compared to the adsorption of p-xylene (39)**

Figure 11 shows the X-ray spectra of H-ZSM-5 before and after the adsorption. The spectra following p-xylene has changed dramatically and is consistent with a change from monoclinic to orthorhombic symmetry even thought the loading was only to 10% of saturation. However, no transformation is found for cyclohexane adsorption at saturation. We concluded that the cyclohexane was only filling the straight channels (probably at the intersections) but not the zig-zag side channels and, thus does not induce the change in symmety. On the other hand, the FTIR seen in Figure 12 shows some changes in the Si-O-Si stretching frequencies with cyclohexane adsorption which reflect changes in bond angles. The straight pores are becoming more eliptical to accomodate the cyclohexane while the overall symmetry is left the same. These studies demonstrate that cyclohexane is anisotropically filling the straight channels and distorting them while the p-xylene accesses the zig-zag channels and causes the whole lattice to deform to orthorhombis symmetry. This occurs during the initial stages of adsorption and the lattice transformation is communicated rapidly throughout the whole structure.

60

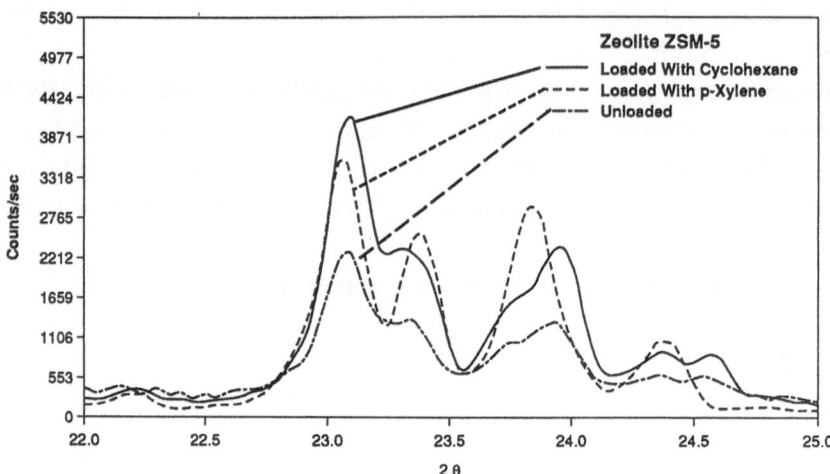

**Figure 11: XRD studies of ZSM-5 before and after adsorption of benzene or p-Xylene (10 % of saturation) (39)**

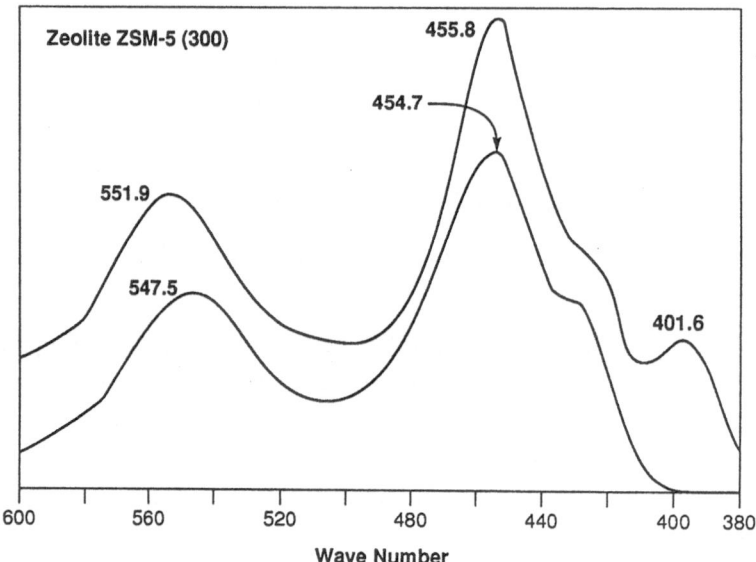

**Figure 12: Infrared spectra of ZSM-5 before (lower curve) and after (upper curve) adsorption of cyclohexane (39)**

It might also be noted that the 401.6 $cm^{-1}$ band in ZSM-5 loaded with cyclohexane is probably a ring stretch that is shifted into the spectra due to the presence of cyclohexane.

### 4.2.3. Comments on the Implications for Adsorption/Catalysis

It is well known that the selectivity of acid catalyzed reactions in the zeolite framework is enhanced with the addition of promoters (metals, oxides or phosphates). The reasons for these selectivity increases are not fully understood, however, pore blockage(40) and pore-mouth constriction(41) have been suggested. None of these explanations are totally satisfactory. In terms of the partial pore blockage or pore mouth constriction models, all estimates of the molecular sizes compared to ring size with any additional atom(s) suggest that if each is rigid there will be no passage past the obstructing atom(s). In terms of the pore blockage and site poisoning models, there should be a concomitant dramatic decrease in overall activity. The argument is that the reaction becomes extremely diffusion controlled. The specific activity for the production of the desired product is not always decreased dramatically; furthermore, a decrease in the reaction activation energies, as would be expected for a transformation to diffusion control, is not found(42).

One of the fundamental assumptions inherent in each of these explanations is that both the zeolitic framework and the sorbing, diffusing, molecules are rigid. In the recent International Zeolite conference, this was one of the "basic assumptions" made by Nowak and Cheetham in their simulation of adsorption and diffusion in zeolites(43). As has recently been documented, zeolites are flexible and change with the inclusion of sorbing organic molecules within the framework structure. It has also been suggested that the Si-O bonds may break-and-reform during adsorption at elevated temperatures ($>300^{o}C$)(44).

### 5. Summary: future needs

It should be apparent that considerable information is available from the accurate measurement of adsorption isotherms over microporous solids. Care must be taken to perform these measurements accurately and assure that equilibrium is achieved. The resultant adsorption isotherms are very sensitive to the pore dimensions, varying by several orders of magnitude in pressure for pores from 3 to 13Å in diameter. $N_2$ or Ar at 77K are both reasonable measurements; however, Ar adsorption at 77K occurs an order of magnitude higher in pressure and may be preferred except for the smallest pores (<4Å). More of these high resolution (and high accuracy) measurements are needed to provide a basis for the theorist to analyze the nature of the interactions that are occurring.

It is further apparent that the state of molecules physisorbed in micropores differs substantially from our current understanding of molecules adsorbing and condensing on solids whose morphology is comprised of larger pore networks. As a consequence, conventional surface area and pore volume measurements should not be used for solids which contain significant microporosity. More *in situ* studies are needed to understand the nature of molecules physisorbed in micropores. The molecules do not condense to form a liquid nor do they progress from a monolayer to form multilayers. The adsorbing species interact strongly with the curved surfaces as well as with each other in the confined voids within micropores. Further, the solid can deform to accommodate the adsorbing species. These complex mutual interactions beg for further experimental study and these insights must be incorporated into future theories that explain physisorption in microporous materials.

## 6. References

1. K. Kaneko, 1994, in *Studies of Surface Science and Catalysis* , J. Rouquerol, F. Rodriguez-Reinoso, K. S. W. Sing, *et al.* Ed. (Elsevier,
2. W. C. Conner, (U.S. patent pending, submitted 1995),
3. W. Conner, S. Christensen, H. Topsøe, A. Pullen, *et al.*, 1994, in *Characterization of Porous Solids III* , J. Rouquerol, F. Rodriguez-Reinoso, K. S. W. Sing, *et al.* Ed. (Elsevier, pp. 151-163.
4. W. C. Conner, in press 1996, *J. Porous Material*
5. E. Maglara, A. Pullen, D. Sullivan and W. C. Conner, 1994, *Langmuir* **10**, 1467-73.
6. S. J. Gregg and K. S. W. Sing, 1982, *Adsorption, Surface Area and Porosity* (Academic Press, London).
7. C. C. Chang and R. J. Kokes, 1973, *The Journal of Physical Chemistry* **77**, 2640-2645.
8. S. A. Zubkov, V. Y. Borovkov, S. G. Gagarin and V. B. Kazansky, 1984, *Chemical Physics Letters* **107**, 337-340.
9. A. Sasse and H. Förster, 1995, *Journal of Molecular Structure* **349**, 97-100.
10. C. C. Chang, L. T. Dixon and R. J. Kokes, 1973, *The Journal of Physical Chemistgry* **77**, 26342640.
11. V. B. Kazansky, V. Y. Borovkov and L. M. Kustov, 1984, in *8th. International Congress on Catalysis Proceedings* , (Verlag Chemie. DECHEMA, Berlin (West)), pp. 3-14.
12. L. M. Kustov and V. B. Kazansky, 1991, *J. Chem. Soc. Faraday Trans.* **87**, 2675-2678.
13. R. Guermeur, C. Jacolin, F. Biquard and C. Blanc, 1990, *Surface Sicence* 157-175.
14. R. Guermeur and C. Jacolin, 1994, *Surface Science* 323-336.
15. W. Conner, J. Fraissard, J. Bonardet, K. Unger, *et al.*, 1994, in *Studies of Surface Science and Catalysis* , Roquerol Ed., Elsevier, pbs.
16. W. C. Conner, M. Ferrero, J. Bonardet and J. Fraissard, 1993, *in press, J. Chem. Soc., Faraday Transactions* .
17. V. R. Choudhary and K. R. Srinivasan, 1986, *J. Catal.* **102**, 316-327.
18. V. R. Choudhary and K. R. Srinivasan, 1986, *J. Catal.* **102**, 328-337.
20. B. Coughlan, W. M. Carroll, P. O'Malley and J. Nunan, 1981, *J. Chem. Soc., Faraday Trans. I* **77**, 3037-3047.
21. C. Förste, A. Germanus, J. Kärger, H. Pfeifer, *et al.*, 1987, *J. Chem. Soc., Faraday Trans. I* **83**, 2301-2309.
22. C. Förste, J. Kärger, H. Pfeifer, L. Riekert, *et al.*, 1990, *J. Chem. Soc. Faraday Trans.* **86**, 881-885.
23. M. Matsumoto, S. Shinoda, H. Takahashi and Y. Saito, 1984, *Bull. Chem.Soc., Jpn.* **57**, 1795-1800.
24. M. Primet, E. Garbowski, M. Mathieu and B. Imelik, 1980, *J.C.S. Faraday I* **76**, 1942-1952.
25. H. P. Wang, T. Yu, B. A. Garland and E. M. Eyring, 1990, *Applied Spectroscopy* **44**, 1070-1073.
26. M. Wilhelm, A. Firouzi, D. E. Favre, L. M. Bull, *et al.*, 1995, *J. Am. Chem. Soc.* **117**
27. D. Hay and H. Jaeger, 1984, *J. Chem. Soc., Chem. Comm.* 1433.
28. E. L. Wu, S. L. Lawton, D. H. Olson, J. Rohrman, A. C., *et al.*, 1979, *The Journal of Physical Chemistry* **83**, 2777-2781.

63

29. G. T. Kokotailo, C. A. Fyfe, G. Kennedy, G. Gobbi, *et al.*, 1986, in *Proceedings of the Seventh International Zeolite Conference*, Tokyo, pp. 361.
30. C. A. Fyfe, H. Gies, G. T. Kokotailo, C. Pasztor, *et al.*, 1989, *J. Am. Chem. Soc.* **111**, 2470-2474.
31. C. Fyfe, G. Kennedy, C. De Schutter and G. Kokotailo, 1984, *J. Chem. Soc., Chem. Commun.* **54**,
32. G. W. West, 1984, *Aust. J. Chem.* **37**, 455-7.
33. J. Parise, T. Gier, D. Corbin and D. Cox, 1984, *J. Phys. Chem.* **88**, 1635-1640.
34. M. Mägi, E. Lippmaa, A. Samason, G. Engelhardt, *et al.*, 1984, *J. Phys. Chem.* **88**, 1518.
35. P. Jacobs, J. Datka, P. Geerlings and W. Mortier, 1985, *J. Phys. Chem.* **89**, 3483-88 and 3488-93.
36. A. W. Chester, E. L. Wu and G. R. Landolt, 1986, in *Proceedings of the Seventh International Zeolite Conference*, Tokyo, pp. 547.
37. E. L. Wu, G. R. Landolt and A. W. Chester, 1986, in *Proceedings of the Seventh International Zeolite Conference*, Tokyo, pp. 547-554.
38. A. Zikanova, M. Bülow and H. Schlodder, 1987, *Zeolites* **7**, 115.
39. J. A. Müller and W. C. Conner, 1993, *J. Phys. Chem.* **97**, 1451-1454.
40. D. Theodorou and J. Wei, 1983, *J. Catal.* **83**, 205,224.
41. M. Niwa, S. Morimoto, K. Masaaki, T. Hattori, *et al.*, 1984, in *Proc. 8th Int. Congress on Catalysis, Berlin, IV*, pp. 701.
42. W. Kaeding, L. B. Young and C. Chu, 1984, *J. Catal.* **89**, 267.
43. A. K. Nowak and A. K. Cheetham, 1988, in *I.C. Zeolites*, pp. 475.
44. J. Rabo, (1987), personnel communications.

# THEORY AND SIMULATION OF ADSORPTION IN MICROPORES

## KEITH E. GUBBINS

*Cornell University*
*School of Chemical Engineering*
*Ithaca, NY 14853-5201, U.S.A.*

ABSTRACT. Methods for the study of adsorption that are based on statistical mechanics are described. The most useful such theory at present is density functional theory (DFT). Both the rigorous basis of the theory, and two approximate forms of it, are described. The most useful methods of molecular simulation, in which the equations of statistical mechanics are solved exactly on a fast computer, are also described, and their strengths and weaknesses for adsorption work are discussed. The remainder of the chapter is devoted to a description of three applications. The first is an application of DFT to determine pore size distributions from adsorption isotherms. Several applications to carbons are shown. The second application is to the study of the selective adsorption of trace gases from a carrier gas stream of nitrogen or air. Both DFT and simulation methods are used to determine the effects of pore size and shape, temperature and other variables on the selectivity, and it is shown that these methods can be used to determine the optimum pore size and pressure for maximum selectivity. The third application involves the use of molecular simulation to investigate the effect of acitive surface sites on the adsorption of water in activated carbons. The density and geometric arrangement of such sites on the carbon surfaces are shown to have a dramatic impact on the adsorption behavior.

## 1. Introduction

In this chapter a review of the most useful theoretical and molecular simulation methods for adsorption will first be given, followed by a number of applications. The methods to be used are based in statistical mechanics. The procedure is to first define the model of the system to be used; this entails writing down equations for the fluid-fluid and fluid-solid intermolecular interactions, and for a complete description of the solid adsorbent - the solid structure (location and species of atoms), pore size and shape, and a detailed molecular description of the surface. The statistical mechanical equations are then solved for this precisely defined model of the system of interest. Such solutions can be of two general kinds. First, an approximate theory can be used to simplify and solve the equations; examples of such approaches are the virial expansion, integral equation approximations, and density functional theory. An alternative, second approach is to solve the equations numerically on a fast computer; in that case no

*J. Fraissard (ed.), Physical Adsorption: Experiment, Theory and Applications, 65–103.*

approximations are necessary, and provided that the runs are long enough, and the simulated system is large enough to properly describe the real system, the calculation should give exact answers within the statistical uncertainty of the simulation. This latter approach is usually called molecular simulation. The use of approximate theories offers the advantage of speed and ease of calculation, so that results can be obtained relatively easily for a wide range of values of the system variables (pressures, temperatures, compositions, pore sizes, etc.). A disadvantage of this approach is that the approximations are likely to lead to errors under some conditions, so that the model is not accurately described; moreover, the approximate theories are difficult to apply to more complex systems, e.g. those involving H-bonding fluids, complex pore shapes, heterogeneous surfaces, etc. The molecular simulation approach can be more readily applied to such complex systems, and in general yields exact results for the model system. Moreover, it can be used to study a wide range of properties, some of which are not accessible to either theory or experiment, e.g. molecular correlation functions, time correlation functions, diffusion rates in various adsorbed molecular layers, distribution of residence times for adsorbate molecules on specific surface sites, and so on. A disadvantage of molecular simulations is that long computing times may be needed for complex adsorption systems; in addition, the real time that can be simulated in a reasonable time is restricted to times of the order 10 nanoseconds, so that dynamical processes that are slow (e.g. desorption from strong binding sites, phase separations at high density in connected pore structures) are not easily studied. This latter problem should be eased by further improvements in computing power.

The methods to be described and used are much more powerful than the older and more empirical methods of analysis, such as the Kelvin equation, Dubinin-Radkevitch equation, etc. These older approaches do not start from a well-defined molecular model, so that the connection between molecular behavior and macroscopic properties is not made clear. The methods of statistical mechanics are not restricted, in general, to pores of a particular size range or shape, and should hold for pore sizes down to zero.

The methods are described in Section 2. This is followed by a discussion of three applications. The first two, to the determination of pore size distributions from adsorption isotherms (Sec. 3), and to the selective adsorption of trace components from mixtures (Sec. 4), involve simple models of porous materials with homogeneous surfaces. Density functional theory is the main tool used in these two applications. The third application is to the study of adsorption of water in activated carbons (Sec. 5), in which the carbon surfaces are heterogeneous and have active chemical sites present.

## 2. Methods

Although integral equation methods and the virial expansion have been used to study adsorption, modern forms of the density functional theory (DFT) are at present the most powerful and useful theoretical approach for equilibrium properties, and it is this approach that is described here. Molecular simulation provides an alternative approach to approximate theories, and is more easily applied to complex fluids and pore geometries.

## 2.1. DENSITY FUNCTIONAL THEORY

The objective in DFT is to calculate the density profile, $\rho(\mathbf{r})$, for all locations $\mathbf{r}$ in the pore. Once $\rho(\mathbf{r})$ is known, the other thermodynamic properties, such as adsorption isotherm, heat of adsorption, free energies, phase transitions, etc., can be easily calculated.

### 2.1.1 Thermodynamics

We consider an open system containing a mixture of components A, B, ....R, characterized by the grand canonical variables $T$, $V$, $\mu_A$,.... $\mu_R$, where $\mu_\alpha$ is the chemical potential of component $\alpha$. The appropriate thermodynamic potential is then the grand potential or grand free energy, $\Omega$, defined by

$$\Omega = A - \sum_\alpha \mu_\alpha N_\alpha \tag{1}$$

where $A$ is the Helmholtz energy and $N_\alpha$ is the number of molecules of component $\alpha$. The exact differential of $A$ is given by the well known thermodynamic expression

$$dA = SdT - PdV + \gamma dS + \sum_\alpha \mu_\alpha dN_\alpha \tag{2}$$

where $S$ is entropy, $P$ is pressure, $V$ is volume, $\gamma$ is surface tension, and $S$ is surface area. By differentiating eqn. (1) and using (2) we obtain the differential of the grand potential,

$$d\Omega = -SdT - PdV + \gamma dS - \sum_\alpha N_\alpha d\mu_\alpha \tag{3}$$

From eqns. (2) and (3) the surface tension is given by

$$\gamma = \left( \frac{\partial A}{\partial S} \right)_{T,V,N} = \left( \frac{\partial \Omega}{\partial S} \right)_{T,V,\mu} \tag{4}$$

where $N=N_A$, $N_B$, ....$N_R$ and $\mu=\mu_A$, ....$\mu_R$. The free energies $A$ and $\Omega$ are homogeneous functions of first order in the variables $V$, $S$ and $N$ (for $A$), and $V$ and $S$ (for $\Omega$), so that equations (2) and (3) can be integrated using Euler's theorem to give

$$A = -PV + \gamma S + \sum_{\alpha} \mu_\alpha dN_\alpha \qquad (5)$$

$$\Omega = -PV + \gamma S \qquad (6)$$

We note that for a homogeneous fluid, $\Omega = -PV$.

### 2.1.2. Rigorous Basis of DFT

We first introduce two free energy functionals, $\Omega[\rho(\mathbf{r})]$ and $A[\rho(\mathbf{r})]$, which have the following properties:
(i) they can be defined even if the system is not at equilibrium, and are a functional of $\rho(\mathbf{r})$, i.e. for a given $\rho(\mathbf{r})$ there is a unique value of $\Omega$ and $A$;
(ii) when $\Omega$ or $A$ is allowed to vary by varying $\rho(\mathbf{r})$, its global minimum corresponds to the equilibrium density profile, $\rho_{eq}(\mathbf{r})$, of the system;
(iii) these minimum values of $\Omega$ and $A$ are the thermodynamic grand free energy and the intrinsic Helmholtz energy (Helmholtz energy in the absence of any external field, such as that due to the pore walls) for the system at fixed chemical potential, volume and temperature; thus $\Omega$ is given by equation (6) and $A$ by (5).

It can be shown rigorously that such a set of conditions can be met [1]. For a pure fluid we introduce the two functionals $\Omega[f]$ and $A[f]$, defined by [1]

$$\Omega[f(\mathbf{r}^N)] = Tr_{cl} f(\mathbf{r}^N)\{U + V - \mu N + kT \ln f(\mathbf{r}^N)\} \qquad (7)$$

$$A[f(\mathbf{r}^N)] = Tr_{cl} f(\mathbf{r}^N)\{U + kT \ln f(\mathbf{r}^N)\} \qquad (8)$$

where $U$ is the total intermolecular potential energy, $V$ is the potential energy due to external fields (in our case these will be due to interaction of the fluid molecules with the pore walls), $f(\mathbf{r}^N) = f(\mathbf{r}_1, \mathbf{r}_2, ... \mathbf{r}_N)$ is the molecular $N$-body distribution function for the (in general, nonequilibrium) state of the system, and $Tr_{cl}$ is the classical trace, given by

$$Tr_{cl}(...) = \sum_N \left(\frac{q_{qu}}{\Lambda_t^3}\right)^N \frac{1}{N!} \int (...) dr_1 dr_2 ... dr_N \qquad (9)$$

where $q_{qu}$ is the quantal part of the molecular partition function, $\Lambda_t$ is the usual de Broglie wavelength (the kinetic energy part of the molecular partition function), and $\mathbf{r}_i$ is the position of the center of molecule i. In equation (9) and what follows we assume spherical molecules for convenience; the extension to nonspherical molecules is straightforward [2]. The N-body distribution function obeys the normalization condition

$$Tr_{cl} f(\mathbf{r}^N) = 1 \qquad (10)$$

For the system at equilibrium, the N-body distribution function, $f_{eq}$ is given by the usual grand canonical expression,

$$f_{eq}(\mathbf{r}^N) = \frac{1}{\Xi}\exp\{-\frac{1}{kT}(U + V - \mu N)\} \tag{11}$$

where $\Xi$ is the grand partition function, given by

$$\Xi = Tr_{cl}\exp\{-\frac{1}{kT}(U + V - \mu N)\} \tag{12}$$

When eqn. (12) is substituted into (7) we obtain

$$\Omega[f] = -kT\ln\Xi = \Omega \tag{13}$$

where $\Omega = -PV + \gamma S$ is the grand potential. Similarly, on using (12) in (8) gives

$$A[f] = -kT\ln\Xi + \mu < N >= A_i \tag{14}$$

where $<N>$ is the average number of molecules in the system, and $A_i$ is intrinsic Helmholtz energy, i.e. the Helmholtz energy in the absence of the external field $V$. Thus, the free energy functionals defined in eqns. (7) and (8) satisfy the third condition given at the beginning of this subsection. It is also easy to show that they satisfy the second condition, that they are each a minimum at equilibrium, when $f=f_{eq}$; this can be shown by using the Gibbs inequality [1,2].

From eqns. (7) and (8), $\Omega[f]$ can be rewritten in the form

$$
\begin{aligned}
\Omega[f] &= A[f] + Tr_{cl}fV - \mu Tr_{cl}fN \\
&= A[f] + \int d\mathbf{r}f(\mathbf{r})v(\mathbf{r}) - \mu < N > \\
&= A[f] + \int d\mathbf{r}f(\mathbf{r})v(\mathbf{r}) - \mu\int d\mathbf{r}f(\mathbf{r})
\end{aligned}
\tag{15}
$$

where $v(\mathbf{r})$ is th external potential acting on a single molecule i, and where

$$= \sum_i v(\mathbf{r}_i) \tag{16}$$

and $f(\mathbf{r})$ is a one-body distribution function proportional to the probability density of finding a molecule at position $\mathbf{r}$, and is given by

$$f(\mathbf{r}_1) = \sum_N \left(\frac{q_{qu}}{\Lambda_t^3}\right)^N \frac{1}{(N-1)!} \int d\mathbf{r}_2 d\mathbf{r}_3 \ldots d\mathbf{r}_N f(\mathbf{r}_2, \mathbf{r}_3, \ldots \mathbf{r}_N) \qquad (17)$$

So far we have treated $\Omega$ and $A$ as functionals of the full N-body distribution function, $f(\mathbf{r}^N)$. It is possible to prove [1] that there is only one unique external potential $v(\mathbf{r})$ that gives rise to a given equilibrium one-body distribution function $f(\mathbf{r})$ and to the corresponding N-body function $f(\mathbf{r}^N)$. It follows that $f(\mathbf{r}^N)$ is a functional of $f(\mathbf{r})$, and we can therefore equally well take $\Omega[f]$ and $A[f]$ to be functionals of the one-body distribution function $f(\mathbf{r})$. For the equilibrium fluid with one-body distribution function given by eqn. (17), $f(\mathbf{r})=\rho_{eq}(\mathbf{r})$, the local number density (number of molecules per unit volume at point $\mathbf{r}$); this follows by applying the normalization property for f(r) to eqn. (17). Thus we can write our free energy functionals as $\Omega[\rho(\mathbf{r})]$ and $A[\rho(\mathbf{r})]$.

Since $\Omega$ is a minimum at equlibrium, we can write

$$\left(\frac{\delta\Omega[\rho(\mathbf{r})]}{\delta\rho(\mathbf{r})}\right)_{\rho=\rho_{eq}} = 0 \qquad (18)$$

where the differentiation is at constant external field $v(\mathbf{r})$ and $(\mu, V, T)$, and since this minimum value is just the grand potential we have

$$\Omega[\rho(\mathbf{r})] = \Omega \qquad (19)$$

From eqns. (15) and (18) we find

$$\mu_i(\mathbf{r}) + v(\mathbf{r}) = \mu \qquad (20)$$

where $\mu_i(\mathbf{r})$ is the equilibrium intrinsic chemical potential (i.e. chemical potential in the absence of the external field), defined by

$$\mu_i(\mathbf{r}) = \left(\frac{\delta A}{\delta\rho(\mathbf{r})}\right)_{\rho=\rho_{eq}} \qquad (21)$$

Equations (20) and (21) are the fundamental ones for the density functional approach to inhomogeneous fluids. Eqn. (20) states that the chemical potential is made up of an intrinsic contribution $\mu_i$ and an external field contribution $v$. Given an expression for $A$, the equilibrium density profile $\rho_{eq}(\mathbf{r})$ can be determined by minimization of $A$; this is usually carried out by iterative methods on a computer. The intrinsic chemical potential and the grand potential can then be obtained from eqns. (21) and (15).

*2.1.3. Approximate Functionals*

To make the theory tractable an explicit expression for $\Omega$ is needed. A generalized van der Waals expression is used, in which $\Omega$ is written as a sum of a contribution from the short range repulsive forces plus a contribution from the longer range attractive forces; the latter contribution is usually treated in a mean field approximation, which is the same as that used in most equations of state for bulk fluids. For simplicity of notation we shall continue to assume that the molecules are spherical; the generalization to nonspherical molecules is straightforward, and is given elsewhere [2]. To formally separate the potential into two parts, we introduce a pair intermolecular potential $u(r;\lambda)$, given by

$$u(r;\lambda) = u_o(r) + \lambda u_1(r) \tag{22}$$

where $u_o(r)$ is the short-range repulsive part and $u_1(r)$ the long-range attractive part of the potential, respectively. Thus $u(r;\lambda)$ becomes the reference potential $u_o(r)$ when $\lambda=0$ and the full potential $u(r)$ when $\lambda=1$. With this potential splitting, and assuming that the potentials are pairwise additive, it is easy to show that the Helmholtz energy functional is given by [2]

$$A[\rho(r)] = \int dr a_o[\rho(r)] + \frac{1}{2} \int_0^1 d\lambda \int dr_1 dr_2 f(r_1, r_2; \lambda) u_1(r_{12}) \tag{23}$$

where the first term on the right side is $A_o[\rho(r)]$, $a_o(r)$ is the Helmholtz energy density at point $r$ for a fluid of molecules with potential $u_o(r)$ and density profile $\rho(r)$, and $f(r_1, r_2; \lambda)$ is the pair distribution function in the fluid with potential $u(r;\lambda)$, and can be written as

$$f(r_1, r_2; \lambda) = \rho(r_1)\rho(r_2)g(r_1, r_2; \lambda) \tag{24}$$

where $g(r_1, r_2; \lambda)$ is the pair correlation function.

*Mean Field Approximation.* In the mean field approximation the pair correlation function $g(r_1, r_2; \lambda)$ is set equal to unity, its limiting value for large $r$, so that in this approximation eqn. (23) becomes

$$A[\rho(r)] = \int dr a_o[\rho(r)] + \frac{1}{2} \int dr_1 dr_2 \rho(r_1)\rho(r_2)u_1(r_{12}) \tag{25}$$

Using this in eqn. (15) gives the mean field approximation to the grand potential functional,

$$\Omega[\rho(\mathbf{r})] = \int d\mathbf{r} a_o[\rho(\mathbf{r})] + \frac{1}{2}\int d\mathbf{r}_1 d\mathbf{r}_2 \rho(\mathbf{r}_1)\rho(\mathbf{r}_2)u_1(r_{12}) + \int d\mathbf{r}\rho(\mathbf{r})v(\mathbf{r}) - \mu\int d\mathbf{r}\rho(\mathbf{r})$$

$$(26)$$

Eqn. (26) is the usual starting point for density functional theories. The mean field approximation leads to some errors, since it neglects effects of attractive forces on the intermolecular correlations, and usually leads to liquid phase densities that are 5-15% too low; it also leads to classical critical exponents. It can be improved upon by using some approximate form for the pair correlation function in eqn. (24). However, more effort has been concentrated on finding accurate methods for calculating $a_o[\rho(\mathbf{r})]$. For simple fluids of spherical molecules it is the $u_o$ term, i.e. the repulsive part of the intermolecular potential, that is the main factor in determining the structure of the fluid.

*Smoothed Density Approximation.* The Helmholtz energy density $a_o[\rho(\mathbf{r})]$ can be written as the sum of an ideal gas part and a residual part, i.e. $a_o = a_o^{id} + a_o^{r}$. The ideal gas part is easily written down and is exactly local, i.e. $a_o^{id}(\mathbf{r})$ only depends on the density at the point $\mathbf{r}$, $\rho(\mathbf{r})$, and not on the densities at neighboring locations, $\rho(\mathbf{r}')$. The residual part, $a_o^{r}(\mathbf{r})$, however, is nonlocal and depends in general on the density profile of the inhomogeneous fluid at all points $\mathbf{r}$, or at least on those in the neighborhood of the point of interest. In the smoothed density approximation, it is assumed that this residual Helmholtz energy density for the actual *inhomogeneous* fluid can be equated to that for a *uniform* fluid (having the same potential $u_o$) at some smoothed density $\bar{\rho}(\mathbf{r})$, i.e.

$$A_o[\rho(\mathbf{r})] = \int d\mathbf{r} a_o^{id}[\rho(\mathbf{r})] + \int d\mathbf{r}\rho(\mathbf{r})\psi_o^{r,un}[\bar{\rho}(\mathbf{r})]$$

$$(27)$$

where $\psi_o^{r} \equiv a_o^{r}/\rho = \psi_o - \psi_o^{id}$ is the residual (non-ideal gas part) of the Helmholtz energy per molecule, $\psi_o^{r,un}$ is the value for the uniform fluid, and $\bar{\rho}(\mathbf{r})$ is some smoothed density that is yet to be defined. This smoothed density is some average over the local densities at points $\mathbf{r}'$ near to $\mathbf{r}$, the point of interest, and is usually written as

$$\bar{\rho}(\mathbf{r}) = \int d\mathbf{r}' \rho(\mathbf{r}')w(|\mathbf{r} - \mathbf{r}'|)$$

$$(28)$$

where $w(r)$ is a normalized isotropic weighting function

$$\int d\mathbf{r} w(r) = 1$$

$$(29)$$

Thus the problem is to decide on an appropriate form for the weighting function, $w(r)$. Once this is known the smoothed density corresponding to any given local density profile $\rho(\mathbf{r})$ can be calculated from eqn. (28). The replacement of the actual residual

Helmholtz energy density for the nonuniform fluid by that for a uniform fluid of some average density is the essential approximation in this form of density functional theory.

In applying the theory, a further approximation is generally made. The Helmholtz energy density of the fluid with the smooth repulsive intermolecular potential $u_o$ is equated to that for a fluid of hard spheres of diameter $\sigma$. This is known to be a very good approximation for uniform fluids of Lennard-Jones molecules, provided the hard sphere diameter is suitably chosen, and it is assumed to be equally valid for the nonuniform fluids considered here. The residual free energy per molecule, $\psi_o^{r,un}$, can then be calculated from an accurate equation of state for uniform hard spheres. The Carnahan-Starling equation [3] is generally used for this purpose, with $\bar{\rho}(r)$ used for the density.

A variety of procedures have been proposed to determine the weighting function $w(r)$. The basic idea in these theories is usually to choose $w(r)$ to give the best possible description of the uniform hard sphere fluid, i.e. to describe both the structure and the thermodynamics of the fluid. This can be largely achieved by requiring that the weighting function gives a good account of the two-body direct correlation function of the uniform hard sphere fluid, $c_o(r_1,r_2)=c_o(r)$; this will give the two-body structure of the fluid correctly and also the thermodynamics, since the equation of state depends only on this function. In density functional theory the direct correlation function is given by a second functional derivative of the residual Helmholtz energy functional [1],

$$c_o(r_1,r_2) = -\frac{1}{kT}\frac{\delta^2 A_o^r}{\delta\rho(r_1)\delta\rho(r_2)} \tag{30}$$

where $A_o^r$ is the last term on the right hand side of eqn. (27). Thus the weighting function needs to be chosen so that the resulting expression for $A_o^r$ yields, via eqn. (30), a direct correlation function for the uniform hard sphere fluid that is as close as possible to the correct one over a range of state conditions.

The simplest choice for the weighting function in the above theory would be to take $w(r)=\delta(r)$, the Dirac delta function. In this case the smoothed density as defined by eqn. (28) would just be the local density, $\rho(r)$. The resulting theory is called the *local form of density functional theory*. Although this gives useful results for fluid-fluid interfaces [2], it is poor for solid-fluid interfaces because of the very large density fluctuations near the wall. Such a simple weighting function, when used with eqn. (30), leads to the following result for the hard sphere direct correlation function,

$$c_o(r,r') = \left(\frac{1}{\rho(r)} - \frac{a_o''[\rho(r)]}{kT}\right)\delta(r-r') \tag{31}$$

where $a_o''$ is the second density derivative of $a_o$. This delta-function result is, of course, unphysical, and leads to poor results for the fluid structure near pore walls.

*Tarazona's Theory* [4]. For a reference fluid of hard spheres the Percus-Yevick (PY) theory gives a good approximation to $c_o(r)$ for the uniform hard sphere fluid over a wide range of density. In Tarazona's theory it is assumed that $w(r)$ has a power series expansion in the smoothed density [5],

$$w(r';\overline{\rho}(\mathbf{r})) = w_o(r') + w_1(r')\overline{\rho}(\mathbf{r}) + w_2(r')\overline{\rho}(\mathbf{r})^2 + ... \tag{32}$$

The coefficients $w_o(r')$, $w_1(r')$, $w_2(r')$, etc. can be estimated by first calculating the expansion of $c_o(r)$ from eqn. (30), and then comparing this expansion term by term with the density expansion of the PY result for $c_o(r)$. Tarazona has carried out such a scheme including terms up to $w_2$ in eqn. (32). Substituting (32) into eqn. (28) gives an integral equation for the smoothed density profile in terms of the local density profile,

$$\overline{\rho}(\mathbf{r}) = \overline{\rho}_o(\mathbf{r}) + \overline{\rho}_1(\mathbf{r})\overline{\rho}(\mathbf{r}) + \overline{\rho}_2(\mathbf{r})(\overline{\rho}(\mathbf{r}))^2 + ... \tag{33}$$

where

$$\overline{\rho}_i(\mathbf{r}) = \int d\mathbf{r}'\,\rho(\mathbf{r}')w_i(|\mathbf{r}-\mathbf{r}'|) \qquad i = 0,1,2 \tag{34}$$

The direct correlation function for the homogeneous hard sphere fluid is obtained from eqn. (30) with $A_o'$ evaluated at the smoothed density given by eqn. (33), using the Carnahan-Starling equation [3] for $a_o'$. The equations for the weighting functions $w_o(r')$, $w_1(r')$ and $w_2(r')$ are then obtained by comparing with the PY result. The equations are [4]:

$$w_o(r) = \frac{3}{4\pi\sigma^3} \qquad\qquad\qquad r^* < 1$$
$$= 0 \qquad\qquad\qquad r^* > 1$$

$$w_1(r) = 0.475 - 0.648r^* + 0.113r^{*2} \qquad\qquad r^* < 1$$
$$= 0.288r^{*-1} - 0.924 + 0.764r^* - 0.187r^{*2} \qquad 1 < r^* < 2 \tag{35}$$
$$= 0 \qquad\qquad\qquad\qquad r^* > 2$$

$$w_2(r) = \frac{5\pi\sigma^3}{144}\left(6 - 12r^* + 5r^{*2}\right) \qquad\qquad r^* < 1$$
$$= 0 \qquad\qquad\qquad\qquad r^* > 1$$

where $r^* = r/\sigma$.

Tarazona's theory to second order gives good results for the hard sphere direct correlation function, the results being only very slightly different from the PY values at a high (liquid-like) reduced density of $\rho\sigma^3=0.8$, while at lower densities the two theories are usually indistinguishable. For a hard sphere fluid in contact with a hard wall the Tarazona theory gives excellent agreement with molecular simulation results for both the density profile and surface tension [4]. It has also been applied successfully to the solid-fluid transition for the hard sphere fluid [4]. There have also been successful applications to Lennard-Jones fluids in pores; examples are shown in Section 3 below.

The Tarazona DFT has the advantages of being both quite accurate and relatively easy to use in numerical calculations for realistic intermolecular potentials such as the Lennard-Jones. It is also able to predict solid-fluid transitions, in contrast to some other forms of DFT. However, it also has several shortcomings:

- it is difficult to use for mixtures, since the extension to mixtures involves some ambiguity in the definition of the weighting functions [6]; as a result, it has not been extended past first order.
- the weighting factor depends on the smoothed density, and so must be calculated afresh for each point in the pore.
- the theory does not give the one-dimensional limit correctly, and so cannot be used for cylindrical pores of diameter below about $4\sigma$ [7].

*Kierlik-Rosinberg-Rosenfeld Theory.* In this theory [8], which was developed by Kierlik and Rosinberg from ideas originally proposed by Rosenfeld [9] and has been shown to be equivalent to the Rosenfeld theory [10], it is assumed that the residual Helmholtz energy of the hard sphere reference fluid, $\psi_o^{r,un}$ of eqn. (27), is a functional of several different smoothed densities, $\overline{\rho}_1, \overline{\rho}_2, \overline{\rho}_3, \dots$, each of which is given by eqn. (28) but with different weighting functions; this is in contrast to the single weighting function used in the Tarazona theory. Kierlik and Rosinberg then seek the minimum set of weighting functions (and hence smoothed densities) that are needed to give the PY equation for the direct correlation function for hard spheres *exactly*. They find that four weighting functions are needed for both pure fluids and binary mixtures, and that these are:

$$w_1(r) = \Theta(r'-r)$$

$$w_2(r) = \delta(r'-r)$$

$$w_3(r) = \frac{1}{8}\delta'(r'-r) \hspace{3cm} (36)$$

$$w_4(r) = -\left(\frac{1}{8}\right)\delta''(r'-r) + \frac{\delta'(r'-r)}{2\pi r}$$

where $\Theta$ is the Heaviside step function ($\Theta$ is 0 for $r'<r$ and 1 for $r'>r$), $\delta$ is the Dirac delta function (0 except at $r'=r$, where it is infinity, and with an itegral over $r'$ of unity), and $\delta`(r)$ and $\delta``(r)$ are the first and second derivatives of $\delta(r)$, respectively. The resulting theory gives very good agreement with simulation results for hard sphere and Lennard-Jones fluids in pores. It is particularly useful for studies of mixtures in pores, since it is usually in quite good agreement with simulation results and is easy to apply, not suffering from the disadvantages of the Tarazona theory. A disadvantage of the theory for some applications is that it does not predict a stable solid phase, and so cannot be used to study melting or freezing transitions. Advantages of the Kierlik-Rosinberg-Rosenfeld theory over that of Tarazona include:

- It gives the PY result for hard spheres exactly.
- The equations for the weighting functions $w_i$ are simple, and independent of density, making calculations easier.
- It is immediately applicable to mixtures, without any ambiguity in defining the weighting functions.
- It gives the one dimensional result exactly.

## 2.2. MOLECULAR SIMULATION

For more complex molecules (e.g. polar, chain molecules, water) DFT becomes difficult to use, and it is generally easier and more accurate to use a numerical method to solve the statistical mechanical equations. Such methods are called *molecular simulation*, and provided they are carefully carried out they should give an exact solution for the model adsorption system of interest [11,12]. As in DFT, one starts from a model of the pore geometry and its atomic structure, and equations for the fluid-fluid and fluid-solid intermolecular forces. Two methods are in general use for solving the statistical mechanical equations: Monte Carlo and molecular dynamics (Fig. 1). In the Monte Carlo (MC) method a random number generator is used to move and rotate the molecules in a random fashion. If the system is at fixed temperature, volume and number of molecules of each species, statistical mechanics tells us that the probability of a particular arrangement of the molecules is proportional to $\exp(-U/kT)$, where $U$ is the total intermolecular energy of the molecular assemblage, $k$ is Boltzmann's constant and $T$ is temperature. In the MC scheme, moves are accepted or rejected according to a recipe that ensures that the various molecular arrangements appear with a probability given by this Boltzmann distribution law. After generating a long sequence of such moves (typically several million), they can be averaged using the equations of statistical mechanics to obtain the various equilibrium properties of the model system.

In molecular dynamics (MD) the motion and velocity of each molecule is followed in time by solving the Newton equations of motion. In this case the properties of the system are obtained, following a suitable equilibration period, by averaging these trajectories and velocities over time. Typically it is possible to simulate a few nanoseconds of real time, which is sufficient to obtain reliable estimates of many properties for most, but not all, systems of interest (see below). Since the simulation is a dynamic one, it is possible to calculate transport properties, as well as equilibrium ones.

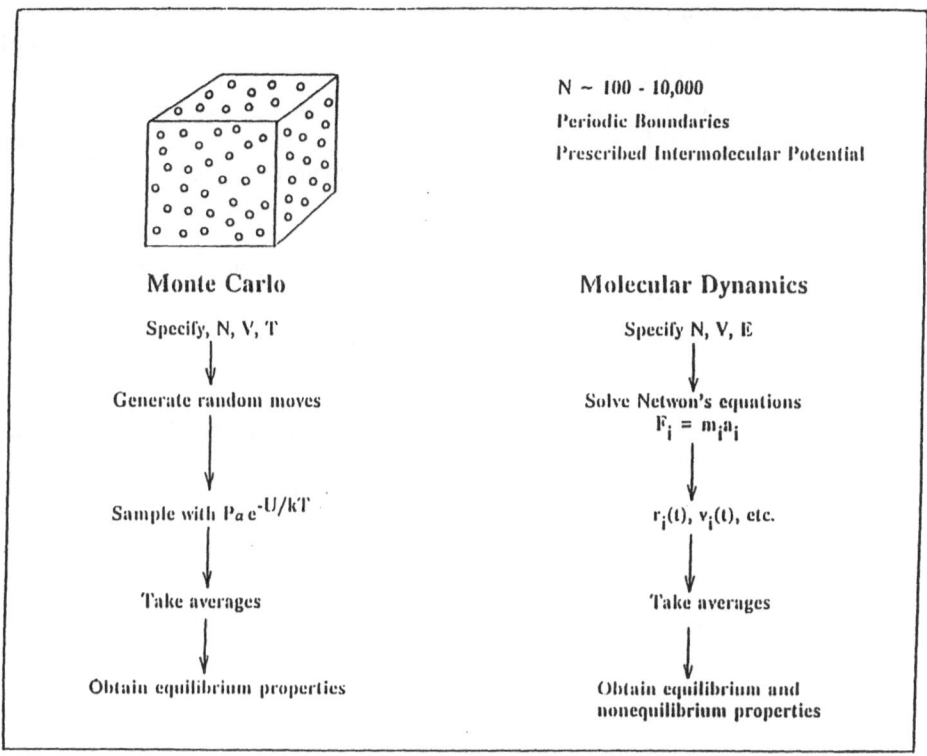

Figure 1. Schematic outline of the Monte Carlo and Molecular Dynamics simulation methods.

The two techniques have several common features. The calculation will give accurate results for the precisely defined model system provided that the simulation is long enough and the system is large enough. In practice the length of the simulation run and the number of molecules are limited by the speed and storage capacity of current computers. Typically the number of molecules in the simulation is a few thousand and for MD the runs may be of the order of one or a few nanoseconds. In order to minimize boundary effects in such small samples, periodic boundary conditions are used. That is, the sample is surrounded on all sides by replicas of itself, each replica having the same molecules in the same positions with the same orientiations and velocities as those in the central sample, so that when a molecule in the central system moves through a boundary and so out of the system, it is replaced by a molecule moving into the sample through the opposite face of the box. Anyone who has played Pacman, Asteroids, or similar video games will be familiar with this periodic boundaries trick.

There are also some significant differences between MC and MD. MC is easy to program, and it is easy to change the state variables from, say $N,V,T$ (the canonical variables of number of molecules, volume and temperature) to $\mu,V,T$ (the grand canonical variables) or $N,P,T$ (isobaric variables). MD is more difficult to program, and it is less straightforward to change variables, although it can be done. MD has two important advantages over MC: it is possible to calculate dynamical properties (e.g. diffusion rates in the pore), and the molecular motions are natural. The latter property of MD is in contrast to MC, where the motions are artificial, and this means that the motions in MD can be observed with computer graphics and will give a picture of the actual motions of the molecules themselves.

While these simulation methods are applicable to a wide range of problems, difficulties arise with some applications because of the limited speed or storage of computers. These include:

- Ionic fluids, where the long-range Coulomb forces require very large systems in order to avoid cutoff errors.
- Systems in which there are long-range spatial correlations, e.g. near a critical point; the system size needed will grow rapidly as the critical point is approached, and it is usually not feasible to study such systems much closer than a K or so from the critical point.
- Very large systems, e.g. a bulk polymer, micelles, systems of large biological molecules, etc.
- Slow processes that require microseconds or longer, e.g. some phase separations and domain growth phenomena.
- Quantum phenomena, such as reacting systems.

The simulation methods used most commonly to study adsorption and phase transitions in small pores are Grand Canonical Monte Carlo (GCMC), Gibbs Ensemble Monte Carlo (GEMC), and molecular dynamics (MD). These each have advantages and disadvantages [13], as summarized in Table 1.

The advantages and disadvantages of these various methods affect the choice of method for a particular application. Methods for overcoming the disadvantages of a given method are often available. The difficulty of molecule insertion at high densities, for example, which is more pronounced for significantly nonspherical molecules, can be overcome by using a biased sampling method [14], which attempts to insert the molecules into 'holes' in the fluid, thus improving the chance of successful insertion; the effects of such biasing on the statistical sampling is removed at a later stage. Similarly, thermodynamic integration methods [7,15] can be used to determine the true phase transition point in cases where hysteresis occurs. Chemical potentials can be determined by using the 'test particle method' [14].

TABLE 1

Advantages and Disadvantages of Various Methods for Adsorbed Systems [13]

| Method | Advantages | Disadvantages |
| --- | --- | --- |
| GCMC | Natural method<br>Know $\mu$, and hence P | Get hysteresis loops<br>Don't see interfaces<br>Fails at high density |
| GEMC | Get true equilibrium point<br>Fast for mixtures | Don't know $\mu$<br>Don't see interfaces<br>Fails for high density |
| MD | See interface, phase splitting<br>Get dynamic information | Don't know $\mu$<br>Need large systems |

## 3. Determination of Pore Size Distributions from Adsorption Isotherms.

Classical methods [16] for analysis of pore size distributions (PSD) from adsorption isotherm measurements have relied on thermodynamic models based on the Kelvin equation, or on semiempirical treatments such as those of Horvath and Kawazoe (HK) [17] for slit shaped pores or of Saito and Foley (ST) [18] for cylindrical pores. The Kelvin-based methods are only applicable to relatively large pores, and tend to fail below a pore width of about 70-80 Å for carbons, and below about 35 Å for silicas and zeolites, for example; the use of the modified Kelvin equation extends this range only slightly. While the HK and ST methods have a wider range of application, they still underpredict pore sizes for the smaller pores. One useful application of DFT has been to develop a more accurate method for obtaining PSD's based on an accurate molecular theory [19,20,21,22,23,24]. In both the classical and DFT approaches, the adsorption $\Gamma(P)$ at pressure $P$ is usually approximated as

$$\Gamma(P) = \int_{H_{min}}^{H_{max}} \Gamma(P,H) f(H) dH \qquad (37)$$

where H is pore size (width for slits, diameter for cylinders), f(H) is the pore size distribution, and $\Gamma(P,H)$ is the adsorption for a material whose pores are all of width H. In the DFT approach the latter quantity is calculated from the DFT. The procedure is to approximate the pores in the material as being of some simple geometry (e.g. slits or cylinders) and then to calculate $\Gamma(P,H)$ for a range of pore sizes covering the range of H of interest. The PSD f(H) is assumed to be described by a function such as the gamma or lognormal distribution. Using experimental data for $\Gamma(P)$ and a fit to the $\Gamma(P,H)$ data from DFT, the above equation is then inverted to obtain f(H).

The method has been successfully applied to nitrogen adsorption in carbons and silicas [24]. In the case of carbons, the pores are modeled as slits of width H with two semi-infinite walls; here H is defined as the distance between the planes through the carbon atoms in the first layer of each wall. The fluid-wall interaction is modeled using the Steele 10,4,3 potential [25], with the usual interaction parameters for carbon. This potential averages over the positions of the carbon atoms in the graphite walls, and so neglects the wall structure. This is not believed to cause significant errors at 77 K because of the close spacing of the C atoms in graphite. The solid-fluid potential well depth parameter, $\varepsilon_{sf}/k$, largely determines the pressure at which the monolayer forms on the pore walls. It was obtained by fitting to adsorption data for a nonporous material having the same chemical structure as the porous material; a Vulcan carbon provided by BP Research was used, and this procedure gave $\varepsilon_{sf}/k=53.22$ K. The nitrogen-nitrogen potential is approximated by a spherical Lennard-Jones interaction.

With these intermolecular potentials, individual nitrogen pore isotherms at 77 K were generated using the Tarazona DFT for pore sizes ranging from $H^*=H/\sigma=1.68$ to 100 (H=6.0 to 357 Å), where $\sigma$ is the Lennard-Jones diameter for nitrogen. Typical results are shown in Figure 2 for micropores, and for a wider range of pore sizes in Figure 3. For $H^*$ values above a critical value of $H_{c1}^*=3.8$ (13.6 Å), capillary condensation is observed and the isotherm is of Type IV. Below $H^*=3.8$ the isotherms show continuous filling, but there is another region of discontinuous filling for pore sizes in the range $H_{c2}^*=3.6$ (12.8 Å) to $H_{c3}^*=2.55$ (9.1 Å); in this latter range a $0\to1$ layering transition, in which an incomplete monolayer on each wall abruptly fills to completion, is occurring. For pore widths below $H^*=2.55$ continuous pore filling is observed. In these ultramicropores, only one layer of adsorbate can be accommodated, so that no sharp transitions can occur at 77 K. The results for very small pores can be understood from Figure 4, where the fluid-wall intermolecular potentials are shown. The two potential wells merge to a single minimum at a pore width of $H^*=2.25$ (8.0 Å), and for $H^*=1.94$ (6.9 Å) the single well reaches its maximum depth of about twice that of its original depth for large pores. Below this pore width a rapid rise in the well depth is seen, as a result of overlap of the repulsive regions of the potential for the two walls (molecular sieve effect).

Gibbs Ensemble Monte Carlo (GEMC) calculations have confirmed the general behavior predicted by DFT for the whole range of pore widths, as seen in Figure 3. In particular the transitions from continuous filling to either layering transitions or capillary condensation are well described by the theory, as is the transition

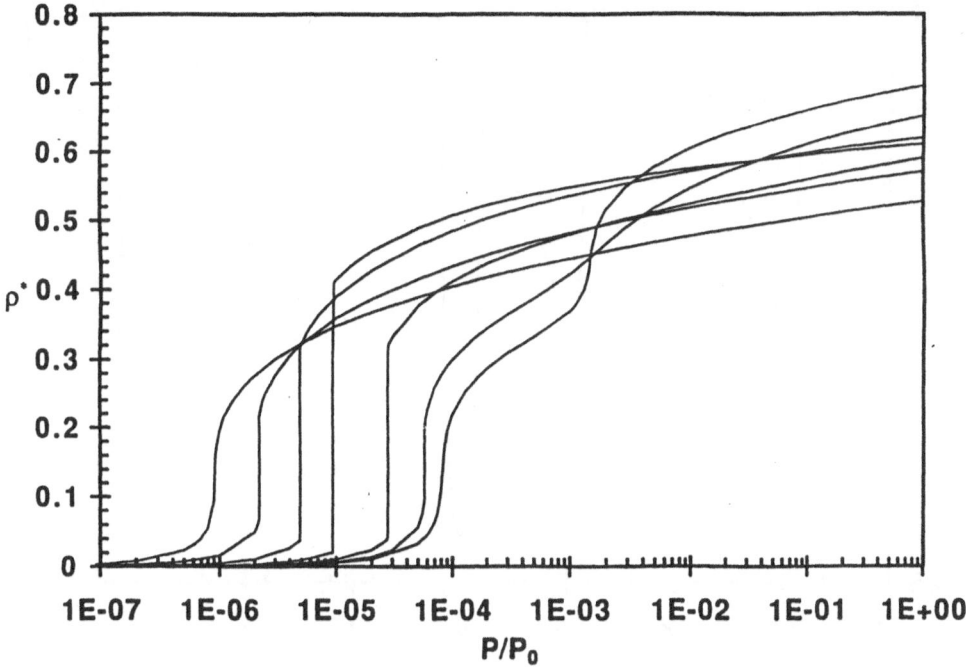

Figure 2. Isotherms for nitrogen at 77 K in slit carbon pores of widths (left to right) H*=2.5, 2.6, 2.75, 3.0, 3.25, 3.5 and 3.75, as calculated from Tarazona DFT [from Lastoskie et. al, ref. 24].

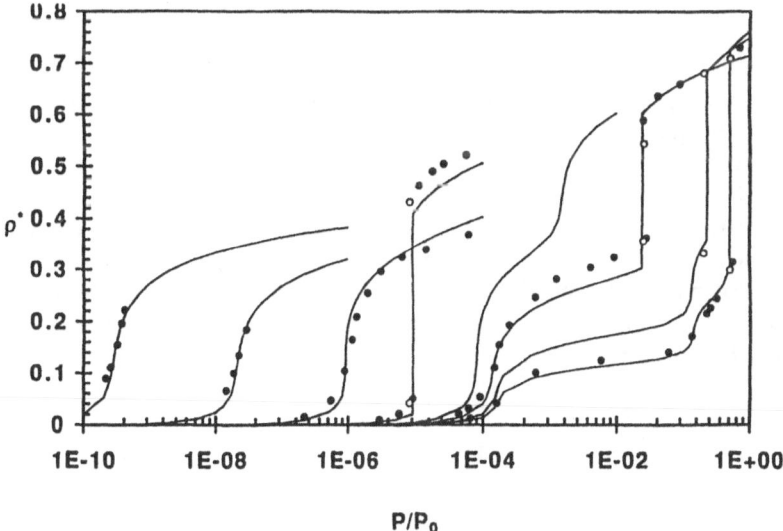

Figure 3. Adsorption isotherms for the system shown in Fig. 2 from DFT (lines) and GEMC (points) for pore widths (left to right) H*=2.0, 2.5, 3.0, 3.75, 5.0, 8.0 and 12.0. Open and closed circles represent pore-pore and pore-bulk fluid simulations, respectively. [From Lastoskie et al., ref. 24].

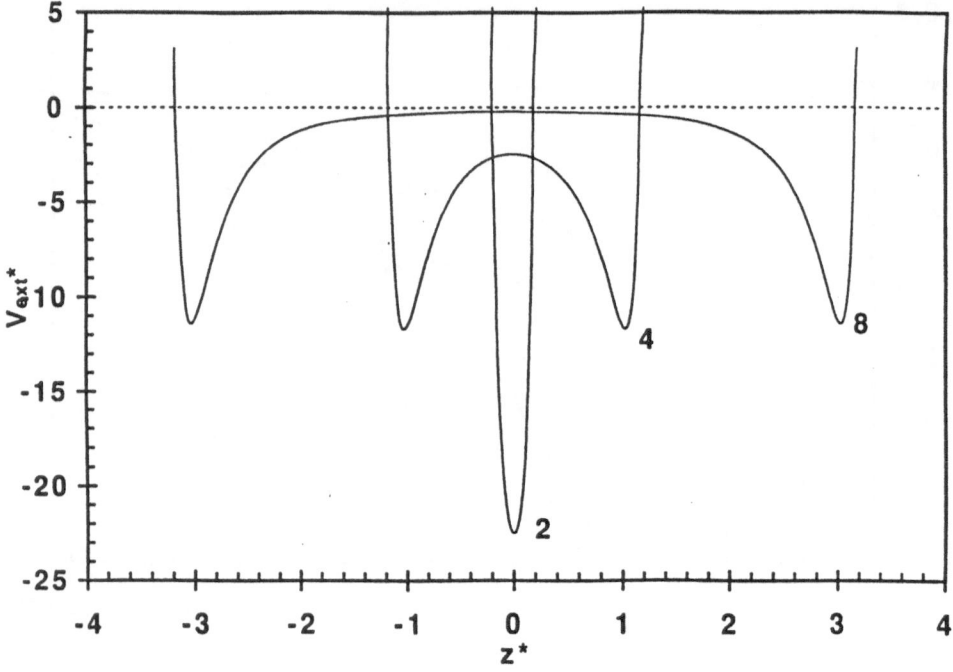

Figure 4. Wall-fluid potentials for nitrogen in graphite slit pores of width H* values shown. Here $V_{ext}^* = V_{ext}/\varepsilon_{ff}$, $z^* = z/\sigma_{ff}$, where $\varepsilon_{ff}$ and $\sigma_{ff}$ are nitrogen parameters. [From Lastoskie et al., ref. 24].

or pore filling pressures. The latter is important, since it is the pore filling pressure that is used as the primary measure of pore size in applying the theory to experimental data.

In applying the theory to obtain PSD's for real materials, the sorbent material is approximated as having an array of non-connected slit pores of a range of sizes, with pore width described by a distribution function $f(H)$. The distribution is assumed to have positive values for all $H$, and $f(H)$ is assumed to be continuous. Functions used to describe $f(H)$ have included the gamma distribution and lognormal distribution, each of which are multimodal functions. The two functions were found to give nearly identical results provided that at least three modes were allowed, and in the results that follow the trimodal gamma distribution was used. The equation for this is:

$$f(H) = \sum_{i=1}^{3} \frac{\alpha_i (\gamma_i H)^{\beta_i}}{\Gamma(\beta_i) H} \exp(-\gamma_i H) \tag{38}$$

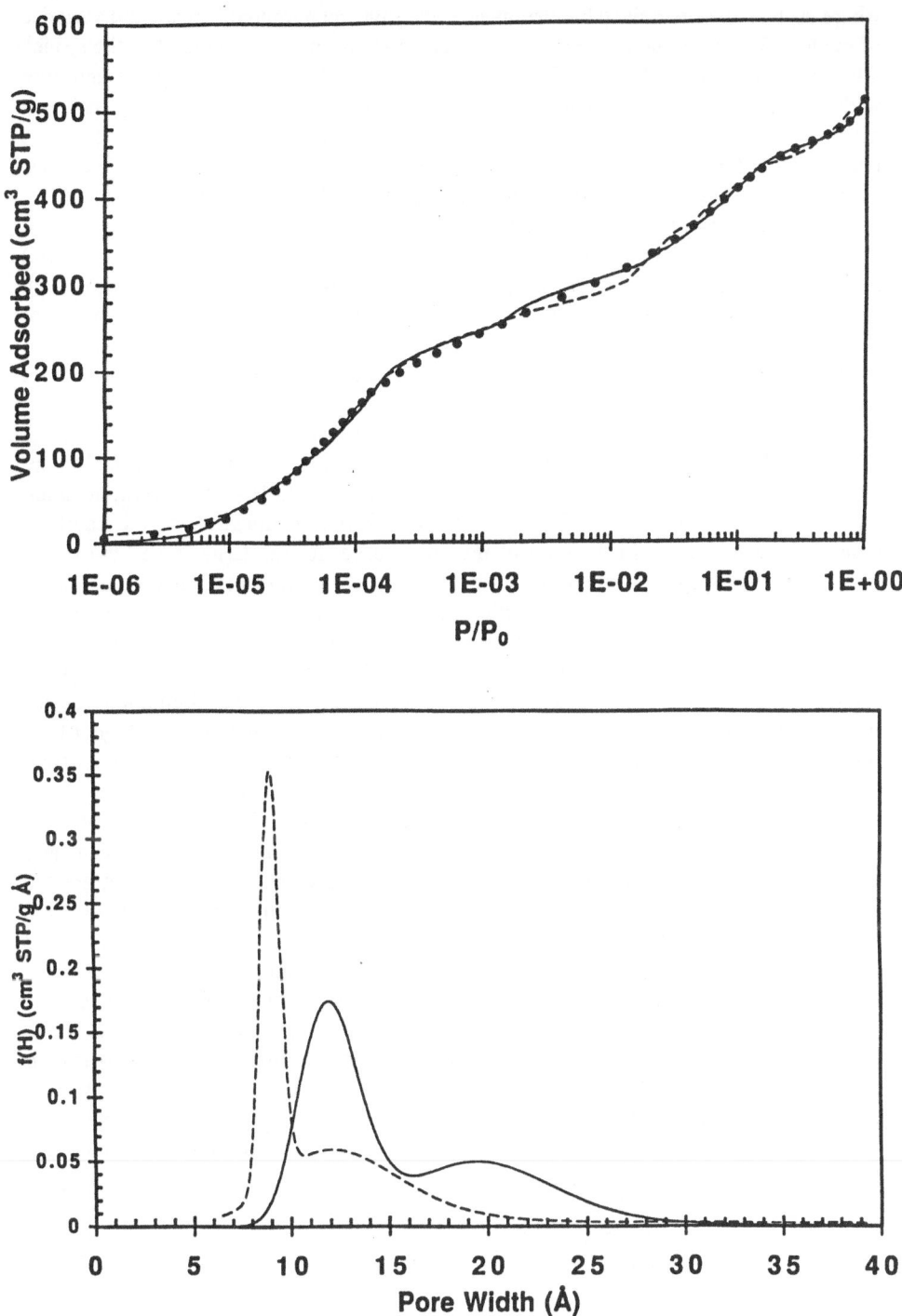

Figure 5. (a) Nitrogen adsorption isotherm, and (b) pore size distribution for microporous carbon AC610. Symbols are experimental data; solid and dashed lines are results of DFT and HK equations, respectively. [From Lastoskie et al., ref. 24].

where $\alpha_i$, $\beta_i$ and $\gamma_i$ are adjustable parameters that give the amplitude, mean and variance of mode i. To determine the PSDs of porous carbons from experimental data the model adsorption isotherms calculated from DFT are correlated in terms of pressure and pore width. The adsorption integral, eqn. (37), is then solved numerically using a minimization routine to determine $\alpha_i$, $\beta_i$ and $\gamma_i$. The number of modes in eqn. (38) has been set to 3, but this can be varied; in general the number of modes should be at least equal to the number of inflection points in the experimental adsorption isortherm.

Examples of such applications for two highly microporous carbons, AC610 and AX21 are shown in Figures 5 and 6. Only the DFT and Horvath-Kawazoe (HK) results are shown, since the Kelvin-based methods give unphysical results for such small pores. The HK theory underpredicts the size of the micropores. Comparisons of the Tarazona DFT method used here with results of the local theory have also been made [19]; the local theory predicts too sharp a peak in the micropore region for both carbons.

In Figure 7 we compare the pore filling pressures as predicted by the Tarazona DFT, the HK equation and the modified Kelvin equation (MK) for nitrogen in slit pore carbons. The MK equation involves replacing the pore width H in the Kelvin equation by a modified width (H-2t), where t is the adsorbed film thickness at the capillary condensation pressure. In the case of the DFT curve, for the larger pores the filling pressure is the capillary condensation pressure, while for small pores it is the point of inflexion on a continuously filling isotherm; cusps in the DFT curve represent critical pore sizes separating regions of continuous filling from capillary condensation or layering transitions. Also shown in Fig. 7 are molecular simulation results from the GEMC method. The DFT is seen to agree very well with the simulation results, whereas the other theories predict pore sizes that are too small for a given filling pressure. The Kelvin equation result (not shown) lies above the MK result [19]. Both the Kelvin and MK equations are valid for slit carbons only for quite large mesopores, and break down for H values below about 70-80 Å. Comparisons of Tarazona DFT with the local form of DFT shows good agreement for the mesopores, i.e. H values above 20 Å [19]. For smaller pores the local theory predicts pore sizes that are too large. From Fig. 7 it can be concluded that adsorption isotherm measurements down to very low pressures, $P_c/P_o < 10^{-9}$, are needed to sample the smallest pores. This is beyond the range of pressure measurement of currently available commercial sorptometers.

So far the applications of the DFT method to materials such as silicas, aluminas, zeolites, etc., in which the pores are to a first approximation cylindrical, have been less extensive, although the results available indicate that the DFT is of similar accuracy and utility for such cases. A comparison of the filling pressures vs. pore diameter for the case of nitrogen in cylindrical oxide pores at 77 K from Tarazona DFT, the Saito-Foley equation (ST), the Kelvin equation (K), and from GEMC simulations is shown in Figure 8. The Kelvin and Saito-Foley equations again underpredict the pore size for small pores. It should be noted that the range of validity of the Kelvin and modified Kelvin (not shown) equations is considerably greater for oxide materials than for carbons, being accurate for pore diameters above about 35 Å. This is expected,

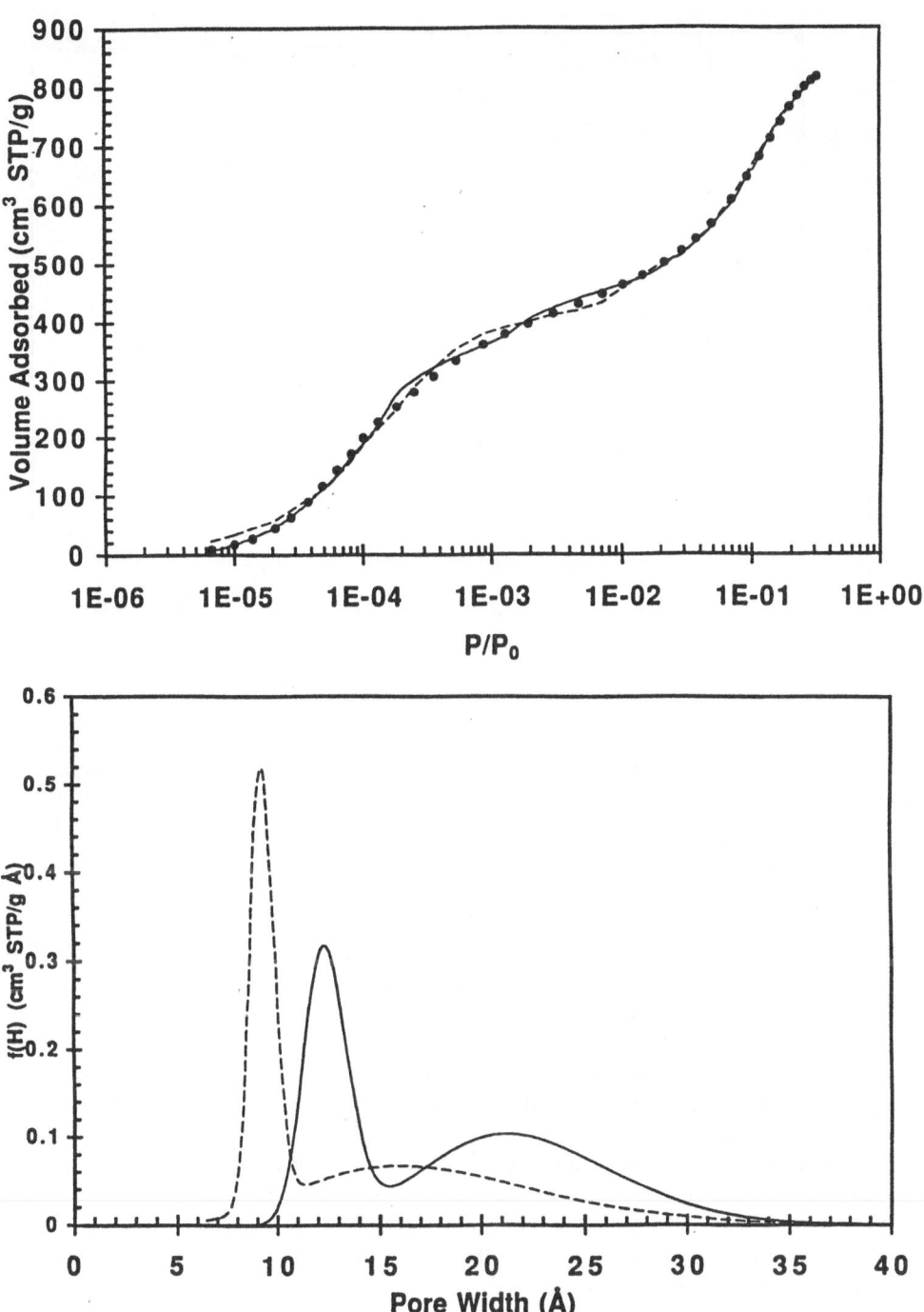

Figure 6.   (a) Nitrogen adsorption isotherm, and (b) pore size distribution for microporous carbon AX21. Key as for Fig. 5.  [From Lastoskie et al., ref. 24].

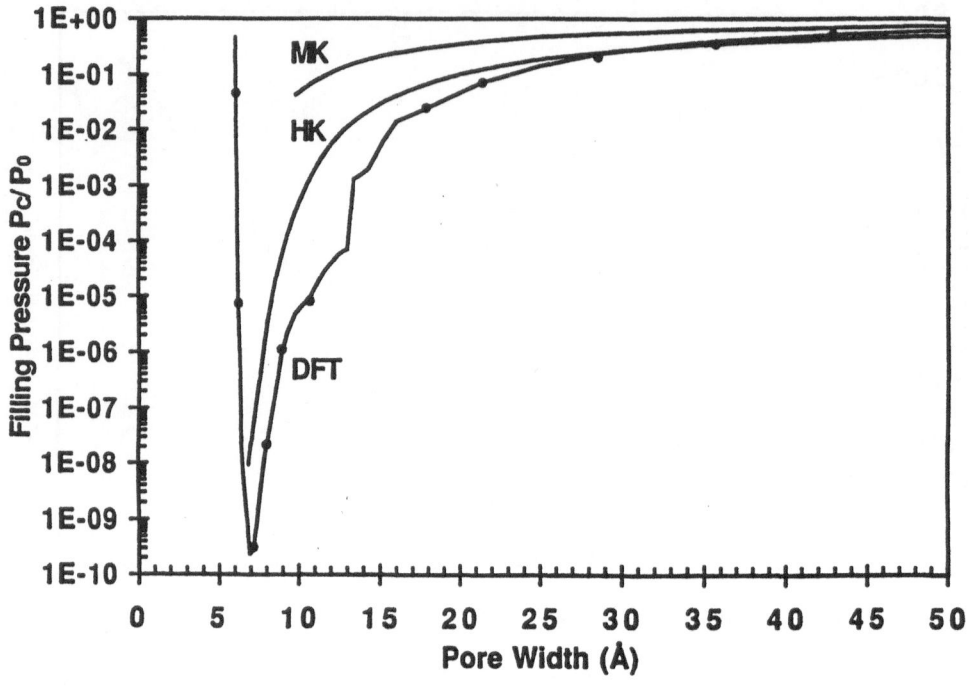

Figure 7. Pore filling pressures $P_c$ for nitrogen in slit pore carbons at 77 K, as predicted by the Tarazona theory (DFT), Horvath-Kawazoe equation (HK), modified Kelvin equation (MK), and from Gibbs Ensemble Monte Carlo simulations (points). [From Lastoskie et al., ref. 24].

since carbons provide a much more strongly adsorbing surface than oxides, a result of the high surface density of the carbon atoms. The DFT is again seen to be in very good agreement with the simulations, indicating that it will provide an accurate interpretation of the single pore isotherms.

The main advantages of the DFT method over the classical ones are (a) its accuracy for small pores, (b) it gives the full isotherm (not just the capillary condensation pressure), (c) it can be used at supercritical pressures, and (d) its basis in fundamental statistical mechanics means that it can be improved in a systematic way. It can be expected to work well for micropores, and can in principle be used to obtain PSDs from data other than the adsorption isotherm, e.g. heats of adsorption. In its present form of application, however, the DFT method still relies on the approximation that the real material can be treated as an *effective porous material*, in which all heterogeneity is approximated by a distribution of pore sizes. Thus, heterogeneity due to chemical groups on the surface, variations in pore shape, pore swelling under pressure, and pore networking and blocking effects are not accounted for explicitly. In addition, the use of the spherical Lennard-Jones potential for nitrogen may lead to errors. The importance of these various effects needs to be investigated by a combination of experiment and modeling. The theoretical treatment could then be further improved by incorporating the most important of these effects. It would also be of interest to apply the DFT analysis method to other experimentally measured properties, such as heats of adsorption, which may be a more sensitive measure of the PSD.

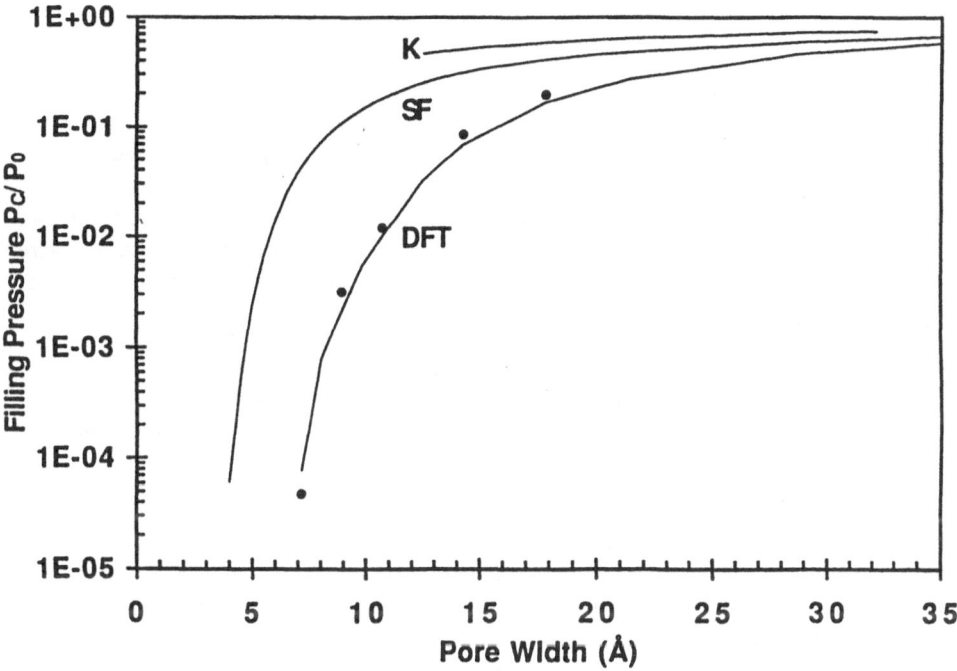

Figure 8. Pore filling pressures $P_c$ for nitrogen in cylindrical oxide pores at 77 K, as predicted by the Tarazona theory (DFT), Saito-Foley equation (ST), Kelvin equation (K), and from Gibbs Ensemble Monte Carlo simulations (points). [From Lastoskie et al., ref. 24].

## 4. Selective adsorption of trace components from mixtures

DFT and simulation methods can be used to investigate the effect of pore size, shape and type of material on selective adsorption of trace pollutants or contaminants from gas or liquid mixtures; these methods can also be used to suggest optimal designs for adsorbent materials for the removal of trace components [26,27]. Few experimental studies have been reported [28,29]. Direct experimental studies are difficult, because of the very low bulk concnetrations of the trace component, and because of the difficulty in finding very well characterised adsorbents for which the pore size can be varied in a controlled way. The Kierlik-Rosinberg-Rosenfeld form of DFT is particularly well suited to these calculations, and comparisons with GCMC simulation show that the theory is very accurate for calculations of the selectivity for trace solutes, as will be shown below.

The selectivity $S_2$ for a trace component X (2) in a carrier stream 1 (e.g. air, methane or water) is defined as

$$S_2 = \frac{(x_2 / x_1)}{(y_2 / y_1)} \tag{39}$$

where $x_i$ and $y_i$ are the mole fractions of component $i$ in the pore and in the bulk phase, respectively; $x_i$ is an average mole fraction over the pore space. In general $S_2$ will depend on the differences between the molecular sizes, shapes and attractive fluid-wall forces for the two components. In the limit of very low pressure (bulk density), where interactions among the adsorbate molecules can be neglected, it is easy to show that [26]

$$\lim_{\rho_b \to 0} S_2 = \frac{\int_0^H \exp[-u_{fw,2}(z)/kT]dz}{\int_0^H \exp[-u_{fw,1}(z)/kT]dz} \tag{40}$$

where $u_{fw,i}(z)$ is the fluid-wall potential for a molecule of species $i$ at distance $z$ from the pore wall, and $H$ is the pore width or diameter. If the pore is relatively large with respect to the molecular size, energetic considerations often dominate, and for simple molecules we might expect that

$$\lim_{\rho_b \to 0} S_2 \approx \exp[(\varepsilon_{sf,2} - \varepsilon_{sf,1})/kT] \tag{41}$$

where $\varepsilon_{fw,i}$ is the well depth of the fluid-wall interaction for species $i$. Equations (40) and (41) indicate that the selectivity, at least at low pressure, depends exponentially on the difference between the molecular interactions with the wall, so that even small differences in these can lead to large selectivities. This general feature holds also for higher bulk densities, and provides the basis for the usefulness of adsorption in separations.

The Kierlik-Rosinberg-Rosenfeld DFT has been used to study the behavior of the selectivity for trace components of simple gases in carrier gases of both methane [26] and nitrogen [27,30]. Here we give only a brief description of some of the results for trace pollutant gases in a nitrogen carrier, but the general features are found to be similar for trace gases in methane. The gases modeled [27,30] included carbon tetrachloride, carbon tetrafluoride and sulfur dioxide, in a slit carbon porous solid. The fluid-wall interaction was of the 10,4,3 type [25]. For the fluid-fluid interactions, both spherical Lennard-Jones (LJ) and site-site Lennard-Jones models were used; the results from these two models were qualitatively similar, although significant quantitative differences in the selectivities were obtained.

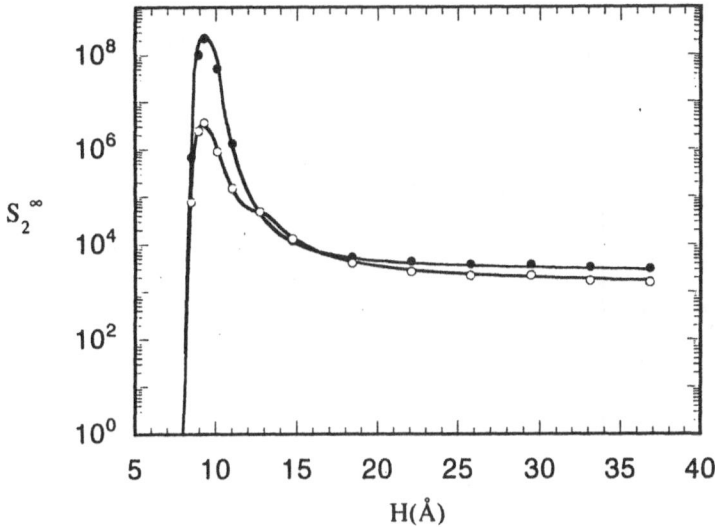

Figure 9. Effect of pore width on selectivity for $N_2(1)/CCl_4(2)$ mixtures in slit carbon pores at 300 K. Here the bulk gas mole fraction of $CCl_4$ is one part per million, $x_2=10^{-6}$. Symbols are GCMC simulation results, lines are DFT. Filled circles are for a pressure of 8.310 kPa ($\rho_b^*=10^{-4}$) and open circles are for 8175 kPa ($\rho_b^*=10^{-1}$). [From Sowers and Gubbins, ref. 27].

The effects of pore width and pressure on the selectivity for carbon tetrachloride in carbon tetrachloride (2)/nitrogen (1) mixtures adsorbed in carbon slit pores is shown in Figures 9 and 10. Here a simple spherical LJ model is used for both nitrogen and carbon tetrachloride. For this system very large selectivities are possible, in excess of $10^8$ for the optimal pore width and pressure, because of the substantial difference in attractive energies of the two components, $\varepsilon_{ff,2}/\varepsilon_{ff,1}=3.57$. The bulk mole fraction of one part per million $CCl_4$ is sufficiently low that the limit of infinite dilution is reached, and so we write the selectivity as $S_2^\infty$. From Figure 9 it is seen that an optimal pore width of H=9.2Å exists for which the selectivity curve exhibits a strong maximum. This is the pore width at which the pore can just accommodate a single layer of CCl$_4$ molecules. This corresponds to the pore width at which the fluid-wall potential wells have coalesced into one deep well, and the well is at its deepest. As the pore width increases the selectivity decreases rapidly due to the decreasing fluid-wall potential, and there is a dramatic decline in selectivity. The optimal pore width can be estimated to good accuracy for a wide range of mixtures from the equation

$$H_1 = \sigma_{cc} + \sigma_{ff,stronger} \tag{42}$$

where $\sigma_{cc}$ and $\sigma_{ff,stronger}$ are the LJ diameters for the carbon-carbon interaction and for the fluid-fluid interaction of the more strongly adsorbed component.

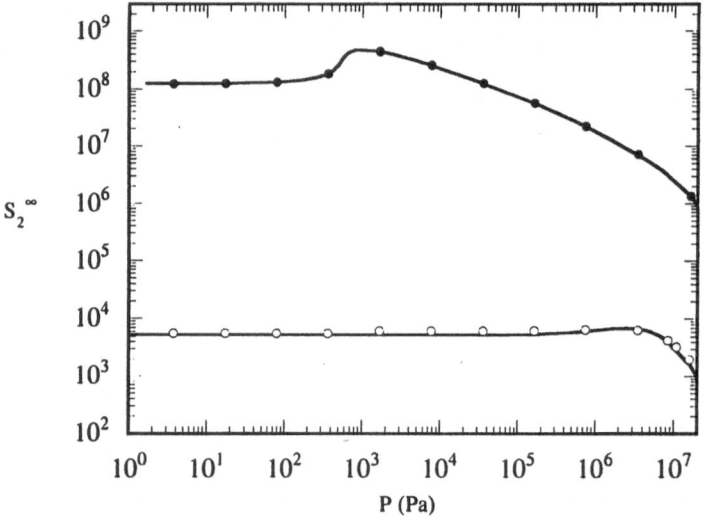

Figure 10. Effect of pressure on selectivity for N$_2$/CCl$_4$ mixtures in a slit carbon pore at 300 K, $x_2$=10$^{-6}$. Symbols are GCMC results, lines are DFT. Filled and open circles are results for H=9.2 and 18.4 Å, respectively. [From Sowers and Gubbins, ref. 27].

At low pressures the selectivity is determined solely by fluid-solid interactions, and in this region the selectivity is constant as seen in Figure 10. At somewhat higher pressures the fluid-fluid interactions become significant and this causes the selectivity to rise. For the optimal pore width of 9.2Å there is a pronounced maximum in the selectivity curve, as seen in Figure 10. Closer examination shows that this corresponds to the formation of a loose monolayer coverage; at this pressure the packing is loose enough that almost all carbon tetrachloride molecules can sit in the well of the potential [27]. At this condition the great majority of the molecules in the pore are CCl$_4$ ones, despite the bulk concentration of only one part per million. At somewhat higher pressures for H=9.2 Å the packing density in the pore increases, and repulsive fluid-fluid interactions lead to some molecules being pushed out of their most favored position at the minimum of the potential well. This leads to a decrease in the selectivity.

It should be noted that the agreement between the DFT and simulation results in these figures is excellent. Similar agreement was obtained for all of the systems studied, where component 2 was highly dilute. The agreement is somewhat poorer for concentrated mixtures.

Results for $N_2/CF_4$ mixtures are shown in Figures 11 and 12. Here the bulk gas concentration is $x_2=10^{-4}$; a lower value was used than in the case of $N_2/CCl_4$ mixtures since it was sufficient to reach the infinite dilution limit for this mixture. In this case the selectivities are lower, with values ranging up to a few hundred, because of the smaller difference in the fluid-fluid interactions of the two components, $\varepsilon_{ff,2}/\varepsilon_{ff,1}=1.67$. Because of these weaker differences in potential, there is no significant maximum in the selectivity on varying the pressure. Results for the $N_2/SO_2$ mixtures in these carbons are similar to those shown for $N_2/CF_4$.

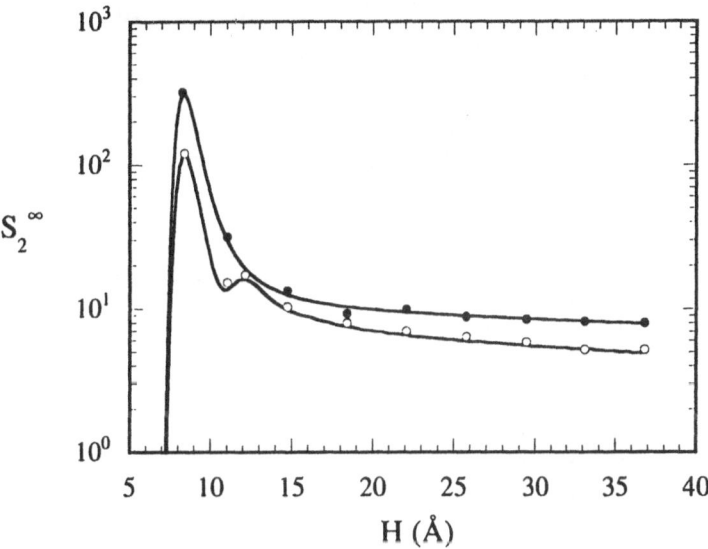

Figure 11. Effect of pore width on selectivity for $N_2/CF_4$ mixtures in slit carbon pores at 300 K and $x_2=10^{-4}$. Points and lines are GCMC and DFT results, respectively. Filled and open circles are for pressures of 8.310 and 8193 KPa, respectively. [From Sowers and Gubbins, ref. 27].

If the pore geometry is changed from slit to cylindrical, even more dramatic selectivities are seen [30]. The cylindrical pore offers greater confinement and deeper potential wells near the walls, leading to these higher selectivities. Typical results for the $N_2/CF_4$ mixture are shown in Figure 13. For the cylinders the bulk mole fraction was lowered to $x_2=10^{-9}$ in order to reach the infinite dilution limit. Changing from slit

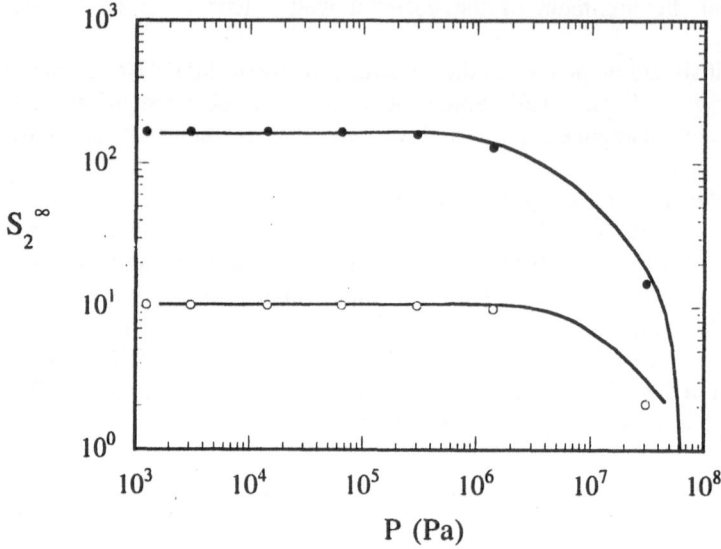

Figure 12. The same as Figure 10, but for the $N_2/CF_4$ mixture. [From Sowers and Gubbins, ref. 27].

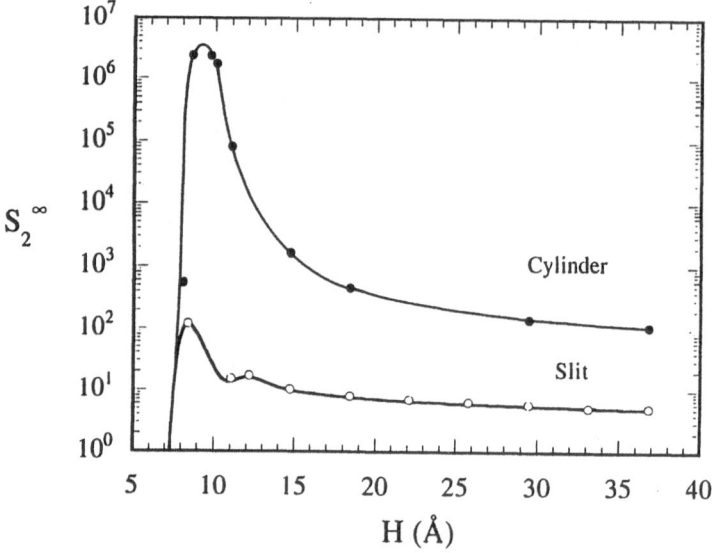

Figure 13. Comparison of effect of pore width on selectivity for the $N_2/CF_4$ mixture in graphitic slit ($x_2=10^{-6}$) and cylindrical ($x_2=10^{-9}$) pores at 300 K. Points are GCMC and lines are DFT results. [From Sowers and Gubbins, ref. 30].

to cylindrical geometry is seen to lead to a somewhat larger optimal pore size, and to increases of selectivity of about 3-4 orders of magnitude, the amount depending on the pressure. The results indicate the potential for the use of buckytubes as adsorbents of very high selectivity, particularly if the pore diameter can be suitably tuned for the mixture of interest.

The above results for optimal selectivity represent an ideal system. Almost any departure from this ideal model will result in a decrease in selectivity. In order to study the effect of a more realistic intermolecular potential, site-site LJ models were used for the adsorbate molecules [30]. For the $N_2/CCl_4$ mixtures in slit pore carbons, at the optimal pore size and pressure, the effect of switching to a site-site model is to lower the selectivity by about two orders of magnitude. For other conditions for this mixture, and for other mixtures, the reduction is smaller than this. Using a carbon that is not monodisperse, but has a distribution of pore sizes will also diminish the selectivity, particularly if the main peak in the PSD is not located at the optimum pore size. For AC610, for example (Figure 5), the PSD is bimodal, with the main micropore peak having a maximum at 11.9 Å. If the selectivity for carbon tetrachloride from a $N_2/CCl_4$ mixture is calculated for this carbon, assuming the PSD shown in Figure 5, a selectivity about 10 times smaller than for monodisperse pores of the optimal size of H=9.2 Å is found [30]. Other factors that will reduce selectivity are pore networking and blocking, and the presence of water vapor, which may be preferentially adsorbed at active sites and so block some of the surface. Pore swelling may also be significant at higher pressures.

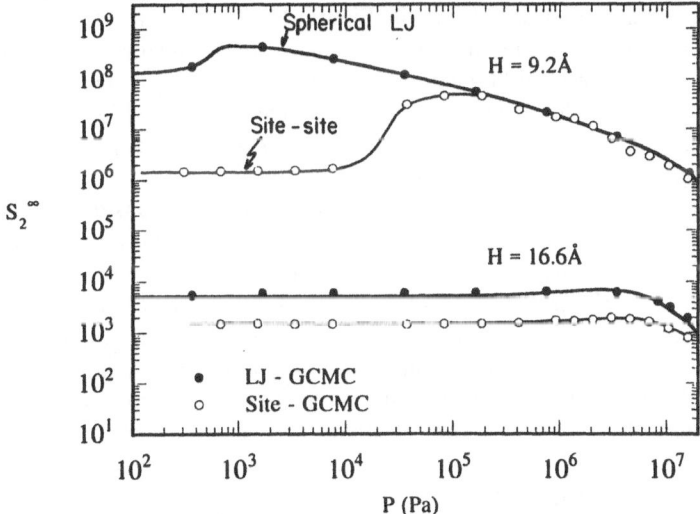

Figure 14. Comparison of selectivity results for $N_2/CCl_4$ mixtures in slit graphitic carbons at 300 K, $x_2=10^{-6}$, calculated from spherical LJ model and site-site LJ model, for two pore widths. [From Sowers and Gubbins, ref. 30].

These calculations serve to show the importance of trying to synthesize monodisperse porous materials with pores of the optimum size and shape, and of operating at the optimum pressure. The use of strong adsorbents, such as buckytubes, should also enhance selectivity. The calculations shown here are of course equilibrium ones. In practice the performance will also depend on attaining satisfactory diffusion rates.

## 5. Adsorption behavior of water and water/hydrocarbon mixtures on activated carbons

The applications described above are for pores with homogeneous surfaces. When surface active sites having a high adsorption energy for one of the adsorbates are present, the adsorption behavior can be dramatically different from that for the simpler cases. In this section we discuss the adsorption of water and water/hydrocarbon mixtures on a model activated carbon [31,32,33]. When the surface is homogeneous, i.e. that of a graphitic carbon, almost no water is adsorbed; the surface is hydrophobic. The physical reason for this behavior is that, in order to adsorb and conform to the surface geometry, hydrogen bonds that would otherwise be present in bulk water will have to be broken. However, when oxygenated groups, such as CO, COOH, OH, etc. are present on the surface the water molecules can hydrogen-bond to these, and this leads to an appreciable adsorption. Such activated carbons are prepared by heating carbonaceous material in the presence of water, oxygen or carbon dioxide. Some typical experimental results are shown in Figure 15. The curves show successive experimental adsorption isotherms obtained after heat treating a carbon with oxygenated sites at successively higher temperatures, in some cases in the presence of hydrogen. It is seen that the adsorption of water is greatly enhanced by the presence of surface oxygenated sites, whereas for graphitic carbon with few if any sites the water uptake is negligible except at high pressures.

The adsorption of water on such activated carbons is of considerable industrial significance, since such carbons are widely used for separations in the chemical, petroleum and pharmaceutical industries for separations and for removal of pollutants from water and air. The presence of water in air or gas streams is known to have a pronounced effect on breakthrough curves (plots of pollutant concentration leaving an adsorption bed versus time). Breakthrough occurs at much earlier times when water vapor is present, apparently because the water molecules adsorb strongly on active sites and so block part of the surface. The selectivity and adsorption of activated carbons used in water treatment also depends strongly on the type and placement of active sites.

Molecular simulation can be used to investigate well-defined models of such water-activated carbon systems [32,33]. The objective of such studies is to examine the effect of active surface sites (e.g. OH, COOH, CO, etc.) on the adsorption of (a) water, (b) water/hydrocarbon mixtures. In particular, we are interested in the effect of the surface density, geometric arrangement and chemical species of the sites on the equilibrium adsorption behavior, including pore filling, phase transitions and selectivity in the case of mixtures.

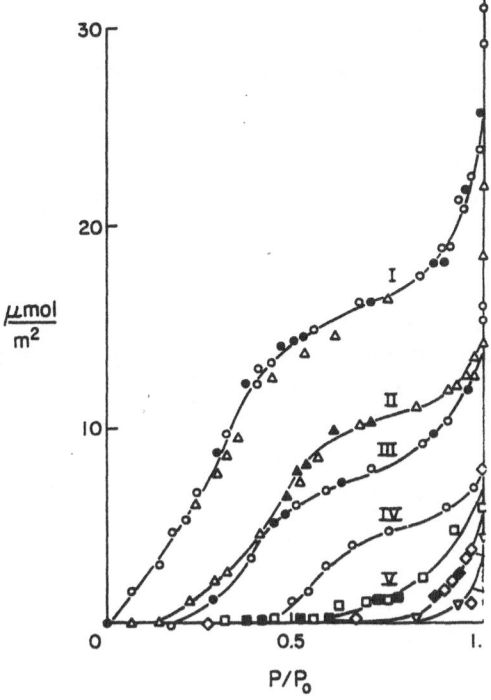

Figure 15. Experimental results for adsorption of water on oxygenated carbons. (I) Heated in vacuo at 200°C, (II) in vacuo at 950°C, (III) in vacuo at 1000°C, (IV) at 1100°C in a hydrogen stream, (V) in hydrogen at 1150°C, (VI) in hydrogen at 1700°C, and (VII) at 3200°C. Solid symbols denote desorption. From Gregg and Sing [16].

The carbon is modeled as made up of slit graphite pores of width H; active oxygenated sites on the carbon surface can be modeled using existing potential models such as OPLS (optimised potentials for liquid state) [34], or using square well attractive sites [33]. Water can also be modeled using either potentials of the OPLS type, or using square well sites. The OPLS model consists of Lennard-Jones site-site potentials, summed over appropriate atom sites in the molecule or group, plus a sum of Coulomb potential terms resulting from point charges placed at strategic points in the molecule to mimic the H-bonding. In the case of water there is a single LJ center which is the O atom; in a surface group such as COOH there would be a number of LJ sites for the various atoms in the group. The square well interaction is

$$u_{HB}(r) = -\varepsilon_{HB} \qquad \textit{if } r < \sigma_{HB}$$
$$= 0 \qquad \textit{otherwise}$$

(43)

where $\sigma_{HB}$ is the diameter of the square well interaction, $\varepsilon_{HB}$ is the depth of the energy well, and r is the distance between two square well sites. Our experience with these two models [32,33] is that they give very similar results for most equilibrium properties, such as adsorption isotherms, heats of adsorption, etc., and so we give below only some results for the simpler square well (SW) model. The SW model captures the strong orientation dependence and short range character of the H-bond. Because it is of short range it is not necessary to resort to Ewald summation methods to account for long range forces, as in using the OPLS model, which involves Coulomb potentials, and so it is possible to use smaller systems and to speed up the calculations. However, it should be noted that for other properties, particularly transport properties and studies of the subtle differences in fluid structure near the wall resulting from different oxygenated groups, the OPLS model would be superior.

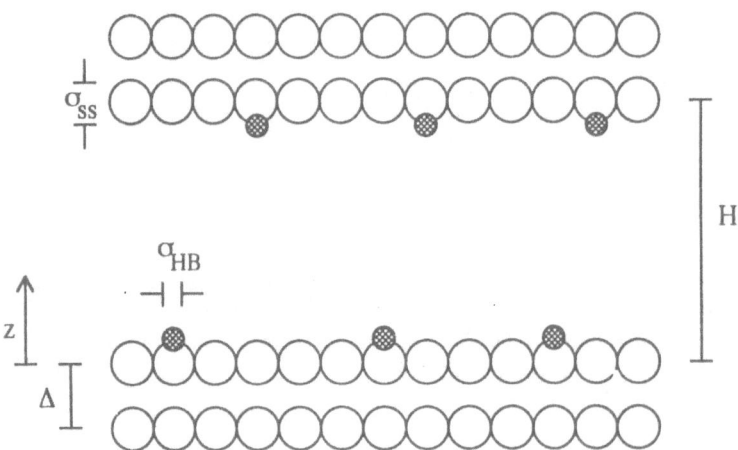

Figure 16. Lateral view of pore geometry for activated carbon model. Open circles are carbon atoms forming the walls of the slit pore, arranged on a graphite lattice. Shaded circles represent the square well sites (activated sites). [From Müller et al., 33].

In the SW model water is modeled using a Lennard-Jones sphere centered on the oxygen, with four square well association sites arranged tetrahedrally; two of these represent the H atoms (which can H-bond to other waters or to surface sites) and two the lone pair electrons [33]. The potential parameters are adjusted to represent the vapor-liquid coexistence properties of water, and in the case of the wall sites, are adjusted to approximately represent COOH groups. The interaction of water with the graphite wall was approximated using the (10,4,3) potential [25]. Thus the interaction of a water molecule with the walls is of the form

$$u_{fw}(z) = u_{wall}(z) + u_{wall}(H - z) + u_{HB} \qquad (44)$$

where $u_{wall}$ is the (10,4,3) interaction between the water molecule and the graphite wall, and $u_{HB}$ is the SW interaction between SW sites on the water molecule and SW sites in the wall; the latter term is only present when an appropriate (H) site on the water molecule overlaps with an active wall site, i.e. when r is less than $\sigma_{HB}$. Grand canonical Monte Carlo simulations were used [33] to study the system for a range of pressures, pore widths, site densities and site arrangements at 300 K. Long runs were needed to ensure ergodicity, i.e. to ensure that phase space is properly sampled. The parameter $\varepsilon_{HB}/kT$ was about 12, where $\varepsilon_{HB}$ is the well depth of the interaction between the water molecules and the surface sites; with such a value it is possible to use conventional GCMC without biasing, but long runs (typically 10-40 million configurations for equilibration and a further 50-100 million for averaging) were needed to ensure that both adsorption and desorption of water molecules onto surface sites were properly sampled.

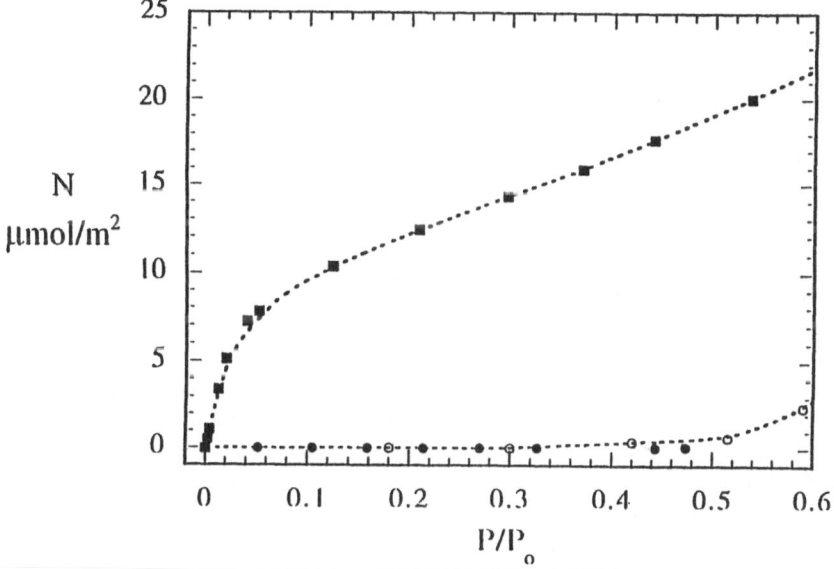

Figure 17. Adsorption isotherms for water and a simple LJ model of propane on a planar graphite surface with no active sites, at 300 K. Closed circles are simulation results for water, open circles are experimental data for water on graphon [35], squares are for propane. Lines are a guide to the eye. [From Müller et al., 33].

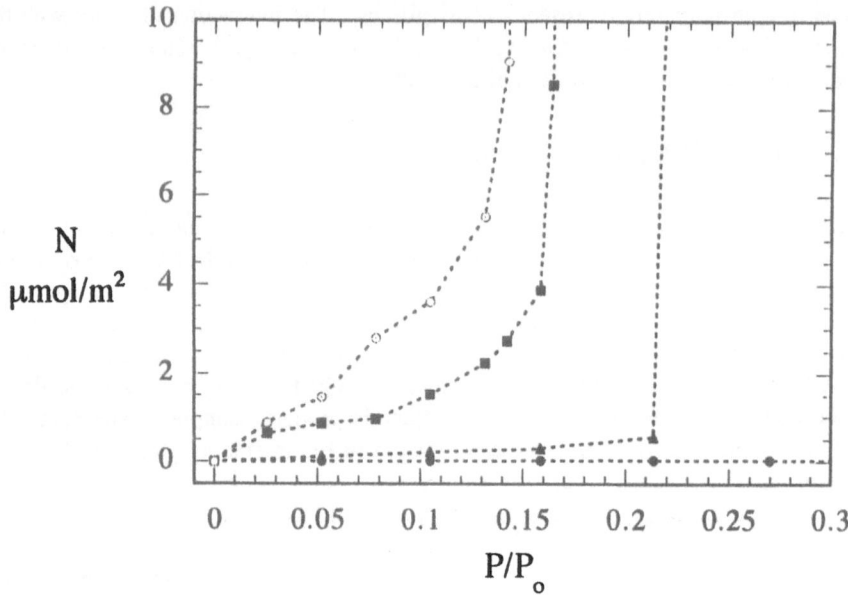

Figure 18. Adsorption isotherms for water on an activated carbon surface at 300 K, from simulation. Closed circles are for n=0 (unactivated graphitic surface), triangles are for n=0.222, squares for n=0.444, and open circles are for n=1 site/nm$^2$; lines are a guide to the eye. [From Müller et al., 33].

The adsorption behavior for pure water is found to depend very strongly on the density and geometric arrangement on the carbon surface of the active sites. When no surface sites are present the pore walls are hydrophobic, and for a pore of width H*=H/σ=20, where H is pore width and σ is the diameter of the water molecule, there is no appreciable adsorption until relatively high pressures, P/P°~0.6, as shown in Figure 17. At somewhat higher pressures capillary condensation occurs. This is in agreement with experimental data for graphitic carbons that have been heated in a reducing atmosphere to remove surface groups [16,35]. The behavior for water is in marked contrast to that for simpler fluids such as nitrogen or hydrocarbons, which adsorb much more strongly at low pressures on graphitic carbons; an adsorption isotherm for a simple model of propane is included in Figure 17 for comparison purposes. When even a low density of active sites are added to the carbon surface, the adsorption behavior is qualitatively changed; the adsorption isotherms at low pressure are shown in Figure 18. Here n is the surface density of active sites, in sites/square nanometer. In these calculations sites were added at random locations on the surface. For reference it should be noted that there are approximately 38 carbon atoms per square nanometer. The effect of a site density of n=0.444 site/nm$^2$ is to greatly increase the adsorption of water at low pressures, and further increases in the density of sites raises the adsorption at low pressures. At site densities of n=0.444 site/nm$^2$ and above, capillary condensation no longer occurs, but is replaced by a continuous filling of the pore. Observation of typical molecular configurations during the course of the simulation [33] shows that pore filling occurs in a completely different way to that for simple fluids such as hydrocarbons, nitrogen, etc. Instead of forming a well-ordered adsorbed layer on the surface, water molecules adsorb onto active sites and then form nuclei for water clusters to form. Where possible the water molecules form 'bridges' between surface sites, and if sites are placed so as to promote this the adsorption is greatly enhanced. During the final stages of pore filling there is no first order phase transition as for simple fluids; instead, the clusters grow until they bridge across between the pore walls, and the vacant spaces in the pore are then filled in. The adsorption at a particular pressure is increased by increasing the site density n (Figure 18), the adsorption levelling out above a site density of about 10-15 sites/nm$^2$.

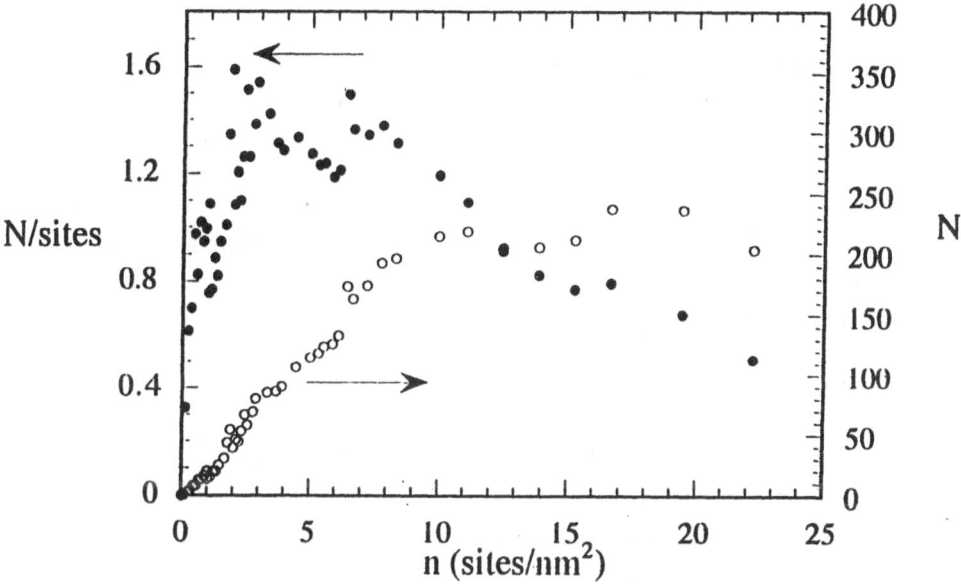

Figure 19. Effect of density of active surface sites, n in site/nm$^2$, on adsorption for a pore of width 2 nm at 300 K, P/P$_o$=0.042. Open circles (right ordinate) correspond to the number of molecules adsorbed, N; closed circles (left ordinate) are the number of molecules per associating site. [From Müller et al., 33].

The adsorption depends strongly on the geometric arrangement of the sites. By grouping the sites at appropriate separation distances from each other the extent of adsorption can be greatly enhanced through the cooperative 'bridging' effect. This is illustrated in Figure 20 and Table 2, where adsorption results for several arrays of sites are compared. Three types of array were considered, as shown in Figure 20. The first is a random array, as was used in the previous results shown in Figures 17-19. The

100

second is a regular array of sites spanning the lattice, and the third takes the same number of sites but compresses them into a small area, keeping the regular square array.

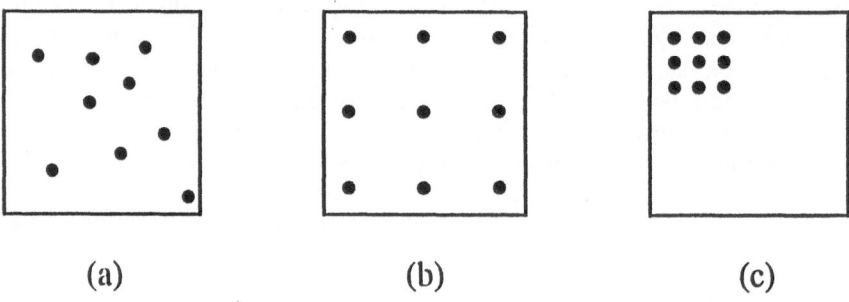

(a)                    (b)                    (c)

Figure 20. Arrangement of surface sites for n=1 site/nm$^2$: (a) random array; (b) regular square array spanning the surface; (c) dense square array over one quarter of the surface. [From Müller et al., 33].

TABLE 2

Number of Adsorbed Water Molecules, N, for Different Site Arrangements and Densities (n=site/nm$^2$), for H=2 nm, T=300 K, P/P°=0.042. [From Müller et al., 33].

| Geometry | N | |
|---|---|---|
| | n=1 | n=1.5 |
| Random 1, (a) | 30±2 | 39±4 |
| Random 2, (a) | 19±2 | 30±3 |
| Random 3, (a) | 15±5 | 28±4 |
| Regular array, (b) | 5±1 | 14±4 |
| Dense array, (c) | 41±2 | 52±1 |

Different random placements of surface sites are seen to give considerable variation in adsorption, depending on whether the surface sites 'land' on the surface with separations that promote bridging or not. In array (b) of Fig. 20 the sites are too far apart for bridging to occur at these rather low site densities. In the case of the dense array, however, the site-site distance is suitable for a strong bridging effect. Thus an appropriate site-site separation is critical in determining the amount of adsorption at these modest pressures, where cooperative bridging effects can increase adsorption by as much as one order of magnitude in some cases.

The main conclusions of these calculations are: (a) the lack of adsorption for graphitic carbons, and the large influence of even a low density of surface H-bonding sites closely parallels the experimental findings; (b) above a rather low threshold density of surface sites, capillary condensation is not observed, but instead a continuous filling takes place; this is because of the dramatically different filling mechanism for water in such materials, which proceeds by adsorption of water molecules onto active sites, followed by the growth of water clusters on these sites, leading eventually to filling via a percolation threshold; (c) the importance of cooperative bridging effects in adsorption, so that not only site density but site arrangement on the surface is very important. More work is needed on these systems to elucidate the influence of type of surface site, effects of temperature variation, etc., as well as on the kinetic phenomena involved.

Recent work [36] on water/methane mixtures in such model activated carbons again shows that the behavior and selectivity depend strongly on the site density and arrangement. For n=0 the pores strongly select methane, with selectivities of over 10 at 300 K. However, as n is increased the selectivity drops rapidly, and an inversion of selectivitiy occurs at a relatively low site density. For a bulk water mole fraction of $x_w$=0.031 and a 2 nm pore at 300 K and a pressure of 2.5 Mpa, the selectivity inversion occurs at a site density of only n=0.13 site/nm$^2$, for example. For higher selectivities water is preferentially adsorbed, with water selectivities of about 30 at n=0.4.

## 6. Conclusions

There have been major advances in the application of molecular simulation and statistical mechanical methods to the study of adsorption over the past decade, with many new workers entering the field. Most of the theoretical studies so far have been liimited to simple fluids, simple pore geometries and homogeneous surfaces, but this situation is now changing and studies of more complex systems are starting to appear. Of particular interest will be the application of these methods to effects of heterogeneous and doped surfaces, and to effects of surface irregularities and pore networking.

The major problem in this area remains the difficulty of effectively coupling the modeling work with well-designed experiments. A large part of the difficulty is in finding porous materials that are sufficiently regular and well characterized, so that the modeler can be confident of knowing the solid structure. Although many small-pore zeolites meet this condition, many other materials do not. Some of the new materials, such as MCM-41 and buckytubes, offer hope of regular, well characterised pores with diameters in excess of 1 nm, but current versions of these materials seem not to be as regular as one would wish.

The selection of topics covered in this chapter is a personal one, chosen because of my own interests and knowledge, and omits a vast amount of interesting and important recent modeling work. It also focusses on equilibrium behavior only, since the theory of transort phenomena in pores is still in a rudimentary state. Much remains to be done on the use of molecular simulation to study such kinetic behavior.

## Acknowledgments

It is a pleasure to thank many colleagues for helpful discussions, particularly Christian Lastoskie, Erich Müller, David Nicholson, Nick Quirke, Luis Rull, Susanne Sowers, and Lourdes Vega. This work was supported by grants (nos. CTS-9508680 and INT-921415) from the National Science Foundation, and by a NATO cooperative research grant (no. 931517). Supercomputing time was provided under a NSF Metacenter grant (no. MCA93S011P).

## References

1.  Evans, R. (1979) *Adv. Phys.* **28**, 143; Evans, R. (1992) in D. Henderson (ed.), *Fundamentals of Inhomogeneous Fluids*, Marcel Dekker, New York, p. 85.
2.  Gubbins, K.E. (1994) in M.G. Velarde and C.I. Christov (eds.), *Fluid Physics, Lecture Notes of Summer Schools*, World Scientific; Gray, C.G., Gubbins, K.E. and Joslin, C.G. (1996), *Theory of Molecular Fluids, Volume 2, Ch. 8*, in press, Clarendon Press, Oxford.
3.  Carnahan, N.F. and Starling, K.E. (1969) *J. chem. Phys.* **51**, 635.
4.  Tarazona, P. (1985) *Phys. Rev. A* **31**, 2672. It should be noted that the equations given in this paper for the weighting functions are incorrect. Correct expressions are given in: Tarazona, P., Marini Bettolo Marconi, U. and Evans, R. (1987) *Mol. Phys.* **60**, 573.
5.  Although the local and smoothed densities depend on the vector $\mathbf{r}$, the weighting function $w(r')$ has spherical symmetry and depends only on the scalar $r'$. We note from eqn. (28) that the radial function $r'$ in $w(r')$ does not correspond to the point $\mathbf{r}$ at which the smoothed density is evaluated, in general.
6.  Tan, Z., Marini Bettolo Marconi, U., van Swol, F. and Gubbins, K.E. (1989) *J. chem. Phys.* **90**, 3704.
7.  Peterson, B.K., Gubbins, K.E., Heffelfinger, G.S., Marini Bettolo Marconi, U. and van Swol, F. (1988) *J. chem. Phys.* **88**, 6487.
8.  Kierlik, E. and Rosinberg, M.L. (1990) *J. chem. Phys.* **A42**, 3382; (1991) *ibid.* **A44**, 5025.
9.  Rosenfeld, Y. (1989) *Phys. Rev. Lett.* **63**, 980.
10. Phan, S., Kierlik, E., Rosinberg, M.L., Bildstein, B. and Kahl, G. (1993) *Phys. Rev. E* **48**, 618.
11. Allen, M.P. and Tildesley, D.J. (1987) *Computer Simulation of Liquids*, Oxford University Press, Oxford.
12. Gubbins, K.E. and Quirke, N., eds. (1996) *Molecular Simulation and Industrial Processes: Methods, Applications and Prospects*, Gordon & Breach, London.

13. Gubbins, K.E., Sliwinska-Bartkowiak, M. and Suh, S.-H. (1996) *Molecular Simulation*, in press.

14. Gubbins, K.E. (1994) Application of Molecular Theory to Phase Equilibrium Predictions, in S.I. Sandler (ed.), *Models for Thermodynamic and Phase Equilibrium Calculations*, Marcel Dekker, New York, pp. 507-600.

15. Peterson, B.K. and Gubbins, K.E. (1987) *Mol. Phys.* **62**, 215.

16. Gregg, S.J. and Sing, K.S.W. (1982) *Adsorption, Surface Area and Porosity*, Academic Press, New York.

17. Horvath, G. and Kawazoe, K. (1983) *J. Chem. Eng. Japan* **16**, 474.

18. Saito, A. and Foley, H.C. (1991) *AIChE Journal* **37**, 429.

19. Lastoskie, C., Gubbins, K.E. and Quirke, N. (1993) *J. Phys. Chem.* **97**, 4786.

20. Lastoskie, C., Gubbins, K.E. and Quirke, N. (1993) *Langmuir* **9**, 2693.

21. Lastoskie, C., Gubbins, K.E. and Quirke, N. (1993) in J. Rouquerol (ed.), *Characterization of Porous Solids III*, Elsevier, Amsterdam, p. 51.

22. Olivier, J.P., Conklin, W.B. and Szombathely, M.von (1994), Determination of Pore Size Distribution from Density Functional Theory: A Comparison of Nitrogen and Argon Results, in J. Rouquerol, F. Rodriguez-Reinoso, K.S.W. Sing and K.K. Unger (eds.), *Characterization of Porous Solids III*, Elsevier, Amsterdam, pp 81-89.

23. Olivier, J.P. (1996) *Adsorption*, in press.

24. Lastoskie, C., Quirke, N. and Gubbins, K.E. (1996) Structure of Porous Adsorbents: Analysis using Density Functional Theory and Molecular Simulation, in W. Rudzinski, W.A. Steele and G. Zgrablich (eds.), *Equilibria and Dynamics of Gas Adsorption on Heterogeneous Solid Surfaces*, Elsevier Pub. Co., Amsterdam, in press.

25. Steele, W.A. (1973) *Surface Sci.* **36**, 317; Steele, W.A. (1974) *The Interaction of Gases with Solids*, Pergamon, Oxford.

26. Jiang, S., Gubbins, K.E. and Balbuena, P.B. (1994) *J. Phys. Chem.* **98**, 2403.

27. Sowers, S.L. and Gubbins, K.E. (1995) *Langmuir* **11**, 4758.

28. Golden, T.C. and Sircar, S. (1994) *AIChE J.* **40**, 935.

29. Joseph, J.C., Myers, A.L., Golden, T.C. and Sircar, S. (1993) *J. Chem. Soc., Faraday Trans.* **89**, 3491.

30. Sowers, S.L. and Gubbins, K.E. (1996) *Adsorption*, in press.

31. Ulberg, D.E. and Gubbins, K.E. (1995) *Mol. Phys.* **84**, 1139.

32. Maddox, M., Ulberg, D.E. and Gubbins, K.E. (1995) *Fluid Phase Eqba.* **104**, 145.

33. Müller, E.A., Rull, L.F., Vega, L.F. and Gubbins, K.E. (1996) *J. Phys. Chem.* **100**, 1189.

34. Jorgensen, W.L. and Tirado-Rives, J. (1988) *J. Am. Chem.Soc.* **110**, 1657, and references therein.

35. Walker, P.L. and Janov, J. (1968) *J. Colloid Interface Sci.* **28**, 449.

36. Müller, E.A. and Gubbins, K.E. (1996), to be published.

# INTERMOLECULAR FORCES AND SIMULATION IN PORES

DAVID NICHOLSON
*Department of Chemistry,*
*Imperial College of Science, Technology and Medicine,*
*London SW7 2AY*
*UK*

## 1. Intermolecular Forces, Simulation and Experiment

The key importance of intermolecular forces in adsorption scarcely needs to be stressed. We can achieve an approximate, but adequate, description of gases and solids by ignoring intermolecular interactions. Although the same is not true for liquids, much of the essence of the liquid state can be captured by a hard sphere description. However, to describe the densified non-uniform fluid states of matter, that form close to solid surfaces, we need to pay attention to both the attractive and repulsive components of the interactions between the molecules.

Computer simulation has helped to focus attention on this aspect of the problem because it enables the statistical mechanics to be done exactly [1,2]. Simulation of adsorbate systems is dealt with in detail in other lectures. Here it will suffice to illustrate the context of intermolecular forces in relation to simulation and experiment (figure 1). The intermolecular force model is the essential

*Figure 1.* Simulation, intermolecular forces and experiment

input to the simulation. Since the statistical mechanics is done exactly, we see that experiment can in principle provide a probe for the potential model. Conversely, if we can get the potential model right, we can interpret experiments at the molecular level.

In an ideal world where we knew everything and possessed an infinite computer, we would

105

*J. Fraissard (ed.), Physical Adsorption: Experiment, Theory and Applications,* 105–131.
© 1997 *Kluwer Academic Publishers.*

always begin by writing a wave function for the whole system of interest. Clearly it is necessary to make compromises in the real world and it is now widely accepted that we can take a chemist's view of matter, identify the individual atoms, and approach the problem of intermolecular forces from the point of view of atom-atom summations.

In physical adsorption we are always dealing with systems that contain more than one atomic species, and involve a condensed phase (the adsorbent). To describe the system fully we need to account for adsorbate-adsorbate interactions as well as adsorbate-adsorbent interactions. Obviously the underlying physics must be the same for both types of interaction, but the treatment of the adsorbate adsorbent interactions has several aspects that make it worthy of special attention. One way of handling this problem is to treat the adsorbent solid as a collective entity [3]. This approach has some advantages, but it is less flexible than the atom-atom method.

Once the intermolecular potentials have been established, they can be fed into a simulation, and the results set against experiment. However agreement with experiment, although always encouraging, doesn't mean that the potential models chosen are correct. Nor does disagreement necessarily imply that they are wrong. For example, the zero coverage heat of adsorption and Henry law constant are frequently used as a criteria for evaluating a potential model. The zero coverage (isosteric) heat of adsorption per molecule, is calculated from

$$q_{st}(0) = kT - \frac{\int\int u \exp(-\beta u) dr d\omega}{\int\int \exp(-\beta u) dr d\omega} = kT - <u>$$  (1)

and the Henry law constant $k_H$ is given by

$$k_H = \int\int [\exp(-\beta u) - 1] dr d\omega$$  (2)

Here $u$ is an adsorbate-adsorbent potential and the integrals are over all positions ($r$) and orientations ($\omega$) at temperature $kT = 1/\beta$. Clearly the integrand in the numerator of equation (1) has a sharp maximum at the position of the potential minimum, especially at low temperatures. This means that $q_{st}(0)$ is not very sensitive to the shape of the potential, but is mainly governed by the well depth or holding potential. Some values of $q_{st}(0)$ and $k_H$ are shown in table 1.

TABLE 1. Some values of $q_{st}(0)/kT$ (top line) and $k_H$ (bottom line) for different adsorbent well depths and Mie potentials.

| $\epsilon_{ad}/kT$ | 12-3 | 9-3 | 9-4 | 10-4 | 9-5 | 9-6 |
|---|---|---|---|---|---|---|
| 12.5 | 12.90 | 12.91 | 12.92 | 12.91 | 12.93 | 12.93 |
| | $1.56 \times 10^4$ | $1.79 \times 10^4$ | $1.54 \times 10^4$ | $1.46 \times 10^4$ | $1.37 \times 10^4$ | $1.24 \times 10^4$ |
| 5.0 | 5.06 | 5.11 | 5.10 | 5.09 | 5.09 | 5.08 |
| | 17.0 | 19.1 | 16.2 | 15.5 | 14.3 | 13.0 |
| 3.3 | 3.24 | 3.28 | 3.24 | 3.25 | 3.19 | 3.15 |
| | 4.77 | 5.24 | 4.49 | 4.33 | 4.03 | 3.71 |

These were calculated using an *n-m* (attractive repulsive) Mie potential having the form:

$$u = \epsilon_{ads} \frac{1}{n-m} \left[ m \left( \frac{z_{min}}{z} \right)^n - n \left( \frac{z_{min}}{z} \right)^m \right] \qquad (3)$$

where $\epsilon_{ads}$ is the well depth at $z=z_{min}$. The values of $\beta\epsilon_{ads}$ in Table 1 cover a range from around liquid nitrogen temperature to ambient temperature for a typical adsorbent.
Some examples of the potentials are illustrated in Fig2.

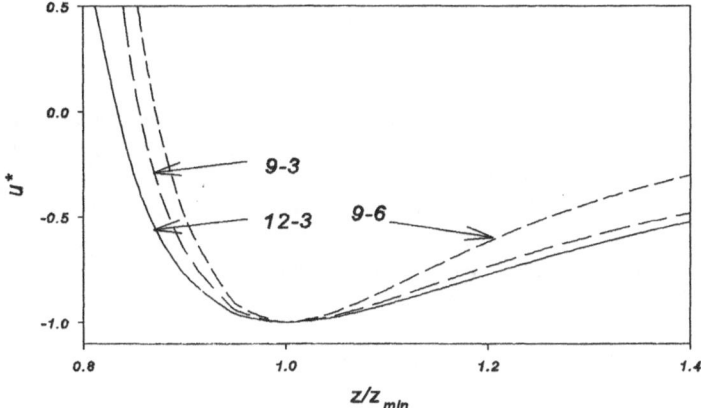

*Figure 2*. Examples of Mie potentials for different attractive and repulsive exponents

In a typical simulation of an adsorption system, we replicate an ensemble of classical particles governed by a Hamiltonian, $H=KE+PE$. The intermolecular interaction model accounts for the potential energy part. Since much of the interest in physical adsorption is in low temperature systems, the potential energy tends to be dominant. At ambient and higher temperature, much of the behaviour of the adsorbate is accounted for by its kinetic energy and experimental behaviour may not be particularly sensitive to the potential model. It is therefore always worthwhile to look at how many $kT$ are in each part of the potential model.

Neither of the preceding points is meant to imply that it is never important to seek the best potential function. However, once this has been achieved, it may be satisfactory to model a complex object by much simpler effective potentials. In the next section I want to examine some of the fundamental aspects of potential functions, and then go on to discuss the construction of potentials for adsorption systems.

## 2. Constructing a potential

### 2.1. SITES - THE TOTAL POTENTIAL ENERGY

If we adopt the chemical compromise we begin by labelling sites in the adsorbate molecules and in the adsorbent. These can be atoms, but a one-to-one correspondence is not mandatory. For example $CH_2$ or $CH_3$ groups in hydrocarbons are often chosen as sites, and the bond between atoms, rather than the nuclei of the atoms, can also be an appropriate choice.

The total potential energy in the Hamiltonian for the interacting sites can be written as

$$U_N = \sum_{i>j}^{N} u_{ij} + \sum_{i>j>k}^{N} u_{ijk} + \ldots \tag{4}$$

representing sums of two-body, three-body etc contributions to the potential energy. $u_{ij}$ is the interaction between a pair of sites labelled $i$ and $j$, also written $u^{(2)}$, $u(r_i, r_j)$, $u(r_{ij})$. Frequently in an adsorption problem the positions of the adsorbent sites are fixed in space (rigid adsorbent model). In a simulation this has the enormous advantage that the adsorbate-adsorbent potential can be calculated once, at the beginning of the simulation, and then stored as an external field whose magnitude at a given position can be found by interpolation. The external field experienced by a given adsorbate site, placed at a position $r$ is then

$$u^{(1)}(r) = \sum_{j} u^{(2)}(r,r_j) + \sum_{j>k} u^{(3)}(r,r_j,r_k) + \ldots \tag{5}$$

where the sums extend over all sites in the adsorbent.

### 2.2. THE MULTIPOLE EXPANSION

Intermolecular forces are entirely electrostatic in origin, it is convenient to treat the interaction between a pair of molecules as if it consisted of two separate parts: (a) A long range part which is mainly attractive [except that electrostatic interaction between ions with the same charge or dipoles oriented head to head (or tail to tail) etc. would be repulsive]. (b) A short range repulsive part due to electron overlap and Pauli exclusion.

A site can be pictured as a collection of positive and negative charges. We anticipate that since many of the charges are carried by electrons, quantum mechanics will be needed. The first step is to consider the potential energy $V_{AB}$ between the charged particles on site $A$ and those on site $B$. This can be written as the sum of all the coulombic interactions between the charges on the two sites. [The sites are now labelled $A$ and $B$ because we want to use $i$ as a summation index for the charges]. The pair interaction energy comes from the quantum mechanical total energy.

The problem of finding $V_{AB}$ can be broken down into two parts: (a) Find the field at any point due to the charges on $A$ (or $B$). (b) Place $B$ (or $A$) in this field and find $V_{AB}$.

*(i) The Field.* Charges, $e_i$ are located at positions $r_i$ with respect to an origin. The electrostatic field at some point $P$, situated at $r$ is:

$$\Phi = \frac{1}{4\pi\epsilon_o} \sum_i \frac{e_i}{|r - r_i|} \tag{6}$$

When $r > r_i$ this can be Taylor expanded and expressed as a power series in $r$ containing components of the multipole moments (the multipole expansion) [4,5].

(ii) *The Interaction.* If a second charge cloud (on $B$) is now placed in the field of the first (on $A$), the interaction energy between them is

$$V_{AB} = \sum_j \Phi_j^{(A)} e_j^{(B)} \tag{7}$$

$V_{AB}$ can be written as a double multipole expansion in which the first few terms are:

$$
\begin{aligned}
V_{AB} &= T q^{(A)} q^{(B)} + T_\alpha (q^{(A)} \mu_\alpha^{(B)} - q^{(B)} \mu_\alpha^{(A)}) \\
&+ T_{\alpha\beta} \left( \frac{1}{3} q^{(A)} \Theta_{\alpha\beta}^{(B)} + \frac{1}{3} q^{(B)} \Theta_{\alpha\beta}^{(A)} - \mu_\alpha^{(A)} \mu_\beta^{(B)} \right) \\
&+ T_{\alpha\beta\gamma} \left( \frac{1}{15} q^{(A)} \Omega_{\alpha\beta\gamma}^{(B)} - \frac{1}{15} q^{(B)} \Omega_{\alpha\beta\gamma}^{(A)} - \frac{1}{3} \mu_\alpha^{(A)} \Theta_{\beta\gamma}^{(B)} + \frac{1}{3} \mu_\alpha^{(B)} \Theta_{\beta\gamma}^{(A)} \right)
\end{aligned} \tag{8}
$$

Here $q$ is the charge on an ion, $\mu_\alpha$ is the component of the dipole moment in the $\alpha (= x, y\ z)$ direction with respect to a laboratory frame of reference, $\Theta_{\alpha\beta}$ is the $\alpha\beta$ component of the quadrupole tensor. By a suitable choice of molecular axes, the 9 components of this array can be reduced to three diagonal ones ($xx$, $yy$, $zz$). For linear molecules ($CO_2$, $N_2$) only one component is needed. $\Omega$ is an octopole tensor, and can have up to 27 components, but again this number can be substantially reduced [5].

Charge distributions with leading dipoles also have quadrupoles, octopoles etc. (in general). Those with leading quadrupoles also have octopoles, etc. The existence or non-existence of multipoles depends on the symmetry of the charge distribution [5]. The $T$ terms are coupling tensors. They are functions of the components of the vector $r$ that joins the origins in the two sites $A$ and $B$. The first few coupling tensors have the components

$$T = (4\pi\epsilon_o)^{-1} r^{-1} \tag{9}$$

$$T_\alpha = (4\pi\epsilon_o)^{-1} \nabla_\alpha r^{-1} = (4\pi\epsilon_o)^{-1} r_\alpha r^{-3} \tag{10}$$

$$T_{\alpha\beta} = (4\pi\epsilon_o)^{-1} \nabla_\alpha \nabla_\beta r^{-1} = 4\pi\epsilon_o^{-1} (3 r_\alpha r_\beta - r^2 \delta_{\alpha\beta}) r^{-5} \tag{11}$$

$$(4\pi\epsilon_o) T_{\alpha\beta\gamma} = \nabla_\alpha \nabla_\beta \nabla_\gamma r^{-1} = [-15 r_\alpha r_\beta r_\gamma + 3 r^2 (r_\alpha \delta_{\beta\gamma} + r_\beta \delta_{\gamma\alpha} + r_\gamma \delta_{\alpha\beta})] r^{-7} \tag{12}$$

Note that $T_\alpha$ goes like $r^{-2}$, $T_{\alpha\beta}$ goes like $r^{-3}$, $T_{\alpha\beta\gamma}$ goes like $r^{-4}$ etc. Since $q^A, q^B$ are ionic charges the first non-zero term in $V_{AB}$ for neutral molecules is $-T_{\alpha\beta}\mu_\alpha^A\mu_\beta^B$. In general this expression has 9-terms. More compact forms of these equations can be written using spherical tensors [4].

## 2.3. THE SUPERMOLECULE

The Hamiltonian for two interacting sites $(A,B)$ can be written in three parts:

$$\hat{H}_{AB} = \hat{H}_A + \hat{H}_B + V_{AB} \tag{13}$$

where the Hamiltonian operators for the isolated $A$ and $B$ sites are:

$$\hat{H}_A = \hat{K}_A + V_A , \qquad \hat{H}_B = \hat{K}_B + V_B \tag{14}$$

and $V_{AB}$ expresses the interaction for the AB pair (the supermolecule). In principle one can now calculate the interaction energy:

$$u_{AB} = E_{AB} - E_A - E_B \tag{15}$$

where $E_{AB}$ etc are total energies from

$$\hat{H}_A\Psi_A = E_A\Psi_A \; ; \; \hat{H}_B\Psi_B = E_B\Psi_B; \; \hat{H}_{AB}\Psi_{AB} = E_{AB}\Psi_{AB} \tag{16}$$

Although in principle $u_{AB}$ can be evaluated using a quantum mechanical package there are several difficulties including:

(i) *r and ω dependence*. A single calculation will give $u_{AB}$ for a particular separation and orientation. To obtain a complete potential surface (eg. to use in a simulation) many calculations need to be carried out.

(ii) *Computing*. SCF (self consistent field) calculations give only the *repulsive* part and electrostatic parts of the potential. For the complete interaction it is essential to include full electron correlation. This requires large computing resources but the calculations are not standard and size inconsistencies may occur.

(iii) *Basis sets*. The results can depend on the set of wave functions chosen. Also the technical problem of basis set superposition error, where one molecule spuriously "borrows" electrons from the orbitals of the other. This can be corrected by carrying out the calculations for the isolated molecules with and without all the orbitals, but in some cases the correction is larger than the correlation energy.

(iv) *Relative magnitudes of the energies*: $E_A + E_B \approx E_{AB}$ therefore $u_{AB}$ is only a very small part of any of these terms ($\sim 10^{-5}$).

(v) *True pair potentials*. If all these problems are overcome we have a set of true pair potentials would result. The total interaction between molecules needs to include three-body and possibly higher order terms.

## 2.4. PERTURBATION THEORY

Since the interaction energy is such a small fraction of the total, perturbation theory provides a valuable adjunct, and is now incorporated into some quantum mechanical packages, but it is also possible to make use of perturbation theory to obtain the long range part of the interaction with the aid of auxiliary experimental data. The essential idea is to establish an expansion using wave functions for the isolated sites in order to form averages involving the interaction Hamiltonian, $V_{AB}$. The long range energy has three parts:

(i) *Electrostatic*. Interaction between permanent multipoles.

(ii) *Induced*. Permanent multipole polarises (ie distorts) another molecule. The multipoles due to this polarization interact with permanent multi poles.

(iii) *Dispersion*. Instantaneous multipoles polarize another molecule.

The perturbation theory expansion of Rayleigh and Schrödinger expresses the total energy of interaction as a series of terms $\varepsilon_1+\varepsilon_2+\varepsilon_3+\dots$. Although its convergence is not assured, this is not important since in practice terms beyond $\varepsilon_3$ have little significance. The first order term is:

$$\varepsilon_1 = <\psi_o^A\psi_o^B| V_{AB} |\psi_o^A\psi_o^B> \tag{17}$$

The second order term is:

$$\varepsilon_2 = -\sum_{n,m} \frac{|<\psi_o^A\psi_o^B| V_{AB} |\psi_n^A\psi_m^B>|^2}{\Delta E^A + \Delta E^B} \tag{18}$$

Here the summation is over all states except the ground state. The symbols are as follows:

$\psi_n^A$    Wavefunction for A in excited state $n$ (electron promoted to orbital above the ground state; $n=0$ is the ground state).

$\Delta E_n^A$    Excitation energy required to promote an electron from the ground state to state $n$ ($= E_n^A - E_o^A$).

$V_{AB}$    The interaction energy between $A$ and $B$. At long range, when the sites are sufficiently far apart ($r_i < r$), the multipole expansion can be used.

$\varepsilon_1$ is the ground state expectation value for $V_{AB}$. It gives the long range part of the electrostatic interaction between permanent multipoles.

$\varepsilon_2$ includes the induction terms, when one site is in its ground state, and the other in its excited state ($n=0$, $m\neq0$); and the dispersion interaction, when both sites are in excited states ($n\neq0$, $m\neq0$).

At shorter range, the Fermionic nature of the electrons and the possibility of electron exchange between overlapping orbitals, need to be taken into account. Formally this can be accomplished by antisymmetrisation of the products of the wave functions. Tang and Toennies [6] showed that a satisfactory correction to the two-body dispersion energy could be achieved if each term was multiplied by a damping factor (see equation (21) below).

The results for the long range parts of the interaction can be reexpressed in terms of frequency dependent polarizabilities and multipole moments, quantities that can be obtained from experiment or quantum mechanical calculation.

### 2.4.1. *Electrostatic Interactions*

The energy given by $\varepsilon_1$ in equation (17) is the sum of the products of multipoles at each site [4,5].

The expressions rapidly become complicated as the order of multipoles increases. However the site-site approximation minimises these complications (see below) and at the same time gives a far more accurate account of electrostatic interactions at close molecular separations.

When molecules and adsorbents are modelled as collections of sites, a first approximation to the electrostatic part of the interaction is to assign a "partial charge" to each site. At the lowest level of approximation it then becomes a simple matter to calculate electrostatic interactions as a sum over Coulombic potentials acting between charged sites in the adsorbent and in the adsorbate. Stone and coworkers [7,8] have shown that such point charge interactions are not always a satisfactory approximation for estimating electrostatic terms. A better representation can be achieved by assigning higher multipoles (dipoles, quadrupoles etc) to the sites and summing over the various multipole interactions, known as distributed multipole analysis (DMA). The multi poles (usually up to quadrupole) are found from electron density distributions centred at each site. In the lowest order, Mulliken analysis gives the point charges, but inclusion of the higher order terms greatly improves the accuracy. The DMA technique has been applied to the interactions between chlorine molecules adsorbed on a graphite surface [9], but no full analysis of adsorbent sites has been made up to now.

### 2.4.2. Induced Interactions

These come from the term $\varepsilon_2$ when one site is in its ground state and the other in an excited state. The excited state is related to the polarizability of the site. In general this property is anisotropic, ie. an electric field in the $x$-direction for example, gives rise to different distortions in the $x,y,z$ directions. Thus the dipole-dipole polarizability is a 9-component tensor $\alpha_{1\alpha\beta}$ where $\alpha\beta$ represents $xx,xy,xz$, etc. For spherical molecules these components are all the same. The components can be expressed in terms of the matrix elements of the dipole operators at the site in question

$$\alpha_{1\alpha\beta} = \sum_{n \neq 0} \frac{\langle\psi_0|\mu_\alpha|\psi_n\rangle\langle\psi_n|\mu_\beta|\psi_0\rangle + \langle\psi_0|\mu_\beta|\psi_n\rangle\langle\psi_n|\mu_\alpha|\psi_0\rangle}{E_n - E_0} \qquad (19)$$

$\alpha_1/(4\pi\epsilon_o)$ is of the order of $1\text{Å}^3$. Higher order polarizabilities are defined in a similar way and involve products of quadrupoles ($\alpha_2$) and octopoles ($\alpha_3$) etc. Polarizabilities involving cross terms vanish when the sites have sufficiently high symmetry [5].

The electrostatic and polarizability contributions to the induction energy can be factored out from $\varepsilon_2$ and the induced interaction due to a dipole polarizability at site A, for example, can be written as

$$u_{ind} = -\frac{1}{2}\sum_{\alpha,\beta} \alpha^A_{1\alpha\beta} F_\alpha F_\beta \qquad (20)$$

where $F_\alpha$ is the field in the $\alpha$ direction at the site $A$ due to the other charges in the system. A common example in adsorption science would be when $A$ is a site on an adsorbate molecule inside a zeolite cavity. The field due to partial charges on the adsorbent atoms would be computed as the sum, at the position of $A$, of the all the coulombic terms due to the partial charges on the constituent atoms of the zeolite [10,11]. Since positive and negative charges tend to be near neighbours, the resultant field is usually quite small and induced interactions do not contribute greatly to the overall potential energy function. A possible exception to this is when there are one or two isolated ions carrying their full charge complement, as in zeolite A. Nevertheless induced interactions tend to be a relatively small part of the total potential energy and are always attractive. The induced interactions arising from higher order polarizabilities can usually be neglected, even when high accuracy is sought.

Note that the induced interactions are not pair additive. Furthermore the induced multipole moments of the polarized molecule set up an additional field, causing further polarization.

### 2.4.3. *Dispersion Interactions*

When both sites are in the excited state, $\varepsilon_2$ in equation (18) yields the two-body dispersion energy. Provided that the sites have been chosen to have high symmetry [5], the dispersion energy can be expressed as a series of even order terms that depend on inverse powers of the site separations

$$u_{disp}^{AB} = -f_6\frac{C_6}{r^6} - f_8\frac{C_8}{r^8} - f_{10}\frac{C_{10}}{r^{10}}...\tag{21}$$

Where $f_6$, $f_8$ and $f_{10}$ account for the exchange force damping and are discussed further below. The coefficients $C_n$ come from the equation (18) and can be expressed in terms of the frequency dependent polarizabilities.

$$C_6 = \frac{r^6}{2\pi}T_{\alpha\beta}T_{\gamma\delta}\int_0^\infty \alpha_{1\alpha\gamma}^A(i\omega)\,\alpha_{1\beta\delta}^B(i\omega)\,d\omega\tag{22}$$

$$C_8 = \frac{r^8}{6\pi}T_{\alpha\beta\gamma}T_{\delta\epsilon\zeta}\int_0^\infty [\alpha_{1\alpha\delta}^A(i\omega)\alpha_{2\beta\gamma,\epsilon\zeta}^B(i\omega) + \alpha_{1\alpha\delta}^B(i\omega)\alpha_{2\beta\gamma,\epsilon\zeta}^A(i\omega)]\,d\omega\tag{23}$$

When $A,B$ are isotropic the coupling tensor product in (22) is $6/r^6$, and the factor in front of the integral becomes $3/\pi$, similarly, the pre-integral factor in (23) reduces to $15/\pi$ for isotropic sites. The $C_{10}$ coefficient for isotropic sites, is

$$C_{10} = \frac{7}{\pi}\int_0^\infty [2\alpha_1^A(i\omega)\,\alpha_3^B(i\omega) + 5\alpha_2^A(i\omega)\,\alpha_2^B(i\omega) + 2\alpha_3^A(i\omega)\,\alpha_1^B(i\omega)]\,d\omega\tag{24}$$

Although the main contribution to dispersion energy comes from the $C_6$ term, $C_8$ and $C_{10}$ terms can be a substantial part of the total dispersion energy, and since their distance dependence differs from that of the leading term, it is important to examine these contributions. Dispersion interactions are always attractive and always present when there are polarizable electrons.

### 2.5. REPULSIVE ENERGY

Repulsive interaction, due to Pauli exclusion and electrostatic repulsion, becomes increasingly important as sites approach each other and eventually dominate the interaction. Clearly a correct form for the repulsive energy is vital if the shape and depth of the potential well are to be faithfully reproduced. Unfortunately this is not easily achieved. Frequently a Mie (inverse *m*th) distance dependence is used; $m=12$ has become hallowed by usage and finds its way into adsorbate-adsorbent interactions as an inverse 9th or 10th power repulsion after summation or integration over the adsorbent atoms. Quantum mechanical calculations and arguments based on quantum mechanics show that to a first approximation, the repulsive energy between a pair of atoms is proportional to the square of the overlap integral between orbitals on the atoms [12,13]. At separations relevant to physical adsorption, the wave functions decay exponentially, and a Born-Mayer form is therefore suitable:

$$u_{rep} \doteq A \exp(-br) \qquad (25)$$

The constants $A$ and $b$ can be found from SCF calculations, but $u_{rep}$ is underestimated by about 30% [14] and empirical adjustment is necessary. The EHT method has also been applied to this problem [12,15]. Currently, the most reliable approach to obtaining $A$ and $b$ appears to be adjustment against experimental data, after the remainder of the potential function has been estimated. Values for rare gas pairs and some other molecules are reasonably well established. These can be used to decompose values obtained for adsorbate adsorbent pairs according to the combination rules [16]:

$$u_{rep}^{A,B}(r) = (A^A A^B)^{1/2} \exp[-2b^A b^B r/(b^A + b^B)] \qquad (26)$$

The parameter $b$ also appears in the damping functions derived by Tang and Toennies [6] (equation (21)). These can be written in the form

$$f_n = 1 - \left( \sum_{k-0}^{n} \frac{(br)^k}{k!} \right) \exp(-br) \qquad (27)$$

where $n$ takes the values 6,8,10 etc. These functions vary smoothly in a sigmoid fashion from 1.0 to zero as the site separation decreases.

## 2.6. THREE-BODY DISPERSION INTERACTIONS

Three-body terms arise from the $\varepsilon_3$ term in the perturbation expansion. This involves triple products of the wave functions such as $\psi^A \psi^A \psi^B$ and both induced and dispersion energies are generated according to whether the wave functions are in ground or excited states, as already discussed for the $\varepsilon_2$ term. There are of course no many-body electrostatic terms, but many-body terms do occur for exchange interactions and some of the deficiencies of two-body repulsive models can be traced to this source [17]. Since two-body induced interactions are quite small, no serious approximation is likely to arise in neglecting this part of the three-body contribution. The three-body long range dispersion interactions can be a significant part of the total energy [18], especially in confined spaces, where overlap effects and non uniformity may be important.

In an adsorption problem, three categories of three-body interaction can occur, (a) those involving adsorbate sites alone, (b) those involving two adsorbate sites and one adsorbent site, and (c) those involving two adsorbent sites [19,20]. The last group contribute to the adsorbate adsorbent potential and can be include in the grid pre-calculated for interpolation when a rigid adsorbent model is used [10]. The first two categories are more difficult to handle in a simulation because they involve configurations that alter at each step, necessitating a lengthy recalculation of the total potential energy [20]. The second category ((b) above) are often referred to as "mediated interactions". In principle the evaluation of three-body interactions is no different from that of the two-body terms, but the equations become more complicated. A general three-body term can be written as [18,21,22]:

$$u^{ABC}(l_1,l_2,l_3) = -Z^{ABC}(l_1,l_2,l_3) \, T_{l_1}^{AB} \, T_{l_2}^{BC} \, T_{l_3}^{CA} \tag{28}$$

Here $l_1,l_2,l_3$ take values of 1 for dipole interactions, 2 for quadrupole etc and the whole three-body interaction includes a summation over all possible combinations of multipole. The electronic component, $Z$ is an integral over components of the relevant polarizabilities [18,23],

$$Z^{ABC}(l_1,l_2,l_3) = \frac{1}{\pi}\int_0^\infty \alpha_{l_1}^A(i\omega)\,\alpha_{l_2}^B(i\omega)\alpha_{l_3}^C(i\omega)\,d\omega \tag{29}$$

and the coupling tensors are given by equation (9-12) for $l_k$=1 and 2. The full expression for the three body interactions contains a very large number of terms and some simplifications are necessary as discussed below.

The fourth order term in the Rayleigh Schrödinger expansion yields a three-body triple-dipole interaction. It was shown some time ago [24] that in uniform bulk fluids, the third order three-body terms involving quadrupoles, cancel with the fourth-order three-body interaction, and this leaves only the three-body triple-dipole interaction to account for. The latter contributes some 10-15% of the total potential and is repulsive. However this cancellation occurs for geometrical reasons, and does not necessarily extend to non-uniform systems since some of the triangles that occur in uniform fluids may be absent.

## 3. Semi-empirical potentials

### 3.1. ISOTROPIC AND ANISOTROPIC SITES

Decomposing the system into sites minimises much of the complicated algebra involved in evaluating internal interactions. It has been seen already that spherical symmetry greatly simplifies all the tensor products. In many cases in adsorption it is reasonable to assume that sites have spherically symmetrical polarizability [25]. One important exception is a graphitic adsorbent where the polarizability is better described as having cylindrical symmetry [26,27].

### 3.2. STATIC POLARIZABILITIES

The equation for the coefficients involving frequency dependent polarizabilities can be evaluated to a good approximation by expressing the frequency dependence as a Lorentzian [6]

$$\alpha_l(i\omega) = \frac{\alpha_l(0)}{1 + (i\omega/\eta_l)^2} \tag{30}$$

where $\eta_l$ is the *lth* pole transition energy, and $\alpha_l(0)$ is a static polarizability. If we impose the restriction that the sites are spherically symmetrical the 2-body dispersion coefficients can be obtained by carrying out the integrations

$$C_6^{AB} = \frac{3}{2} \frac{\eta_1^A \eta_1^B}{\eta_1^A + \eta_1^B} \alpha_1^A \alpha_1^B \tag{31}$$

$$C_8^{AB} = \frac{15}{4} \left[ \frac{\eta_1^A \eta_2^B \alpha_1^A \alpha_2^B}{\eta_1^A + \eta_2^B} + \frac{\eta_2^A \eta_1^B \alpha_2^A \alpha_1^B}{\eta_2^A + \eta_1^B} \right] \tag{32}$$

$$C_{10}^{AB} = 7 \left[ \frac{\eta_1^A \eta_3^B \alpha_1^A \alpha_3^B}{\eta_1^A + \eta_3^B} + \frac{\eta_3^A \eta_1^B \alpha_3^A \alpha_1^B}{\eta_3^A + \eta_1^B} + \frac{5}{2} \frac{\eta_2^A \eta_2^B \alpha_2^A \alpha_2^B}{\eta_2^A + \eta_2^B} \right] \tag{33}$$

for sites with cylindrical symmetry, equation (31) is

$$C_6^{'AB} = \frac{\eta_1^A \eta_1^B r^6}{4(\eta_1^A + \eta_1^B)} T_{\alpha\beta} T_{\beta\alpha} \alpha_{1\alpha\alpha}^A \alpha_{1\beta\beta}^B \tag{34}$$

( the convention of summation over repeated subscripts is implied). The anisotropy can be expressed in terms of the normal (N) and parallel(P) components of the polarizability tensor by the factor

$$\gamma = (\alpha_{1N} - \alpha_{1P})/3\overline{\alpha} \tag{35}$$

where the trace of the polarizability tensor is

$$\overline{\alpha} = (2\alpha_{1N} + \alpha_{1P})/3 \tag{36}$$

For example $\gamma$ for graphite is positive and has a value of ~0.26 [27]. The coefficient in equation (31) is modified by a factor of $(1-\frac{1}{2}\gamma)$ thus for an isotropic adsorbate $A$ on anisotropic graphite,

$$C_6^{A,Gr} = \frac{3}{2}\left(\frac{\eta_1^A \eta_1^{Gr}}{\eta_1^A + \eta_1^{Gr}}\right)\bar{\alpha}_1^A \bar{\alpha}_1^{Gr}(1 - \frac{1}{2}\gamma) \tag{37}$$

Anisotropic higher order terms have received little attention.

The three-body dispersion terms, even within the limitation of spherical symmetry are still rather complicated. When the Lorentzian approximation is used, the electronic factor, $Z$ can be written as a function of the static polarizability and excitation energies [18,23]

$$Z^{ABC}(l_1,l_2,l_3) = \frac{1}{2}\alpha_{l_1}^A \alpha_{l_2}^B \alpha_{l_3}^C \eta_{l_1}^A \eta_{l_2}^B \eta_{l_3}^C$$
$$\times \frac{\eta_{l_1}^A + \eta_{l_2}^B + \eta_{l_3}^C}{(\eta_{l_1}^A + \eta_{l_2}^B)(\eta_{l_2}^B + \eta_{l_3}^C)(\eta_{l_3}^C + \eta_{l_1}^A)} \tag{38}$$

In the triple dipole 4th order term the electronic factor is [18]

$$Z_i^{(4)} = -\frac{45}{32}[\alpha_1^{(i)}]^2 \alpha_1^{(j)}\alpha_1^{(k)}\eta_1^{(i)}\eta_1^{(j)}\eta_1^{(k)}$$
$$\frac{2(\eta_1^{(i)})^2 + (\eta_1^{(j)} + \eta_1^{(k)})(\eta_1^{(j)}\eta_1^{(k)}) + 2\eta_1^{(i)}(\eta_1^{(j)} + \eta_1^{(k)})(2\eta_1^{(i)} + \eta_1^{(j)} + \eta_1^{(k)})}{(\eta_1^{(i)} + \eta_1^{(j)})^2(\eta_1^{(i)} + \eta_1^{(k)})^2(\eta_1^{(j)} + \eta_1^{(k)})^2} \tag{39}$$

where $\{i,j,k\}$ are permuted over the three interacting species.

With the assumption of spherical symmetry, the tensor factors in equation (28) can be expressed as functions of the angles at the vertices of the triangle forming the interacting triplet and its side lengths,

$$W(DDD) = 3r_{12}^{-3} r_{23}^{-3} r_{31}^{-3} (1 + 3\cos\phi_1 \cos\phi_2 \cos\phi_3) \tag{40}$$

$$W(DDQ) = \frac{3}{16} r_{12}^{-3} r_{23}^{-4} r_{31}^{-4} [(9\cos\phi_3 - 25\cos 3\phi_3)$$
$$+ 6\cos(\phi_1 - \phi_2)(3 + 5\cos 2\phi_3)] \tag{41}$$

$$W(QQD) = \frac{15}{64} r_{12}^{-5} r_{23}^{-4} r_{31}^{-4} [3(\cos\phi_3 + 5\cos 3\phi_3) +$$
$$20\cos(\phi_1 - \phi_2)(1 - 3\cos 2\phi_3) + 70\cos 2(\phi_1 - \phi_2)\cos\phi_3] \tag{42}$$

$$W(QQQ) = \frac{15}{128}r_{12}^{-5}r_{23}^{-5}r_{31}^{-5} \left[ -27 + 220 \cos \phi_1 \cos \phi_2 \cos \phi_3 \right.$$
$$+ 490 \cos 2\phi_1 \cos 2\phi_2 \cos 2\phi_3$$
$$\left. + 175 \left( \cos 2(\phi_1 - \phi_2) + \cos 2(\phi_2 - \phi_3) + \cos 2(\phi_3 - \phi_1) \right) \right] \tag{43}$$

$$W^{(4)}(DDD) = Z_2^{(4)}r_{12}^{-6}r_{23}^{-6}[1+\cos^2\phi_2] + Z_3^{(4)}r_{23}^{-6}r_{31}^{-6}[1+\cos^2\phi_3]$$
$$+ Z_1^{(4)}r_{31}^{-6}r_{12}^{-6}[1+\cos^2\phi_1] \tag{44}$$

Again only the triple dipole term has been studied for anisotropic sites [22]. There are no analytical expressions in this case, and no full scale study incorporating such terms has yet been attempted, although the results obtained for a graphite substrate [22,28] suggest that site anisotropy in the adsorbent can have a significant effect. When anisotropy is neglected the DDD (triple-dipole) term is repulsive and the ratio of the DDD to DD energy remains nearly constant at about 0.6 as the adsorbate moves away from the surface. When anisotropy, typical of graphite, is included in the calculation, the relative magnitude of these terms is smaller, but the three body term changes from repulsive to attractive (Figure 3) when the adsorbate is moved away from the surface.

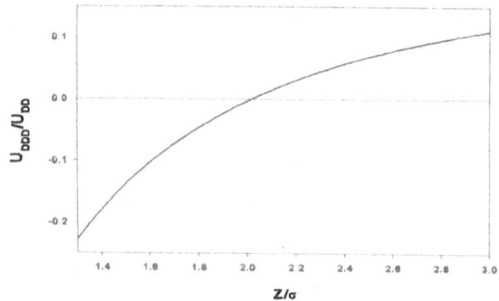

*Figure 3.* The relative contribution from the triple-dipole adsorbate-adsorbent-adsorbent dispersion energy interaction compared to the two-body dipole dispersion energy, for Ar over anisotropic graphite. The anisotropy $\gamma=0.26$ [27]. If the graphite is assumed to be isotropic, the graph is a horizontal line at ~-0.55.

## 3.3. PARAMETER DETERMINATION

At this stage we can, in principle, construct a usable adsorbate-adsorbent potential, based on site-site summation, at least for spherically symmetrical sites, if it is possible to provide values for three orders of polarizability and excitation energies, and the repulsive parameters, $A$ and $b$. Of course it needs to be born in mind that the long range parameters are required for in-lattice species. This has been achieved for some adsorbates in silicalite [29] and $AlPO_4$-5 pores [30].

### 3.3.1 Dipole Parameters
As a first step the excitation energies themselves can be re-expressed in terms of their corresponding polarizabilities and sums over oscillator strengths

$$\eta_l = [S_l/\alpha_l]^{1/2} \tag{45}$$

where $S_\ell$ is a sum over the oscillator strengths $f_n^\ell$

$$S_\ell = \sum f_n^\ell \tag{46}$$

According to time dependent perturbation theory, $S_l$ should be the total number of electrons, this assumption can be tested. Coupled Hartree Fock calculations have been made [31] for a number of small atoms and for ions in lattices [33-37]; from these calculations the frequency dependent polarizabilities can be found, and hence accurate evaluation of $C_6$ for these atoms. These calculations confirm that the polarizable electrons are found in the outer shells, and that $S_l$ is equivalent to an effective number of electrons, $N_{eff}$, rather than the total number. With the aid of the quantum mechanical data it is possible to arrive at an expression for $N_{eff}$ for neutral atoms, which can be expressed in terms of the total number $N$, in the outer shell [10],

$$N_{eff}^o = aN^2 + bN + c \tag{47}$$

some values for the coefficients a and b are given in Table 2[10].

TABLE 2. Parameters for the calculation of $N^o_{eff}$ in equation (47).

|          | a       | b      | c      |
|----------|---------|--------|--------|
| H, He    | -0.109  | 0.9330 | 0      |
| Li to Ne | -0.202  | 0.7075 | 0.0857 |
| Na to Ar | -0.0293 | 0.9966 | 0.0062 |
| K to Kr  | -0.118  | 1.0059 | 0.0011 |
| Rb to Xe | -0.0237 | 1.1767 | 0      |

In-lattice species generally carry partial charges and both $N_{eff}$, and polarizability will differ from the neutral atom values on account of this. We therefore write

$$N_{eff} = N_{eff}^o \pm |q| \tag{48}$$

To calculate site-site interactions involving adsorbents we need to have polarizabilities for the in lattice species. Apart from the CHF calculations referred to above, two other approaches are possible: (i) Clausius Mosotti equations [38,39] and (ii) Auger data [29]. The first method is more useful for single species adsorbents, since it relies on measurements of dielectric constant, $\epsilon$, for the whole solid. The Clausius-Mosotti equation gives the polarizability from

$$\frac{4\pi}{3V}\sum_i \alpha_1^i = \frac{\epsilon - 1}{\epsilon + 2} \tag{49}$$

where the sum is over all species, and $V$ is the volume per formula unit in the solid. An extension of this method using the Kramers-Krönig equations has been used to extract the normal and transverse components of the polarizability of carbon in graphite [27]. The Auger method starts from the assumption that the charge left at a site in the lattice following the ejection of an Auger electron, causes an instantaneous relaxation of surrounding sites that can be related directly to the polarizability [29]. This method can be used where more than one lattice species is involved, and supplemented by the Clausius-Mosotti equation, together with the assumption that the total polarizability is additive. The method has been applied to determine the polarizability of O and Si species in silicalite and $AlPO_4$-5[30]. It has been shown that the polarizabilities of anions in a lattice differ considerably from that of the corresponding free ion and are highly dependent on environment. The polarizabilities of lattice cations by contrast vary very little with environment and are virtually the same as those of the free species [38]. Thus oxide adsorbents for example can be expected to have quite different holding potentials, largely as a consequence of the differences in oxide polarizability and in the numbers of effective electrons associated with the ions.

### 3.3.2. Higher Order Parameters

The higher order sum rules, $S_2$, $S_3$ needed in equations (45) and (46) are easily determined since they are related by a hierarchy to $S_1$ [40]:

$$S_2 = 3(S_1\alpha_1)^{1/2} \; ; \; S_3 = 7.5(S_2\alpha_2)^{1/2} \tag{50}$$

Coupled HF calculations for isolated neutral species again provide a valuable data base for the determination of the higher order polarizabilities. With the aid of these data, and equations due to Kiselev [41] we arrive at the following expressions for $\alpha_2$ and $\alpha_3$

$$\alpha_2 = \frac{K_8^2 \alpha_1}{144}[1 + X]^2 \tag{51}$$

where

$$X = \left[1 + \frac{24\sqrt{3}}{K_8}\left(\frac{\alpha_1}{N_{eff}}\right)^{1/4}\right]^{1/2} \tag{52}$$

$$\alpha_3 = \frac{K^2 \alpha_1^2}{4S_3}[1 + Y]^2 \tag{53}$$

where

$$Y = \left[1 + \frac{4\,\alpha_1\,S_3}{K\,N_{\text{eff}}}\right]^{1/2} \qquad (54)$$

and

$$K = \frac{1.6406\,K_{10}}{\eta_1} - \frac{0.625}{\eta_2}\left(\frac{\alpha_2}{\alpha_1}\right)^2 \qquad (55)$$

$K_8$ and $K_{10}$ are found by comparing $C_8$ and $C_{10}$ from (32) and (33) with quantum mechanical data, and were shown to be simply related to $N_{\text{eff}}$ [10]. Thus $K_8$ and $K_{10}$ are known if $N_{\text{eff}}$ is known.

In summary, the parameters needed for the evaluation of two-body and three-body dispersion interactions up to and including the quadrupole terms are $\alpha_1$, $N_{\text{eff}}$, $K_8$ and $K_{10}$. Some typical values for this set of parameters are given in Table 3. Fuller tabulations are given elsewhere [10]. The dispersion coefficients for calculated by this method are in good agreement with *ab initio* data, where these are available.

TABLE 3. Parameters for the calculation of dispersion potentials and C6/C8 coefficients for self interactions (in atomic units).

| | q/e | $N_{\text{eff}}$ | $\alpha_1/\text{Å}^3$ | $K_8$ | $K_{10}$ | $C_6$ | $C_6$ * | $C_8$ | $C_8$ * |
|---|---|---|---|---|---|---|---|---|---|
| O(MgO) | -2.0 | 5.656 | 1.59 | 3.171 | 8.654 | 62.7 | 59.1 | 1427 | 1412 |
| O(silicalite) | -1.0 | 4.656 | 1.20 | 2.89 | 7.00 | 37.3 | - | 586 | - |
| O(AlPO4) | -0.9 | 4.556 | 1.12 | 2.87 | 6.85 | 33.2 | - | 352 | - |
| Cℓ(NaCℓ) | -1.0 | 6.55 | 2.96 | 3.40 | 9.81 | 171 | 160 | 5086 | 5404 |
| Na | +1 | 4.455 | 0.148 | 2.84 | 6.71 | 1.59 | 1.51 | 11 | 12 |
| Si | +2.0 | 1.52 | 0.38 | 2.00 | | 3.80 | - | - | - |
| N | - | 3.18 | 1.10 | 2.55 | 5.28 | 27.0 | 27.5 | 502 | 443 |
| Ar | 0 | 6.11 | 1.64 | 3.29 | 9.51 | 68.3 | 67.2 | 750 | 1060 |

* *Ab initio*

### 3.3.3 Repulsive Parameters

Quantum mechanical calculations on simple species, can give a guide to the relative magnitude of the parameters $A$ and $b$, if it is assumed that the attractive part of the interaction is given with sufficient accuracy by the terms discussed in the foregoing sections, then the depth of the potential well for the adsorbate adsorbent interaction is readily determined with the aid of experimental zero coverage heats or Henry law data, and $A$ and $b$ can be optimised on this basis. Some guidance on anisotropic effects can be gained from extended Hückel calculations [15].

## 4. Potential Corrugation

Site-site summations to produce an adsorbate-adsorbent potential inside a zeolite cavity yield a complex potential surface that reflects the crystal structure of the framework. To gain general insights into the nature of adsorption in porous materials, these are often modelled as an assembly of "unit pores". The units typically have planar or cylindrical geometry; the graphite slit pore being a popular choice and typical example.

An inverse $n^{th}$ power interaction with a planar array of atoms can be expressed as a Fourier expansion by exploiting the underlying periodicity of the array. The expansion is made for a two dimensional layer and its coefficients expressed as Fourier transform

$$\frac{1}{r^n} = u_o(z) + u_1(z)f_1(x,y) + \dots \qquad (56)$$

where the array of adsorbent sites is in the $x,y$ plane and $\mathbf{r}=(x,y,z)$. The function $f_1$ expresses the periodicity of the corrugation in the $xy$ plane. The attenuation of this corrugation as $z$ increases away from the surface is close to being exponential and is given by $u_1(z)$.

The series in equation (56) turns out to be rapidly convergent and it is rarely necessary to go beyond the first two terms. The first term is equivalent to that which would be obtained if the sheet of adsorbent sites was treated as a continuum. When the 12-6 potential is used as the site site pair interaction, and a summation is made over several layers of the adsorbent, the popular 10-4-3 potential results:

$$u_{sf}(z) = 2\pi\rho_s\epsilon_{sf}\sigma_{sf}^2\Delta\left[\frac{2}{5}\left(\frac{\sigma_{sf}}{z}\right)^{10} - \left(\frac{\sigma_{sf}}{z}\right)^4 - \frac{\sigma_{sf}^4}{3\Delta(0.61\Delta + z)^3}\right] \qquad (57)$$

Recently Girardet and co-workers [42] have developed an ingenious extension of this technique to the calculation of potentials inside cylindrical spaces.

The Fourier expansion demonstrates that the corrugation is a relatively small perturbation, superimposed on the continuum background. Nevertheless corrugations can exert profound effects on adsorbate properties in some circumstances, especially at low temperatures. Amongst these may be mentioned the location, or even the existence, of submonolayer transitions in the adsorption isotherm, heat curves etc, which are often associated with changes from commensurate to incommensurate structures in the adsorbate, and the form of the adsorption isotherm in micropores [43, 44]. Transport properties in small pores are also be influenced by the effects of "windows" and "cages" created in this way [45].

When the interaction is between a permanent multipole in the adsorbate and an array of multipoles, the Fourier expansion technique can yield an interesting insight. The first order (continuum) term vanishes [46, 47] and the effect of interaction between multipoles and a regular array is then to modify the amplitude of the surface corrugation [46, 48].

## 5. Potentials in pores

Many of the extraordinary properties of fluids adsorbed in porous materials are the direct consequence of two factors: the enhancement of the adsorbate-adsorbent potential caused by "overlap" of attractive interactions from surrounding walls, and the constraints imposed by the repulsive part of the adsorbate adsorbent interaction on packing and freedom of molecular movement i the adsorbate molecules. Of course the significance of these factors varies greatly as pore width is altered. Pores with widths of only one or two diameters (micropores) are especially intriguing since adsorbates can not only exhibit the most remarkable anomalies in comparison to bulk material, but such pores are also of great importance in many industrial processes. It is worth mentioning that the significant parameter here is the ratio of pore width to molecular size, rather than the absolute pore size - a bulky organic molecule for example may experience repulsive interactions from all the surfaces of a pore that would accommodate more than a single layer of Ar.

The variation of pore potential with pore width has been illustrated many times. Fig 4 shows potential functions for a spherical adsorbate described by a 12-6 potential inside a smooth walled (10-4-3 cf equation (57)) graphite slit pore. The slit width is measured between the atom centres of the opposite pore walls and expressed in units of the adsorbate hard sphere diameter, $\sigma$. The internal space, $H'$, available to the adsorbate (sometimes known as the chemical width) is not a precisely defined quantity, but it has been suggested [49] that a reasonable estimate is

$$H' = H - 0.149\sigma - 0.851\sigma_s \qquad (58)$$

where $\sigma_s$ is the hard sphere diameter of the solid adsorbent. The potential minima at opposite walls merge to become a single minimum as the pore size decreases. For smaller pore widths that than those in Fig. 4, repulsive terms begin to dominate the interaction an the depth of the single minimum decreases. Eventually the pore becomes too small to admit any adsorbate under a given set of conditions [50].

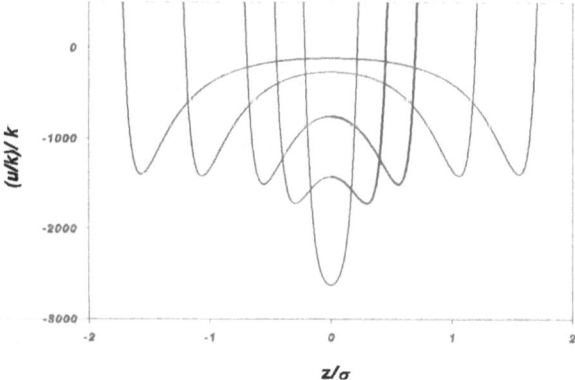

*Figure 4.* Adsorbate adsorbent potentials for a 12-6 methane adsorbate in a graphitic slit pore with smooth walls. The slit widths (in units of the adsorbate size $\sigma$) are 2.0, 2.5, 3.0, 4.0, and 5.0

To illustrate how a full adsorbate-adsorbent potential can behave, and how the various components of the potential contribute to the total, some examples have been chosen from the Ar silicalite and Xe silicalite system [51]. Since the construction of a full scale adsorbate-adsorbent potential can be a lengthy procedure if the three-body terms are included, it will also be useful to examine whether a simple two-body effective potential will suffice as a representation of the more elaborate model. Silicalite is a particularly useful adsorbent for these purposes. Its crystal structure is well defined and there are extensive experimental data for adsorption. It is a reasonable approximation to assume that the O and Si are isotropically polarizable and the partial charges on the framework species ($O^-$, $Si^{2+}$) have been well established [52,53]. The basic set of attractive parameters for Ar and Xe adsorption $\{\alpha_1, N_{eff}, K_8, K_{10}\}$ is available [10], and all the parameters needed to calculate dispersion and induction intercations can be found from these four parameters. The repulsive parameters $\{A,b\}$(equation (20)) have to be determined for the relevant pairs of sites with the aid of experimental data [54,55] for $k_H$ and $q_{st}$, using quantum mechanical calculations as a guide to the approximate magnitude of the parameters. The values needed for the system under discussion are given in Table 4, other values can be found in refs [10], [30] and [51].

TABLE 4. Repulsive parameters for rare gas-silicalite interactions

|  | $A/E_h$ | $b/a_o$ |
| --- | --- | --- |
| Ar | 328.9 | 1.918 |
| Xe | 1026 | 1.728 |
| $O^-$(silicalite) | 1544 | 2.190 |
| $Si^{2+}$ | 6163 | 2.395 |

$$E_h = 4.3598 \times 10^{-18} \text{ J}; a_o = 5.2918 \times 10^{-11}\text{m}$$

With the aid of these data and combining rules (equation (26)) the required cross coefficients can be found.

Figure 5 shows the potential functions at the centre of the straight channel through silicalite. Since there is only a single minimum (cf figure 4), it is clear that this is a very restricted environment for both adsorbates. The graphs show the full two-body potentials (including $C_8$ and $C_{10}$ terms) and repulsive interactions, and the same potentials with the addition of many-body interactions. It appears from fig. 5 that the latter could be safely neglected for Ar in this system, since they amount to no more than 3% of the total interaction, but that many-body terms account for about 15% of the Xe interaction. However it must be kept in mind that the channel geometry varies with position. It has been demonstrated elsewhere, that many-body interactions can have quite an important impact on thermodynamic properties [20]. Moreover, the present calculations are for the monoclinic crystal structure of silicalite [51] (stable below 320K). In the high temperature, orthorhombic structure, many-body terms account for 6% of the total interaction in the same cross section. It can be concluded that, although it may be safe to neglect many-body terms in some systems, no general conclusions can be made, and that these should be investigated where a full scale potential function is the aim. The many body terms are examined more closely below.

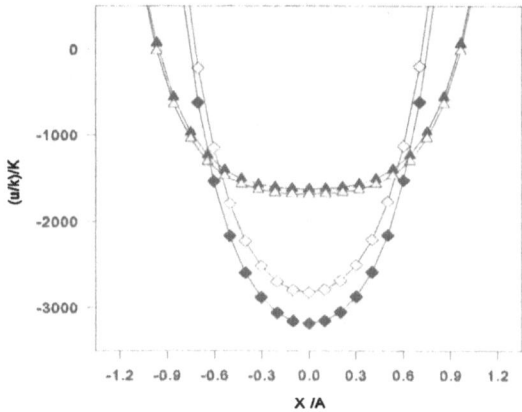

*Figure 5.* Potential energy of Ar (triangles) and Xe (diamonds) in the central cross section of the straight channel in silicalite. Filled points are the full scale potential; open points, the two-body + repulsive terms.

The nature of the various two-body and repulsive terms in these example systems is illustrated in Figures 6 and 7. Several points can be made with regard to these figures: (i) The higher order dispersion interactions are by no means insignificant compared to the inverse 6th power term. (ii) Because of their different $r$-dependence, these higher order terms exert an influence on the shape of the potential. (iii) Oxide adsorbents are often modelled by the O sites alone, but the cation sites can make a contribution which is not negligible if accurate modelling is the goal.

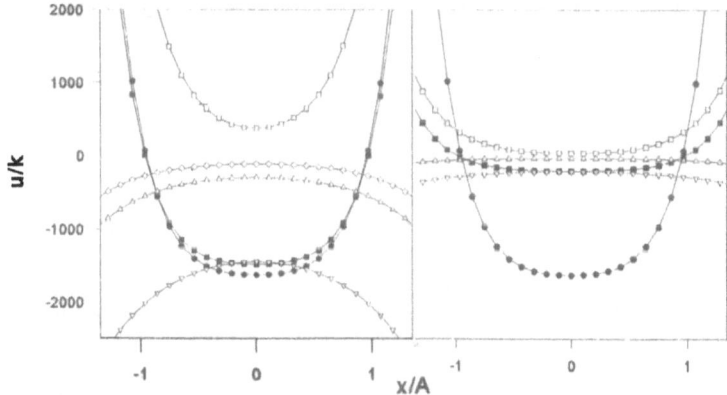

*Figure 6.* Contributions from two body terms to the potential energy of Ar in the straight channel of silicalite. The right hand panel is the interaction with lattice oxygen, and the left hand panel the interaction with lattice Si. Total potential (●) (see also fig 5), total two body + repulsive interaction (■). The open points are the $C_6(\vee)$, $C_8(\triangle)$, $C_{10}(\diamond)$ and repulsive(□), two-body long range terms respectively.

126

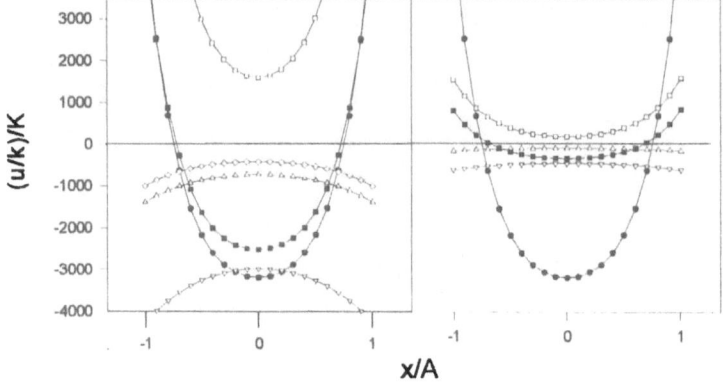

*Figure 7.* Contributions from two body terms to the potential energy of Xe in the straight channel of silicalite.The symbols are as for Figure 6.

The three-body contributions to these systems are shown in Figure 8. As already seen in Figure 5, these contributions are repulsive overall for Ar, but attractive for Xe.

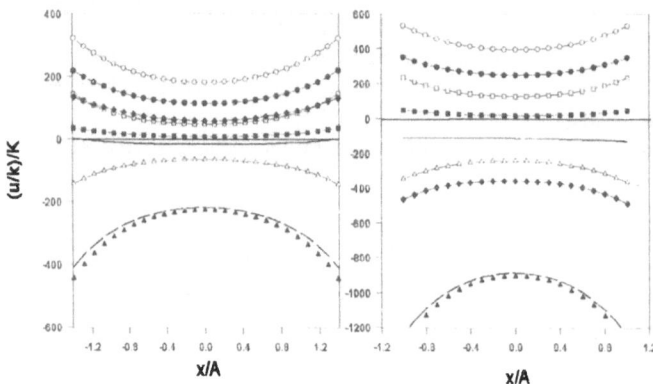

*Figure 8.* Many-body contributions to the interaction potential energy between Ar and silicalite (left hand panel) and Xe and silicalite (right hand panel). The total many-body term is shown as filled diamonds. The remaining filled symbols are the adsorbate-Si-O terms; the open symbols are the adsorbate-O-O terms. Circles denote the DDD, the total of the terms involving quadrupoles (DDQ, DQD, QDD, QQD, QDQ, DQQ and QQQ) terms are shown as triangles. The total contribution from the sum of fourth order + quadrupole terms is shown as a heavy full line for the adsorbate O-O terms, and as a broken line for the adsorbate-Si-O terms.

The reason for this is partly geometrical, and partly a consequence of the placement of the highly polarizable Xe or the much less polarizable Ar. It was mentioned above that in uniform fluids and solids, the three-body term arising from the fourth order perturbation theory almost exactly cancels the higher order third order terms involving quadrupoles. This does not always happen in these heterogeneous nonuniform systems. In fact, it is these terms for the adsorbate-O-Si triangles that play a major part in reducing the Ar three-body terms to a negligible contribution, and that give rise to an attractive many-body contribution from Xe. One may conclude that even more exotic many-body terms may be rather important in porous materials. At the same time it must be kept in mind that the so-called mediated interactions, involving two adsorbate sites and one adsorbent site, may help to counterbalance the many-body terms discussed in these examples as adsorption proceeds to higher coverages [20], since the total many-body interaction in a dense adsorbate approaches more closely to a uniform environment.

The construction of a full scale potential, including many-body terms, requires a lengthy calculation and it is therefore interesting to know how successful an effective two-body potential might be if its parameters are fitted empirically. Unfortunately it is not possible to offer a general answer but we can explore this question in the preceding examples. A 12-6 effective potential can be defined as

$$u = -C\sum r^{-6} + B\sum r^{-12} \qquad (59)$$

where the sums extend over adsorbent sites and $B$ and $C$ can be found by fitting the attractive and repulsive parts of the potential to the full scale potential in the range $-0.13 < x < 0.13$nm [51], figure 9. As might be expected, neither term is an accurate representation of the full potential function.

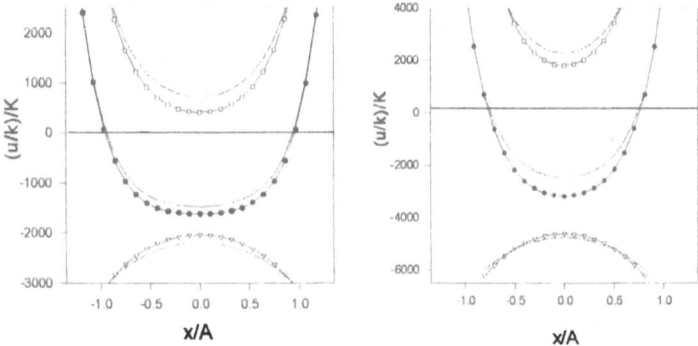

*Figure 9.* A comparison of fitted 12-6 equations and full scale potential functions. The left hand panel is for Ar in the centre of the straight pore of silicalite, the right hand panel is for Xe. The full potential is shown as filled circles. The open squares are the Born Mayer repulsive terms, and the open triangles the total contribution from the attractive terms. The lines are summed 12-6 functions fitted to these two components of the potential. See text.

However, for Ar, the compensation between the two terms results in a total potential which comes quite close to the full scale potential. The compensation does not work as well for Xe; although the range over which the effective potential is fitted could be adjusted to obtain closer agreement. An important point is that the attractive term derived from an inverse 6th potential cannot be accurately fitted to the full potential. It is perhaps more obvious that the inverse 12th power term cannot be an accurate fit to the repulsive Born-Mayer term. One may conclude that because of the compensation effect, the 12-6 potential turns out to be a better basis for an

128

effective potential than the 6-exp Buckingham potential. In micropores, the overlap of the repulsive terms means that these terms are important over the whole pore width and this increases the difficulty of selecting an effective potential.

## 6. Conclusions

Site-site summation provides a useful approach to the calculation of adsorbate potentials at the atomic scale. The construction of a full-scale potential requires several contributing terms. Perturbation theory still provides the most useful starting point for the discussion and evaluation of these contributions and, coupled with results from modern quantum mechanical calculations a systematic procedure can be established for determining the necessary parameters. The short range repulsive part of the potential continues to pose a major problem, and the degree of uncertainty may be enhanced in very small pores. A combination of experimental data, accurate long range terms and guidance from quantum mechanical SCF calculations seems to provide the optimum working strategy at the present time. An advantage of the full scale potential is that transferability can be established, and in principle it should be possible to develop, extend and refine parameter sets.

Although many-body terms appear to make a minor contribution to the overall interaction in some cases, a close analysis shows that this may be the consequence of fortuitous cancellation. The non uniform multi species environment of the adsorption problem means that these terms cannot be safely ignored if accurate potential modelling is desired. Up to now contributions beyond $\varepsilon_4$ in the perturbation expansion have not received detailed study in the adsorption problem.

Surfaces raise particular issues which have not been dealt with in depth here, such as the "rumpling" of the surface due to displacement of atoms, and anisotropy of polarizability, consequent upon broken symmetry. Anisotropy of polarizability is important in the ubiquitous carbon based adsorbents, but a full scale potential, incorporating this feature has not been established.

It is clear that, despite the progress which has been made, there are still challenging problems to be solved in developing high quality potential function for adsorption simulations. Fortunately much can be learned with simple effective potentials that have been tailored for a specific purpose.

## 7. References

1. Allen M. and Tildesley, D.J.(1986) *Computer Simulation of Liquids*, OUP, Oxford.
2. Nicholson, D. and Parsonage, N.G.(1982) *Computer Simulation and the Theory of Physical Adsorption*, Academic Press, London.
3. Dzyaloshinskii, I. E., Lifshitz, E. M. and Pitaevskii, L.P. (1961) *Adv. Phys.*, **10**, 165.
4. Gray, C. G. and Gubbins, K.E., (1984) *Theory of Molecular Fluids*, OUP, Oxford.
5. Buckingham, A.D. (1978) Basic theory of intermolecular forces: Applications to small molecules, *Intermolecular Interactions from Diatomics to Biopolymers*, Ed. B. Pullman., Wiley, New York.
6. Tang, K.T. and Toennies, J.P. (1984) An improved simple model for the van der Waals potentials based on universal damping functions and the dispersion coefficients, *J. Chem. Phys.* **80,** 3726.

7. Stone, A.J. and Alderton, M. (1985) Distributed multipole analysis. Methods and applications, *Mol. Phys.* **56**, 1047-1051.

8. Stone, A.J. and Price, S.L. (1988) Some new ideas in the theory of intermolecular forces: Anisotropic atom-atom potentials, *J. Phys. Chem.* **92**, 3325-3335.

9. Hammonds, K.D., McDonald, I.R., and Tildesley, D.J. (1993) Computational studies of the structure of monolayers of chlorine physisorbed on the basal plane of graphite, *Mol.Phys.* **78**, 173-189.

10. Pellenq, R.J-M. and Nicholson, D.(1994) Intermolecular potential function for the physical adsorption of rare gases in silicalite, *J.Phys.Chem.* **98**, 13339-13349.

11.Hutson, J.M. and Fowler, P.W. (1986) The atom surface interaction potential for He-NaCl: a model based on pairwise additivity, *Surf. Sci.* **173**, 337-350.

12. Vidal-Madjar, C. and Minot, C. (1987) Adsorption potential of alkanes on graphite, *J.Phys.Chem.* **91**, 4004-4011.

13. Murrell, J.N.(1974) in N.H.March (ed.), *Orbital Theories of Molecules and Solids*, OUP, Oxford.

14.. Ahlrichs, R., Bohm, H.J., Brode, S., Tang, K.T., and Toennies, J.P. (1988), Interaction potentials for alkali ion-rare gas and halogen ion-rare gas systems, *J.Chem.Phys.* **88**, 6290-6302.

15. Pellenq, R.J., Pellegatti, A., Nicholson, D., and Minot, C. (1995) Adsorption of argon in silicalite. A semiempirical quantum mechanical study of the repulsive interaction, *J.Phys.Chem.* **99**, 10175-10180.

16. Bohm, H.J. and Ahlrichs, R. (1982) A study of short range repulsions, *J. Chem. Phys.* **77**, 2028-2034.

17.Meath, W.J. and Aziz, R.A. (1984) On the importance and problems in the construction of many body potentials, *Mol.Phys.* **52**, 225,-243.

18. McRury, T.B. and Linder, B. (1972) Three-body nonadditive free energy of gases adsorbed on solids of arbitrary electron delocalisation, *J. Chem. Phys.* **56**, 4368-4377.

19. Nicholson, D. (1991) Fundamentals of equilibria in adsorption, in A. Mersmann and S. Scholl (eds.), *Fundamentals of Adsorption III*, Engineering Foundation, New York, pp3-24.

20. Fernandez-Alonso, F., Pellenq, R.J., and Nicholson, D. (1996) The role of three-body interactions in the adsorption of argon in silicalite-1, *Mol.Phys.* **86**, 1021-1030.

21. Bell, R.J. (1970) Multipolar expansion for the nonadditive third order interaction energy of three atoms, *J. Phys B*, **3**, 751-762.

22. Nicholson, D. (1987) Triple dipole dispersion interactions between a linear adsorbate and an anisotropic adsorbent, *Surf. Sci.* **184**, 255-272.

23. Doran, M B. and Zucker, I.J. (1971) Higher order multipole three body van der Waals interactions and stability of rare gas solids, *J. Phys. C: Solid state*, **4**, 307-312.

24. Barker, J.A. and Pompe, A. (1968) Atomic interactions in argon, *Aust J.Chem.* **21**, 1683-1694.

25. Hutson, J.M. and Fowler, P.W. (1986) The atom surface interaction potential for He NaCl: a model based on pairwise additivity, *Surf. Sci.* **173**, 337-350.

26. Carlos, W.E. and Cole, M.W. (1980) Interaction between a He atom and a graphite surface, *Surf. Sci.* **91**, 339-357.

27. Nicholson, D. (1987) Graphite polarizability, *Surf. Sci.* **181**, L189-L192.

28. Kim, H., Cole, M.W., Toigo, F., and Nicholson, D. (1988) Dispersion interaction between adsorbed linear molecules, *Surf. Sci.* **198**, 555-570.

29. Pellenq, R.J. and Nicholson, D. (1993) In-framework ion dipole polarizabilities in non-porous and porous silicates and aluminosilicates, determined from Auger electron spectroscopy data,

*J. Chem. Soc. Faraday Trans.* **89**, 2499-2508.

30. Boutin, A., Pellenq, R.J., and Nicholson, D. (1994) Molecular simulation of the stepped adsorption isotherm of methane in $AlPO_4$-5, *Chem. Phys. Letts.* **219**, 484-490.

31. Spackman, M.A. (1991) Time dependent Hartree Fock second order molecular properties with a moderately sized basis set. II Dispersion coefficients, *J. Chem. Phys.* **94**, 1295-1305.

32. Cambi, R., Cappelletti, D., Liuti, G., and Pirani, F. (1991) Generalized correlations in terms of polarizability for van der Waals interaction potential parameter calculations, *J. Chem. Phys.* **95**, 1852-1861.

33. Pyper, N.C. (1986) Relativistic *ab initio* calculation of the properties of ionic solids, *Phil. Trans Roy. Soc.* **A320**, 107-137.

34. Fowler, P.W. and Madden, P.A. (1983) The in-crystal polarizability of the fluoride ion, *Mol.Phys.* **49**, 913-923.

35. Fowler, P.W. and Madden, P.A. (1984) In crystal polarizability of alkali and halide ions, *Phys Rev.* **B29**, 1035-1041.

36. Fowler, P.W. and Madden, P.A. (1985) In crystal polarizability of $O^{2-}$, *J. Phys. Chem.* **89**, 2581-2584.

37. Fowler, P.W. and Tole, P. (1988) Polarizability of sulphide in MgS, *Chem. Phys. Letts.* **149**, 273-278.

38. Fowler, P.W. and Hutson, J.M. (1986) A semi-empirical model for atom-surface dispersion coefficients, *Surf. Sci.* **165**, 289-302.

39. Fowler, P.W. (1990) A users guide to polarizabilities and dispersion coefficients for ions in crystals, *Molecular Simulation*, **4**, 313-330.

40. Koutselos, and Mason, E. A (1986) Correlation and prediction of dispersion coefficients for isoelectronic systems, *J.Chem.Phys.* **85**, 2154-2160.

41. Kiselev, A.V. and Poskus, D. P. (1958) *Zur. Fiz. Chim.* **32**, 2854-2857.

42. J., Breton, J. and Girardet, C. (1993) Atom confinement in helicoidal cavities, *J. Chem. Phys.* **98**, 3389-3394.

43. Nicholson, D. (1994) Simulation study of nitrogen adsorption in parallel-sided micropores with corrugated potential functions, *.J. Chem. Soc. Faraday Trans.* **90**, 181-185.

44. Diestler, D.J., Schoen, M., Curry, J.E., and Cushman, J.H. (1994) Thermodynamics of a fluid confined to a slit pore with structured walls, *J.Chem.Phys.* **100**, 9140-9146.

45. Schoen, M., Cushman, J.H., and Diestler, D.J. (1994) Anomalous diffusion in confined monolayers films, *Mol.Phys.* **81**, 475-490.

46. van den Berg, T.H.M.. and van der Avoird, A. (1989) Analytical two- and three- dimensional lattice sums for general multipole interactions, *Chem. Phys.Letts.* **160**, 223-227.

47. Cracknell, R.F., (1990) Thesis, University of London.

48. Vernov, A. and Steele, W.A. (1992) The electrostatic field at a graphite surface and its effect on molecule solid interactions, *Langmuir*, **8**, 155-159.

49 Kaneko, K., Cracknell, R.F., and Nicholson, D. (1994) Nitrogen adsorption in slit pores at ambient temperatures: Comparison of simulation and experiment, *Langmuir*, **10**, 4606,-4609.

50. Lastoskie, C., Gubbins, K.E., and Quirke, N. (1993) Pore size heterogeneity and the carbon slit pore: A density functional theory model, *Langmuir*, **9**, 2693-2702.

51. Nicholson, D., Boutin, A. And Pellenq, R. J-M., (1996) Intermolecular potential functions for adsorption in zeolites: State of the art and effective models, *Mol. Simulation* (in press).

52. van Genechten, K.A. and Mortier, W. (1987) Intrinsic framework electronegativity: A novel concept in solid state *chemistry, J. Chem. Phys.* **86**, 5063-5071.

53. White, J.C. and Hess, A.C. (1993) An examination of the electrostatic potential of silicalite

using periodic Hartree-Fock *theory, J. Phys. Chem.* **97,** 8703-8706.

54. Llewellyn, P., Coulomb, J.P., Grillet, Y., Patarin, J., Lauter, H.J., Reichart, H., and Rouquerol, J. (1993) Adsorption by MFl-type zeolites examined by isothermal microcalorimetry and neutron diffraction. 1. Ar, Kr and methane, *Langmuir*, **9,** 1846-1851.

55. Reichart, H., private communication see also ref 10.

# PORE SIZE CALCULATION IN THE CASE OF DIFFERENT GEOMETRICAL SHAPES

G. HORVATH[1] and M. SUZUKI[2]
[1] *Department of Chemical Engineering, University of Veszprem*
*8201 Veszprem P.O.B. 158, Hungary*
[2] *Institute of Industrial Science, University of Tokyo*
*7-22-1 Roppongi, Minato-ku Tokyo 106 Japan*

## 1. Introduction

The quantitative evaluation of microstructures is an important aspect of the processes using such materials for catalysis or adsorption. In the case of microporous adsorbents is often used as one of the measures the internal surface area based on the well known B.E.T. adsorption isotherm. For these materials the micropore volume and the pore size distribution may be more informative than the surface area. Even if the B.E.T. plot is applied mechanically the physical meaning may be reduced.

Several methods for calculating pore size distribution from adsorption data have been proposed. The main problem of the calculation is similar to the fundamental problem of chemical thermodynamics namely in some sense, the chemical thermodynamics is "contradictio in adjecto". Partly the objects of study are considered to be a continuous phase, partly those consist of individual molecules having a given structure and properties. The building an acceptable and well acting bridge is not too easy task.

The forces occurring in adsorption are not different from those involved in any other interatomic or intermolecular interactions. The rising problem is that they participate in the atoms of solid affected by the whole structure of adsorbent.

The common methods for the determination are the mercury penetration, nitrogen or argon adsorption methods and the so-called molecular probe method. The experimental data obtained from the former methods, do not allow to direct calculation of pore size distributions. Usually a given geometry is taken into consideration, sometimes differing from the real pore structure. In this paper we do not deal with the problems of mercury porosimetry.

The molecular sieve materials usually are supposed - in models - to have slit like or cylindrical pores. The IUPAC classified the pores as micro <2nm, meso 2-50 nm and macro >50 nm.

The most traditional method based on the Kelvin equation. A pore loses its condensed liquid adsorbate at a pressure related to the pore radius. The basis of this

133

*J. Fraissard (ed.), Physical Adsorption: Experiment, Theory and Applications,* 133–149.
© 1997 *Kluwer Academic Publishers.*

134

method is the equilibrium across the meniscus of a capillary condensate. The difficulty in the calculation arises from the fact, that even when the pore has been emptied of condensed liquid, multi-layers of adsorbed molecules remain on the inner surface of pores. The measured desorption is made up of removal of condensed liquid from some pores, plus the adsorbate lost from the surface.

## 2. Dollimore-Heal method

Dollimore and Heal gave a review [1] and criticism of previous methods used for the pore size calculation. Gregg and Sing [2] summarised the historical background and the application of their method.

The nitrogen desorption is carried out at liquid nitrogen temperature (77,34 K). A desorption step can be seen in FIGURE 1. The amount lost in the desorption step is the sum of multi-layer desorption and pore desorption.

$$\Delta V = \Delta V_m + \Delta V_c \qquad (1)$$

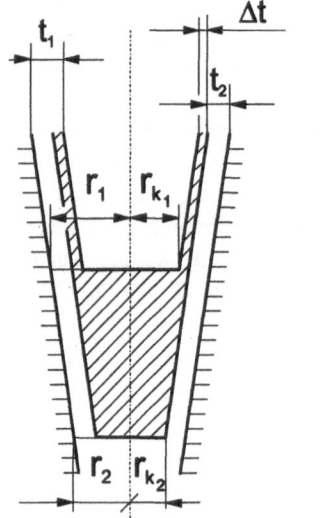
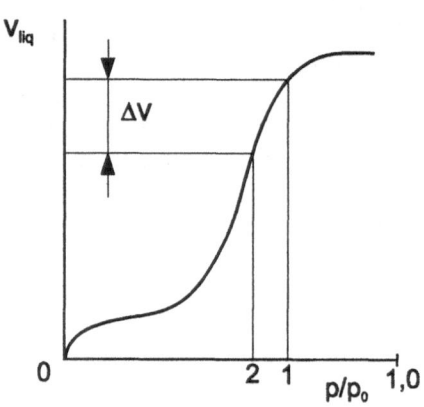

*Figure 1.* Desorption step

The model uses two fundamental relations; namely the Halsey equation for the calculation of adsorbed layer thickness in nm

$$t = 0,43[5/\ln(p_o/p)]^{1/3} \qquad (2)$$

where $p_o$ is the saturation pressure, and the Kelvin equation for the capillary condensate radius, in nm

$$r_k = 0,953/\ln(p_o/p) \qquad (3)$$

supposing that the contact angle equals zero.

The two radii and the actual pore radii are related

$$r_{K1} = r_1 - t_1 \qquad (4)$$
$$r_{K2} = r_2 - t_2 = r_2 - t_1 + \Delta t \qquad (5)$$

Wheeler defined a pore length distribution function $L(r)$. Using this function after the actual desorption step (see FIG. 1) the wall adsorption volume in the pores larger pores than $r_2$

$$V_m = \int_{r_2}^{\infty} \pi[r^2 - (r-t)^2] L(r) dr =$$

$$= \int_{r_2}^{\infty} \pi[2rt - t^2] L(r) dr = t\, F(>r_2) - \pi t^2 L(>r_2) \qquad (6)$$

for one step

$$\Delta V_m = F(>r_2)\Delta t - 2\pi t L(>r_2)\Delta t \qquad (7)$$

where $F$ is the total wall area of pores larger than $r_2$. Substituting into equation (1), the pore desorption is

$$\Delta V_c = \Delta V - F(>r_2)\Delta t + 2\pi t L(>r_2)\Delta t \qquad (8)$$

The corresponding pore volume

$$\Delta V_p = r^2 \pi\, L(r)\Delta r \qquad (9)$$

and the actual capillary volume

$$\Delta V_c = (r-t)^2\, \pi\, L(r)\Delta r \qquad (10)$$

give the following equation

$$\Delta V_p = \frac{r^2}{(r-t)^2} \Delta V_c \tag{11}$$

$\Delta V$ and $p$ are measured, $t$ and $r$ can be calculated from equations (2), (3), (5). After the $i^{th}$ desorption step

$$\Delta V_{p_i} = \frac{r_i^2}{(r_i - t_i)^2} \left[ \Delta V - (t_{i-1} - t_i) \sum_{n=1}^{i-1} F_n + 2\pi t_i (t_{i-1} - t_i) \sum_{n=1}^{i-1} L_n \right] \tag{12}$$

$$F_i = \frac{2\Delta V p_i}{r_i} \tag{13}$$

and

$$L_i = \frac{F_i}{2\pi r_i} \tag{14}$$

can be obtained.

As for the slit like pores a small modification is necessary. The actual $t$ values can be calculated from equation (2). Taking into consideration the slab geometry, the adequate equations are

$$r_k = 0,476 / \ln( p_o / p ) \tag{15}$$

$$\Delta V_m = \Delta t \ F( > r_2 ) \tag{16}$$

$$\Delta V_c = \Delta V - \Delta t \ F( > r_2 ) \tag{17}$$

$$\Delta V_p = r \ \Delta F \tag{18}$$

$$\Delta V_c = (r - t )\Delta F \tag{19}$$

$$\Delta V_p = \frac{r}{r-t} \Delta V_c \tag{20}$$

$$\Delta V_{p_i} = \frac{r_i}{r_i - t_i} \left[ \Delta V_i - (t_{i-1} - t_i) \sum_{n=1}^{i-1} F_n \right] \tag{21}$$

$$F_i = \frac{\Delta V p_i}{r_i} \tag{22}$$

The authors in their original work [1] measured and calculated in 10-1,7 nm pore intervals for cylindrical shapes. This method is well applicable with the help of master sheets or suitable numerical method fitted to the measuring apparatuses. The pore size distribution function in this case is

$$\frac{dVp}{dr} \qquad (23)$$

For microporus adsorbents Dollimore's method is not advisable since the Kelvin equation is no longer valid when pore size approaches the molecular dimensions.

## 3. Horvath-Kawazoe method

In the micropore domain the Kelvin equation is not valid, therefore new theoretical background has to be introduced for the establishment of pore calculation [3]. Let us consider the molar integral change of the free energy (Gibbs function for $T$=constant):

$$\Delta G^{ads} = \Delta H^{ads} - T\Delta S^{ads} \qquad (22)$$

The molar integral change of enthalpy on adsorption is

$$\Delta H^{ads} = -q^{diff} - RT + K(T\beta / \theta)(\partial \pi / \partial T)_\theta \qquad (23)$$

The calorimetric definition of $q^{diff}$ is

$$q^{diff} = \Delta H^{vap} - RT - (\partial h_f / \partial n_a)_T \qquad (24)$$

The molar integral change of entropy is

$$\Delta S^{ads} = \Delta S^{tr} + \Delta S^{rot} + \Delta S^{vib} \qquad (25)$$

Except for $\Delta S^{tr}$, the terms of equation (25) are nearly constant, thus one can write

$$\Delta S^{ads} = \Delta S^{tr}(w/w_\infty) + \Delta S_0 \qquad (26)$$

The free energy change can be calculated from the gas-phase pressure. Combining this with the previous equation:

$$\Delta G^{ads} = RT \ln(p/p_o) = \Delta H^{ads} - T(\Delta S^{tr}(w/w_\infty) + \Delta S_0) \qquad (27)$$

If we consider the limit of $\Delta G^{ads}$ as $p$ approaches $p_o$,

$$\lim_{w \to w_\infty} \Delta S^{tr}(w/w_\infty) = 0$$
$$\lim_{p \to p_o} RT \ln(p/p_o) = \Delta H_0^{ads} - T\Delta S_0 = 0 \qquad (28)$$

then

$$\Delta H_0^{ads} = T\Delta S_0 = -\Delta H^{vap} = -T\Delta S_0^{vap} \tag{29}$$

We can also write

$$RT \ln(p/p_o) = -q^{diff} - RT + K(T\beta/\theta)(\partial\pi/\partial T)_\theta - T(\Delta S^{tr}(w/w_\infty) + \Delta S_0) \tag{30}$$

Supposing that $K(T\beta/\theta)(\partial\pi/\partial T)_\theta \cong RT$, the expression can be simplified to

$$RT \ln(p/p_o) - \Delta H^{vap} = -q^{diff} - T\Delta S^{tr}(w/w_\infty) \tag{31}$$

In the range of $p/p_o \geq p_e/p_o > 0$, the above general equation is approximated into Eq. (32), since the adsorbed phase is considered similar to the liquid phase and then

$$\left| T\Delta S^{tr}(w/w_\infty) \right| << \left| q^{diff} \right| \tag{32}$$

$$-q^{diff} = RT \ln(p/p_o) - \Delta H^{vap} \tag{33}$$

Having considered the definition of $q^{diff}$ on the molecular level [4]

$$-q^{diff} = K(U_0 + P_a - \Delta E^{vib} - \Delta E^{tr} - \Delta E^{rot}) \tag{34}$$

we used the same logic to obtain:

$$RT \ln(p/p_o) = K(U_o + P_a) \tag{35}$$

where
  $U_o$ is the expression of the adsorbent-adsorbate interactions,
  $P_a$ is an implicit function of the adsorbate-adsorbate/adsorbent interactions, and
  $K$ is the Avogadro's number.
Some authors [5,6,7] have determined potential functions over a carbon layer and inside a pore having either slit-like or cylindrical shape. They suggest that the so-called (10:4) potential is likely more realistic than the (9:3) potential, but it is not possible to choose between these models on the basis of the precision with which they represent the adsorption data. Our aim was to develop a suitable method for the calculation and not to decide which is the most adequate model among those mentioned above.
The micropores of molecular-sieve materials are usually considered to be slits between two graphitized carbon layer planes [8]. Supposing that the potential fields of these layers can be approximated by a potential of the graphite layer, simple mathematical expressions can be used.
The potential function [9] over a graphite surface (interaction energy between a gas molecule and an infinite graphite layer) is

$$\Phi = \Phi^* \left[ -(\sigma/r)^4 + (\sigma/r)^{10} \right] \tag{36}$$

$$\Phi^* = (N_a A_a)/(2\sigma^4) \qquad (37)$$

where $\sigma$ is the distance between a gas atom and the surface at zero interaction energy, $r$ is the distance from surface, $N_a$ is the number of atoms per unit area of surface, $A_a$ is a constant given by the Kirkwood-Müller, London, etc., equations.

The potential function between two parallel layers in case of one adsorbate molecule is:

$$\Phi = \left[(N_a A_a)/(2\sigma^4)\right]\left[-(\sigma/r)^4 + (\sigma/r)^{10} - \{\sigma/(l-r)\}^4 + \{\sigma/(l-r)\}^{10}\right] \qquad (38)$$

Where $l$ is the distance between the nuclei of the two layers FIGURE 2. A similar expression is used for the present case of adsorption on two parallel surfaces. The potential, originated from the interaction of the adsorbate molecules in a pore, increases the interaction energy. The potential function between the two carbon layers filled with adsorbates is

$$\Phi = \frac{N_a A_a + N_A A_A}{2\sigma^4}\left[-\left(\frac{\sigma}{r}\right)^4 + \left(\frac{\sigma}{r}\right)^{10} - \left(\frac{\sigma}{l-r}\right)^4 + \left(\frac{\sigma}{l-r}\right)^{10}\right] \qquad (39)$$

where $N_A$ is the number of molecules per unit area of the adsorbate and Kirkwood-Müller

$$A_a = \frac{6mc^2\alpha_a\alpha_A}{\dfrac{\alpha_a}{\chi_a} + \dfrac{\alpha_A}{\chi_A}} \qquad (40)$$

$$A_A = \frac{3mc^2\alpha_A\chi_A}{2} \qquad (41)$$

or London expressions

$$A_a = \frac{3}{2}\alpha_a\alpha_A \frac{I_a I_A}{I_a + I_A} \qquad (42)$$

$$A_A = \frac{3}{4}\alpha_A^2 I_A \qquad (43)$$

where $m$ is the mass of an electron, $c$ is the velocity of light, $\alpha_a$ is the polarizability and $\chi_A$ is the magnetic susceptibility of an adsorbent atom, $\alpha_A$ and $\chi_A$ are the polarizability and magnetic susceptibility of an adsorbate molecule, $I_a$ and $I_A$ are the corresponding first ionisation energies.

140

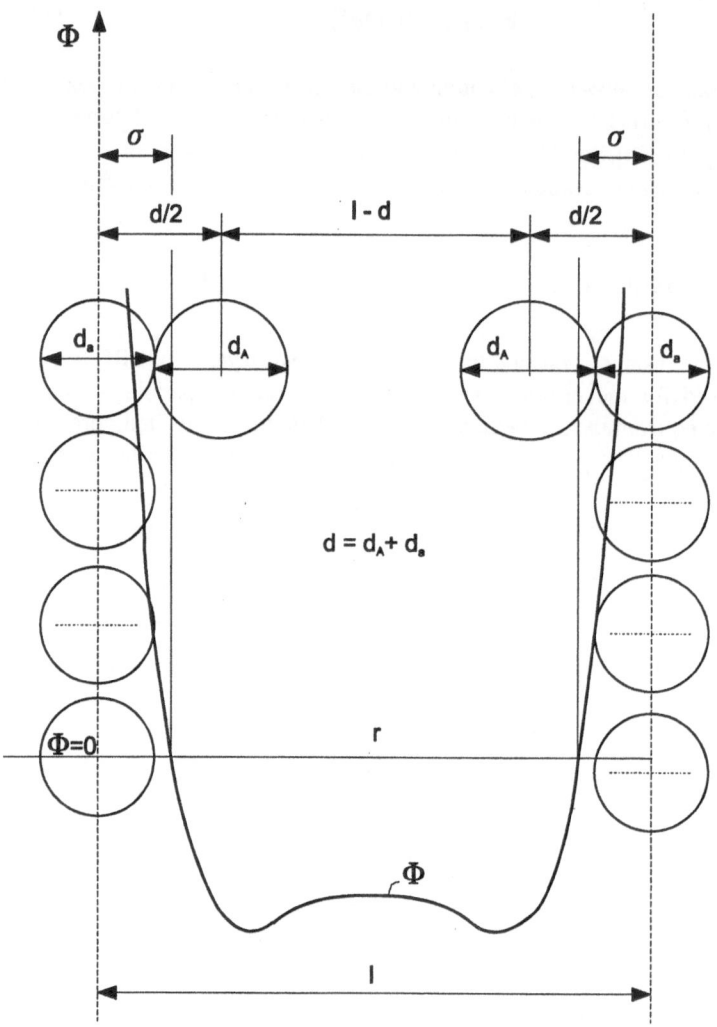

*Figure 2.* Potential function in a slit like pore

Equation (39) consists of two parts. One of them, $N_a A_a / 2\sigma^4$ multiplied by the expression in parentheses, corresponds to $U_o$ (Eq. (35)); the other corresponds to $P_a$ (Eq. (35)).

Having taken into account Eq. (35), an average potential value that depends on the absolute values of distances between the two layers can be calculated:

$$RT \ln(p/p_o) = K \frac{N_a A_a + N_A A_A}{2\sigma^4(l-d)} \int_{d/2}^{l-d/2} \left[ -\left(\frac{\sigma}{r}\right)^4 + \left(\frac{\sigma}{r}\right)^{10} - \left(\frac{\sigma}{l-r}\right)^4 + \left(\frac{\sigma}{l-r}\right)^{10} \right] dr \qquad (44)$$

where

$$d = d_a + d_A \qquad (45)$$

$d_a$ is the diameter of an adsorbent atom, and $d_A$ is the diameter of the adsorbate molecule. Integration gives the following result:

$$RT \ln(p/p_o) = K \frac{N_a A_a + N_A A_A}{\sigma^4(l-d)} \left[ \frac{\sigma^4}{3(l-d/2)^3} - \frac{\sigma^{10}}{9(l-d/2)^9} - \frac{\sigma^4}{3(d/2)^3} + \frac{\sigma^{10}}{9(d/2)^9} \right] \qquad (46)$$

where $l$ should be larger than $d$. When $l=d$, Eq. (44) gives a finite value of $p/p_o$, $p_c/p_o$. So the above equation is considered to be valid for $p/p_o \geq p_c/p_o$, which satisfies the condition assumed for deriving Eq. (33).

We now have a function of $p$ with respect to $l$. Having used the data of a nitrogen isotherm at liquid nitrogen temperature, where $w/w_\infty = q(p)$, the expression

$$w/w_\infty = f(l-d_a) \qquad (47)$$

can be obtained. This function gives us the effective pore size distribution since $w$ is considered as the mass of the nitrogen adsorbed into pores smaller than $(l-d_a)$, and $w_\infty$ is the maximum amount of nitrogen adsorbed into the pores.

The necessary physical data can be seen in TABLE 1.

TABLE 1. Physical properties for calculation of model parameters

| | Carbon | Reference | Nitrogen | Reference |
|---|---|---|---|---|
| Diameter [nm] | 0.34 | [5] | 0.3 | [6] |
| Liquid density [g/cm$^3$] | | | 0.808 | |
| Polarizability [cm$^3$] | $1.02 \times 10^{-24}$ | [6] | $1.46 \times 10^{-24}$ | [10] |
| Susceptibility [cm$^3$] | $13.5 \times 10^{-29}$ | [6] | $2 \times 10^{-29}$ | [11] |
| Ionisation potential [eV] | 11.26 | [12] | 14.53 | [12] |
| Surface density [Molecule/ cm$^2$] | $3.845 \times 10^{15}$ | [9] | $6.7 \times 10^{14}$ | |

Everett and Pawl [5] gave an equation for the calculation of $\sigma$ values:

$$\sigma = 0.858 d/2 \qquad (48)$$

In our case,

$$\sigma = 0.275nm$$

From the numerical data substituted into Eq. (44) the following expressions can be obtained, where $l$ is in nm.

In case of Kirkwood-Müller expression

$$ln\frac{p}{p_o} = \frac{61,23}{(l-0,64)}\left[\frac{1,895.10^{-3}}{(l-0,32)^3} - \frac{2,7087.10^{-7}}{(l-0,32)^9} - 0,05014\right] \qquad (49)$$

and for London expression

$$ln\frac{p}{p_o} = \frac{18,37}{(l-0,64)}\left[\frac{1,895.10^{-3}}{(l-0,32)^3} - \frac{2,7087.10^{-7}}{(l-0,32)^9} - 0,05014\right] \qquad (50)$$

From the theoretical part it follows that if we know the $p$-$l$ functions, we know the $p-(l-d_a)$ functions as well (TABLE 2). In the same table pore size calculated according to Dollimore and de Boer can be found. The two slit model curves (FIGURE 3) intersect at a pore size of 1.34 nm and a relative pressure of $p' = 5 \times 10^{-2}$. At this pressure the slopes of the two curves are nearly equal. At the point of intersection the two methods could be combined. Beyond $p'$ the Dollimore method can be used while below $p'$ the new method may be used.

TABLE 2. Comparison of pore sizes

| Relative pressure | Pore sizes [nm] | | | | |
|---|---|---|---|---|---|
| | Horváth-Kawazoe | | Dollimore-Heal | | de-Boer |
| | KM parameter | L parameter | Slit | Cylindrical | |
| $p/p_o$ | $l$-$d_a$ | $l$-$d_a$ | | | $2t$ |
| $1,46.10^{-7}$ | 0,4 | 0,42 | - | - | - |
| $6,47.10^{-7}$ | 0,43 | 0,45 | - | - | - |
| $1,19.10^{-6}$ | 0,46 | 0,49 | - | - | - |
| $1,05.10^{-5}$ | 0,5 | 0,53 | - | - | - |
| $1,54.10^{-4}$ | 0,6 | 0,68 | - | - | - |
| $8,86.10^{-4}$ | 0,7 | 0,8 | - | - | - |
| $2,95.10^{-3}$ | 0,8 | 0,94 | - | - | - |
| $2,95.10^{-3}$ | 1,1 | 1,42 | 1,16 | 1,51 | - |
| $2,05.10^{-2}$ | 1,3 | 1,95 | 1,32 | 1,71 | - |
| $4,43.10^{-2}$ | 1,5 | 2,60 | 1,46 | 1,90 | 0,76 |
| $7,59.10^{-2}$ | 3,0 | - | 2,23 | 3,15 | 1,0 |
| $3,15.10^{-1}$ | 10,0 | - | 5,09 | 8,50 | 1,8 |
| $7,24.10^{-1}$ | | | | | |

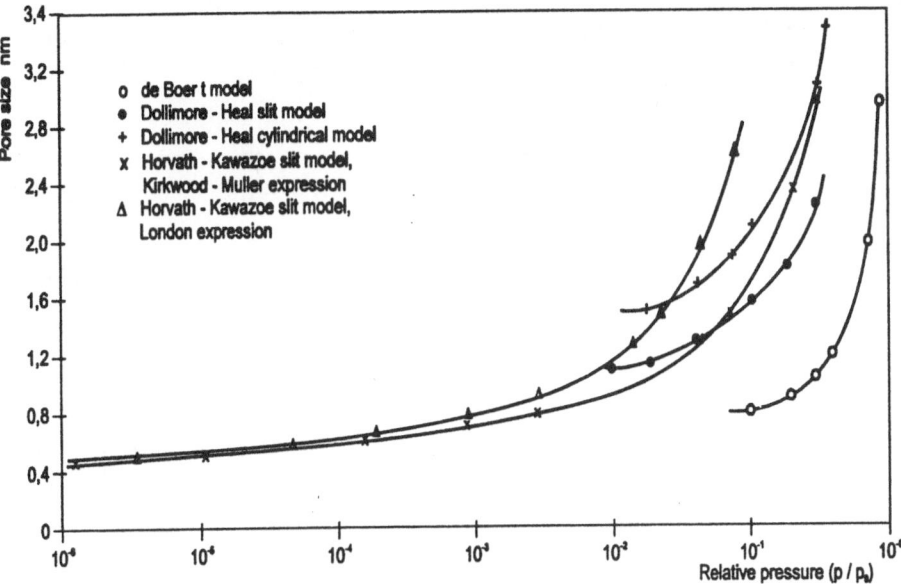

*Figure 3.* Pore sizes according to different models for carbon

## 4. Saito-Foley method

Saito and Foley (13) developed the cylindrical variation of the former method. Two types, a line averaged and an area averaged model, were examined. The equation of cylindrical potential was taken from the works of Everett and Powl [5].(For explanation see FIGURE 4.)

$$\Phi(r) = \frac{5}{2}\pi\Phi^* \left[ \frac{21}{32}\left(\frac{d_o}{r_p}\right)^{10} \sum_{k=0}^{\infty} \alpha_k \left(\frac{r}{r_p}\right)^{2k} - \left(\frac{d_o}{r_p}\right)^4 \sum_{k=0}^{\infty} \beta_k \left(\frac{r}{r_p}\right)^{2k} \right] \qquad (51)$$

$$\alpha_k^{0.5} = \frac{\Gamma(-4.5)}{\Gamma(-4.5-k)\Gamma(k+1)}$$

$$\beta_k^{0.5} = \frac{\Gamma(-1.5)}{\Gamma(-1.5-k)\Gamma(k+1)}$$

The line averaged potential for this cylindrical model

$$\overline{\Phi}_{LA} = \frac{\int_{o}^{r_p-d_o} \Phi(r)\,dr}{r_p - d_o} \qquad (52)$$

144

$$ln\left(\frac{P}{P_0}\right)=\frac{3}{4}\frac{\pi K}{RT}\frac{(N_aA_a+N_AA_A)}{d_o^4}\sum_{k=0}^{\infty}\left[\frac{1}{2k+1}\left(1-\frac{d_o}{r_p}\right)^{2k}\left\{\frac{21}{32}\alpha_k\left(\frac{d_o}{r_p}\right)^{10}-\beta_k\left(\frac{d_o}{r_p}\right)^4\right\}\right] \quad (53)$$

The area averaged model

$$\overline{\Phi}_{AA}=\frac{\int_0^{r_p-d_o}\Phi(r)2\pi rdr}{\int_0^{r_p-d_o}2\pi rdr} \quad (54)$$

$$ln\left(\frac{P}{P_0}\right)=\frac{3}{4}\frac{\pi K}{RT}\frac{(N_aA_a+N_AA_A)}{d_o^4}\sum_{k=0}^{\infty}\left[\frac{1}{k+1}\left(1-\frac{d_o}{r_p}\right)^{2k}\left\{\frac{21}{32}\alpha_k\left(\frac{d_o}{r_p}\right)^{10}-\beta_k\left(\frac{d_o}{r_p}\right)^4\right\}\right] \quad (55)$$

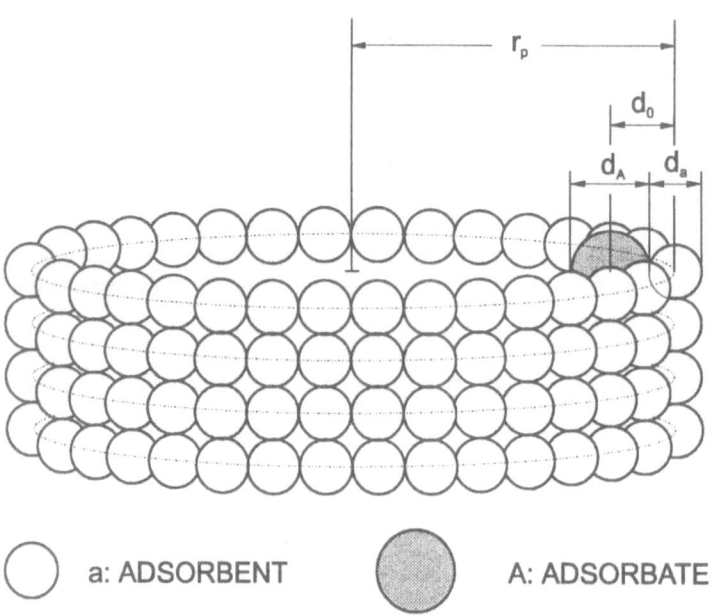

a: ADSORBENT          A: ADSORBATE

*Figure 4.* Explanation to Saito-Foley model

Substituting the physical parameters, a relationship between the relative pressure and pore sizes can be obtained (FIGURE 5.). Their opinion is that in case of zeolites, the area averaged model seems to be more realistic.

Ruthven [14] discussed in detail the comparison of experimental heat values of sorption with theoretical values. To find the appropriate interaction function including some uncertainty in the choice of physical parameter values is one of the keys of pore size determination. The calculated and measured heat values have to be equal.

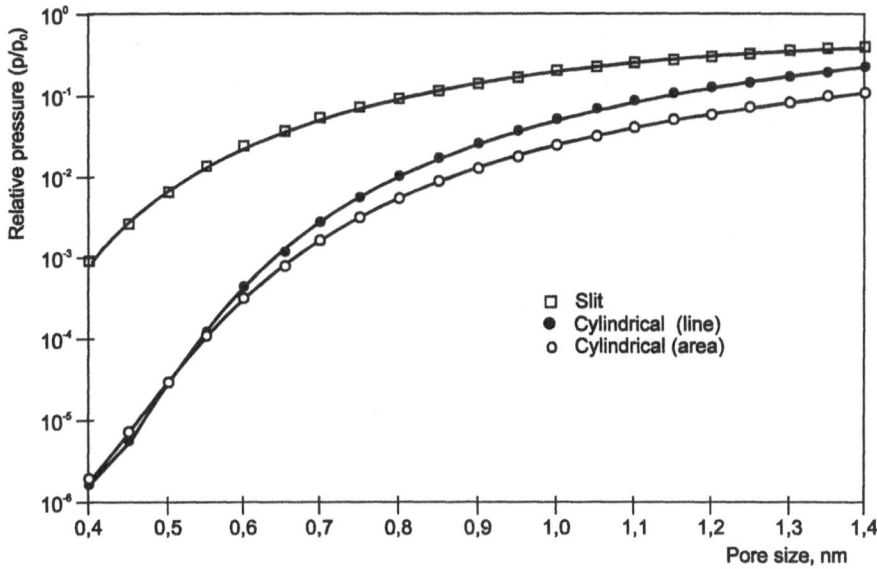

*Figure 5.* Pore sizes for zeolite according to Saito-Foley

## 5. Molecular probe method

For small pores which have molecular sieving abilities sometimes it is not possible to determine pore size distribution by nitrogen or argon adsorption method. In this case the possible way of direct determination of effective pore size is the molecular probe method. Adsorption from saturated pressure of different molecules, some point of pore size distribution can be obtained.

Beside the experimental energy parameters, the molecular probe method is the other proof for the numerical values of pore size calculation in a given material system.

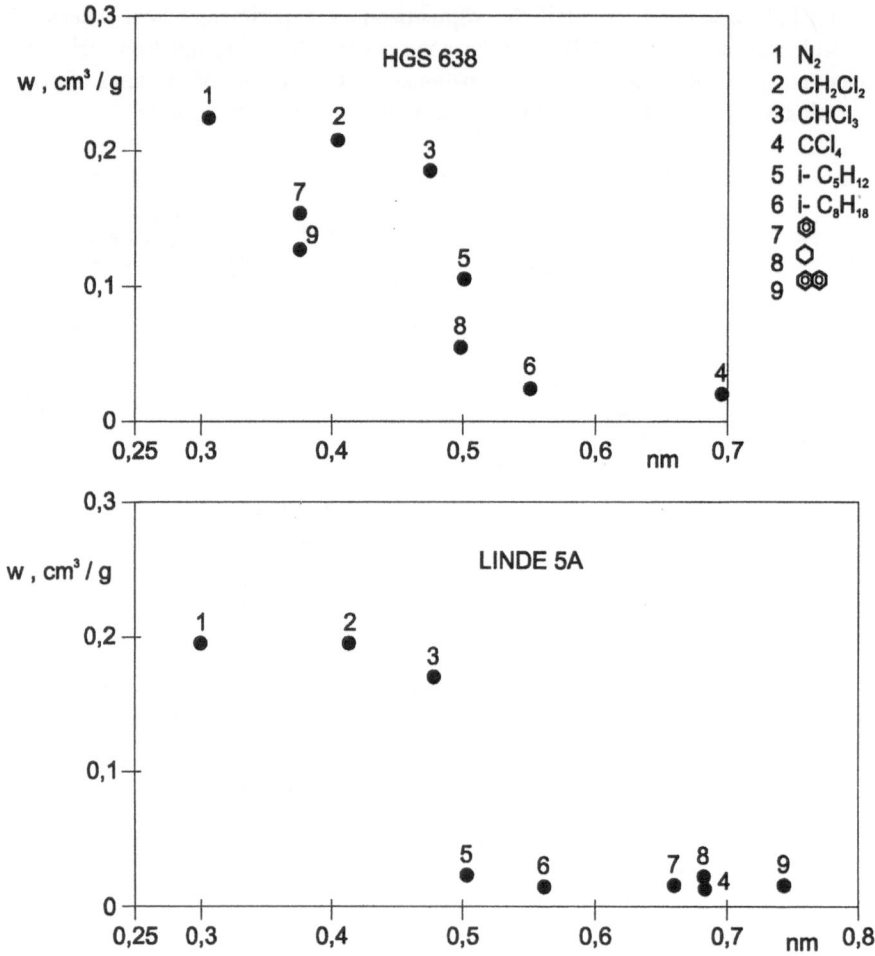

*Figure 6.* Molecular probes

## 6. Conclusion

The former methods show that even in the most simple case many parameter effects the pore size measurements as well as the calculations of them. The parameter sensitivity proof justifies that the diameter of adsorbent atom has a large effect the others have only moderate influences. It can also concluded that the more realistic geometrical shape provides the more adequate results. As for the energetic interactions an independent proof is necessary to choose the numerically realistic variation, namely the calculation and measurement of $q^{diff}$. The third adsorptive possibility the molecular probe method. When one harmonises these independent proofs has a good chance to find adequate numerical methods for a given material system.

## 7. Notation

| | | |
|---|---|---|
| $A_a$ | constant in Lennard-Jones | [J/molec.] |
| $A_A$ | constant in Lennard-Jones | [J/molec.] |
| $d_a$ | diameter of an adsorbent atom | [nm] |
| $d_A$ | diameter of an adsorbate molecule | [nm] |
| $d_0$ | arithmetic mean of diameters of adsorbate and adsorbent atoms | [nm] |
| $E^{vib}$ | vibrational energy | [J/mol] |
| $E^{tr}$ | translational energy | [J/mol] |
| $E^{rot}$ | rotational energy | [J/mol] |
| $F$ | wall area | [m$^2$] |
| $G$ | free energy | [J/mol] |
| $H$ | enthalpy | [J/mol] |
| $H^{ads}$ | enthalpy of adsorption | [J/mol] |
| $H_0^{ads}$ | limited enthalpy of adsorption | [J/mol] |
| $H^{vap}$ | enthalpy of vaporisation | [J/mol] |
| $h_f$ | residual heat of immersion | [J/g] |
| $I_a$ | ionisation potential of adsorbent atom | [eV] |
| $I_A$ | ionisation potential of adsorbate atom | [eV] |
| $K$ | Avogadro's number | [molec./mol] |
| $l$ | distance between nuclei of two layers | [nm] |
| $L(r)$ | pore length distribution function | |
| $mc^2$ | kinetic energy of an electron | [J] |
| $N_a$ | number of atoms per unit area of adsorbent | [atom/cm$^2$] |
| $N_A$ | number of molecules per unit area of adsorbate | [molec/ cm$^2$] |
| $n_a$ | adsorbed amount | [mol/g] |
| $P_a$ | potential energy of adsorbate-adsorbate-adsorbent interaction | [J/mol] |
| $p$ | pressure of $N_2$ | [Pa] |
| $p_c$ | pressure corresponding to $l=d(=d_a+d_A)$ | [Pa] |
| $p_o$ | saturation pressure of $N_2$ | [Pa] |
| $q^{diff}$ | differential heat of adsorption | [J/mol] |
| $q^{st}$ | isosteric heat of adsorption | [J/mol] |
| $R$ | gas constant | [J/(mol·K)] |
| $r$ | distance from the nuclei of adsorbent layer | [nm] |

| | | |
|---|---|---|
| $r_K$ | Kelvin radius | [nm] |
| $r_p$ | maximal value of r | [nm] |
| $S$ | entropy | [J/(mol·K)] |
| $S^{ads}$ | entropy of adsorption | [J/(mol·K)] |
| $S^{rot}$ | entropy of rotation | [J/(mol·K)] |
| $S^{tr}$ | entropy of translation | [J/(mol·K)] |
| $S^{vap}$ | entropy of vaporisation | [J/(mol·K)] |
| $S^{vib}$ | entropy of vibration | [J/(mol·K)] |
| $S_0$ | limited entropy value of adsorption | [J/(mol·K)] |
| $t$ | adsorbed layer thickness | [nm] |
| $T$ | temperature | [K] |
| $U_0$ | adsorptive potential | [J/mol] |
| $V$ | volume | [cm$^3$] |
| $V_c$ | pore desorption | [cm$^3$] |
| $V_m$ | multi-layer adsorption | [cm$^3$] |
| $V_p$ | pore volume | [cm$^3$] |
| $w$ | adsorbed amount of $N_2$ | [g/g] |
| $w_\infty$ | maximum amount of $N_2$ adsorbed into the pores | [g/g] |
| $w_\infty^i$ | maximum adsorbed amount of $i$-th compound | [g/g] |
| $\alpha_a$ | polarizability of adsorbent | [cm$^3$] |
| $\alpha_A$ | polarizability of adsorbate | [cm$^3$] |
| $\alpha_k$ | parameter | [-] |
| $\beta$ | surface needs for one molecule | [m$^2$/molec.] |
| $\beta_k$ | parameter | [-] |
| $\theta$ | fraction of surface occupied by molecules | [-] |
| $\pi$ | two-dimensional pressure | [N/m] |
| $\rho$ | density of liquid $N_2$ | [g/cm$^3$] |
| $\rho^i$ | density of $i$-th compound | [g/cm$^3$] |
| $\sigma$ | distance between a gas atom and the nuclei of the surface at zero interaction energy | [nm] |
| $\Phi$ | interaction energy | [J/molec.] |
| $\Phi^*$ | characteristic interaction energy | [J/molec.] |
| $\Phi_{LA}$ | line averaged $\Phi$ | [J/molec.] |
| $\Phi_{AA}$ | area averaged $\Phi$ | [J/molec.] |
| $\chi_a$ | magnetic susceptibility of an adsorbent atom | [cm$^3$] |
| $\chi_A$ | magnetic susceptibility of an adsorbate molecule | [cm$^3$] |

## 8. References

1. Dollimore, D. and Heal, G.R. (1964) An Improved Method for the Calculation of Pore Size Distribution from Adsorption Data, *J. Appl. Chem.* **14,** 109-114.
2. Gregg, S.J., Sing, U.S.W. (1982) *Adsorption, Surface Area and Porosity*, Academic Press, London.
3. Horvath, G. and Kawazoe, K. (1983) Method for the Calculation of Effective Pore Size Distribution in Molecular Sieve Carbon, *J. Chem. Eng. Japan,* **16,** 470-475.
4. Ross, S. and Olivier, J.P. (1964) *On Physical Adsorption,* Interscience pub., New York.
5. Everett, D.H. and Powl, J.C. (1976) Adsorption in Slit-like Cylindrical Micropores in the Henry's Law Region, *J. Chem. Soc. Faraday Trans.* **1, 72,** 619-636.
6. Sams, J.R., Constrabaris, G. and Halsey. (1960) Second Virial Coefficients of Neon, Argon, Krypton and Xenon with a Graphitized Carbon Beach, *J. Phys. Chem.* **64,** 1689-1695.
7. Stoeckli, F., (1974) The Gas-Solid Interface Calculations of Adsorption Potentials in Slot-line Pores of Molecular Dimensions. *Helv. Chim. Acta,* **57,** 2195-2199.
8. Chihara, K., Suzuki, M. and Kawazoe, K. (1978) Adsorption Rate on Molecular Siering Carbon by Chromatography. *AIChE J.* **24,** 237-246.
9. Walker, P.L. (1966) *Chemistry and Physics of Carbon Dekker*, New York.
10. Kagaku Binran (1966) *Maruzen,* Tokyo.
11. Samsonov (1968) *Handbook of the Phys. Chem. Properties,* Plenum, New York
12. *Handbook of Chemistry and Physics,* (1975), CRC Press, Cleveland.
13. Saito, A. and Foley, H.C. (1991) Curvature and Parametric Sensitivity in Models for Adsorption in Micropores, *AIChE J.,* **37,** 429-436.
14. Ruthven, D.M. (1984) *Principles of Adsorption and Adsorption Processes,* John Wiley & Sons, New York.

# HIGH-PRESSURE PHYSISORPTION OF GASES ON PLANAR SURFACES AND IN POROUS MATERIALS

G.H. FINDENEGG and M. THOMMES

*Stranski-Institut für Physikalische und Theoretische Chemie, Technische Universität Berlin, D 10623 Berlin, Germany*

## 1. Introduction

Physical adsorption of gases on solid surfaces is due to the attractive van der Waals interactions of gas molecules with the solid. The gas-solid interaction potential $u_s(z)$ exhibits a minimum close to the surface and tends to zero for large distances from the surface ($z \rightarrow +\infty$) (see Figure 1). The interaction energy at the minimum of the gas-solid potential, $-\varepsilon_s$, is typically a factor 10 greater than the thermal energy $k_B T$, where $k_B$ is the Boltzmann constant and $T$ the absolute temperature. Accordingly, gas molecules will accumulate in the vicinity of the surface. At sufficiently low temperatures (typically at and below the normal boiling temperature of the adsorptive) a dense monolayer of molecules is formed at pressures far below the saturated vapor pressure $p_o(T)$ and a multilayer adsorbed film of increasing thickness and liquid-like density builds up on strongly adsorbing substrates as $p_o$ is approached. In this low-temperature region the adsorption of vapors can be analysed in terms of a two-phase model in which an *adsorbed phase* (superscript a) coexists with the bulk phase (superscript b). Accordingly, one writes

$$n^g = n^a + n^b \tag{1}$$

$$V^g = V^a + V^b \tag{2}$$

where $n^g$ is the amount of gas and $V^g$ the overall macroscopic volume accessible to the gas molecules; $n^a$ represents the amount and $V^a$ the volume of the adsorbed phase

*J. Fraissard (ed.), Physical Adsorption: Experiment, Theory and Applications, 151–179.*
© *1997 Kluwer Academic Publishers.*

152

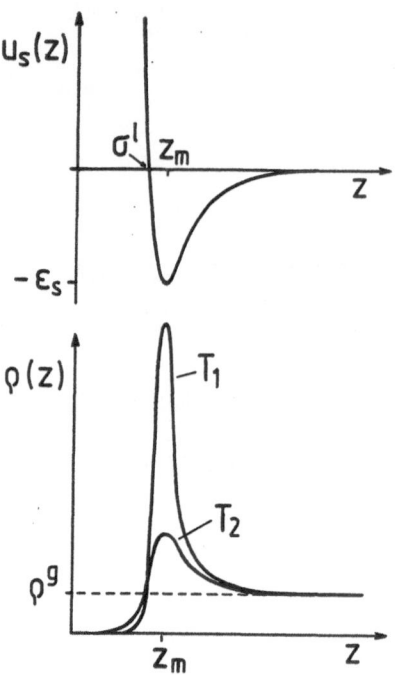

*Figure 1.* Gas-solid interaction potential $u_s(z)$ (upper graph) and density profile $\rho(z)$ of an ideal adsorbed gas at a flat structureless surface for two temperatures $(T_2 > T_1)$ (lower graph).

(*adsorption space*). For an adsorbent with a planar surface of surface area $A_s$ this volume corresponds to a thickness $z^\alpha$ of the adsorbed layer, viz. $V^\alpha = z^\alpha A_s$.

At higher temperatures the model of an adsorbed phase becomes progressively unrealistic for two reasons: (i) the tendency of the molecules to accumulate near the surface becomes less pronounced, i.e., the peak in the local density $\rho(z)$ of the fluid near the surface decreases sharply with increasing temperature (see Fig.1); (ii) due to the weaker physisorption at elevated temperatures, higher pressures have to be applied in order to reach a significant surface coverage. In this case the density of the bulk gas is no longer negligible against the density near the surface, and the profile of the local density $\rho(z)$ exhibits a smooth transition from the surface into the bulk gas. In this case any definition of the adsorption space becomes arbitrary and other concepts to quantify the adsorbed amount are required. The most general concept, first introduced by Gibbs [1], is to express adsorption by the *surface excess amount* of the fluid.

## 1.1. ADSORPTION AND SURFACE EXCESS AMOUNTS

Experimentally, the adsorbed amount is commonly determined by gas volumetry [2-4], microbalance techniques [5-7], or by a combination of both [8]. In the former case, a known amount $n^g$ of gas is allowed into the volume $V^g$ in contact with the solid adsorbent. The adsorbed amount is defined operationally as the excess amount over and above the amount which would be present if the density of the gas remained constant and equal to that of the bulk fluid up to the surface [9]. Accordingly, the experiments yield not $n^a$ but the *surface excess amount* $n^\sigma$

$$n^\sigma = n^g - \rho^g V^g \tag{3}$$

where $\rho^g$ is the (molar) density of the bulk gas at the experimental temperature and pressure. From eq. (1) - (3) one finds

$$n^\sigma = n^a - \rho^g V^a = (\rho^a - \rho^g) V^a \tag{4}$$

where the term $\rho^g V^a$ represents the amount of gas in a volume equal to $V^a$ somewhere in the bulk phase and $\rho^a = n^a / V^a$ is the mean density of the adsorbed phase. At sufficiently low pressures and low temperatures, the gas density is negligably small against the density near the surface ($\rho^g << \rho^a$) and thus $n^\sigma \approx n^a$. However, it is important to realize that in principle any volumetric or gravimetric method for measuring the adsorbed amount yields the surface excess amount $n^\sigma$ rather than the total amount $n^a$ which can be derived only indirectly on the basis of eq. (4) and relies on a model assumption about the extension of the adsorption space.

The concept of surface excess quantities was introduced first for fluid interfaces by Gibbs [1] and requires a convention concerning the exact location of the mathematical surface dividing the total volume V into the volumes of the two coexistent bulk phases. In the case of gas/solid interfaces it is convenient to define a Gibbs dividing surface by adopting a *reference volume* for the solid, $V_s^* = \rho_s^* m_s$, where $m_s$ is the mass and $\rho_s^*$ the *reference mass density* of the solid (the asterisk indicates that $\rho_s^*$ and thus $V_s^*$ represent reference values to be determined by some specified prescription). Accordingly, the volume of the gas phase is $V^g = V - \rho_s^* m_s$ and the surface excess amount of gas defined relative to the solid adsorbent becomes

$$n^\sigma = n^g - \rho^g(V - \rho_s^* m_s) \tag{3a}$$

Eq. (3a) represents an operational definition for $n^\sigma$ as all quantities on the r.h.s. are experimentally accessible. Prescriptions for the determination of $\rho_s^*$ will be given later.

An understanding of the different definitions of adsorption on a molecular level is based on a consideration of the density profile of the fluid, $\rho(z) = <\delta n / \delta V>$, where $\delta n$ represents the amount of gas in a volume element $\delta V$ of infinitesimal width $\delta z$ at a distanze $z$ from the surface. Density profiles cannot be measured experimentally but can be calculated by means of the density functional theory (and other theoretical roots), or $\rho(z)$ may be computed by molecular simulation techniques, as outlined in other articles of this volume. The density profile for an ideal gas in contact with an attractive surface is illustrated in Figure.1. Here we are interested in the way in which the surface excess amount depends on the choice of the Gibbs dividing plane. As explained above, the surface excess amount of the adsorbed gas is defined by comparing the real system with a hypothetical reference system in which the local density of the gas changes stepwise from $\rho(z) = 0$ at $z < z_0$ to $\rho(z) = \rho^b$ at $z > z_0$, where $z_0$ represents the location of the Gibbs dividing surface. Accordingly, the surface excess amount per unit surface area $A_s$ of the solid, $\Gamma = n^\sigma / A_S$ is given by [9]

$$\Gamma = \int_{-\infty}^{z_o} \rho(z)dz + \int_{z_o}^{+\infty} [\rho(z) - \rho^b]dz \tag{5}$$

Here the lower integration limit ($z \to -\infty$) is far inside the solid substrate, where $\rho(z)$ is zero for non-swellable inert substrates, and the upper integration limit ($z \to +\infty$) is far in the bulk fluid, where $\rho(z) = \rho^b$. Let us calculate $\Gamma$ for the adsorption of a hard-sphere fluid at a hard attractive wall. In this model the molecules of the fluid can approach the surface no closer than to a distance $z_m$ measured from the surface layer of interaction centers of the solid, which is defined as the $z = 0$ plane (see Figure 2). We consider two limiting cases of adsorption:

(a)     *High-pressure limit* at low temperatures, when a layer of molecules exists at the surface while the bulk density is negligably small in comparison with the density in this surface layer.

In this limiting case ($\rho^b = 0$) eq. (5) yields

$$\Gamma = \int_{-\infty}^{+\infty} \rho(z)dz = n^a / A_S \tag{6}$$

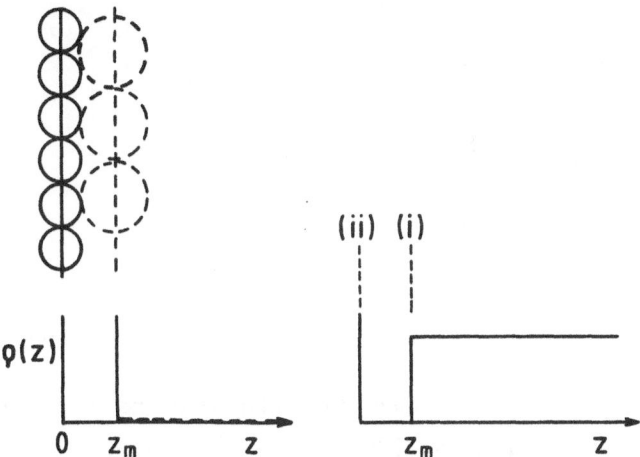

*Figure 2*. Schematic density profiles $\rho(z)$ of a fluid for two limiting cases:
(a) low-pressure adsorption at low temperature $(T \ll T_c)$ (b) high-pressure adsorption at high temperatures $(T > T_c)$ with two choices (i) and (ii) for the location of the Gibbs surface (see text).

where $n^a$ represents the total amount of the fluid in the adsorption space. No precise definition of the adsorption space is required as long as the bulk density of the fluid is sufficiently small. Indeed, in this low-pressure region the result for $\Gamma$ is independent of the locationof the Gibbs deviding surface, as the density of the fluid is effictively zero everywhere except in a narrow layer near the substrate surface.
(b)      *High-pressure limit* at high temperatures, when the density of the bulk fluid becomes similar to the density of the adsorbed layer.
In terms of our simple hard-sphere model, this limiting case implies a single-step density profile, viz. $\rho(z) = 0$ for $z < z_m$ and $\rho(z) = \rho^b$ for $z \geq z_m$, where $z_m$ is the distance of closest approach of the fluid molecules to the surface. In this case the result for $\Gamma$ depends on an arbitrary definition of the precise location of the Gibbs surface $z_o$ relative to $z_m$. Let us consider two alternative choices for $z_o$:

(i)      $z_o = z_m$ (center of the first layer of molecules of the fluid): In this case the integrand of eq. (5) is zero in the entire range and thus

$$\Gamma^{(i)} = 0$$

156

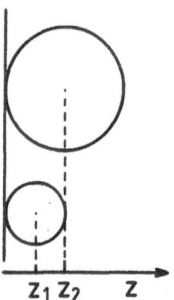

$$z_1\ z_2 \qquad z$$

*Figure 3.* Influence of the size of adsorbate molecules on the excluded volume

(ii)    $z_o = 0$ (center of surface atoms of the solid): Here, the first integral of eq. (5) is again zero but the second integral yields

$$\Gamma^{(ii)} = \int_0^{z_m} [\rho(z) - \rho^b]\,dz = -\rho^b z_m$$

This simple calculation illustrates that in high-pressure adsorption the magnitude of $\Gamma$ becomes strongly dependent on the choice of $z_o$; even negative values of $\Gamma$ are possible if the Gibbs dividing plane is placed somewhere in the region not accessible to the *centers* of the fluid molecules. On the basis of the simple hard-sphere model considered above, choice (i) appears to be the most natural choice for the Gibbs surface, as it yields $\Gamma = 0$ if the density of the adsorbed layer is equal to the bulk density of the fluid.

As was already mentioned in the context of eq. (3a), in experimental adsorption studies the location of the Gibbs dividing plane is defined implicitly by adopting a value of $V_s^*$, the nominal volume of the solid. Commonly $V_s^*$ is defined as the volume of a *nonadsorbable fluid* which is *displaced* by the solid adsorbent. Helium is commonly chosen for these displacement measurements as it exhibits weaker gas-solid interactions than any other fluid. The nominal density $\rho_s^* = m_s / V_s^*$ of a solid determined by displacement measurements with helium gas is called the *helium density of the solid.* There are indications that the adsorption of helium gas on typical adsorbents like activated carbon it is not entirely negligible at room temperature, and thus heliumdisplacement measurements should be performed preferably at higher temperatures (300-400°C) [10,11]. However, in this case it is necessary to apply a correction for the thermal expansion of the solid, which implies that the thermal expansion coefficient of the adsorbent in the relevant temperature range is available from independent experiments (e.g., X-ray densities). A more fundamental problem of defining the "true" volume of the solid arises from the *different size* of fluid molecules (cf. Figure 3): If the volume $V_s^*$ is defined as the volume that is inaccessible to the centers of the fluid molecules, $V_s^*$ becomes dependent on the size of the test molecules because large molecules will exclude a somewhat greater volume than small molecules. In the examle of Fig. 3 this volume increment is $(z_2 - z_1)A_s$. This *excluded-volume effect* becomes particulary important in the case of microporous solids, where ultra-narrow pores may be accessible to helium atoms but not to larger atoms or molecules [12,13].

The considerations outlined above can be summarized as follows:

1. Generally, the *adsorbed amount* represents a *surface excess amount* and its magnitude depends on a convention concerning the volume of the solid phase. In low-pressure adsorption (i.e. low bulk densities of the fluid as compared with the density of the adsorbed film) the surface excess $n^\sigma$ is nearly equal to $n^a$, the total amount of fluid in the adsorption space near the surface, and $n^\sigma$ becomes independent of the chosen reference volume of the solid.

2. In high-pressure adsorption the magnitude of the surface excess for given experimental conditions (i.e. given density profile $\rho(z)$ of the adsorbed fluid) is a function of the chosen nominal volume of the solid; this effect becomes more pronounced as the density of the bulk fluid approaches the mean density in the layer next to the solid.

3. Generally, the volume of finely divided or porous solid adsorbents cannot be determined on an absolute scale but depends on an operational definition (e.g., displacement of helium gas). Accordingly, the surface excess amount of dense fluids against solid surfaces represents a *relative quantity*, too, in the sense that we may analyse its dependence on temperature and pressure but cannot argue about the absolute magnitude of $n^\sigma$ or $\Gamma$, particulary in the region of very high bulk densities of the fluid [4].

## 1.2. CHARACTERISTIC TEMPERATURE RANGES

Temperature has a strong influence on the physisorption of gases on solid surfaces, and pronounced differences in the high-pressure adsorption behavior exist depending on whether the experimental temperature is below or above the critical temperature $T_c$ of the adsorptive. These differences result from the fact that at temperatures below $T_c$ the physisorption isotherms are bounded on the high-pressure side by the respective saturation pressure $p_o(T)$ of the vapor/liquid coexistence curve, and the density of the bulk fluid remains rather low in this regime (except at temperatures close to $T_c$. This low-temperature situation is illustrated by isotherm 1 in Figure 4. On the other hand, at high temperatures the strength of adsorption (expressed, for example, by the Henry's Law constant) is much weaker than at low temperatures, but above $T_c$ the pressure range of the adsorption isotherms is not limited (see isotherm 2 in Figure 4).

158

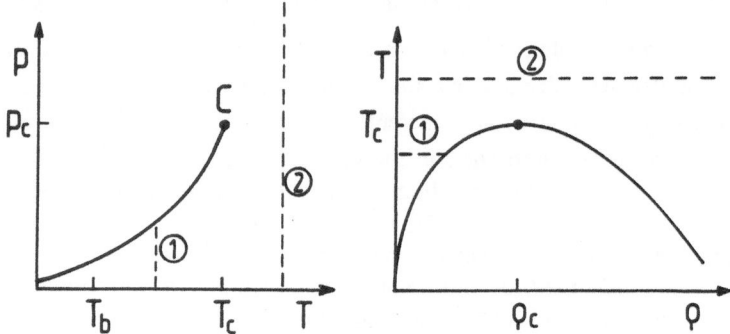

*Figure 4.* Phase diagram of a pure fluid in the $p$-$T$-plane (left) and in the $T$-$\rho$-plane (right); path 1 and path 2 represent isotherms at a subcritical temperature $(T_1 < T_c)$ and at a supercritical temperature $(T_2 > T_c)$. Point C is the critical point, $T_b$ the normal boiling temperature of the fluid.

Accordingly, the density of the bulk fluid can exceed the critical density $\rho_c$ and densities typical for the liquid state can be reached at sufficiently high pressures, as indicated in Figure 4b. This has an immediate effect on the pressure dependence of the measured adsorption excess amount $n^\sigma$ (see eq. 4): Whereas $n^a$, the total amount of fluid in a given volume $V^a$ near the surface, approaches some limiting value at large pressures, the background term $\rho^g V^a$ becomes more significant as the pressure is raised, and sooner or later it will increase more strongly than $n^a$. Accordingly, the high-pressure surface excess isotherms $n^\sigma = n^\sigma(p)$ must exhibit a *maximum* at some pressure $p_{max}$ at $T > T_c$. Note that this pecularity of high-pressure physisorption is due to the properties of both the adsorbed layer and the bulk fluid.

Below the critical temperature, multilayer adsorption of the vapor will occur at sufficiently attractive solid substrates: As the vapor pressure approaches the saturation pressure $p_o(T)$ a multilayer adsorbed film of liquid-like density is formed on flat or weakly curved surfaces, and the thickness $\ell$ of this film increases sharply for $p \to p_o$, signifying *complete wetting* of the substrate by the fluid at vapor/liquid coexistence. At temperatures well below $T_c$ the density of the saturated vapor is negligably small as compared with that of the adsorbed film, and thus the low-pressure approximation $n^\sigma = n^a$ applies. Closer to $T_c$ however, the density of the saturated vapor is no longer negligable against that of the coexistent liquid phase ultimately; at the critical point, the densities of the two phases become identical (see Fig. 4).

For a discussion of the high-pressure adsorption behavior it is appropriate to consider three distinct temperature regimes [14]:

(i)     low-temperature region $(T < T_c)$
(ii)    transition region (supercritical region)
(iii)   high-temperature region $(T > 1.1 \ T_c)$

The adsorption behavior in the transition region (ii) at temperatures just above $T_c$ and fluid densities around $\rho_c$ is of special interest. As will be explained in the subsequent Sections, the criticality of the bulk fluid causes interesting adsorption phenomena both at flat surfaces and for fluids in pores of mesoscopic size. In Section 2 we will discuss high-pressure adsorption phenomena on flat (or weakly curved) surfaces and in microporous materials in the high-temperature region. In Section 3 we turn to multilayer adsorption on nonporous substrates in the low-temperature region, emphasizing the behavior just below $T_c$. Finally, in Section 4 pore condensation phenomena in mesoporous materials will be considered; here again, the pecularities of the pore filling behavior in the region between the normal boiling temperature and the critical temperature of the fluid will be emphasizes (cf. Fig.4).

## 2. Fluid adsorption in the high-temperature region

In this Section we present results on the high-pressure adsorption of simple fluids on carbon adsorbents with patchwise homogeneous surfaces and in microporous carbon materials at temperatures well above the critical temperature $T_c$ of the adsorptive.

The adsorption of argon, krypton and methane on *Graphon* has been studied over a wide temperature range and pressures up to 150 bar, using a microbalance technique [5,15]. *Graphon* is a graphitized carbon black with a specific surface area of ca. 80 m$^2$/g; the exposed surface of this non-porous colloidal material consists of patches of graphite basal plane. Adsorption isotherms of gases on this uniform material are appropriate for comparison with theoretical studies of adsorption on flat adsorbing walls. Figure 5 shows adsorption isotherms of krypton on Graphon at several temperatures, corresponding to reduced temperatures $T_r = T / T_c$ from 1.2 to 1.8, and reduced pressure $p_r = p / p_c$ up to 2.7. In the graph at the left, the surface excess mass

160

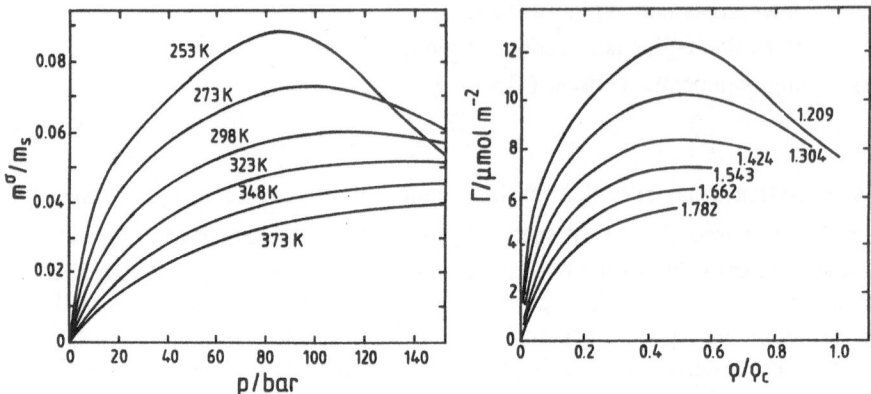

*Figure 5.* Adsorption isotherms of Kr on Graphon in the temperature range 253-373K (1.2 $T_c$ to 1.8 $T_c$): Specific surface excess mass $m^\sigma/m_s$ as a function of pressure $p$ (left) and surface excess amount per unit area, $\Gamma$, as a function of reduced density $\rho/\rho_c$ (right).

$m^\sigma$ of krypton per unit mass of adsorbent is plotted as a function of pressure. The isotherms at temperatures up to 298 K exhibit a distinct maximum in the experimental pressure range and $p_{max}$, the position of the maximum, is shifted to higher pressures with increasing temperature. As a consequence, the isotherm obtained at the lowest temperature $T_1$ intersects those for higher temperatures at pressures $p > p_{max}$ $(T_1)$, as can be seen for the 253 K isotherm. This temperature inversion of the adsorption isotherms at high pressures becomes more pronounced as the temperature is decreased and approaches $T_c$ [5b]. However, no such crossing of the isotherms occurs if the surface excess isotherms are plotted as a function of the density of the gas, as shown in the graph at the right-hand side. Indeed, from the definition of the surface excess amount (eq. 3) it is obvious that surface excess amounts obtained at different temperatures should be compared at the same density of the bulk gas: changes in $\Gamma$ (or $m^\sigma$) with temperature at constant $\rho^g$ imply corresponding changes in the total amount of fluid near the surface. Accordingly, it is desirable to measure $\Gamma$ at constant bulk density.

Adsorption isotherms for krypton in a molecular sieve carbon (MSC) with BET area 700 $m^2/g$ and 0.67 $cm^3/g$ total pore volume (39 % micropore volume < 1 nm, 9 % mesopore volume 1-100 nm) are presented in Figure 6 [16]. As to be expected for a microporous material, the adsorption isotherms rise much more steeply than those on Graphon at low pressures and level off at higher pressures. This influence of the microporosity of the MSC material can be seen more directly in the graph on the right-hand side of Fig.6, where the ratio of specific surface excess mass of krypton on the two

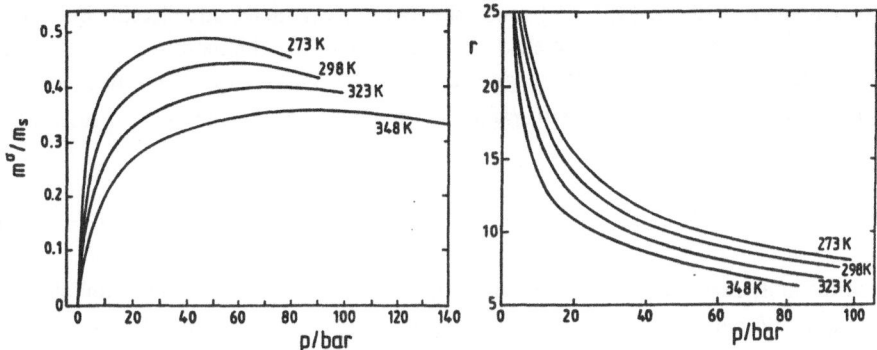

*Figure 6.* Adsorption isotherms (specific surface excess mass ) $m^\sigma / m_s$ of Kr in a carbon molecular sieve (MSC) in the temperature range 273-348 K (left) and the ratio $r$ of the specific surface excess mass in MSC and on Graphon (right) at pressures up to 150 bar.

materials, $r = m^\sigma$ (MSC)/ $m^\sigma$ (GRA), is plotted as a function of pressure. The ratio of the BET surface areas of the two materials is 700/80 ≈ 9. The large values of $r$ at low pressures reflect the high adsorption energy of the gas molecules in the micropores. With increasing pressure the ratio $r$ then falls off to values below 9, as it becomes increasingly difficult to add further molecules into the micropores, whereas unlimited adsorption is possible at a free surface. High-temperature adsorption isotherms of gases in microporous adsorbents are commonly correlated by a modified Dubinin-Astakhov equation or related treatments [3, 17]. However, the physical justification of pore-filling models based on the Polanyi potential theory is questionable at supercritical temperatures where the *adsorption potential* $\Delta\mu \equiv \mu_o - \mu \approx RT \ln(p_o / p)$ can be defined only by an arbitrary extrapolation of the vapor pressure curve $p_o(T)$ to temperatures $T > T_c$. Alternative treatments based on statistical mechanics (e.g. nonlocal density functional theory [8, 20] or integral equation theories [18]) have been applied more recently to the analysis of high-pressure adsorption of gases in activated carbons in the high-temperature region.

A simple semiempirical equation for correlating high-pressure adsorption data in this high-temperature region can be obtained from a model of hard spheres in contact with an attractive hard wall. For such a model the first three virial coefficients of the virial expansion for $\Gamma$,

$$\Gamma = B\rho + C\rho^2 + D\rho^3 + ... ,$$

162

were calculated rigorously [19] and it was found that this expansion starts approximately as a geometric series with alternating signs, i.e. $q = -C/B \approx -D/C > 0$. It was suggested that the virial expansion may then be approximated over a wider density range (up to $\rho \approx 1/q$) by the sum of an infinite series, i.e.

$$\Gamma \approx B\rho/(1+q\rho) \tag{7}$$

which formally resembles the Langmuir equation. For the system argon/Graphon $1/q$ corresponds to densities and pressures of 2.0 mol dm$^{-3}$ (40 bar) and 3.5 mol dm$^{-3}$ (90 bar) at temperatures of 253K and 323 K, respectively [19]. To correlate the adsorption data over an even wider range of densities and pressures (including the maximum of the surface excess isotherms), eq.(7) may be augmented by a Padé approximation, to give

$$\Gamma = \frac{B\rho - k\rho}{1+q\rho} \approx \Gamma^a - \rho z^a \tag{8}$$

where $\Gamma^a$ is given by the r.h.s. of eq.(7) with $B$ replaced by $B' = (B + k/q)$, and $z^a = k/q$. The physical significance of eq.(8) is seen by inspection of eq.(4): the quantity $\Gamma^a$ in eq.(8) represents the *total* amount of adsorbate per unit surface area in the nominal adsorption space of thickness $z^a$, and the term $\rho z^a$ is the amount in an equivalent volume of the bulk phase. Eq.(8) was found to fit experimental surface excess data for argon/Graphon at pressures up to 150 bar at reduced temperatures from 1.7 to 2.1 (253-323 K) within experimental accuracy. The isotherms for Krypton/MSC presented in Fig.6 can also be fitted almost within experimental accuracy by eq.(8)[19]. The applicability of this simple equation suggests that at high reduced temperatures these systems behave almost like hard-sphere fluids near an attractive hard wall. Thus in this high-temperature region the surface excess isotherms conform to a Langmuir-type model for $\Gamma^a$ and the layer thickness $z^a$ appears to be independent of pressure along a given isotherm. In this case the simple model of an *adsorbed phase* introduced in the Introduction represents a reasonable description of the system. Considerable progress has been made in the past 15 years in the statistical mechanics of fluid adsorption based on first principles [20], and the maximum of the experimental surface excess isotherms can be reproduced by several theoretical treatments [21-23]. An analysis of the high-pressure adsorption of gases in activated carbon based on an explicit consideration of the energetic heterogeneity of the surface has also been presented [24].

## 3. Multilayer adsorption at elevated pressures

Below the critical temperature the pressure range of vapor adsorption isotherms is bounded by the saturation pressure $p_o(T)$ and unlimited multilayer adsorption is expected to occur on sufficiently attractive flat substrates in the temperature range from below the normal boiling temperature $T_b$ to the critical temperature $T_c$ (cf. Fig.4). Figure 7 shows the density profile for a multilayer adsorbed film as calculated by means of the Born-Yvon-Green theory for a Lennard-Jones fluid at a structureless flat wall interacting with the fluid molecules by a 9:3 potential [25]. This and other theoretical studies indicate that the adsorbed film has a density similar to that of the saturated liquid, except for the layer next to the surface where the film may be compressed to a somewhat higher density. In the classical Frenkel-Halsey-Hill (FHH) theory of multilayer adsorption the adsorbed film is treated as a slab of liquid of uniform density $\rho_o^\ell$ , and thus its thickness $\ell$ is related to the measured surface excess by $\Gamma = \ell(\rho_o^\ell - \rho^g)$. Adsorption isotherms for a given system measured at several temperatures below or near $T_b$ commonly tend to coincide in the multilayer region when either the adsorption $\Gamma$ or the film thickness $\ell$ is plotted as a function of the relative pressure $p/p_o$. In this low-temperature region $\Gamma$ and $\ell$ are essentially equivalent, as $\Gamma \approx \rho_o^\ell \ell$ and $\rho_o^\ell$ is a weak function of temperature. At temperatures approaching $T_c$, on the other hand, $\rho_o^\ell$ decreases appreciably with increasing temperature and thus $\ell = \Gamma / (\rho_o^\ell - \rho^g)$ is no longer simply proportional to $\Gamma$. The question then arises which of the two quantities is to be considered in an analysis of adsorption isotherms in this temperature range $(T_b < T < T_c)$. This question has been investigated by studying the multilayer region of the system propane/Graphon in a wide temperature range up to $T/T_c = 0.98$ [28]. It was found that the surface excess $\Gamma$ for given values of the relative fugacity $f/f_o$ decreases in a pronounced manner with increasing temperature. On the other hand, the nominal film thickness $\ell = \Gamma / (\rho_o^\ell - \rho^g)$ derived from the experimental values of $\Gamma$ is nearly independent of temperature at given $f/f_o$ over a wider temperature range, as shown in Figure 8, where $\ell$ is plotted vs. $\ln(f/f_o)$ on a double logarithmic scale. This observation is consistent with the FHH equation which can be written in the form [26-28]

$$\ln(f/f_o) = -\alpha / \ell^n \qquad (9)$$

where $\alpha$ represents the energy density of the gas/solid interaction which is expected to be only a weak function of temperature. For non-retarded van der Waals interactions one expects n = 3[26,27]. From a linear regression of the film thickness isotherms

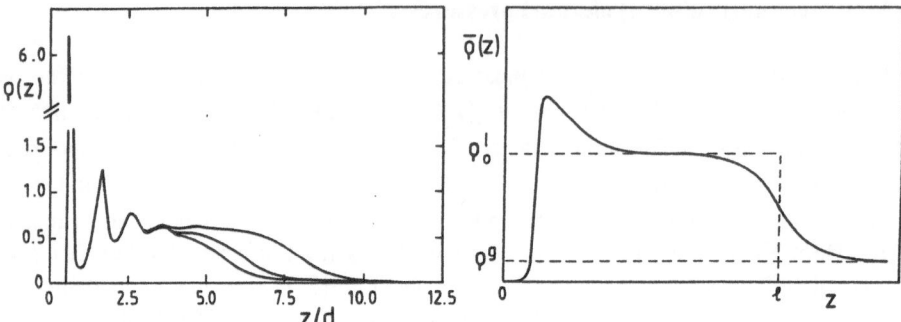

*Figure 7.* Density profiles of multilayer adsorbed films of a Lennard-Jones fluid at a strongly adsorbing surface. Left: Theoretical results for a Lennard-Jones fluid (argon) on graphite at a reduced temperature $T/T_c = 0.74$ [25]: profiles for different gas densities close to the saturation pressure a re shown. Right: Sketch of the density profile for a multilayer adsorbed film at a near-critical temperature where the attractive potential causes a compression of the fluid in the region next to the surface (the layering of the film next tot the surface is omitted in this sketch).

plotted in Figure 8 in a range of fugacities $f/f_o$ from 0.75 to 0.97 (corresponding to a film thickness from 1.0 to 2.5 nm) we find n=2.55±0.3 for 9 isotherm in the temperature range $0.74 \leq T/T_c \leq 0.982$ [28]. This value of the exponent n agrees with results for hydrocarbons and other gases on graphite at lower reduced temperatures. From this agreement it has been concluded that multilayer adsorption data obtained at temperatures approaching $T_c$ can be correlated in terms of eq. (9) with $\ell$ given by $\ell = \Gamma / (\rho_o^\ell - \rho^g)$. Only at temperatures very close to $T_c$ this simple slab model appears to be insufficient, as indicated in Figure 8 by the fact that the film thickness isotherm at the highest experimental temperature ( $T/T_c$ =0.982) falls well above those for $T/T_c \leq 0.93$. This deviation may be explained by a compression of the fluid in the layer next to the surface on strongly attractive substrates like graphite, as shown in the sketched density profile in Fig.7. This effect is expected to become important close to the critical point, where the density $\rho_o^\ell$ decreases but the compressibility of the liquid increases markedly as one approaches $T_c$ from below. Experimental evidence for the existence of such a compressed layer next to the substrate comes from a volumetric study of the boundary layer of liquid propane against graphite, which was performed over a wide temperature range up to nearly the critical point [28,29]. When this compressed layer is taken into account by assuming a two-step density profile instead of the simple slab model, somewhat lower values of the film thickness are obtained and the film thickness isotherms of propane/Graphon based on such a two-step density profile of the multilayer adsorbed film were found to nearly coincide - including the

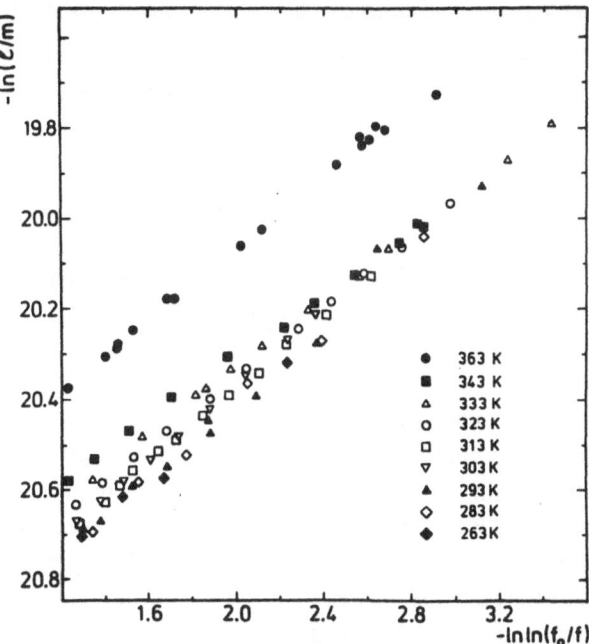

*Figure 8.* Experimental test of the FHH equation for the system propane/Graphon for a temperature range $T / T_c$ from 0.74 to 0.98. The film thickness $\ell$ is derived from the experimental surface excess $\Gamma$ on the basis of a one-step density profile [28].

isotherm at the highest temperature, $T / T_c = 0.98$. The best-fit values of the exponent n are not significantly changed by this modification of the assumed density profile [28]. The main conclusion from this analysis is that the simple FHH equation (eq. 9) remains applicable up to nearly the critical point, but the exponent n becomes significantly smaller than the ``theoretical`` value n=3, even for strongly attractive adsorbents like graphite. Values of n as low as 1.5 are found for weaker adsorbents like silica in this temperature range [30].

## 4.     Pore condensation and phase behavior of fluids in mesopores

Pore condensation is the phenomenon whereby a gas condenses to a liquid-like phase in pores at a pressure $p$ less than the saturation pressure $p_o$ of the bulk liquid. It represents an example of a *shifted bulk transition* under the influence of the attractive fluid-wall interactions [31]. For pores of uniform shape and width (e.g., ideal slit-like or cylindrical mesopores) pore condensation can be treated on the basis of the classical Kelvin equation, which relates the shift of the chemical potential $\mu$ relative to that of the bulk transition, $\Delta\mu = \mu - \mu_o \approx - k_B T \ln (p/p_o)$, to macroscopic properties of the fluid, viz. the densities of coexistent liquid and gas, $\rho^{\ell}$ and $\rho^g$, the surface tension $\gamma$, and the contact angle $\theta$ of the liquid against the wall material. Condensation in cylindrical pores (radius $R$) is expected when [32]

$$\ln(p/p_o) = -2\gamma \cos\theta / k_B T R \, (\rho^{\ell} - \rho^g) \qquad (10)$$

For slit-like pores $R$ is replaced by the pore width $D$. Depending on the relative strength of fluid-wall and fluid-fluid interactions, a multilayer adsorbed film forms at the pore wall prior to pore condensation (complete-wetting case; $\cos\theta = 1$), or pore condensation takes place inspite of weak adsorption (partial wetting case; $0 < \cos\theta < 1$); however, neither adsorption nor pore condensation occurs at $p < p_o$ if the fluid-wall interaction is too weak (non-wetting case; $-1 < \cos\theta < 0$). For systems which exhibit partial wetting at low temperatures one expects that the contact angle decreases with increasing temperature and a transition from incomplete to complete wetting occurs at the so-called *wetting temperature* $T_w$. In the context of adsorption at elevated pressures we are mainly interested in pore condensation phenomena in the complete-wetting regime between $T_w$ and the critical temperature $T_c$.

From the classical Kelvin equation (eq.10) one expects that pore condensation persists up to $T_c$, and that the pore condensation pressure $p$ approaches the respective saturation pressure $p_o$ (i.e., the shift of the chemical potential $\Delta\mu$ approaches zero) for $T \rightarrow T_c$. This follows immediately from the temperature dependence of the surface tension $\gamma$ and of the density increment $\rho^{\ell} - \rho^g$, which sufficiently close to $T_c$ can be represented by power laws [33]

$$\gamma = \gamma_o t^{\mu}; \qquad \rho^{\ell} - \rho^g = B t^{\beta}$$

where $t = (T_c - T)/T_c$ is the distance from the critical temperature in reduced units, $\mu$ and $\beta$ are *critical exponents* which have universal (i.e. system-independent) values ($\mu = 1.26$, $\beta = 0.31$), while $\gamma_o$ and $B$ are non-universal constants. From eq. (10) with $\cos\theta = 1$ and replacing vapor pressures $p$ by fugacities $f$ one thus expects

$$\ln(f/f_o) \propto -t^{\mu-\beta} \rightarrow 0 \qquad \text{for } t \rightarrow 0.$$

However, the classical Kelvin equation neglects adsorption at the pore wall which, as mentioned above, competes with pore condensation in the complete-wetting regime. The presence of such a multilayer adsorbed film at the pore walls leads to a modified

Kelvin equation valid for large pore sizes [31]. For cylindrical pores one obtains instead of eq.(10)

$$\ln(f/f_o) = -2\gamma / k_B T(R-1.5\ell)(\rho^\ell - \rho^g) \qquad (11)$$

where $\ell$ is the film thickness at the pore-condensation point. For wide slit pores one expects that $\ell$ is given by eq.(9), as for adsorption at a single flat wall. As discussed in Section 3, the interaction parameter $\alpha$ in eq.(9) is a weak function of temperature; this implies that $\ell$ is a nearly temperature-independent function of $f/f_o$. In contrast, the ratio $\gamma / (\rho^\ell - \rho^g)$ in the Kelvin equation is a decreasing function of $T$ for $T \to T_c$. Accordingly, in eq. (11) the film thickness $\ell$ at which pore condensation occurs is implicitly a function of temperature, and thus eq. (11) does not allow to calculate the pore condensation line in a straightforward way as in the case of eq.(10). The analogous problem for cylindrical pores has been treated by Saam and Cole [34]. The Saam-Cole theory yields a map of the regions of stability, metastability, and instability of the multilayer adsorbed film at the pore wall in terms of the parameters $\alpha$, $\gamma$, and $\rho^\ell - \rho^g$. Some predictions of this theory have been experimentally confirmed [35]. However, like other classical treatments of pore condensation, this theory does not account for the pecularities of the critical region.

Modern scaling theories of the criticality of fluids in pores attribute the critical point shift to the confinement of the space in the pore. In bulk fluids, the correlation length $\xi$ of density fluctuations can grow without limit as one approaches the critical point: when $T_c$ is approached along bulk coexistence from below $(T < T_c)$, the correlation length in the two coexistent fluid phases grows by a power law $\xi = \xi_o t^{-\nu}$ for $t \to 0$, where $\xi_o$ is a characteristic length (of the order of magnitude of the molecular diameter $\sigma$), and $\upsilon$ is another universal critical exponent ( $\upsilon = 0.63$). For a fluid confined between two parallel walls, $\xi$ can grow unlimited for $t \to 0$ in the two directions parallel to the walls but not perpendicular to the walls, where it cannot grow beyond the wall-to-wall distance $D$. From finite-size scaling arguments one expects pore criticality at a temperature increment $\Delta T_c$ below bulk criticality, such that $\xi(\Delta T_c) = cD$, where $c$ is a factor depending on the strength of solid-fluid interactions. Accordingly, for slit pores of width $D$ [36]

$$\Delta T_c / T_c = (cD / \xi_o)^{-1/\upsilon} \qquad (12)$$

For cylindrical pores a different situation arises, at least from a theoretical point of view. As the correlation length can diverge in only one direction (along the pore axis), one does not expect to observe true pore criticality but instead a rounding of the first-order phase transition caused by finite size effects. Since the broadening of the phase transition is expected to be proportional to $\exp[(-R/\sigma)^2]$, where $R$ ist the pore radius and $\sigma$ the molecular diameter of the fluid, these rounding effects can be suppressed by chosing materials with rather wider pores and narrow pore size distribution [37].

In the past only few experimental studies of pore condensation have been made using well-characterized porous materials, and have been focussed at the phase behavior of the fluid in the pores. Commonly in these studies the pore critical point was estimated by analysing sorption isotherms at temperatures up to the bulk critical temperature. Specifically, the sorption of Xe and $CO_2$ in Vycor glass [38], Ar and Xe in silica gel [39], and cyclopentane in MCM-41 [40] has been analysed in such a way. Below we discuss in some detail how the phase diagram of the pore fluid can be derived from experimental sorption data [41].

## 4.1 SORPTION ISOCHORES AND PORE CONDENSATION LINE

Figure 9 shows a schematic temperature-density diagram with the gas/liquid coexistence curve $T_o(\rho)$ of the bulk fluid and the *pore condensation line* $T_p(\rho)$ near the bulk critical point C (graph at the left). This graph illustrates the case when the pore condensation line terminates in a pore-critical point at a temperature $T_{cp} < T_c$ at a density $\rho_k$ less than the critical density $\rho_c$ of the bulk fluid. Pore condensation occurs whenever the pore condensation line is crossed along either an isothermic or isochoric path. To locate the pore critical point, it is advantageous to take isochoric temperature scans (*sorption isochores*) rather than isotherms, because of the flat top of the coexistence curve in the vicinity of the critical point C. As the temperature is lowered at a density $\rho_i < \rho_k$, the sorption isochore will exhibit a vertical step at the pore condensation temperature $T_p(\rho_i)$, the sorption hysteresis is expected to become more and more pronounced as $T_p(\rho_i)$ moves away from the pore-critical temperature (see Fig.9, graph at the right). Isochores at densities $\rho_i > \rho_k$ will not exhibit a vertical step but a continuous increase of the sorption isochore down to the temperature of bulk gas/liquid coexistence. The condition for pore criticality is thus the disappearance of a step in the sorption isochore, as illustrated in Fig. 9. However, a truly vertical step in the sorption isochores (or sorption isotherms) is expected only for pores of uniform size and shape but not for real porous material with finite pore-size distribution. A less stringent criterion for pore criticality, based on the disappearance of the transition from multilayer adsorption to pore condensation may be adopted [41]. From such an analysis of the sorption isochores, and from the known pore volume, one can also determine the mean densities of the coexisting gas-like and liquid-like states of the pore fluid.

Sorption isotherms [42] and sorption isochores [41] of $SF_6$ in controlled-pore silica glasses of different mean pore sizes have been measured by gravimetric and volumetric techniques. For illustration we summarize some results for controlled-pore glasses with wide mesopores (24 and 31 nm) designated as CPG-240 and CPG-350,

169

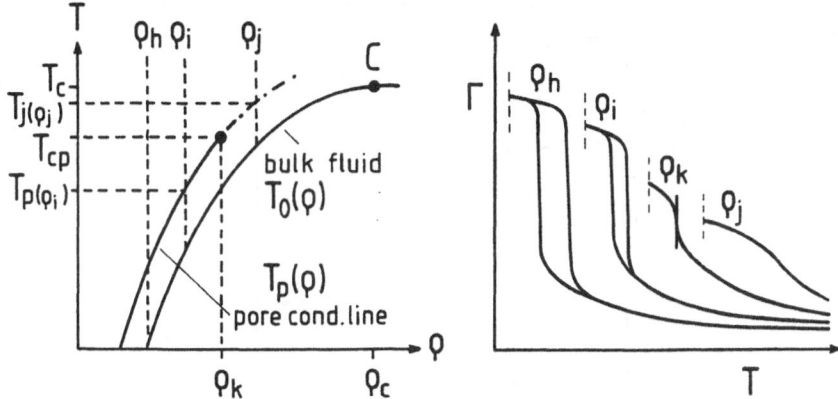

*Figure 9.* Schematic representation of the $T-\rho$ phase diagram of a fluid showing the bulk coexistence curve with the critical point C and the pore condensation line which terminates at the pore critical temperature $T_{cp}$ (left); sorption isochores corresponding to four different bulk densities are sketched in a $\Gamma - T$ diagram (right).

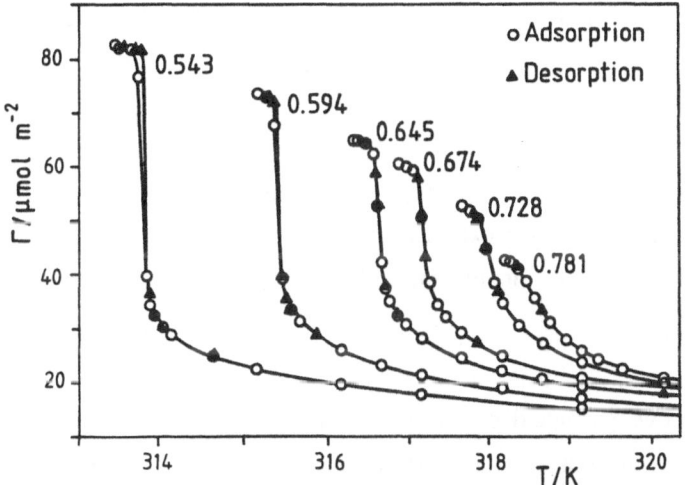

*Figure 10.* Sorption isochores for $SF_6$ in controlled-pore glass CPG-350 at reduced densities $\rho/\rho_c$ from 0.543 to 0.781. Each isochore ends at the respective bulk coexistence temperature $T_o(\rho)$ (from [41]).

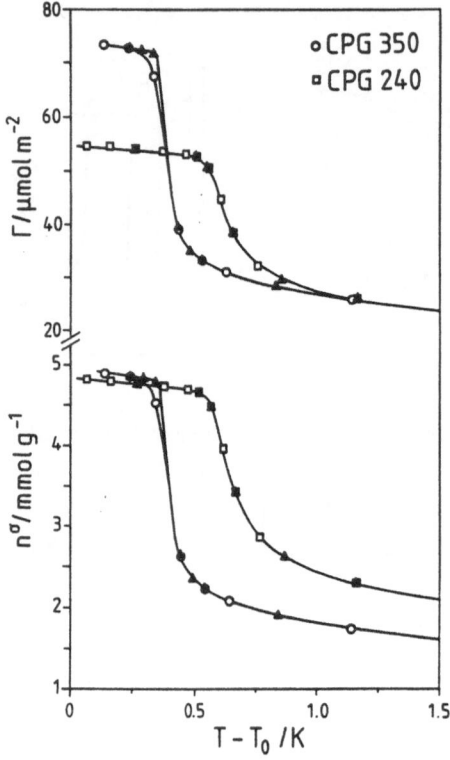

*Figure 11.* Sorption isochores of $SF_6$ in CPG-240 and CPG-350 at $\rho/\rho_c \approx 0.60$. Results expressed as surface excess amount per unit area $\Gamma$ (upper graph) and as specific excess amount $n^\sigma$ per unit mass of adsorbent (lower graph) (from 41).

respectively [41]. Sorption isochores of $SF_6$ in these materials have been measured for reduced densities $\rho/\rho_c$ between 0.5 and 0.8 in a temperature range from 333 K down to the respective gas-liquid coexistence temperature $T_o(\rho)$. Figure 10 shows the sorption excess amount per unit surface area, $\Gamma = n^\sigma / m_s a_s$, for $SF_6$ in CPG-350 plotted as a function of temperature, from the respective coexistence temperature $T_o$ up to ca. 320 K. The experimental isochores in this figure are consistent with the behavior sketched in Figure 9: The isochores at relatively low bulk densities ($\rho/\rho_c < 0.6$)

exhibit an almost vertical section in a temperature range ca. 0.5 K above $T_o$ due to pore condensation, and a narrow hysteresis loop is detected at these densities. The isochore at the highest experimental density $(\rho / \rho_c = 0.78)$ exhibits a smooth and fully reversible increase of $\Gamma$ for $T \to T_o$, indicating that the fluid no longer undergoes pore condensation.

Figure 11 shows a comparison of the isochores of $SF_6$ in CPG-240 and CPG-350 at nearly the same bulk density of the fluid $(\rho/\rho_c = 0.59)$. As to be expected from the Kelvin equation, pore condensation in CPG-350 occurs closer to the bulk coexistence temperature $T_o$ than in CPG-240. Hysteresis is still detectable in the former but not in the latter material, indicating that the fluid in CPG-240 is closer to pore criticality than in the wider pores of CPG-350. At temperatures below the pore condensation temperature $T_p$, when the pores are filled with liquid, the surface excess amount per unit mass of CPG, $n^\sigma / m_s$, reaches similar plateau values in the two materials, as to be expected from the similar specific pore volumes $v_p$ of the two glasses. On the other hand, at temperatures well above $T_p$, where the fluid forms an adsorbed film at the pore wall, the surface excess amount per unit surface area, $\Gamma = n^\sigma / m_s a_s$, becomes equal in the two materials. These regularities of the low-temperature and high-temperature sorption behavior in the two glasses are found at all densities at which pore condensation occurs.

The pore condensation temperature $T_p$ may be determined from the isochores by locating the point at which systematic positive deviations from the high-temperature adsorption behavior occur. In the spirit of the Frenkel-Halsey-Hill model of multilayer adsorption, one expects that close to gas/liquid coexistence the thickness $\ell$ of the adsorbed film and hence the surface excess $\Gamma$ should follow a power law in $T - T_o$, viz

$$\Gamma \propto (T - T_o)^{-x} \tag{13}$$

Figure 12 shows experimental isochores for $SF_6$ in CPG-350 plotted as $\Gamma$ vs $(T - T_o)/T_o$ on a logarithmic scale. The solid lines represent a fit of eq.(13) to the data in the high-temperature region. Linear regression of the data on the individual isochores yields values of the exponent $x$ of $0.37 \pm 0.03$ for CPG-240, and $0.35 \pm 0.02$ for CPG-350. These values are close to the $x = 1/3$ dependence expected on the basis of the FHH treatment. The pronounced deviations from the power law at temperature increments $(T - T_o)/T_o < 2.10^{-3}$ (see Figure 12) are attributed to the onset of pore condensation (positive deviations) or to beginning saturation at densities above the pore critical point (negative deviations). As to be seen in the figure, pore condensation has vanished at $\rho / \rho_c = 0.781$. The hysteresis loop vanishes (or becomes too narrow to be detected

172

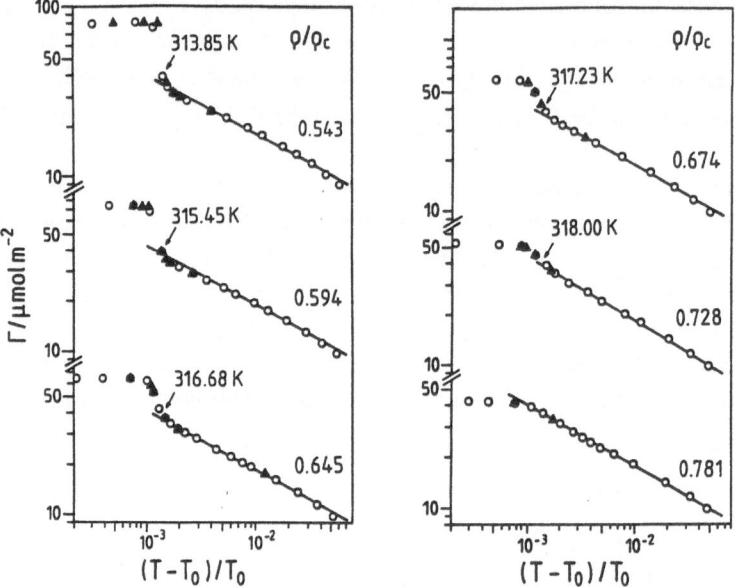

*Figure 12.* Analysis of the sorption isochores of Fig.10 for locating the pore-critical point: The solid lines represent a fit of eq. (13) to the data. The arrows indicate the pore condensation temperature $T_p$ at the respective density of the gas.

experimentally) at even lower densities, at which pore condensation is still seen clearly.

Figure 13 shows the pore condensation line for $SF_6$ in the two CPG materials together with the bulk coexistence curve. For both materials the pore condensation line ends in a pore-critical point at a temperature $T_{cp}$ below the critical temperature $T_c$ of the bulk fluid. The pore-critical temperature can be estimated from the sorption isochores as follows (cf. Figure 9): Let $\rho_i$ be the highest experimental density at which pore condensation was observed, and $\rho_j$ the lowest experimental density at which no pore condensation is exhibited; in this case the pore condensation line must terminate at a density $\rho_k$ somewhere between $\rho_i$ and $\rho_j$ The pore critical temperature $T_{cp}$ is then located between the pore condensation temperature $T_p(\rho_i)$ of the isochore at $\rho_i$ and the temperature $T_j(\rho_j)$ at which the extrapolation of the pore condensation line intersects the isochore at $\rho_j$. For $SF_6$ in CPG-240 we obtain $T_p(0.678) = 317.55 \pm 0.05$ K and $T_j(0.726) = 318.02$ K and thus $T_{cp} = 317.78 \pm 0.23$ K; hence in CPG-240 the critical temperature of $SF_6$ is shifted from the bulk value ( $T_c = 318.70 \pm 0.01$ K) by an increment $\Delta T_c \equiv T_c - T_{cp} = 0.92 \pm 0.24$ K. Similarly, for $SF_6$ in CPG-350 we find

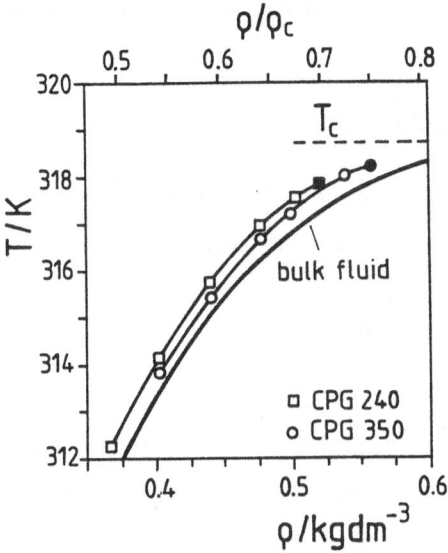

*Figure 13.* Pore-condensation lines of $SF_6$ in CPG-240 and CPG-350 in the weakly undersaturated region close to bulk coexistence. The pore condensation lines terminate at a temperature $T_{cp}$ below $T_c$ (from [41]).

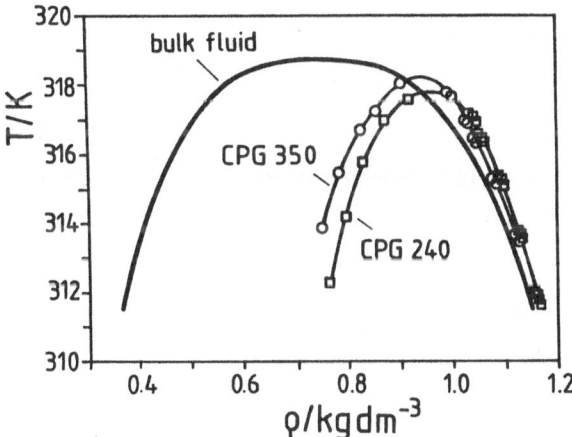

*Figure 14.* $T - \rho$ phase diagram of $SF_6$ in bulk and as a pore fluid in CPG-240 and CPG-350, where the mean density of gas-like and liquid-like states as derived from the sorption isochores is shown (from [41]).

$\Delta T_{cp} = 0.48 \pm 0.23$ K. The quoted uncertainty in $\Delta T_{cp}$ is determined by the gap between the experimental densities $\rho_i$ and $\rho_j$ (see Fig. 9) [41].

## 4.2. PHASE DIAGRAM OF THE PORE FLUID

The mean densities $\rho_p$ of the liquid-like and gas-like states of the pore fluid are obtained from the surface excess amount $n^\sigma$ by the relation

$$\rho_p = n_p / v_p = n^\sigma / (v_p m_s) + \rho^g \tag{14}$$

where $n_p$ is the amount of fluid in the pore volume $v_p$ of unit mass of CPG, and $\rho^g$ is the bulk density of the chosen isochore. The mean density of the saturated pore gas, $\rho_p^g$ is derived from the surface excess amount at the pore condensation temperature $T_p(\rho)$, i.e. the value of $n^\sigma$ at which the experimental data begin to deviate from the linear region in the logarithmic plots of Figure 12. Values of the mean density of the coexisting pore liquid, $\rho_p^l$, are derived from the surface excess amounts in the "plateau" region of the isochore near the coexistence temperature $T_o(\rho)$. Inspection of Figures 10 and 11 shows that the surface excess isochores do not reach a constant plateau below $T_p$ but continue to increase weakly as $T$ approaches $T_o$, because the density of the pore liquid is increasing as $T$ decreases. A $\rho - T$ phase diagram exhibiting the mean densities of gas-like and liquid-like SF$_6$ in CPG-240 and CPG-350 is shown in Figure 14 together with the coexistence curve of the bulk fluid. The points on the gas-like branch of the pore fluid represent $\rho_p^g$ values at $T_p$, the groups of points on the liquid-like branch represent values at temperatures close to $T_o$ of the respective isochores at $\rho < \rho_k$.

## 4.3 CRITICAL POINT SHIFT

The analysis of the sorption isochores presented in Section 4.1 yields a quantitative determination of the pore condensation line of a fluid near its critical region in materials with a narrow pore size distribution. The analysis shows that the pore condensation line does not extend up to the bulk critical temperature, as to be expected from the Kelvin equation or the Saam-Cole theory, but terminates in a point distinctly below $T_c$. This end-point of the pore condensation line is called the *pore critical point*, although its true nature is not clear in the present case of an *interconnected network* of *cylindrical* pores.

In a strict thermodynamic sense, only in ideal slit pores (i.e., a fluid between parallel walls of infinite extension) pore condensation represents a first-order phase transition and only in this special case true pore criticality will occur. The pore coexistence curve $\Delta \rho_p(T)$ is expected to have a critical exponent $\beta = 1/8$ (i.e., the value of the 2D Ising model) in ideal slit pores [33]. Narrow cylindrical pores represent quasi-one-dimensional systems, as the correlation length can diverge in only one dimension of space (along the cylinder axis). Accordingly pore condensation in an ideal cylindrical pore is not first order but represents a continuous phase transition. In a system of interconnected cylindrical pores with large overall porosity, as for CPG materials, however, the pore system will behave essentially three dimensional, especially in the vicinity of the critical point (when the correlation length exceeds the nominal pore diameter). This argument is supported by studies by Chan et al. [43] who interpreted the superfluid transition of $^4$He in Vycor glass as three-dimensional-like.

The results presented above are consistent with the expectation that the shift of the critical temperature $\Delta T_c$ increases with decreasing pore diameter. Theoretical studies suggest that $\Delta T_c$ will vary with the pore radius $R = D/2$ as a power law

$$\Delta T_c \propto R^{-y} . \tag{15}$$

Receently, the study of pore-criticality of $SF_6$ in Controlled-Pore-Glass has been extended to a material with smaller pore size (ca. 7.5 nm pore diameter). In this case adsorption isotherms (instead of isochores) were analysed in a similar way as outlined in Section 5.1. From this analysis we obtain y = 2.1 for $SF_6$ in CPG materials with pore diameters from 7.5 nm to 31 nm [42].

Theoretical studies [37] and results of computer simultations [44-46] indicate that the critical density of the (inhomogeneous) pore fluid and the width of the pore coexistence curve also depend strongly on pore size and on the strength of the attractive fluid-solid potential. For $SF_6$ in CPG, the entire pore coexistence curve is shifted to higher mean densities relative to the coexistence curve of the bulk fluid. At the lowest experimental temperatures, the densities of the liquid-like states of $SF_6$ in CPG-350 and CPG-240 become equal; they appear to be somewhat higher than the density of the bulk liquid, but the deviations from the latter are not significant in view of the uncertainty in the specific pore volumes $\upsilon_p$. Note that the values of $\upsilon_p$ used in this analysis are based on the conjecture that at temperatures far below $T_c$ the density of the pore liquid becomes equal to that of the bulk liquid (see Section 4.2). Close to the pore-critical temperature, the mean density of the pore liquid is distinctly greater than the corresponding density $\rho_0^\ell$. This can be explained by the fact that in the neighborhood of the pore walls the liquid will be compressed in the attractive potential of the substrate to a density higher than in the core of the pore (which is believed to be nearly equal to the bulk liquid density). The thickness of this compressed boundary layer is expected to increase proportional to the correlation length $\xi$ of the bulk fluid on approaching the pore-critical temperature. This effect explains the observed enhancement of the mean densities of the liquid-like states of the pore fluid near the pore-critical temperature (Figure 14). The mean densities of the gas-like states of $SF_6$ in CPG-240 are higher than in CPG-350, and both are much higher than the density of the bulk gas at the given temperature. This is due to the fact that the pore gas represents a highly inhomogeneous fluid, consisting of an adsorbed film with nearly liquid-like density at the pore wall and a core in which the density is approximately equal to that of the bulk gas.

The results for the pore coexistence curve of $SF_6$ in CPG materials are in qualitative agreement with earlier results for $CO_2$ in Vycor glass reported by Burgess et al. [38a] who obtained the pore coexistence curve over a wider temperature range ($50K < T_c - T < 100$ K) but did not study the near-critical range of the pore fluid. From an analysis of the hysteresis loops of the experimental sorptions isotherms, they obtained a coexistence curve of the pore fluid shifted to lower temperatures ($\Delta T_c \approx 39$ K) and higher mean densities relative to the bulk coexistence curve. As for the present systems, the mean density of the gas-like states of the pore fluid was much higher than that of the bulk gas, while the mean density of the liquid-like states of the pore fluid agreed with the bulk liquid density over almost the entire experimental range.

Phase diagrams of pore fluids have also been determined by light scattering and thermodynamic methods. Wong and Chan [47] measured the pore coexistence curve of $^4$He in aerogel in a narrow temperature range close to the critical temperature ($T_c - T \leq 0.06$ K) by heat capacity and vapor pressure measurements. They found a narrow coexistence curve of the pore fluid located entirely in the two-phase region of the bulk fluid. The shift of the critical temperature was found to be very small ($\Delta T_c = 31$ mK), and the pore critical density was higher than the critical density of the bulk fluid. Wong et al. [48] studied the phase behavior of $N_2$ in the same aerogel material in a temperature range $T_c - T \leq 2$ K using light scattering. A qualitatively similar but wider pore coexistence curve than for $^4$He was found, shifted to somewhat lower temperatures ($\Delta T_c = 0.84$ K). Except for the lower mean densities of the pore liquid these results [47,48] are also in qualitative agreement with the findings for $SF_6$ in CPG. However, the interpretation of the aerosol results is not straightforward, because aerogel consists of a network of pores with a wide pore-size distribution, whereas for CPG a well-defined mean pore size can be assigned to each material.

In conclusion the phase behavior of $SF_6$ in controlled pore glass shows that the criticality of pure fluids in mesopores is affected by confinement, in accordance with recent theoretical work and computer simulations studies. For the first time it has been possible to locate the pore condensation line and the pore critical point in materials with a narrow pore-size distribution and it is shown that the shift of the pore-critical temperature $\Delta T_c$ is correlated with the mean pore radius $R$. The results suggest a power law $\Delta T_c \propto R^{-y}$ with $y \approx 2$, but measurements on further CPG materials are required to check this high value of the exponent $y$. It is also desirable to study fluids with stronger fluid-wall interactions than $SF_6$ in the same CPG materials, in order to test theoretical predictions on the influence of the strength of fluid-wall interactions on the criticality of fluids in mesopores.

Acknowledgement. The authors wish to acknowledge helpful discussions with S. Dietrich, R. Evans, D. Everett and K. Gubbins. This work was supported by a grant from the Deutsche Agentur für Raumfahrtangelegenheiten (DARA).

## 5. References

1.  Gibbs, J.W. (1957) *The Collected Works*, Vol. 1, Yale University Press, New Haven, p. 219-331.
2.  Menon, P.G. (1968) *Chem. Rev.* **68**, 277; (1969) *Advances in High Pressure Pressure Res.* (R.S. Bradley, Ed.), Vol.3, p. 313, Academic Press, New York.
3.  Wakasugi, Y., Ozawa, S., and Ogino, Y. (1981) *J. Colloid Interf. Sci.* **79**, 399-409.
4.  Malbrunot, P., Vidal, D., and Vermesse, J. (1992) *Langmuir* **8**, 577-580; Vermesse, J., Vidal, D., and Malbrunot, P. (1996) *Langmuir* **12**, 4190-4196.
5.  Specovius, J., and Findenegg, G.H. (1978) *Ber. Bunsenges. Phys.Chem.* **82**, 174-180; (1980) *Ber. Bunsenges. Phys. Chem.* **84**, 690-696.
6.  Agarval, R.K., and Schwarz, J.A. (1988) *Carbon* **26**, 873-887.
7.  Staudt, R., Saller, G., Tomalla M., and Keller, J.U. (1993) *Ber. Bunsenges. Phys. Chem.* **97**, 98-105.
8.  Jiang, S., Zollweg, J.A., and Gubbins, K.E. (1994) *J. Phys. Chem.* **98**, 5709-5713.
9.  Everett, D.H. (1972) IUPAC Definitions, Terminology and Symbols in Colloid and Surface Chemistry, Part I, *Pure and Appl. Chem.* **31**, 579-638.
10. Sing, K.S.W., Everett, D.H., Haul, R.A.W., Moscou, L., Pierotti, R.A., Rouquerol, J., and Siemieniewska, T. (1985) *Pure Appl. Chem.* **57**, 603-619.
11. Malbrunot, P., Vidal, D., Vermesse, J., Bose, T.K., and Chahine, R. (1996) submitted to *Langmuir* (cited in ref. 4b).
12. Kuwabara, H., Suzuki, T., and Kaneko, K. (1991) *J. Chem. Soc. Faraday Trans.* **87**, 1915.-1916.
13. Kaneko, K., Setoyama, N., Suzuki, T., and Kuwabara, H. (1993) in: *Fundamentals of Adsorption*, Vol. 4 (M. Suzuki, Ed.), Kodansha, Tokyo, p.315-322.
14. Findenegg, G.H. (1984) in: *Fundamentals of Adsorption*, Vol.1 (A.L. Myers, G. Belfort, Eds.), American Institute of Chemical Engineers, New York, p.207-218.
15. Blümel, S., Köster, F., and Findenegg, G.H. (1982) *J. Chem. Soc. Faraday Trans.2* **78**, 1753-1764.
16. Findenegg, G.H., Körner, B., Fischer, J., and Bohn, M. (1983) *Ger. Chem. Eng.* **6**, 80-84.

17. Kaneko, K., Shimizu, K., and Suzuki, T. (1992) *J. Chem. Phys.* **97**, 8705-8711; Kaneko, K., Murata, K., Shimizu, K., Camara, S. and Suzuki, T. (1993) *Langmuir* **9**, 1165-1167.

18. Fischer, J., Bohn, M., Körner, B., and Findenegg, G.H. (1982) *Chem.-Ing.-Techn.* **54**, 763-764; (1983) *Ger. Chem. Eng.* **6**, 84-91.

19. Fischer, J., Specovius, J., and Findenegg, G.H. (1978) *Chem.-Ing. Techn.* **50**, 41-42.

20. Ted Davis, H. (1996) *Statististical Mechanics of Phases, Interfaces, and Thin Films*, Chap.12, VCH Publishers, New York.

21. Sokolowski, S. (1982) *J. Chem. Soc. Faraday Trans.2* **78**, 255-264.

22. Rangarajan, B., Lira, C.T., and Subramanian, R. (1995) *AICHE Journal* **41**, 838-845; Subramanian, R., Pyada, H., and Lira, C.T. (1995) *I&EC Res.* (Preprint).

23. Aranovich, G.L., and Donohue, M.D. (1996) *J. Colloid Interface Sci.* **180**, 537-541.

24. Jagiello, J., Sanghani, P., Bandosz, T.J., and Schwarz, J.A. (1992) *Carbon* **30**, 507-512.

25. Wendland, M., Salzmann, S., Heinbuch, U., and Fischer, J. (1989) *Molec. Phys.* **67**, 161.

26. Steele, W.A. (1974) *The Interaction of Gases with Solid Surfaces*, Chap.5, Pergamon, Oxford.

27. Steele, W.A. (1980) *J. Colloid Interface Sci.* **75**, 13.

28. Findenegg, G.H., and Löring, R. (1984) *J. Chem. Phys.* **81**, 3270-3276.

29. Löring, R., and Findenegg, G.H. (1981) *J. Colloid Interface Sci.* **84**, 355-361.

30. Lawnik, W.H., Goepel, U.D., and Findenegg, G.H. (1995) *Langmuir* **11**, 3075-3082.

31. Evans, R. (1990) *J. Phys.: Condens. Matter* **2**, 8989-9007.

32. Gregg, S.J., and Sing, K.S.W. (1982) *Adsorption, Surface Area and Porosity,* Chap.3, Academic Press, New York

33. Rowlinson, J.S., and Widom, B. (1982) *Molecular Theory of Capillarity*, Chap.9, Oxford University Press, Oxford.

34. Saam, W.F., and Cole, M.W. (1975) *Phys. Rev. B.* **11**, 1086-1105; Cheng, E., and Cole, M.W. (1989) *Langmuir* **5**, 616-625.

35. de Keizer, A., Michalski, T., and Findenegg, G.H. (1991) *Pure Appl. Chem.* **63**, 1495-1502.

36. Fisher, M.E., and Nakanishi, H. (1981) *J. Chem. Phys.* **75**, 5857-5863; Nakanishi, H., and Fisher, M.E. (1983) *J. Chem. Phys.* **78**, 3279-3293.

37. Evans, R. (1990) *Liquids at Interfaces* (Les Houches Session XLVIII, J. Charvolin, J. Joanny, and J. Zinn-Justin, Eds.), Elsevier, Amsterdam

38. Burgess, C.G.V., Everett, D.H., and Nuttall, S. (1989) *Pure Appl. Chem.* **61**, 1845; (1990) *Langmuir* **6**, 1734-1738.

39. Machin, W.D. (1994) *Langmuir* **10**, 1235-1240.

40. Rathousky, J., Zukal, A., Franke, O., and Schulz-Ekloff, G. (1995) *J. Chem. Soc. Faraday Trans.* **91**, 937.

41. Thommes, M., and Findenegg, G.H. (1994) *Langmuir* **10**, 4270-4276.
42. Michalski, T., Thommes, M., and Findenegg, G.H. (1997) *Langmuir*, in print.
43  Chan, M.H.W., Blum, K.I., Murphy, S.Q., Wong, G. K.S., and Reppy, J.D. (1988) *Phys. Rev. Lett.* **61**, 1950-1953
44. Kozak, E., and Sokolowski, S. (1991) *J. Chem. Soc. Faraday.Trans.* **87**, 3415-3422.
45. Balbuena, P.B., and Gubbins, K.E. (1993) *Langmuir* **9**, 1801-1814.
46. Gubbins, K.E., Sliwinska-Bartkowiak, M., and Suh, S.H. (1996) *Molecular Simulation* **17**, 333-367.
47. Wong, A.P.Y., and Chan, M.H.W. (1990) *Phys. Rev. Lett.* **65**, 2567-2580.
48. Wong, A.P.Y., Kim, S.B., Goldburg, W.I., and Chan, M.H.W. (1993) *Phys. Rev. Lett.* **70**, 954-957.

# FUNDAMENTALS OF SINGLE–GAS AND MIXED–GAS ADSORPTION ON HETEROGENEOUS SOLID SURFACES.

W. RUDZIŃSKI
*Department of Theoretical Chemistry*
*Maria Curie–Skłodowska University*
*Lublin 20–031, POLAND*

## Introduction

It is now generally realized that the actual (really existing) solid surfaces are energetically heterogeneous to a greater or lesser extent[1,2]. In terms of localized adsorption, it means the variation of adsorption energy when going from one site to another across the surface.

The generally accepted quantitative measure of the energetic heterogeneity of the actual solid surfaces is the differential distribution of a number of adsorption sites among the corresponding values of adsorption energy. This function, used usually in its form normalized to unity is commonly called — "the adsorption energy distribution". That quantitative information is sufficient for a complete thermodynamic description of adsorption equilibria in the systems where one adsorbed molecule occupies one adsorption site and no interactions exist between adsorbed molecules (Langmuir model).

However, in the systems where the interactions between adsorbed molecules cannot be ignored, or when one adsorbed molecule occupies more than one site, another important physical factor comes into play. This is the way in which adsorption sites characterized by different adsorption energies are distributed on a heterogeneous solid surface. In other words, this is the topography of a heterogeneous solid surface. Yang and coworkers[3–5], and Zgrablich and coworkers[6–10] have shown that even in the simplest case of Langmuirian adsorption the surface topography affects strongly the surface diffusion of adsorbed molecules.

So far, two extreme models of surface topography have, almost exclusively, been considered in theoretical works on adsorption. They are shown schematically in Figures 1 and 2.

*J. Fraissard (ed.), Physical Adsorption: Experiment, Theory and Applications,* 181–240.
© 1997 *Kluwer Academic Publishers.*

Figure 1. A schematic representation of random topography of an energetically heterogeneous surface composed of two kinds of adsorption sites.

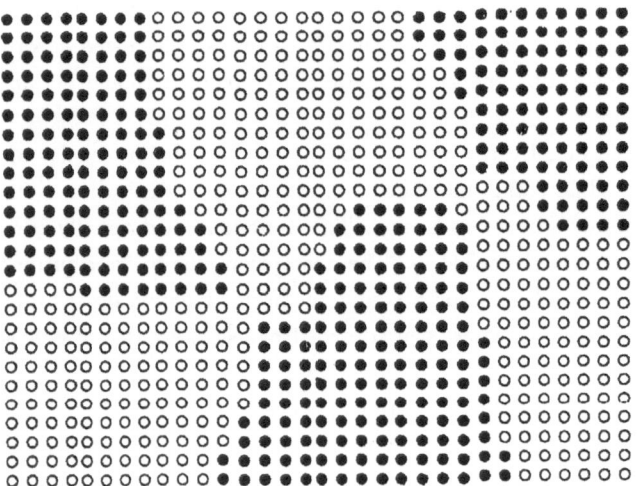

Figure 2. A schematic representation of patchwise topography of an energetically heterogeneous surface composed of two kinds of adsorption sites.

The first one was the "random" model, introduced in literature by Hill[11]. It assumes that adsorption sites characterized by different adsorption energies are distributed over a solid surface completely at random. In other words, no spatial correlations exist between adsorption sites having the same energy of adsorption. A schematic picture of that kind of surface topography is shown in Figure 1.

The other extreme model of surface topography was the "patchwise" model, reported in literature by Ross and Olivier[12]. It assumes, that adsorption sites having the same adsorption energy are grouped on a heterogeneous surface into "patches". These patches are large enough so that the states of an adsorption system in which two interacting molecules are adsorbed on different patches, could be neglected. Thus, the adsorption system can be considered as a collection of independent subsystems, being only in a material and thermal contact. On the contrary, adsorption systems exhibiting random surface topography should be considered as a thermodynamic entity. A schematic picture of a patchwise topography is shown in Figure 2.

The theoretical description of one–site–occupancy adsorption equilibria for both patchwise and random topography was elaborated in the works by Steele[13], Pierotti and Thomas[14], Rudziński et al.[15-18], and by Zgrablich and co–workers[19-20].

Figure 3. A schematic representation of mediate surface topography when the heterogeneous surface is composed of two kinds of adsorpion sites.

Of course, it goes without saying, that the actual solid surfaces will exhibit a surface topography which is "mediate" between patchwise and random. That kind of surface topography is shown schematically in Figure 3, and is sometimes also called — the "partially correlated" surface.

Theoretical solutions for adsorption on surfaces characterized by the "mediate" surface topography have been developed by Tovbin[21,22] and by Rippa and Zgrablich[23]. Every theoretical description of the systems characterized by "mediate" surface topography will be based on a concept of a quantitative measure of the degree of spatial correlations between adsorption sites characterized by the same adsorption energy.

## 1. Principles of Adsorption–Desorption Kinetics: The Traditional Approach.

In the theories of adsorption-desorption kinetics, the mass balance of the adsorbate over the entire heterogeneous solid surface is usually written in the following form,

$$R_d = -M \sum_i X_i \frac{\partial \theta_i}{\partial t} \tag{1.1}$$

where $M$ is the total number of sites on the solid surface; $\theta_i$ means the fractional coverage of sites of $i$-th type; $X_i$ is the fraction of these sites on the solid surface and $t$ is the time. Most frequently, $\frac{\partial \theta}{\partial t}$ is taken to be the expression offered by the Theory of Activated Adsorption–Desorption, (TAAD)

$$\frac{\partial \theta}{\partial t} = K_a p (1 - \theta)^s \exp \left\{ \frac{-\epsilon_a}{kT} \right\} - K_d \theta^s \exp \left\{ \frac{-\epsilon_d}{kT} \right\} \tag{1.2}$$

where $s$ is the number of adsorption sites involved in an elementary adsorption-desorption process, $p$ is the pressure in the gas phase, $\epsilon_a$ and $\epsilon_d$ are the activation energies for adsorption and desorption respectively, $K_a$, $K_d$ are slightly temperature dependent parameters. Further, $T$ and $k$ are the absolute temperature and the Boltzmann constant respectively.

The TAAD approach was developed by applying Absolute Rate Theory which is based on a particular model of reactions. Before being adsorbed or desorbed, molecules must pass through appropriate "activation states" for adsorption or desorption. The activation energy for adsorption, $\epsilon_a$, is defined as the energy difference between a free, gas molecule and the molecule in the "activated state", preceding adsorption. Then, the activation energy for desorption, $\epsilon_d$, is defined as the difference in energy between the adsorbed molecules and those in the activated state preceding desorption. Clark[24] has clearly described the development of the Absolute Rate Theory expressions for the adsorption and desorption rates.

The attempt to apply eq. (1.2) to describe either kinetics of adsorption or desorption usually led to large discrepancies between theory and experiment. One class of improvements was empirically based. Coverage dependence was simply introduced into the activation energies, and empirical formulae were proposed for the coverage dependence of the preexponential factors. Kisliuk[25] used the idea of the "precursor state". They assumed that adsorption was a two step process in which the molecules first formed a weakly bond precursor phase before proceeding to the adsorbed phase. The "precursor state" approaches did not take into account energetic heterogeneity of the actual solid surfaces. So far that important physical factor was taken into consideration by the scientists applying the TAAD approach.

Now, let us consider for simplicity physisorption and the case $s = 1$. So, at the equilibrium when $\frac{\partial \theta}{\partial t} = 0$, eq.(1.2) yields the Langmuir isotherm equation

$$\theta^{(e)}(p, T) = \frac{Kp \exp\left\{\frac{\epsilon}{kT}\right\}}{1 + Kp \exp\left\{\frac{\epsilon}{kT}\right\}} \tag{1.3}$$

where $K = \frac{K_a}{K_d}$, $\epsilon = (\epsilon_d - \epsilon_a)$, and where the superscript $(e)$ refers to the equilibrium. At the equilibrium, the assumption of a discrete distribution of the fraction $X$ of adsorption sites among corresponding values of $\epsilon$, expressed in eq.(1.1), leads to the following expression,

$$\theta^{(e)}(p, T) = \sum_i X_i \frac{K_i p \exp\left\{\frac{\epsilon_i}{kT}\right\}}{1 + K_i p \exp\left\{\frac{\epsilon_i}{kT}\right\}} \tag{1.4}$$

where now $\theta_t^{(e)}$ means the "total", (average) fractional occupancy of all adsorption sites.

In the case of the actual (real) solid surfaces one usually deals with a dense spectrum of adsorption energies which should be represented rather by a continuous function $\chi(\epsilon)$, so that,

$$\theta_t^{(e)}(p, T) = \int_\Omega \frac{Kp \exp\left\{\frac{\epsilon}{kT}\right\}}{1 + Kp \exp\left\{\frac{\epsilon}{kT}\right\}} \chi(\epsilon) d\epsilon \tag{1.5}$$

where $\chi(\epsilon)$ fulfils the normalization condition

$$\int_\Omega \chi(\epsilon) d\epsilon = 1 \tag{1.6}$$

and $\Omega$ is the physical domain of $\epsilon$. For the mathematical convenience, $\Omega$ is frequently assumed to be the interval $(-\infty, +\infty)$, or $(0, +\infty)$. It was shown that replacing the true physical domain $\Omega \in (\epsilon_l, \epsilon_m)$, ($\epsilon_l$ and $\epsilon_m$ meaning the lowest and the maximum values of $\epsilon$ for a heterogeneous solid surface), by $(-\infty, +\infty)$ or by $(0, +\infty)$ does not affect much the behaviour of the calculated isotherm $\theta_t^{(e)}(p, T)$, provided that extremely low or high surface coverages are not considered[1].

## 2. Integral Representation for Equilibrium Adsorption Isotherms.

In this and next sections we will deal with adsorption equilibria, so, meanwhile we drop the superscript $(e)$ in appropriate expressions.

Let $N_t$ denote the total number of the molecules adsorbed at the pressure $p$ and the temperature $T$ on a heterogeneous solid surface, and $M$ denote the total number of sites on that surface. Let further $\theta(\epsilon, p, T)$ denote, the "local" fractional coverage of sites the adsorption energy of which is equal to $\epsilon$. The experimentally monitored $N_t$ value is then expressed by the following integral equation,

$$\theta_t(p, T) = \frac{N_t}{M} = \int_{\epsilon_l}^{\epsilon_m} \theta(\epsilon, p, T)\chi(\epsilon)d\epsilon \tag{2.1}$$

That equation is used either to calculate $\theta_t(p, T)$ when $\theta_l(\epsilon, p, T)$ and $\chi(\epsilon)$ are known, or, to calculate $\chi(\epsilon)$ when $\theta_t(p, T)$ and $\theta_l(\epsilon, p, T)$ are known. The "local" adsorption isotherm may be Langmuir isotherm like in eq. (1.5), or another one (BET, Bragg–Williams, Hill–de Boer, etc.). The actual (really existing) adsorption energy distributions $\chi(\epsilon)$ are expected to have a pretty complicated form, with a number of local maxima and minima. Figure 4 shows an example of the function $\chi(\epsilon)$ calculated by using an advanced numerical method.

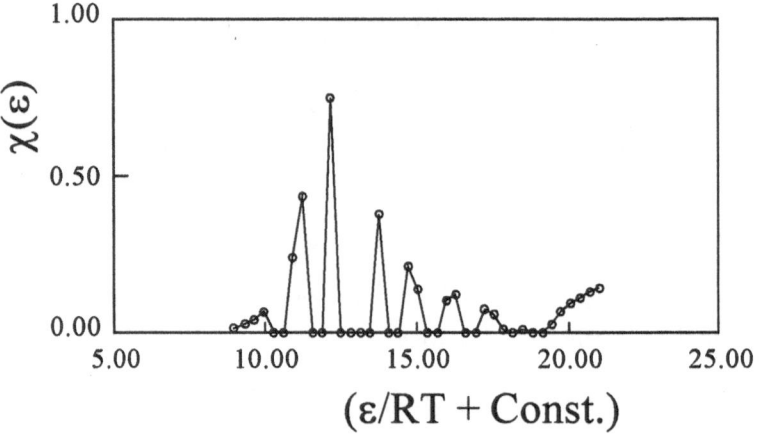

Figure 4. The adsorption energy distribution of $N_2$ adsorbed on carbon black 'Spheron', calculated by Koopal and Vos by using their CAESAR numerical method. (After Koopal and Vos[26].

However, to a certain degree of accuracy, the really existing function can, for practical purposes, be approximated by some "smoothed" functions, the shape of which is described by a small number of parameters.

The following functions have, most frequently, been used to represent the "smoothed" form of the actual adsorption energy distributions.

1. The rectangular function,

$$\chi(\epsilon) = \begin{cases} \frac{1}{\epsilon_m - \epsilon}, & for \ \epsilon \in (\epsilon_l, \epsilon_m); \\ 0, & elsewhere \end{cases} \tag{2.2}$$

**2.** the Gaussian–like function,

$$\chi(\epsilon) = \frac{\frac{1}{c} \exp\left\{\frac{\epsilon - \epsilon_o}{c}\right\}}{\left[1 + \exp\left\{\frac{\epsilon - \epsilon_o}{c}\right\}\right]^2} \tag{2.3}$$

centered at $\epsilon = \epsilon^o$, the variance of which is given by $\frac{\pi \cdot c}{\sqrt{3}} = 1.8c$.

**3.** The Dubinin–Astakhov function

$$\chi(\epsilon) = \frac{r(\epsilon - \epsilon_1)^{r-1}}{(E)^r} \exp\left\{-\left[\frac{\epsilon - \epsilon_1}{E}\right]^r\right\} \tag{2.4}$$

the variance of which is equal to $E$. Depending on the shape parameter $r$, it is a pretty Gaussian–like function for $r = 3$, right hand widened for $r < 3$, and left hand widened for $r > 3$.

Figures 4,5,6 show the form of the above mentioned adsorption energy distributions.

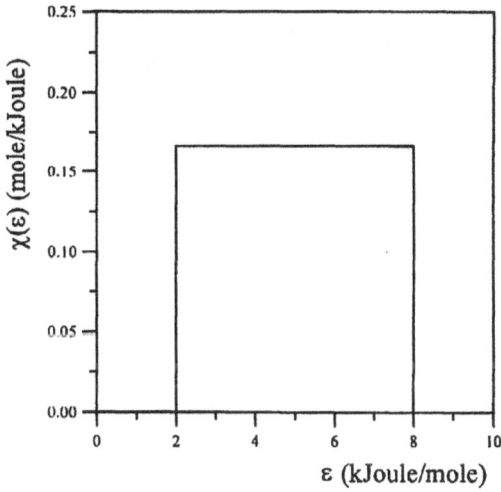

Figure 5: The "rectangular" adsorption energy distribution (2.3) for which $\epsilon_l = 2$, and $\epsilon_m = 8$.

188

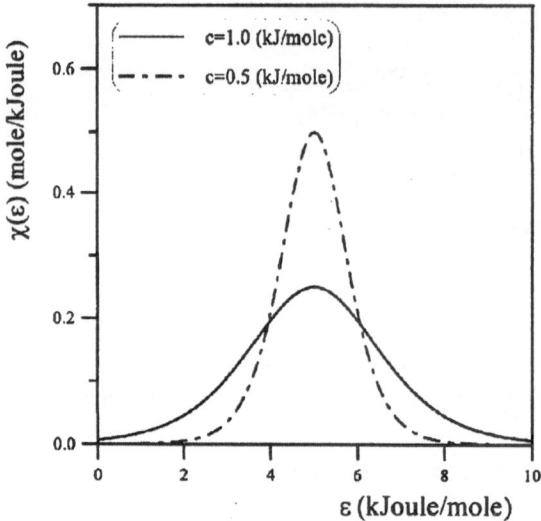

Figure 6. The Gaussian like adsorption energy distribution given by eq. (2.3) drawn by assuming that $\epsilon_0 = 5$ kJ/mole, for two values of the heterogeneity parameter $c$.

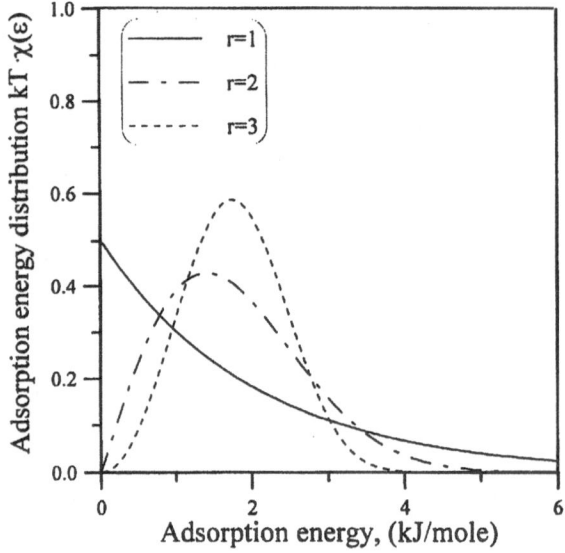

Figure 7. Effect of $r$ on the form of the adsorption energy distribution $\chi(\epsilon)$ calculated from eq. (2.4), by assuming that $\epsilon_l = 0$. The value of $\bar{E}=2$ kJ/mole.

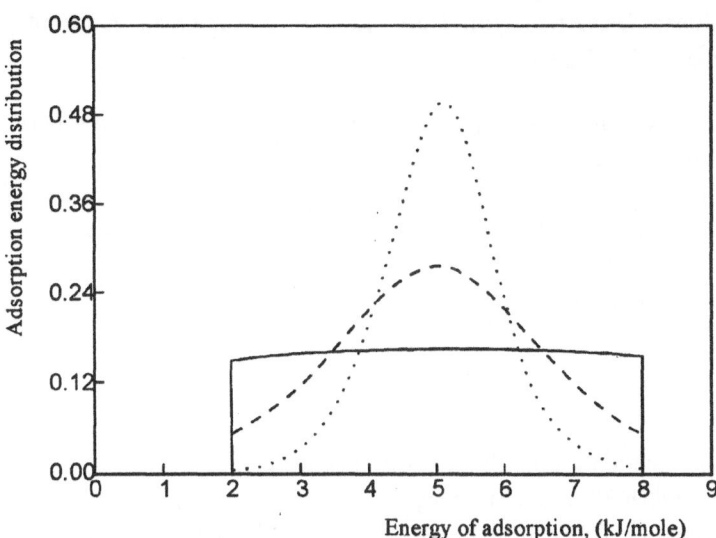

Figure 8. The effect of the heterogeneity parameter $c$ on the shape of the adsorption energy distribution $\chi(\epsilon)$, given in eq. (2.5). As it can be deduced from this figure, $\epsilon_l = 2$kJ/mole, $\epsilon_0 = 5$kJ/mole, and $\epsilon_m = 8$kJ/mole. The values of the parameter $c$ are: 0.5kJ/mole (- - - -), 1.0kJ/mole (– – –), and 5kJ/mole (—). One can see that as the heterogeneity (parameter $c$) increases, $\chi(\epsilon)$ tends to rectangular adsorption energy distribution.

For obvious physical reasons, there must be a certain minimum, and a maximum value of the adsorption energy $\epsilon$, on a heterogeneous solid surface, $\epsilon_l$ and $\epsilon_m$. Thus, the gaussian–like function (2.3) should, for various values of the parameter $c$, be viewed correctly in the way shown in Fig. 7.

All the functions shown in Fig. 7 are normalized to unity, and given by the equation,

$$\chi(\epsilon) = \frac{1}{F_N} \frac{\frac{1}{c}\exp\{\frac{\epsilon-\epsilon_0}{c}\}}{\left[1+\exp\{\frac{\epsilon-\epsilon_0}{c}\}\right]^2} \tag{2.5}$$

where the normalization factor $F_N$ reads,

$$F_N = \left[1+\exp\{\frac{\epsilon_l-\epsilon_0}{kT}\}\right]^{-1} - \left[1+\exp\{\frac{\epsilon_m-\epsilon_0}{kT}\}\right]^{-1} \tag{2.6}$$

Figure 7 shows that when $c \to \infty$, the function in eq (2.5) becomes a rectangular (constant) energy distribution. It means, the rectangular function (2.2) is likely to be a limiting form of all adsorption energy distributions when the surface is strongly heterogeneous.

## 3. The "Condensation Approximation" in the Theories of Adsorption on Heterogeneous Surfaces.

In this section we explain the principles of the theoretical method which is most frequently used to calculate the integral in eq. (2.1). We will do it by considering the simplest case when the "local" isotherm $\theta(\epsilon, p, T)$ is the Langmuir isotherm under the integral sign in eq. (1.5). To that purpose we rewrite the Langmuir isotherm in the following way:

$$\theta(\epsilon, p, T) = \frac{Kp \exp\left\{\frac{\epsilon}{kT}\right\}}{1 + Kp \exp\left\{\frac{\epsilon}{kT}\right\}} = \frac{\exp\left\{\frac{\epsilon - \epsilon_c}{kT}\right\}}{1 + \exp\left\{\frac{\epsilon - \epsilon_c}{kT}\right\}} \tag{3.1}$$

where

$$\epsilon_c = -kT \ln Kp \tag{3.2}$$

The features of the function $\theta(\epsilon, \epsilon_c, T)$ are shown in Figure 8.

While calculating $\theta_t(p, T)$ or $\chi(\epsilon)$ from eq. (2.1) a certain approximation called "Condensation Approximation" is frequently used. It is based on the assumption that adsorption on a heterogeneous surface proceeds in an ideally "stepwise" fashion in the sequence toward decreasing adsorption energies. It means, the true kernel $\theta(\epsilon, p, T)$ in eq.(1.5) is replaced by the following step function, $\theta_c(\epsilon, p, T)$,

$$\theta(\epsilon, p, T) \rightarrow \theta_c(\epsilon, p, T) = \begin{cases} 0, & for \quad \epsilon < \epsilon_c; \\ 1, & for \quad \epsilon \geq \epsilon_c \end{cases} \tag{3.3}$$

Figure 9 shows the derivative $(\partial\theta/\partial\epsilon)$. When $T \rightarrow 0$, the derivative

$$\frac{\partial\theta}{\partial\epsilon} = \frac{\frac{1}{kT} \exp\left\{\frac{\epsilon - \epsilon_c}{kT}\right\}}{\left[1 + \exp\left\{\frac{\epsilon - \epsilon_c}{kT}\right\}\right]^2} \tag{3.4}$$

tends to Dirac delta function $\delta(\epsilon - \epsilon_c)$.

Application of the CA (Condensation Approximation) makes it possible to simplify the calculations in eq.(2.1). Then

$$\theta_t(p, T) = \int_{\epsilon_c}^{\epsilon_m} \chi(\epsilon)d\epsilon = \mathcal{X}(\epsilon_m) - \mathcal{X}(\epsilon_c) \tag{3.5}$$

where $\mathcal{X}(\epsilon)$ is the integral form of $\chi(\epsilon)$

$$\mathcal{X}(\epsilon) = \int \chi(\epsilon)d\epsilon \tag{3.6}$$

When $\chi(\epsilon)$ is, the Gaussian–like function (2.3) for instance, then

$$\mathcal{X} = -\left[1 + \exp\left\{\frac{\epsilon - \epsilon_0}{c}\right\}\right]^{-1} + const \tag{3.7}$$

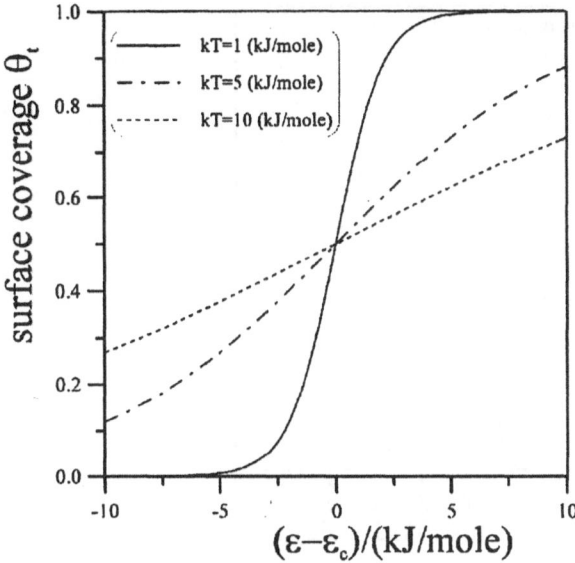

Figure 9: The features of the function $\theta(\epsilon, \epsilon_c, T)$ shown for three different temperatures.

so, the function $\theta_t(p, T)$ in eq. (3.5) becomes the well–known Langmuir–Freundlich isotherm equation,

$$\theta_t(p, T) = \frac{K^\circ(p)^{\frac{kT}{c}}}{1 + K^\circ(p)^{\frac{kT}{c}}} \tag{3.8}$$

in which $K^\circ = K\ exp\{\epsilon_o/kT\}$.

At small adsorbate pressures $\theta_t(p, T)$ in eq.(3.8) reduces to the Freundlich isotherm $\theta_t(p, T) = K^\circ(p)^{\frac{kT}{c}}$.

When $\chi(\epsilon)$ is the rectangular adsorption energy distribution (2.2) then,

$$\theta_t = \frac{\epsilon_m - \epsilon_c}{\epsilon_m - \epsilon_l} = \frac{\epsilon_m}{\epsilon_m - \epsilon_l} + \frac{kT}{\epsilon_m - \epsilon_l}\ln Kp \tag{3.9}$$

i.e. becomes the well-known Temkin's isotherm,

$$\theta_t = C_1 + C_2 \ln p \tag{3.10}$$

where $C_1 = (\epsilon_m + kT \ln K)/(\epsilon_m - \epsilon_l)$ and $C_2 = kT/(\epsilon_m - \epsilon_l)$.

When $\chi(\epsilon)$ is the adsorption energy distribution given in eq. (2.4), then

$$\theta_t = \exp\left\{-\left[\frac{\epsilon_c - \epsilon_l}{E}\right]^r\right\} \tag{3.11}$$

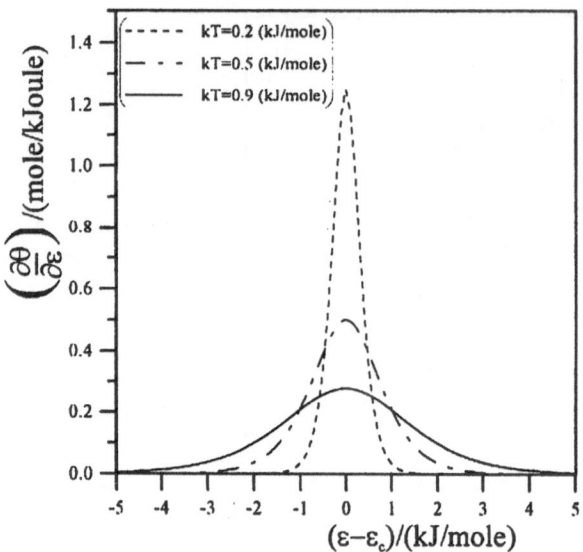

Figure 10. The form of the derivative $(\partial\theta/\partial\epsilon)$ for three values of the temperature $\tau = kT$ (kJ/mole).

or, in another way,

$$\theta_t = \exp\left\{-\left[\frac{kT}{E}\ln\frac{p_l}{p}\right]^r\right\} \tag{3.12}$$

where

$$\ln p_l = -\ln K - \frac{\epsilon_1}{kT} \tag{3.13}$$

Eq. (3.12) is just the well-known Dubinin-Astakhov isotherm, or the Dubinin–Radushkevich isotherm for the particular case when $r = 2$.

Now, let us consider the practical side of the problem, when we have at disposal an experimental adsorption isotherm $N_t(p) = M\theta_t(p)$, and we would like to check which theoretical isotherm equation describes our experimental data best. For that purpose, it is recommended to write equation (3.8) in the following linear form,

$$\ln\frac{N_t/M}{1 - N_t/M} = \ln\frac{\theta_t}{1 - \theta_t} = \frac{kT}{c}\ln K^0 + \frac{kT}{c}\ln p \tag{3.14}$$

A suitable (computer) choice of the parameter $M$ will make the calculated function $\ln\frac{N_t/M}{1-N_t/M}$ linear function of $\ln p$. From the slope and intercept of that linear plot one can easily determine the parameters $c$ and $K^0$.

In the case of the Dubinin–Astakhov isotherm (3.12) the following linear plot is to be considered.

$$\ln N_t = \ln M - \left[\frac{kT}{E}\right]^r\left[\ln\frac{p_l}{p}\right]^r \tag{3.15}$$

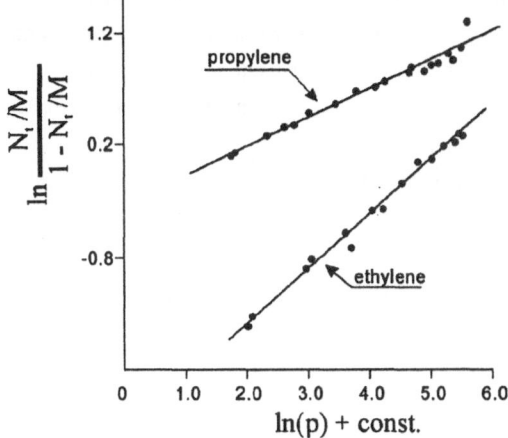

Figure 11. The linear regression (3.14) for the experimental isotherms of ethylene and propylene adsorption on the carbon molecular sieve at 274.85K, (Nakahara and coworkers[27]), made by using the parameters collected in Table 1.

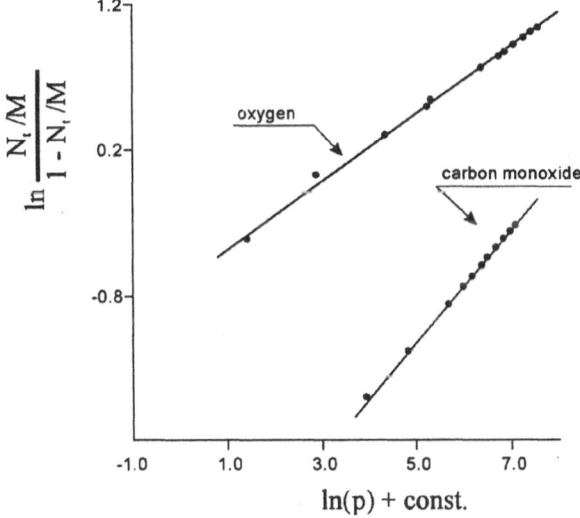

Figure 12. The linear regression (3.14) for the experimental isotherms of oxygen and carbon monoxide adsorption in the molecular sieve 10X at 172.04K, (Nolan et. al.[28]), made by using the parameters collected in Table 1.

When the parameters $p_l$, $r$ are correctly chosen, $\ln N_t$ should be a linear function of $\left[\ln \frac{p_l}{p}\right]^r$ in the whole range of the surface coverages. It is a frequent practice to choose $p_l$ as the saturated vapour pressure $p_s$, but it is not a correct choice, in general.

Temkin's isotherm (3.10) applies when the plot $N_t$ (whichever are the units for the adsorbed amount) is a linear function of $\ln p$.

Figures 10 and 11 show the examples of correlating experimental adsorption isotherms by the Langmuir-Freundlich equation (3.14).

| $T$ $(K)$ | $M$ $(mmol/g)$ | $\frac{kT}{c}$ | $\ln K^0$ $mmHg^{-1}$ |
|---|---|---|---|
| Ethylene on the carbon molecular sieve | | | |
| 274.85 | 4.28 | 0.46 | -2.30 |
| Propylene on the carbon molecular sieve | | | |
| 274.85 | 3.55 | 0.25 | -3.27 |
| Carbon monoxide on the molecular sieve 10X | | | |
| 172.04 | 138.6 | 0.45 | -2.29 |
| Oxygen on the molecular sieve 10X | | | |
| 172.04 | 379.2 | 0.76 | -6.97 |

Table 1. The parameters found by computer while fitting the experimental isotherms for ethylene and propylene adsorption on the carbon molecular sieve MSC-5A reported by Nakahara et. al.[27], and by fitting the isotherms of carbon monooxide and oxygen adsorption on the molecular sieve 10X reported by Nolan et. al.[28], by the linear form of the Langmuir–Freundlich equation (3.14).

## 4. The Rudzinski–Jagiello (RJ) Approach, and the Existence of Henry Law.

The integral in eq. (2.1) is evaluated by using integration by parts

$$\theta_t(p, T) = \theta(\epsilon, p, T)\mathcal{X}(\epsilon)|_{\epsilon_1}^{\epsilon_m} - \int_{\epsilon_1}^{\epsilon_m} \left(\frac{\partial \theta}{\partial \epsilon}\right) \mathcal{X}(\epsilon)d\epsilon \qquad (4.1)$$

Until the surface coverage is not close to zero, or to unity, the first term on the r.h.s. of eq. (4.1) can be neglected. (We will discuss it soon). The second term on the r.h.s. of eq. (4.1) is evaluated by expanding $\mathcal{X}(\epsilon)$ into its Taylor series, around the point $\epsilon = \epsilon_c$, at which the derivative $\left(\frac{\partial\theta}{\partial\epsilon}\right)$ reaches its shap maximum, i.e.,

$$\left(\frac{\partial^2\theta}{\partial\epsilon^2}\right)_{\epsilon=\epsilon_c} = 0 \qquad (4.2)$$

For Langmuir adsorption isotherm, $\epsilon_c$ calculated from eq. (4.2) is the same as $\epsilon_c$ in eq. (3.2).

$$\theta_t(p, T) = -\mathcal{X}(\epsilon_c) - \sum_{l=1}^{\infty} \frac{(kT)^l}{l!} C_l \left[\frac{\partial^l \mathcal{X}}{\partial\epsilon_l}\right]_{\epsilon=\epsilon_c(p,T)}, \qquad (4.3)$$

$C_l$-s are temperature-dependent coefficients given by,

$$C_l = \int_{(\epsilon_l-\epsilon_c)/kT}^{(\epsilon_m-\epsilon_c)/kT} \frac{t^l e^t}{(1+e^t)^2} dt, \qquad (4.4)$$

The terms under the sum in eq. (4.3) are correction terms, whereas the first leading term on the r.h.s. of eq. (4.3), $-\mathcal{X}(\epsilon_c)$, is just what is given by the Condensation Approximation. When only the first correction term ($l = 1$) is retained under the sum in eq. (4.3), and $\theta(\epsilon, p, T)$ is the Langmuir isotherm, we have,

$$\theta_t(p, T) = -\mathcal{X}(\epsilon_c) - \frac{\pi^2}{6}(kT)^2 \left(\frac{\partial\chi}{\partial\epsilon}\right)_{\epsilon=\epsilon_c} \qquad (4.5)$$

As already discussed, the failure of the LF, DR, DA, and of other isotherm equations to reduce correctly to Henry's Law, has its source in assuming the infinite energy limits $(-\infty, +\infty)$, or $(0, +\infty)$ while developing these isotherm equations. For obvious physical reasons, there must exist finite integration limits, $(\epsilon_l, \epsilon_m)$. When RJ expansion is used to represent $\theta_{ti}(p_i, T)$, the function $\epsilon_c(p_i, T)$ is defined in the interval $(-\infty, +\infty)$, and $\chi(\epsilon)$ is to be defined as follows:

$$\chi(\epsilon) = \begin{cases} 0 & \text{for } 0 \leq \epsilon < \epsilon_l, \\ \chi & \text{for } \epsilon_l \leq \epsilon < \epsilon_m, \\ 0 & \text{for } \epsilon_m \leq \epsilon. \end{cases} \qquad (4.6)$$

Accordingly, the function $\mathcal{X}(\epsilon)$ is to be defined now as follows:

$$\mathcal{X}(\epsilon) = \begin{cases} \mathcal{X}(\epsilon_l) & \text{for } 0 \leq \epsilon < \epsilon_l, \\ \mathcal{X}(\epsilon) & \text{for } \epsilon_l \leq \epsilon < \epsilon_m, \\ \mathcal{X}(\epsilon_m) & \text{for } \epsilon_m \leq \epsilon, \end{cases} \qquad (4.7)$$

and the normalization condition states that

$$\mathcal{X}(\epsilon_m) - \mathcal{X}(\epsilon_l) = 1.$$

As the functions describing real physical systems are continuous of an infinite order, it means that $\chi(\epsilon)$ defined in eq. (4.6) and all its derivatives are continuous at the points $\epsilon_l$ and $\epsilon_m$ — this is the existence condition for expansion (4.3). When the first term on the r.h.s. of eq. (4.1) is not neglected, eq. (4.1) takes the more general form

$$\theta_t(p,T) = [\theta(\epsilon_m)\mathcal{X}(\epsilon_m) - \theta(\epsilon_l)\mathcal{X}(\epsilon_l)] - \sum_{l=0}^{\infty} \frac{(kT)^l}{l!}C_l \left(\frac{\partial^l \mathcal{X}}{\partial \epsilon_l}\right)_{\epsilon=\epsilon_c}. \tag{4.8}$$

While carrying out our considerations we take the function (2.31) for illustration. In the limit of a strongly heterogeneous surface, when $kT/c \to 0$ (practically when $kT/c < 0.9$), it is sufficient to retain only the first leading term under the sum in eq. (4.8), which then takes the following form,

$$\theta_t = \theta(\epsilon_m)[\mathcal{X}(\epsilon_m) - \mathcal{X}(\epsilon_c)] + \theta(\epsilon_l)[\mathcal{X}(\epsilon_c) - \mathcal{X}(\epsilon_l)]. \tag{4.9}$$

At very low adsorbate pressures such that $\epsilon_c > \epsilon_m$, from eqs. (4.7) and (4.9), it follows that $\theta_t = \theta(\epsilon_l)$. It means that adsorption is running like on a homogeneous solid surface characterized by the adsorption energy $\epsilon = \epsilon_l$. That apparently strange result was predicted even earlier in the theoretical works by Rudnitzky and Alekseev[25]. 

Let us suppose that half of the surface is covered, and that function (2.31) is still symmetrical. Then,

$$|\mathcal{X}(\epsilon_m) - \mathcal{X}(\epsilon_c)| = |\mathcal{X}(\epsilon_c) - \mathcal{X}(\epsilon_l)| = \frac{1}{2}, \tag{4.10}$$

and eq. (4.9) reduces to

$$\theta_t = \theta(\epsilon_m)[\mathcal{X}(\epsilon_m) - \mathcal{X}(\epsilon_c)] \tag{4.11}$$

because $\theta(\epsilon_m) \gg \theta(\epsilon_l)$. We choose now the integration constant in such a way that $\mathcal{X}(\epsilon_m) = 0$. When the surface is strongly heterogeneous, i.e. when $\epsilon_l \to -\infty$ and $\epsilon_m \to +\infty$, then

$$F_N \to 1, \quad \theta(\epsilon_m) \to 1, \quad \text{and} \quad \theta_t \to -\mathcal{X}(\epsilon_c), \tag{4.12}$$

Then eq. (4.11) reduces to CA, i.e. to the LF isotherm. When $\epsilon_m$ is large, the reduction takes place even at very small surface coverages (adsorbate pressures). This is why linear $\log \theta_t$-vs.-$\log p$ plots are commonly observed at low adsorbate pressures. (The Freundlich isotherm is the low-pressure limit of the function $-\mathcal{X}(\epsilon_c)$ when $\chi(\epsilon)$ is given by (2.3).)

This also means that there is a transition within a certain range of pressures, around $p = p_m = (1/K)\exp(-\epsilon_m/kT)$, from the LF isotherm to the Langmuir isotherm $\theta(\epsilon_l)$. As expected at low adsorbate pressures, Langmuir behaviour means simply Henry's Law, $\theta_t \to pK \exp(\epsilon_l/kT)$. This means that there is a transition region, where the tangent of the plot $\ln \theta_t$ vs $\ln p$ increases from a value smaller than unity (typically between 0.5 and 0.9) to values close to unity when $p \to 0$.

5. **The Cooperative Effect of the Energetic Surface Heterogeneity and of the Interactions Between Adsorbed Molecules on Various Adsorption Observables (isotherms, heats of adsorption, etc.).**

Let us consider the simplest generalization of Langmuir equation taking into account the lateral interactions between the adsorbed molecules. This is the Bragg–Williams isotherm

$$\theta(\epsilon, \mathrm{p}, \mathrm{T}) = \frac{Kp \exp\left\{\frac{\epsilon}{kT} + \frac{\omega}{kT}\theta_1(\epsilon)\right\}}{1 + Kp \exp\left\{\frac{\epsilon}{kT} + \frac{\omega}{kT}\theta_1(\epsilon)\right\}} \qquad (5.1)$$

The term $\frac{\omega}{kT}\theta_l(\epsilon)$ in the exponents represents the additional potential field acting on an adsorbed molecule due to its interactions with neighbouring admolecules. The interaction parameter $\omega$ is the product of the number of the nearest neighbours – adsorption sites, and the interaction energy between two molecules adsorbed on two neighbouring sites.

For a homogeneous solid surface $\theta_l(\epsilon)$ is just the same $\theta(\epsilon, p, T)$ as on the l.h.s. of eq.(5.1). This is also true for a heterogeneous surface characterized by patchwise topography. For surfaces with random topography, $\theta_l(\epsilon)$ is to be replaced by $\theta_t(p, T)$. Due to random distribution of adsorption sites, every local microscopic coverage of surface is the same across the whole surface, and is equal to the average surface coverage $\theta_t(p, T)$. It means the additional force field acting on an adsorbed molecule, from the neighbouring molecules will depend on the average coverage of the neighbouring sites which is equal to $\theta_t(p, T)$. The CA approach can still be applied, but the function $\epsilon_c$ calculated from condition (4.2) takes the following form:

$$\epsilon_c = -kT \ln Kp - \omega\theta_t \qquad (5.2)$$

After inserting this function into eq.(3.5), from eq.(3.7) and assuming $\epsilon_m = +\infty$ we arrive at the following generalization of the Bragg–Williams equation for a heterogeneous surface characterized by the Gaussian–like adsorption energy distribution (2.3), and random topography,

$$\theta_t(\mathrm{p}, \mathrm{T}) = \frac{K^\circ p^{\frac{kT}{c}} \exp\left\{\frac{\omega\theta_t}{c}\right\}}{1 + K^\circ p^{\frac{kT}{c}} \exp\left\{\frac{\omega\theta_t}{c}\right\}} \qquad (5.3)$$

For strongly heterogeneous surfaces, i.e., when $c \to \infty$ $exp\left\{\frac{\omega\theta_t}{c}\right\} \to 1$. That means, the energetic surface heterogeneity "suppresses" the effects originating from the lateral interactions between adsorbed molecules. That suppressing is even stronger for the surfaces characterized by patchwise topography. There, the function $\epsilon_c$ calculated from eq.(4.2) takes the form

$$\epsilon_c = -kT \ln Kp - \frac{\omega}{2} \qquad (5.4)$$

and the corresponding generalization of the Bragg–Williams isotherm takes the form,

$$\theta_t(p, T) = \frac{K^\circ p^{\frac{kT}{c}} \exp\left\{\frac{\omega}{2c}\right\}}{1 + K^\circ p^{\frac{kT}{c}} \exp\left\{\frac{\omega}{2c}\right\}} \tag{5.5}$$

Eq.(5.5) behaves essentially like the Langmuir–Freundlich isotherm, developed for the systems in which no interactions exist between the adsorbed molecules. (The effect of interactions between adsorbed molecules will show up in the correction term in eq. (4.5)).

On a homogeneous solid surface, characterized by adsorption energy $\epsilon$, eq.(5.1) predicts a 2–dimensional condensation to occur below a certain critical temperature $T_c = \omega/4k$.

According to our discussion the surface energetic heterogeneity suppresses the effects of the interactions between the adsorbed molecules. Thus it must suppress also phase transitions in the adsorbed phase. As the surface heterogeneity is so common for the real solid surfaces, suppressing phase transitions must be common too. And this is why they are so rarely observed in the reported experimental adsorption isotherms.

In the case of the Dubinin–Astakhov isotherm (3.12), its generalized form taking into account the interactions between adsorbed molecules reads,

$$\theta_t(p, T) = \exp\left\{-\left[\frac{kT}{E} \ln \frac{p_l}{p} - \frac{\omega}{E}\theta_t\right]^r\right\} \tag{5.6}$$

Figures 13 and 14 show the behaviour of $\theta_t$ v.s. $\ln(p/p_l)$ for various values of of the heterogeneity $(kT/E)$ and the interaction parameter $(\omega/kT)$.

A convenient way to correlate the experimental data by using eq. (5.6) is to use the logarithmic form,

$$\ln \theta_t = \ln \frac{N_t}{M} = -\left(\frac{kT}{E}\right)^r \left[\ln \frac{p_l}{p} - \frac{\omega}{kT} \frac{N_t}{M}\right]^r \tag{5.7}$$

where $N_t$ is the amount adsorbed, expressed in certain units, and $M$ is the maximum adsorbed amount. By chosing suitably the parameters: $M$, $p_l$, $\omega$, $r$, one should get a linear plot of $\ln \frac{N_t}{M}$ vs. the variable $\left[\ln \frac{p_l}{p} - \frac{\omega}{kT} \frac{N_t}{M}\right]^r$.

Usually the term $\frac{\omega}{kT} \frac{N_t}{M}$ within the square bracket is a correction term compared to $\ln \frac{p_l}{p}$. Thus, in a first step one can find good approximate values of $p_l$, $r$, $M$, by making the plot $\ln \frac{N_t}{M}$ vs. $\left[\ln \frac{p_l}{p}\right]^r$ linear. That first step should be done for the region of small adsorbate pressures, where the correction term is expected to be small. In a next step the region of higher adsorbate pressures is to be taken into analysis too, and by using the previously estimated values of the parameters $M$, $p_l$, $r$, one has to make the plot $\ln \frac{N_t}{M}$ vs. $\left[\ln \frac{p_l}{p} - \frac{\omega}{kT} \frac{N_t}{M}\right]^r$, linear in the whole region of pressures.

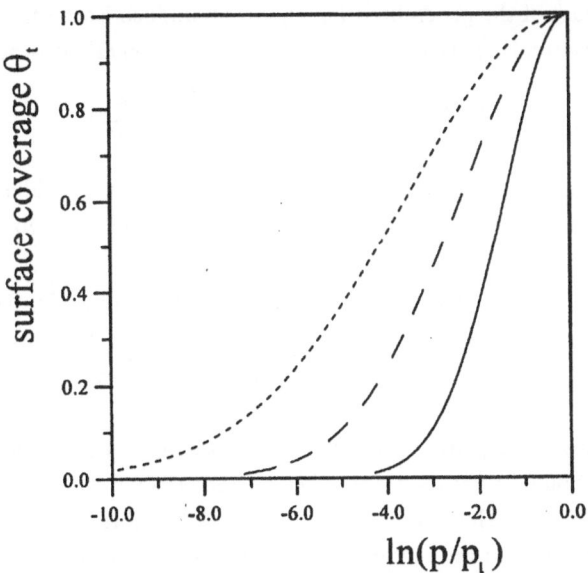

Figure 13. The effect of the degree of surface heterogeneity on the adsorption isotherm $\theta_t$ v.s. $\ln(p/p_l)$, plated for the three values of the heterogeneity parameter $kT/E$; 0.5(—), 0.3(- - -) and 0.2(- - - -). The value of $r = 2$, i.e. we consider Dubinin–Radushkevich isotherm.

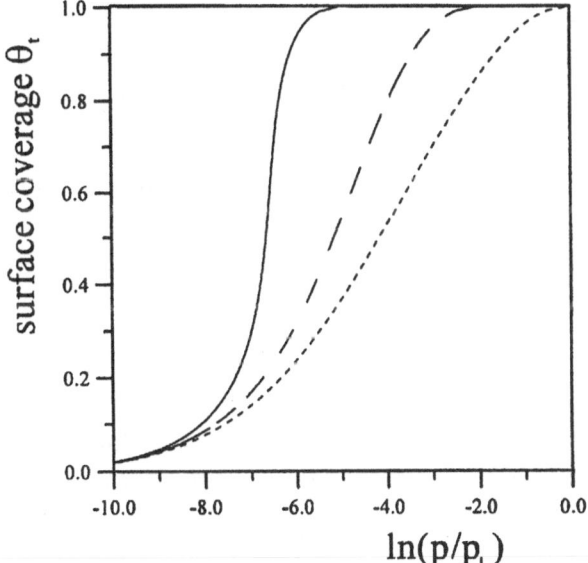

Figure 14. The effect of the interaction between the adsorbed molecules on the adsorption isotherm $\theta_t$ v.s. $\ln(p/p_l)$, plated for the three values of the interaction parameter $\omega/kT$; 0.0(- - - -), 2.0(- - -) and 5.0(—). The value of $r = 2$, i.e. we consider Dubinin–Radushkevich isotherm.

Similarly, in the case of the generalized Langmuir–Freundlich isotherm (5.3), the following linear representation is convenient for fitting the experimental adsorption isotherms.

$$\ln \frac{N_t/M}{1 - N_t/M} = \frac{kT}{c} \ln K^0 + \frac{kT}{c} \ln p + \frac{\omega}{c} \frac{N_t}{M} \tag{5.8}$$

Here only two parameters: $M$ and $\frac{\omega}{kT}$ are to be found by computer, while making linear the plot $\ln \frac{N_t/M}{1-N_t/M}$ vs. $\frac{kT}{c} \left[ \ln p + \frac{\omega}{kT} \frac{N_t}{M} \right]$. In the first step one can try to make the plot $\ln \frac{N_t/M}{1-N_t/M}$ vs. $\ln p$ linear.

Although for $r = 3$, the function (2.4) will look pretty symmetric, its behaviour is still different from that observed in the function (2.3). It means, that even if $r = 3$, the Dubinin–Astakhov isotherm will not simulate exactly the behaviour of the Langmuir--Freundlich isotherm, and vice versa.

Now let us consider the equation related to the rectangular adsorption energy distribution (2.2) which should apply to strongly heterogeneous solid surfaces. Below, we are going to show, that even in the case of random topography, this isotherm equation will not predict now any effects of interactions between the adsorbed molecules to be seen.

From eqs. (2.2), (3.4) and (5.2) we obtain,

$$\theta_t = \frac{\epsilon_m}{\epsilon_m - \epsilon_l} + \frac{kT}{\epsilon_m - \epsilon_l} \ln Kp + \frac{\omega}{\epsilon_m - \epsilon_l} \theta_t \tag{5.9}$$

or, in another way,

$$\theta_t = \frac{1}{1 - \frac{\omega}{\epsilon_m - \epsilon_l}} \left[ \frac{\epsilon_m + kT \ln K}{\epsilon_m - \epsilon_l} \right] + \frac{1}{1 - \frac{\omega}{\epsilon_m - \epsilon_l}} \frac{kT}{\epsilon_m - \epsilon_l} \ln p \tag{5.10}$$

Thus, again, we have arrived at eq. (3.10) developed previously by accepting the Langmuir equation as the local isotherm, except that the coefficients $C_1$ and $C_2$ have now slightly different meaning,

$$C_1 = \frac{1}{1 - \frac{\omega}{\epsilon_m - \epsilon_l}} \frac{\epsilon_m + kT \ln K}{\epsilon_m - \epsilon_l} \tag{5.11}$$

$$C_2 = \frac{1}{1 - \frac{\omega}{\epsilon_m - \epsilon_l}} \frac{kT}{\epsilon_m - \epsilon_l} \tag{5.12}$$

Now, let us consider the behaviour of the isosteric heat of adsorption $Q_{st}$, predicted by the various isotherm equations, developed for various adsorption energy distributions.

$$Q_{st} = -k \left( \frac{\partial \ln p}{\partial \frac{1}{T}} \right)_{\theta_t} \tag{5.13}$$

This is because the heats of adsorption are much more sensitive to the nature of an adsorption system than adsorption isotherms.

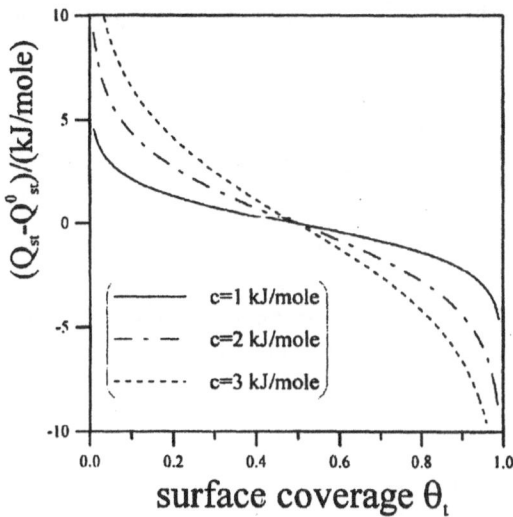

Figure 15. The behaviour of the heat of adsorption curve predicted by eq. (5.14), when $\omega = 0$, for various values of the heterogeneity parameter $c$.

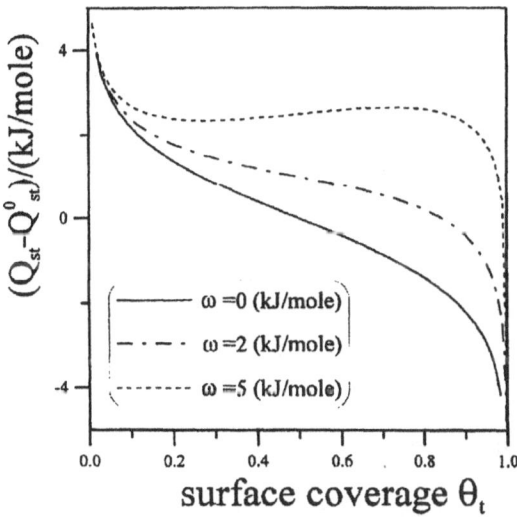

Figure 16. The behaviour of the heat of adsorption curve predicted by eq. (15.4) when $c = 1\ kJ/mole$, for various values of the interaction parameter $\omega$.

For the Langmuir–Freundlich isotherm (5.3), the related equation for the heat of adsorption reads,

$$Q_{st} = Q_{st}^0 + \omega\theta_t + c\ln\frac{1-\theta_t}{\theta_t} \qquad (5.14)$$

where

$$Q_{st}^0 = k\frac{d\ln K^0}{d\frac{1}{T}} \qquad (5.15)$$

Figures 15 and 16 show the behaviour of the heat of adsorption curves predicted by eq. (5.14).

Figure 17 shows the examples of the heat of adsorption curves taken from literature. They are similar to the theoretical curves presented in Figure 16.

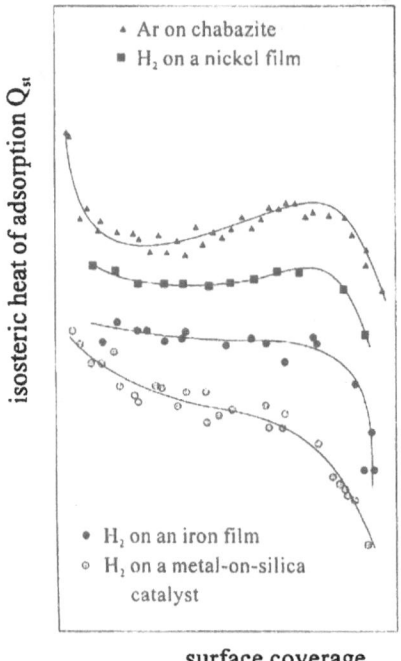

Figure 17. The examples of the heat of adsorption curves published by various authors. The data for $Ar$ on habazite are those reported by Kington and McLeod[30], $H_2$ on a nickel film data were reported by Wedler and Broecker[31], $H_2$ on an iron film data were reported by Beeck[32], whereas the $H_2$ on metal-on-silica catalyst were reported by Schuit and van Reijen[33]. The curves were shifted on Y-scale in such a way that their individual shapes could be clearly seen. Their individual X (coverage) scales were compressed in such a way that the highest reported coverages were identified with $\theta_t = 1$.

In the case of the generalized Dubinin–Astakhov isotherm (5.6), $Q_{st}$ is given by,

$$Q_{st} = Q_{st}^l + \omega\theta_t + E\left[\ln\frac{1}{\theta_t}\right]^{\frac{1}{r}} \tag{5.16}$$

where the non–configurational contribution $Q_{st}^l$ is now given by

$$Q_{st}^l = -k\frac{d\ln p_l}{d\frac{1}{T}} = k\frac{d\ln K}{d\frac{1}{T}} + \epsilon_l \tag{5.17}$$

In the case of Temkin isotherm (5.11), the isosteric heat of adsorption is given by the following expression

$$Q_{st} = Q_{st}^{in} - (\epsilon_m - \epsilon_l - \omega)\theta_t \tag{5.18}$$

where $Q_{st}^{in}$ is the "initial" heat of adsorption when $\theta_t = 0$,

$$Q_{st}^{in} = -k\frac{d}{d\frac{1}{T}}\left(\frac{\epsilon_m + kT\ln K}{kT}\right) \tag{5.19}$$

Thus, it is to be expected that in the case of strongly heterogeneous surfaces, (Temkin isotherm), the experimental isotherm $N_t(p)$ will grow linearly with the logarithm of the equilibrium bulk pressure, whereas the isosteric heat of adsorption will decrease linearly with the adsorbed amount.

## 6. Simultaneous description of adsorption kinetics and equilibria.

At small surface coverages, the second term on the r.h.s. of eq. (1.2) can be neglected. Many papers on the experimental studies of adsorption kinetics[34] were published. but the reported data did not obey the Langmuirian kinetics represented by the first term on the r.h.s. of eq. (1.2). Thus, various empirical laws were formulated to correlate the experimental data for adsorption kinetics. The first attempts by Roginski and Zeldovich[35,36] to provide a theoretical explanation for these empirical laws employed the concept of adsorption on an energetically heterogeneous solid surface. Later on that concept was more thoroughly elaborated by Aharoni and coworkers[37-39], and more recently by Tovbin[40] and Cerofolini[41].

Studies of desorption kinetics were carried out even more extensively. They were stimulated by the wide application of the Temperature Programmed Desorption (TPD) experiments to study the energetic properties of catalysts and catalyst supports[42]. It was observed, that the activation energy for desorption $\epsilon_d$ changed with the surface coverage, so, the theoretical analyses of TPD desorption spectra started with the following equation,

$$-\frac{\partial\theta}{\partial t} = K_d\theta^s\exp\left\{\frac{-\epsilon_d(\theta)}{kT}\right\} \tag{6.1}$$

The dependence of $\epsilon_d$ on $\theta$ was commonly explained as being due to the energetic heterogeneity of the actual solid surfaces, characterized by the dispersion of the activation energies for desorption.

Although every adsorption process is accompanied by a simultaneous desorption process, and vice versa, the studies of adsorption-desorption kinetics proceeded historically along two separate routes. One group of scientists studied kinetics of adsorption at low surface coverages, and neglected desorption phenomena in their studies. This group of scientists treated the surface energetic heterogeneity as a dispersion of the activation energies for adsorption, across a solid surface. The second group of scientists investigating desorption at high surface coverages, mostly in TPD experiments, treated the surface heterogeneity as the dispersion of activation energies for desorption.

While accepting the TAAD approach, one should consider the energetic surface heterogeneity as a simultaneous variation of $\epsilon_a$ and $\epsilon_d$ values, from one adsorption site to another.

The generalization of eq. (1.2) for a heterogeneous solid surface takes then the following form,

$$\frac{\partial \theta_t}{\partial t} = \int\limits_{\Omega_a} \int\limits_{\Omega_d} \left\{ \left[ K_a p (1 - \theta) \exp \left\{ \frac{-\epsilon_a}{kT} \right\} \right] - \left[ K_d \theta \exp \left\{ \frac{-\epsilon_d}{kT} \right\} \right] \right\} \chi(\epsilon_a, \epsilon_d) d\epsilon_a d\epsilon_d \quad (6.2)$$

where $\chi(\epsilon_a, \epsilon_d)$ is a two-dimensional differential distribution of the fraction of the surface sites among corresponding pairs of the values $\{\epsilon_a, \epsilon_d\}$.

Statistical theories of adsorption equilibria provide us with the functions $\theta = \theta(\epsilon_d - \epsilon_a)$. Thus, in order to make use of the statistical theories of adsorption, one must establish the relationship between $\epsilon_a$ and $\epsilon_d$. Seeking for the relationship between $\epsilon_d$ and $\epsilon_a$ on different adsorption sites seems to be a difficult fundamental problem. Finally, the lack of applicability of Langmuirian adsorption kinetics was reported for typical physisorption systems in which the sense of the activation energy for adsorption seems to be difficult for interpretation.

All the above mentioned difficulties disappear when, as the starting point, one applies the Statistical Rate Theory of Interfacial Transport, developed by Ward and Findlay[43]. Recently Rudzinski has generalized their new approach to describe adsorption–desorption kinetics on energetically heterogeneous solid surfaces. (See the chapter by W. Rudzinski in the international monograph: "Equilibria and Dynamics in Gas Adsorption on Heterogeneous Solid Surfaces"[44].

The starting point of our consideration is the equation developed by Ward and Findlay[43].

$$\frac{\partial \theta}{\partial t} = K'_{gs} \left[ \exp\{\frac{\mu^g - \mu^s}{kT}\} - \exp\{\frac{-(\mu^g - \mu^s)}{kT}\} \right] \tag{6.3}$$

where $\mu^g$, $\mu^s$ are the chemical potentials of the gaseous and adsorbed (surface) molecules respectively, and $K'_{gs}$ is a constant. They assumed next that the transient configurations of adsorbed molecules are close to the equilibrium ones, corresponding to the same surface coverage, $\theta$.

Now, let us consider the Langmuir model of adsorption , i.e., one–site occupancy (localized) adsorption when no interactions exist between the adsorbed molecules. Then,

$$\frac{\mu^s}{kT} = \ln \frac{\theta}{q^s(1 - \theta)} \tag{6.4}$$

where $q^s$ is the molecular partition function of the adsorbed molecules. Accepting still the ideal-gas approximation for $\mu^g$, we have

$$\frac{\mu^g}{kT} = \frac{\mu_0^g}{kT} + \ln p \tag{6.5}$$

Now, let us consider the region of low surface coverages, i.e., neglect the second term within the square brackets in eq (6.3). As the adsorption kinetics is essentially a non-equilibrium process, we introduce the superscript $(n)$ at the surface coverages appearing in the equations for adsorption kinetics. Then,

$$\frac{\partial \theta^{(n)}}{\partial t} = K'_{gs} \exp \frac{\mu^g - \mu^s}{kT} = K'_{qs} p q^s \exp\{\frac{\mu_0^g}{kT}\} \frac{1 - \theta^{(n)}}{\theta^{(n)}} \tag{6.6}$$

The main effect of the energetic heterogeneity of the actual (really existing) solid surfaces is related to the dispersion of the minima in the gas-solid potential function across a solid surface. In the case of localized adsorption, these local minima are called "adsorption sites", and the value of the gas-solid potential at these local minima, taken with a reverse sign, is called – "the adsorption energy", and usually denoted by $\epsilon$.

So, while considering the kinetics of adsorption on a heterogeneous solid surface we will write the molecular partition function $q^s$ as the following product,

$$q^s = q_0^s \exp\{\frac{\epsilon}{kT}\} \tag{6.7}$$

where $q_0^s$ is the same for all the adsorption sites, and $\epsilon$ varies from one site to another. We will denote further by $K$ the following product,

206

$$K = q_0^s \exp\left\{\frac{\mu_0^g}{kT}\right\} \tag{6.8}$$

and rewrite eq (6.6) to the following form,

$$\theta^{(n)} = \left[1 + \frac{(\partial\theta^n/\partial t)}{K'_{gs}pK}\exp\{\frac{-\epsilon}{kT}\}\right]^{-1} \tag{6.9}$$

The above equation describes the rate of adsorption on the adsorption sites having adsorption energy equal to $\epsilon$. The experimentally measured surface coverage $\theta_t^{(n)}$, and the observed mean rate of adsorption $(\partial\theta_t^{(n)}/\partial t)$, are the values $\theta^{(n)}$ and $(\partial\theta^{(n)}/\partial t)$ defined in eq (6.9), averaged with a differential distribution of the number of adsorption sites among corresponding values of adsorption energy, $\chi(\epsilon)$.

Thus, the experimentally determined values $\theta_t^{(n)}$ and $(\partial\theta_t^{(n)}/\partial t)$ are to be related by the following equation;

$$\theta_t^{(n)} = \int_{-\infty}^{+\infty} \left[1 + \frac{(\partial\theta^{(n)}/\partial t)}{K'_{gs}pK}\exp\{\frac{-\epsilon}{kT}\}\right]^{-1} \chi(\epsilon)d\epsilon \tag{6.10}$$

where the superscript $(n)$ is used again to denote "nonequilibrium" adsorption isotherm.

Now, let us remark that eq. (6.9) can be rewritten again to the Fermi-Dirac like form,

$$\theta^{(n)} = \frac{\exp\left\{\frac{\epsilon-\epsilon_c^{(n)}}{kT}\right\}}{1 + \exp\left\{\frac{\epsilon-\epsilon_c^{(n)}}{kT}\right\}} \tag{6.11}$$

where $\epsilon_c^{(n)}$ is given by

$$\epsilon_c^{(n)} = kT\ln\frac{(\partial\theta^{(n)}/\partial t)}{K'_{gs}pK} \tag{6.12}$$

The behaviour of the function $\theta^{(n)}(\epsilon, \epsilon_c^{(n)}, T)$ will be exactly the same as the behaviour of the function $\theta(\epsilon, \epsilon_c, T)$ defined in eq. (3.1) and shown in Figure 9.

Thus, when $T \to 0$, $\theta^{(n)}$ tends to the step function $\theta_c^{(n)}$,

$$\lim_{T\to 0} = \theta_c^{(n)} = \begin{cases} 0, & \text{for } \epsilon < \epsilon_c^{(n)}; \\ 1, & \text{for } \epsilon \geq \epsilon_c^{(n)}. \end{cases} \tag{6.13}$$

and then

$$\theta_t^{(n)} = \int_{\epsilon_c^{(n)}}^{\infty} \chi(\epsilon)d\epsilon \tag{6.14}$$

Replacing the true kernel $\theta^{(n)}$, in eq. (6.10) by its corresponding step function (6.13) means assuming that the adsorption proceeds gradually on various adsorption sites in the sequence of decreasing adsorption energies $\epsilon$. At a given temperature $T$, pressure $p$, and $\left(\frac{\partial \theta^{(n)}}{\partial t}\right)$, the adsorption "front" is on the sites whose energy $\epsilon$ is equal to $\epsilon_c$ given in eq. (2.12). But then, the overall adsorption rate $\left(\frac{\partial \theta^{(n)}}{\partial t}\right)$ is, in fact, governed by the local rate of adsorption, on the sites whose adsorption energy is equal to $\epsilon_c$, through the obvious relation

$$\frac{\partial \theta_t^{(n)}}{\partial t} = \text{Const. } \chi(\epsilon_c) \left(\frac{\partial \theta^{(n)}}{\partial t}\right)_{\epsilon = \epsilon_c^{(n)}} \tag{6.15}$$

When $\epsilon = \epsilon_c^{(n)}$, $\theta^{(n)} = \frac{1}{2}$, and from eqs. (6.9) and (6.15) we obtain,

$$\epsilon_c^{(n)} = kT \ln \frac{\left(\frac{\partial \theta_t^{(n)}}{\partial t}\right)}{\chi(\epsilon_c^{(n)}) \tilde{K}_{gs} pK} \tag{6.16}$$

The picture of a sharp "adsorption front" will be a very good representation for the true isotherm (6.9) at not too high temperatures, or, when the variance of $\chi(\epsilon)$ is considerably larger than the variance of $(\partial \theta^{(n)}/\partial \epsilon)$. In the case of the rectangular function (2.2), for instance, the picture of the "sharp adsorption front" will be applicable when $kT/(\epsilon_m - \epsilon_l)$ is small. Then,

$$\theta_t^{(n)} = \frac{\epsilon_m - \epsilon_c^{(n)}}{\epsilon_m - \epsilon_l} \tag{6.17}$$

so, the corresponding equation for adsorption kinetics reads,

$$\frac{\partial \theta_t^{(n)}}{\partial t} = K_{gs} pK \exp\{\frac{\epsilon_m}{kT}\} \exp\{-\frac{\epsilon_m - \epsilon_l}{kT} \theta_t^{(n)}\} \tag{6.18}$$

where

$$K_{gs} = \tilde{K}_{gs} (\epsilon_m - \epsilon_l)^{-1} \tag{6.19}$$

Eq (6.18) is just the Elovich equation which is probably the most popular one to correlate the experimental data for adsorption kinetics. Originally, it was launched as an empirical equation. Later on, it was associated with a constant, (rectangular), distribution of adsorption energies, but next, several additional assumptions had to be made, to arrive at the Elovich kinetic equation. This made all these deviations obscure and the nature of Elovich equation remained still a half-empirical one.

And now, it is important to realize that in addition to the assumption of a rectangular adsorption energy distribution, there is still another essential asumption leading us to the experimentally observed Elovich equation. This is the asumption that the transient configurational states of adsorbed molecules are close to the equilibrium ones corresponding to the same surface coverage. The wide applicability of Elovich equation may be considered as a proof that the latter assumption is usually true in most of the experiments which are carried out in practice. Such a conclusion has received an impressive support in the recent computer simulations of adsorption kinetics reported by Talbot et. al[45].

In such a case equations for adsorption kinetics can, for a heterogeneous solid surface characterized by a certain function $\chi(\epsilon)$, easily be obtained from the corresponding equilibrium adsorption isotherms $\theta_t$.

## 7. Multi–Site–Occupancy Adsorption on Heterogeneous Solid Surfaces.

The measured quantity is the number of adsorbed molecules $N_t$, which, in terms of our theoretical treatment is to be related to the following average $\theta_t$,

$$\theta_t(p, T) = \frac{N_t n}{M} = \int_\Omega \theta(\epsilon_n, p, T)\chi_n(\epsilon_n)d\epsilon_n \tag{7.1}$$

where $M$ is the total number of adsorbed sites, $\Omega$ is the physical domain of $\epsilon_n$, and $\theta(\epsilon_n, p, T)$ is the fractional occupancy of the configurational surface states, characterized by the adsorption energy $\epsilon_n$. This is simply the generalization of the "local adsorption isotherm" known widely in the theories of one–site occupancy adsorption on heterogeneous solid surfaces.

In the case of the patchwise surface topography, $\theta(\epsilon_n, p, T)$ can then be just one of the equations used to represent multi–site occupancy adsorption on the homogeneous solid surfaces.

At the same time, there was no paper proposing an expression for the "local adsorption isotherm" $\theta(\epsilon_n, p, T)$ for the surfaces characterized by random surface topography, until Nitta published the first solution of that problem[47] in 1984. Below we give a brief sketch of his considerations, for the readers' convenience.

When $N_t$ molecules are adsorbed on a heterogeneous surface, the system partition function $Q(N_t, M, T)$, may be written as,

$$Q = q_s^{N_t} \sum_{\{N_{ij}\}} g(N_t, M, \{N_{ij}\}) \exp\left\{\sum_i \sum_j N_{ij}\frac{\epsilon_{ij}}{kT}\right\}, \tag{7.2}$$

where $\epsilon_{ij}$ is the energy of adsorption of $j$th group (mer) on $i$th kind of sites. $N_{ij}$ is the number of adsorption–pairs of sites $i$ and a group $j$; $\{N_{ij}\}$ is a distribution of these adsorption pairs, the subscript $i0$ being used for the empty site, $g(N_t, M, \{N_{ij}\})$ is a combinatorial factor, expressing the number of the distinguishable ways of distributing the $N_t$ molecules on $M$ sites under the condition of a special distribution $\{N_{ij}\}$; $q_s$ is the product of internal and vibrational partition functions of an adsorbed molecule.

As emphasized by Nitta, it is difficult to find a rigorous expression for $g(N_t, M, \{N_{ij}\})$ by taking into consideration the mutual correlations between neighbouring sites and neighbouring groups in a molecule. A simple expression is obtained by assuming that all pairs of site-mer $\{ij\}$ are independent, under the constraints imposed by the distribution $\{N_{ij}\}$. That assumption and the procedure used by Nitta are similar to the quasi-chemical approximation presented by Guggenheim for the molecule-molecule interactions. The expression for $g(N_t, M, \{N_{ij}\})$ derived in this way by Nitta reads,

$$\ln g(N_t, M, \{N_{ij}\}) = \ln g_0(N_t, M) - \sum_{i=0}^{W} \sum_{j=0}^{s} \ln \left( \frac{N_{ij}!}{N_{ij}^*!} \right) , \qquad (7.3)$$

where $g_0$ is the combinatorial factor for a homogeneous surface. While quoting Nitta; "the superscript * stands for the random distribution corresponding to the homogeneous surface". The expression used to represent $g_0$ was following,

$$g_0(N_t, M) = \frac{M!}{N_t!(M - nN_t)!} \cdot \frac{\zeta^{N_t}}{M^{(n-1)N_t}} , \qquad (7.4)$$

where $\zeta$ is the constant relating to the flexibility and symmetry number of a molecule; $n$ is the number of sites occupied by a molecule, given by the sum of all $n_j$'s $(j = 1, 2, \ldots, n)$. The combinatorial factor $g_0$ is for the polymer-like molecules adsorbed on a homogeneous solid surfaces. It will still be the same when the surface containing $M$ adsorption sites is composed of homogeneous patches which are so large that the states of the whole system in which an adsorbed molecule is located partially on another patch can be neglected. The above quoted Nitta's original statement should be understood in such a way that the superscript * refers to the homogeneous surface patches.

The expression $N_{ij}^*$ is calculated by assuming random distribution of mers on a homogeneous surface patch.

$$N_{ij}^* = \left( \frac{M_l}{M} \right) n_j N \quad \text{and} \quad N_{i0}^* = \left( \frac{M_l}{M} \right) (M - nM) \qquad (7.5)$$

Substituting equations (7.3), (7.4) into equation (7.2) and replacing the summation by the maximum term on the right-hand side of equation (7.2), one obtains the following expression for $\ln Q$,

$$\ln Q = N_t \ln q_s + \ln g_0 + \sum_{i=0}^{W} \sum_{j=0}^{s} \left( \ln \frac{N_{ij}^*!}{N_{ij}!} + \frac{N_{ij} \epsilon_{ij}}{kT} \right) , \qquad (7.6)$$

There are constraint conditions for the numbers of adsorption pairs $\{N_{ij}\}$,

$$\sum_{i=1}^{W} N_{ij} - n_j N_t = 0 , \quad \text{and} \quad \sum_{j=1}^{s} N_{ij} - M_i = 0 \quad (7.7)$$

The problem to be solved may now be expressed as that of determining the adsorption-- pairs distribution $\{N_{ij}\}$ which minimizes the Helmholtz energy of the system (or maximizes $\ln Q$) with these constraints. The chemical potential, spreading pressure, internal energy, etc. are derived by the differentiation of $\ln Q$ with respect to $N_t$, $M$ or $T$. Then, the method of undetermined multipliers is used to determine the distribution $\{N_{ij}\}$. From the condition of the maximum with respect to variable $\{N_{ij}\}$, the relations are obtained,

$$- \ln \left( \frac{N_{ij}}{M_i - \sum_{j=1} N_{ij}} \right) + \frac{\epsilon_{ij}}{kT} = \beta_j \quad (i = 1,\dots,W; \; j = 1,\dots,s) \quad (7.8)$$

where $\beta_j$ is the Lagrange multiplier relating to a group $j$ originated from equation (24).

From the condition of the maximum term with respect to $\{N_{ij}\}$, Nitta arrives at the following expression for the chemical potential of the adsorbed molecules, $\mu^s$,

$$\frac{\mu^s}{kT} = - \left( \frac{\partial \ln Q}{\partial N_t} \right)_{M,T} =$$

$$= - \ln(q_s \zeta) + \ln \left( \frac{N_t}{M} \right) - n \ln \left( \frac{M - nN_t}{M} \right) - \sum_{j=1}^{s} n_j \ln \left( \frac{n_j N_t}{M - nN_t} \right) + \quad (7.9)$$

$$+ \sum_{j=1}^{s} n_j \left[ \ln \left( \frac{N_{ij}}{M_i - \sum_{j=1} N_{ij}} \right) - \frac{\epsilon_{ij}}{kT} \right] .$$

To evaluate $\theta_t$, three kinds of surface coverages, $\theta_t$, $\{\theta_{tj}\}$ and $\{\theta_{ij}\}$ are to be defined;

$\theta_t = \frac{nN_t}{M}$,     the fraction of sites occupied;

$\theta_{tj} = \frac{n_j N_t}{M}$,   $(j = 1,\dots,s)$,     the fraction of sites occupied by the segment of type $j$;

$\theta_{ij} = \frac{N_{ij}}{M_i}$,   $(i = \dots,W; \; j = 1,\dots,s)$,   the fraction of sites of type $i$ occupied by the segments of type $j$.

The problem of evaluating $\theta_t$ is not trivial, and Nitta proposed a complicated numerical procedure, in which only a discrete distribution of adsorption energy can be applied. However, that problem becomes easy when all the segments of an admolecule are of the same kind.

This, for instance, is the case of hydrocarbon homologous series, the adsorption of which by activated carbons is so frequently studied experimentally. Nitta's equation (7.9) takes then the following form,

$$\mu^s = -\epsilon_{ni} - kT \ln(q_s \zeta n) - (n-1)kT \ln \theta_t + nkT \ln \frac{\theta_i}{1-\theta_i} \; , \tag{7.10}$$

where

$$\theta_i = \frac{N_i}{M} \tag{7.11}$$

and

$$\epsilon_{ni} = n\epsilon_{1i} \tag{7.12}$$

When the surface becomes homogeneous, i.e., $\theta_i \to \theta_t$, equation (7.10) reduces to the well–known Flory's isotherm for multi–site occupancy adsorption on homogeneous solid surfaces,

$$\mu^g = \mu^s = -\epsilon_n - kT \ln(q_s \zeta n) + nkT \ln \frac{\theta}{(1-\theta)^n} \; , \tag{7.13}$$

The averaging of $\theta_i$ in equation (7.13) with respect to the dispersion of $\epsilon_{ni}$ defined in (7.12) can be carried out easily by applying the Rudziński–Jagiełło method. Thus,

$$\theta_t = \int_{\epsilon_{nc}}^{\infty} \chi_n(\epsilon_n)d\epsilon_n + Corr. \; , \tag{7.14}$$

where $\epsilon_{nc}$ is found from the condition

$$\left( \frac{\partial^2 \theta}{\partial \epsilon_n^2} \right)_{\epsilon_n = \epsilon_{nc}} = 0 \; , \tag{7.15}$$

and Corr. is a correction term which can, safely, be neglected in the case of strongly heterogeneous surfaces.

After performing the differentiation (7.15) in equation (7.10) we arrive at the following expression for $\epsilon_{nc}$,

$$\epsilon_{nc}^{(r)} = -\mu^s - kT \ln(q_s \zeta n) - (n-1)kT \ln \theta_t \; , \tag{7.16}$$

where the superscript "r" refers to random topography. After replacing $\mu^s$ by $\mu^g$ of an ideal gas,

$$\frac{\mu^g}{kT} = \ln \left[ \frac{p\Lambda^3}{q_g kT} \right] \tag{7.17}$$

eq. (7.16) takes the following form,

$$\epsilon_{nc}^{(r)} = -kT \ln(nK'p) - (n-1)kT \ln \theta_t \; , \tag{7.18}$$

where

$$K' = \frac{q_s \zeta \Lambda^3}{q_g kT} \; . \tag{7.19}$$

$\Lambda$ is the thermal Broglie wavelength and $q_g$ is the internal molecular partition function of the molecules in the bulk gas phase.

In the case of patchwise topography $\theta_i$, and $\epsilon_{ni}$ in eq. (7.15) are to be identified with $\theta$ and $\epsilon_n$ in equation (7.13). The obtained equation for $\epsilon_{nc}^{(p)}$ takes then the following form,

$$\epsilon_{nc}^{(p)} = -kT \ln \left( \frac{n^{1+\frac{n}{2}}}{\left(n^{\frac{1}{2}}+1\right)^{n-1}} K'p \right) \qquad (7.20)$$

In the case of multi–site occupancy adsorption the growing value of $n$ may affect, in general, both the spread, (the second central moment $\mu_{2n}$), and the shape, ($\mu_{3n}$ and higher central moments), of the function $\chi_n(\epsilon_n)$. One of the simplest analytical approximations that is flexible enough to reproduce such behaviour is the following function,

$$\chi_n(\epsilon_n) = \frac{r\left(\epsilon_n - \epsilon_{nl}\right)^{r-1}}{(E_n)^r} \exp\left\{-\left[\frac{\epsilon_n - \epsilon_{nl}}{E_n}\right]^r\right\} \qquad (7.21)$$

where $\epsilon_{nl}$ is the lowest value of the adsorption energy $\epsilon_n$ on a given heterogeneous surface, $E_n$ is related to the width of that function, and $r$ governs its shape.

The generalized form of the Dubinin–Astakhov equation for the case of patchwise surface topography is obtained by inserting $\epsilon_{nc}^{(p)}$ defined in eq. (7.20), into eq. (7.14), when $\chi_n(\epsilon_n)$ is represented by the function defined in eq. (7.21). Doing so we arrive at the following equation for $\theta_t$,

$$\theta_t = \exp\left\{-\left[\frac{kT}{E_n} \ln \frac{p_1}{p}\right]^r\right\} \qquad (7.22)$$

where

$$\ln p_0 = -\ln \left[\frac{n^{1+\frac{n}{2}}}{\left(n^{\frac{1}{2}}+1\right)^{n-1}} K'\right] - \frac{\epsilon_{nl}}{kT} \qquad (7.23)$$

In a similar way we obtain the expression for $\theta_t$ in the case of random topography, by inserting $\epsilon_{nc}^{(r)}$ defined in eq. (7.18) into eq. (7.14),

$$\theta_t = \exp\left\{-\left[\frac{kT}{E_n} \ln \frac{p_1}{p\theta_t^{(n-1)}}\right]^r\right\} \qquad (7.24)$$

where

$$\ln p_0 = -\ln(nK') - \frac{\epsilon_{nl}}{kT} \qquad (7.25)$$

Thus, in the case of patchwise topography, the experimental data should be well correlated by the linear plot

$$\ln N_t \quad \text{vs.} \quad \left[\ln \frac{p_1}{p}\right]^r \qquad (7.26)$$

While carrying out this linear regression the two parameters; $p^0$ and $r$ are to be adjusted by computer.

Then in the case of random topography, eq. (7.22) suggests the following linear regression to be made,

$$\ln N_t = \ln \frac{M}{n} - \left(\frac{kT}{E_n}\right)^r \left[\ln \frac{p_l}{p\theta_t^{(n-1)}}\right]^r \tag{7.27}$$

in which the three parameters; $p_0$, $r$, and $M$ are to be adjusted by computer.

Surface topography affects also the way in which the interactions between adsorbed molecules influence the adsorption on heterogeneous surfaces.

As a starting point we have to consider the contribution to the surface chemical potential $\mu^s$. Even when accepting the simplest mean–field (Bragg–Williams) approximation, a simple expression for $\mu^s_{in}$ is obtained only if one neglects the "excluded interactions" by the existing chemical bonds between the adsorbed segments. While accepting this simplification, we have

$$\mu^s_{in} = \omega\theta \tag{7.28}$$

where $z$ is the number of the nearest neighbours adsorption sites, decreased by the average number of chemical bonds linking the adsorbed mer to the rest of the admolecule. Further, $u_{12}$ is the interaction energy between two segments adsorbed on the neighbouring adsorption sites.

As the condition (7.15) is fulfilled when $\theta = \frac{1}{2}$, $\epsilon_{nc}^{(p)}$ takes the more general form,

$$\epsilon_{nc}^{(p)} = -kT \ln \left(\frac{n^{1+\frac{n}{2}}}{\left(n^{\frac{1}{2}}+1\right)^{n-1}}K'p\right) - \frac{1}{2}\omega \tag{7.29}$$

and the final equation (7.22) is still valid, except that $p_l$ is to be interpreted now as follows,

$$\ln p_l = -\ln\left[\frac{n^{1+\frac{n}{2}}}{\left(n^{\frac{1}{2}}+1\right)^{n-1}}K'\right] - \frac{1}{2}\left(\frac{\omega}{kT}\right) - \frac{\epsilon_{nl}}{kT} \tag{7.30}$$

Thus, in the case of patchwise topography and strong surface heterogeneity the experimental data will still be correlated by the linear plot (7.26), even in the presence of interactions between adsorbed molecules.

In the case of random surface topography, the interactions between adsorbed molecules will be a source of a new configurational term in the isotherm equation.

Now, the starting point is equation (7.10) which we will write as follows,

$$\mu^s = -\epsilon_{ni} - kT \ln(q_s\zeta n) - (n-1)kT \ln\theta_t + nkT \ln\frac{\theta_i}{1-\theta_i} + \omega\theta_t \tag{7.31}$$

Accordingly, the generalized function $\epsilon_{nc}^{(r)}$ takes the following form,

$$\epsilon_{nc}^{(r)} = -\mu^s - kT \ln(q_s \zeta n) - (n-1)kT \ln \theta_t - \omega \theta_t \tag{7.32}$$

or, in another form

$$\epsilon_{nc}^{(r)} = -kT \ln(nK'p) - (n-1)kT \ln \theta_t - \omega \theta_t \tag{7.33}$$

Thus, the generalized form of eq. (7.27) taking into account interactions between adsorbed molecules reads,

$$\ln N_t = \ln \frac{M}{n} - \left(\frac{kT}{E_n}\right)^r \left[\ln \frac{p_1}{p\theta_t^{(n-1)}} - \frac{\omega}{kT}\theta_t\right]^r \tag{7.34}$$

where $p_0$ is still given by eq. (7.25).

To make use of the equations for $N_t(p, T)$ developed for patchwise and random topography one must know how the parameters $E_n$ and $r$ change with the changing number of mers $n$. It turns out that they change in a different way for the surfaces characterized by patchwise and random topography. That problem was studied in detail by Marczewski et. al[47]. Below we repeat briefly their results.

So, let us assume that every admolecule adsorbs on the surface in such a way that each "mer" interacts strictly with a single site and in the same way. The total adsorption energy of a chosen admolecule is, $\epsilon_n$,

$$\epsilon_n = \sum_{l=1}^{n} \tilde{\epsilon}_{i(l)} , \qquad (i = 1, 2, ..., W); \tag{7.35}$$

where the subscript $l$ denotes the local "mer" position in an admolecule, $\tilde{\epsilon}_{i(l)}$ is the interaction energy of the $l$th "mer" with the $i$th type of adsorption site, and $W$ is the number of site types. When the molecule size equals the site size $(n = 1)$,

$$\tilde{\epsilon}_1 = \tilde{\epsilon}_{i(1)} = \tilde{\epsilon}_i . \tag{7.36}$$

In our further consideration the character " ~ " will be used to denote a quantity related to one mer.

We define next $\chi_n(\epsilon_n)$ as the differential distribution of the number of the surface configurational states of an adsorbed molecule, into corresponding values of $\epsilon_n$. This is simply the generalization of the "adsorption energy distribution" for multi-site occupancy adsorption.

For practical purposes, it is useful to consider the effect of $n$ on $k$th central moment $\mu_{kn}$ of the adsorption energy distribution $\chi_n(\epsilon_n)$,

$$\mu_{kn} = \int\limits_{\Delta\epsilon_n} (\epsilon_n - \overline{\epsilon_n})^k \chi_n(\epsilon_n) d\epsilon_n , \qquad (7.37)$$

where $\Delta\epsilon_n$ is the physical domain of $\epsilon_n$. The adsorption energy distribution $\chi_n(\epsilon_n)$ is now the differential distribution of the number of the surface (configurational) states of an adsorbed molecule into the corresponding values of $\epsilon_n$. The second central moment $\mu_2$ is related to the variance $\sigma$,

$$\mu_2 = \sigma^2 , \qquad (7.38)$$

whereas higher central moments are related to the form of $\chi_n(\epsilon_n)$. For instance, the skewness $\beta_1^*$,

$$\beta_1^* = \frac{\mu_3}{\sigma^3} \qquad (7.39)$$

shows whether $\chi_n(\epsilon_n)$ is symmetrical ($\beta_1^* = 0$), a left hand widened ($\beta_1^* < 0$), or a right hand widened ($\beta_1^* > 0$) function.

Let us consider first the patchwise model where adsorbent surface is divided into energetically homogeneous patches, which are sufficiently large to neglect the number of sites placed in the patch boundaries. This means, that if a given site is of $S_i$ type, then its neighbour is also of $S_i$ type.

Thus, in the case of the patchwise surface

$$\epsilon_n = n\tilde{\epsilon}_i , \qquad (7.40)$$

and

$$\overline{\epsilon_n} = n\overline{\tilde{\epsilon}} \qquad (7.41)$$

For the considered patchwise surface topography

$$\mu_{kn} = n^k \tilde{\mu}_{k1} \qquad (7.42)$$

The most important effect of $n$ on $\chi_n(\epsilon_n)$ will, of course, be that on the second central moment, i.e. the effect of $n$ on the variance $\sigma_n$. From equation (8) we have,

$$\sigma_n = n\tilde{\sigma}_1 = (\mu_{2n})^{\frac{1}{2}} \qquad (7.43)$$

It may also be easily shown that, for patchwise topography considered for the moment,

$$\chi_n(\epsilon_n) = \frac{1}{n}\chi_1(\tilde{\epsilon}_1) . \qquad (7.44)$$

From equations (10) and (11) it follows, that the average energy and the width of the distribution function increase with the increasing number of adsorption sites which are occupied by an adsorbed molecule. However, the other parameters determining the shape of the distribution function $\chi_n(\epsilon_n)$, like the skewness $\beta_1^*$ for instance, remain for the patchwise topography invariant of $n$.

A somewhat different situation occurs in the case of random surface topography. In this case,

$$\overline{\epsilon_{n+1}} = \overline{\epsilon_n} + \overline{\tilde{\epsilon}} \tag{7.45}$$

$$\chi_{n+1}(\epsilon_{n+1}) = \int_{\Delta\epsilon_n} \chi_n(\epsilon_n)\chi_1(\epsilon_{n+1} - \epsilon_n)d\epsilon_n , \tag{7.46}$$

where

$$(\epsilon_{n+1} - \epsilon_n) \in \Delta\tilde{\epsilon} \tag{7.47}$$

Further,

$$\mu_{2n} = n\tilde{\mu}_{21} , \quad \text{i.e.} \quad \sigma_n^2 = n\tilde{\sigma}_1^2 \tag{7.48}$$

$$\mu_{3n} = n\tilde{\mu}_{31} \tag{7.49}$$

but

$$\mu_{4n} = n\tilde{\mu}_{41} + 3\tilde{\mu}_{21}^2 n(n-1) \tag{7.50}$$

and

$$\beta_{1n}^* = \frac{\tilde{\beta}_{11}^*}{n^{\frac{1}{2}}} \tag{7.51}$$

Thus, in contrast to patchwise topography, in the case of random surface topography the increasing value of $n$ affects also the shape of the adsorption energy distribution. With the increasing value of $n$ the distribution function $\chi_n(\epsilon_n)$ becomes more and more similar to a Gaussian function.

Now let us draw attention to the fact that eq. (7.12) has the same form as eq. (7.40) for the surface characterized by patchwise topography. Thus, Rudzinski and Everett who first considered the generalization of Langmuir–Freundlich isotherm for multi-site-occupancy adsorption, proposed to apply the relation (7.43) for both patchwise and random surfaces.

In the next papers published by Rudzinski et. al[48,49] it was stated that the above assumption may not be perfectly true in the case of multi-site-occupancy adsorption on the surfaces characterized by random topography. This is because eq. (7.5) in Nitta's consideration neglects the effect of connectivity of mers on the calculated $N_{ij}^*$ values. So, they attempted to check whether the relation (7.48) might not be more appropriate for the surface characterized by random topography. The connectivity condition is the essential one for the effect of the number of mers on $\chi_n(\epsilon_n)$ while considering patchwise and random topography. Thus, in their two recent publications[48,49] on the extensions of the Dubinin-Radushkevich isotherm for the case of multi-site-occupancy adsorption, Rudzinski et. al. assumed that $E_n = \sqrt{n}E_1$.

For the purpose of illustrative calculations, we will consider here the case when the chemical bonds between mers exclude only a small portion of the adsorption sites being the nearest neighbours – sites of an adsorbed mer. So, we will accept the relation $E_n = nE_1$, to see effects of surface topography on multi-site-occupancy adsorption on heterogeneous solid surfaces.

Further, we will use equations (7.22–7.25) to show the effect of $n$, (the number of mers), on the behaviour of adsorption isotherms. In order to study the behaviour of adsorption isotherms in much different pressure regimes, we will draw the fractional coverage $\theta_t$ as the function of $\ln p$.

Then, a particular choice of the value of $\left(-\ln K' - \frac{\epsilon_{nl}}{kT}\right)$ in the expressions (7.23) and (7.25) will shift only the function $\theta_t(\ln p)$ toward higher or lower values of $\ln p$, without changing its shape. Thus, we will put the parameter $\left(-\ln K' - \frac{\epsilon_{nl}}{kT}\right)$ equal to zero. Finally, we will plot $\theta_t$ as the function of the dimensionless quantity $\ln \frac{p}{p_l}$.

Figures 18 and 19 show the effect of $n$ for two differrent values of $r$.

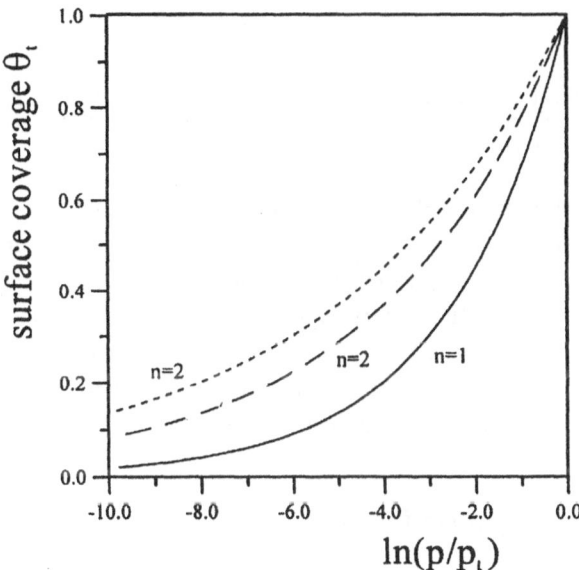

Figure 18. The effect of the number of mers, $n$, in the adsorbed molecule on the behaviour of adsorption istherm when $r = 1$, i.e. when $\chi_n(\epsilon_n)$ is a strongly nonsymmetrical, right hand widened adsorption energy distribution. The parameter $kT/E_1 = 0.4$ is typical for methane adsorption on carbonaceous adsorbents. It is still assumed that no interactions exist between the adsorbed moelcules, so, the interaction parameter $\omega = 0$. The solid line (—) is for $n = 1$, the slightly broken line (- - -) is for $n = 2$ and random topography, whereas the strongly broken line (- - - -) is for $n = 2$ and patchwise topography.

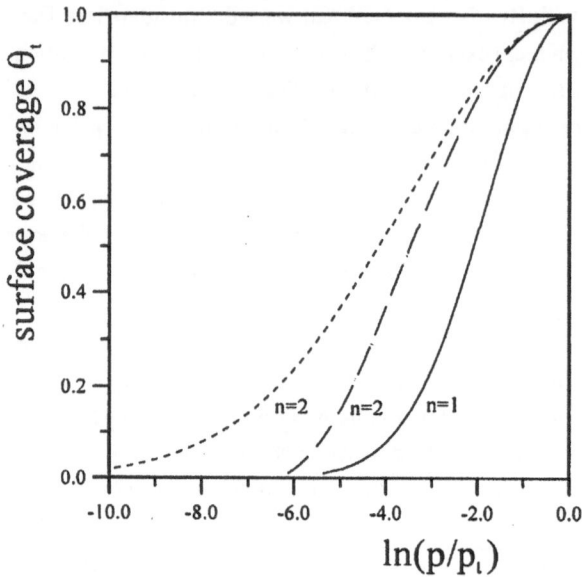

Figure 19. The effect of the number of mers, $n$, when $r = 2$, i.e. $\chi_n(\epsilon_n)$ is a slightly nonsymmetrical, right hand widened adsorption energy distribution. Other parameters and the meaning curves are the same as in Figure 16.

The comparison of Figures (18,19) show that in the case of patchwise topography the energetic surface heterogeneity affects the multi–site–occupancy adsorption isotherms, stronger than in the case of random topography. Thus, the effect of surface topography here is very similar to that of the surface topography in the case of one-site-occupancy collective adsorption, considered in Section 5. The explanation of this similarity is rather obvious. The chemical bonds between mers may be considered to some extent as very strong interactions between the adsorbed mers.

## 8. The Mixed-Gas Adsorption Equilibria: The Current State of the Research.

Nowadays the role of the energetic heterogeneity of the actual solid surfaces is almost commonly taken into account by the authors proposing various theoretical approaches to mixed-gas adsorption. They may be classified into three groups[50,51]:

1. the 'molecular' approaches employing methods of statistical thermodynamics;
2. the 'thermodynamic' approaches based on the methods of phenomenological thermodynamics;
3. the computer simulations of mixed-gas adsorption.

We will focus our attention on the two approaches that were most commonly used in the hitherto theoretical studies of mixed–gas adsorption.

Of all the "molecular" approaches, the most commonly used is the "Integral Equation" approach introduced into literature by Hoory and Prausnitz[52] at the beginning of the seventies. That approach was later developed by the adsorption group in the Department of Theoretical Chemistry, MC Sklodowska University in Lublin.

The other fundamental approach is the "Ideal Adsorbed Solution" approach, also introduced into literature at the beginning of the seventies by Myers and Prausnitz[53]. That approach was used and developed further in numerous papers on mixed–gas adsorption.

## 9. The Integral Equation Approach: Theoretical principles.

The first approach ever used to predict mixed-gas isotherms from the single component data is known as the integral-equation (IE) approach, and is based on the integral representation for $\theta_{ti}$,

$$\theta_{ti}(\mathbf{p}, T) = \int_{\Omega_1} \cdots \int_{\Omega_n} \theta_i(\epsilon, \mathbf{p}, T) \chi_{(n)}(\epsilon) d\epsilon_1 \cdots d\epsilon_n, \tag{9.1}$$

where: $\theta_{ti}(\mathbf{p}, T)$ is the total surface coverage by the component $i$ at the set of the partial pressures $\mathbf{p} = \{p_1, p_2, ..., p_n\}$, $\theta_i(\epsilon, \mathbf{p}, T)$ is the fractional coverage by the component $i$ ($i = 1, 2, ..., n$) of a certain class of adsorption sites, characterized by a set of the adsorption energies $\epsilon = \{\epsilon_1, \epsilon_2, \ldots, \epsilon_n\}$ for the single components; $\chi_{(n)}(\epsilon)$ is the $n$-dimensional normalized differential distribution of the number of the adsorption sites among various sets $\epsilon$,

$$\int_{\Omega_1} \cdots \int_{\Omega_n} \chi_{(n)}(\epsilon) d\epsilon_1 \cdots d\epsilon_n = 1, \tag{9.2}$$

and $\Omega_i$ is the $n$-dimensional physical domain of $\epsilon_i$. For the adsorption isotherms of single components, we have

$$\theta_{ti}(\mathbf{p}, T) = \int_{\Omega_i} \theta_i(\epsilon_i, \mathbf{p}, T) \chi_i(\epsilon_i) d\epsilon_i, \tag{9.3}$$

where

$$\chi_i(\epsilon_i) = \int_{\Omega_1} \cdots \int_{\Omega_{i-1}} \int_{\Omega_{i+1}} \cdots \int_{\Omega_n} \chi_{(n)}(\epsilon) d\epsilon_1 \cdots d\epsilon_{i-1} d\epsilon_{i+1} \cdots d\epsilon_n. \tag{9.4}$$

The integral (9.3) can be easily evaluated using the Rudziński–Jagiełło method.

When $T \to 0$, or when the variance of $\chi(\epsilon)$ is much larger than that of the derivative $\partial\theta/\partial\epsilon$,

$$\theta_{ti}(\mathbf{p}, T) = -\mathcal{X}_i(\epsilon_{ic}) , \quad \text{where} \quad \frac{d}{d\epsilon_i}\mathcal{X}_i(\epsilon_i) = \chi_i(\epsilon_i) \tag{9.5}$$

The features of the adsorption model are coded in the function $\epsilon_{ci}(p, T)$. When $\theta(\epsilon, p, T)$ is the Langmuir equation,

$$\epsilon_{ci}(p, T) = -kT \ln K_i p. \tag{9.6}$$

One of the fundamental problems of mixed-gas adsorption is a very common one, when the monolayer capacities estimated from the single-gas adsorption isotherms are different for different components. One faces that problem even in the case of mixed-gas adsorption on a hypothetical homogeneous solid surface. In such a case even the generalization of Langmuir equation for mixed-gas adsorption is difficult. So far it has been done only for the case when $\forall j(j \neq i) : \mathcal{N}_{sj} = \mathcal{N}_{si}$[54]:

$$\theta_i(\epsilon, \mathbf{p}, T) = \frac{K_i p_i \exp\left(\dfrac{\epsilon_i}{kT}\right)}{1 + \displaystyle\sum_{j=1}^{n} K_j p_j \exp\left(\dfrac{\epsilon_j}{kT}\right)}, \tag{9.7}$$

where $n$ denotes the number of the components in the adsorbed mixture. Provided that $\mathcal{N}_{si}$ and $\mathcal{N}_{sj}(j \neq i)$ are not much different, the next problem is to define the multidimensional adsorption energy distribution $\chi_{(n)}(\epsilon)$, which would reduce to $\chi_i(\epsilon_i)$ after $n-1$ integration steps, as outlined in eq. (9.4).

A general strategy is to reduce the $n$D integral (9.1) to a 1D integral by using various physical arguments. Most commonly, it is done by considering the correlations between the adsorption energies $\epsilon_i$ and $\epsilon_j(j \neq i)$, with $i, j = 1, 2, \cdots, n$. Two physical situations have been considered so far:

1. The adsorption energies $\epsilon_i$ and $\epsilon_j$ are not correlated at all for $j \neq i$;
2. Functional relationships exist:

$$\epsilon_i = f_{ij}(\epsilon_j), \quad i \neq j = 1, 2, \cdots, n. \tag{9.8}$$

## 10. The Case when Adsorption Energies are Strongly Correlated, and the Nature of the Gas-Solid Interactions is Similar for Various Components.

In such a case one can expect that the local minima of the gas-solid potential function for one of the components, will simultaneously be the local minima for other components. Then, until the sizes of the adsorbed molecules are not much different, we may assume that all the components are adsorbed on the same lattice of energetically different adsorption sites. Next, it is to be expected that the higher will be the adsorption energy for a certain component on one of the adsorption sites, the higher will also be its adsorption energy for other components. It means, the function $\epsilon_i = f_{ij}(\epsilon_j)$ in eq. (9.8) will be a one-to-one function.

Now, let us consider the extreme case of high correlations existing between the adsorption energies of various components, represented by the condition on every adsorption site[44,51]:

$$\epsilon_j = \epsilon_i + \Delta_{ji}, \quad j, i = 1, 2, \cdots, n, \tag{10.1}$$

where $\Delta_{ji}$-s are the constants. Then, $\theta_i(\epsilon_i, \mathbf{p}, T)$ can be rewritten to the following form:

$$\theta_i(\mathbf{p}, \epsilon_i, T) = \frac{K_i p_i}{\sum_j K_j p_j \exp\left(\frac{\Delta_{ji}}{kT}\right)} \frac{\sum_j K_j p_j \exp\left(\frac{\Delta_{ji}}{kT}\right) \exp\left(\frac{\epsilon_i}{kT}\right)}{1 + \sum_j K_j p_j \exp\left(\frac{\Delta_{ji}}{kT}\right) \exp\left(\frac{\epsilon_i}{kT}\right)}, \tag{10.2}$$

and the averaged function $\theta_{ti}(\mathbf{p}, T)$ is given by the 1D integral

$$\theta_{ti}(\mathbf{p}, T) = \frac{K_i p_i}{\sum_j K_j p_j \exp\left(\frac{\Delta_{ji}}{kT}\right)} \int_{\Omega_i} \frac{\sum_j K_j p_j \exp\left(\frac{\Delta_{ji}}{kT}\right) \exp\left(\frac{\epsilon_i}{kT}\right)}{1 + \sum_j K_j p_j \exp\left(\frac{\Delta_{ji}}{kT}\right) \exp\left(\frac{\epsilon_i}{kT}\right)} \chi(\epsilon_i) d\epsilon_i. \tag{10.3}$$

The above integral can also be evaluated by using the RJ approach. The function $\epsilon_c$ is defined now as follows:

$$\epsilon_c(\mathbf{p}, T) = -kT \ln \left( \sum_j K_j p_j \exp\left(\frac{\Delta_{ji}}{kT}\right) \right). \tag{10.4}$$

Equation (10.3) is again a kind of a master equation from which various isotherm equations can be developed corresponding to various adsorption energy distributions $\chi_i(\epsilon_i)$. For the Gaussian-like function (2.3) leading to the LF isotherm for single gas adsorption, $\theta_{ti}$ takes the following form

$$\theta_{ti}(\mathbf{p}, T) = \frac{K_i p_i \exp\left(\frac{\epsilon_{0i}}{kT}\right)}{\sum_j K_j p_j \exp\left(\frac{\epsilon_{0j}}{kT}\right)} \frac{\left[\sum_j K_j p_j \exp\left(\frac{\epsilon_{0j}}{kT}\right)\right]^{kT/c_i}}{1 + \left[\sum_j K_j p_j \exp\left(\frac{\epsilon_{0j}}{kT}\right)\right]^{kT/c_i}}, \tag{10.5}$$

because $(\epsilon_{0i} + \Delta_{ji})$ is the most probable value of $\epsilon_j$. (The numerator and denominator before the integral sign in eq. (10.3) have been multiplied by $\exp(\epsilon_{0i}/kT)$.)

When the adsorption isotherms of single components are well correlated by the DA equation (3.12), the equations for mixed-gas adsorption take the following form

$$\theta_{ti}(\mathbf{p}, T) = \frac{p_i/p_{il}}{\sum_{j=1}^{n} p_j/p_{jl}} \exp\left(-\left[\frac{kT}{E} \ln\left(\frac{1}{\sum_{j=1}^{n} p_j/p_{jl}}\right)\right]^{r_i}\right). \tag{10.6}$$

Both eq. (10.5) and (10.6) seem to be promising for correlating the mixed-gas adsorption data for the components having a similar chemical character. Another obvious condition is that the molecules of the different components should have similar sizes.

However, even in the case of such chemically similar molecules, the functional relationship (9.8) may have a form more complicated than that in eq. (10.1). It can be deduced from the low-temperature adsorption isotherms of single components.

At low temperatures, the adsorption will proceed in a fairly stepwise fashion, and the experimentally measured isotherm $\theta_{ti}(p, T)$ will be given by eq. (9.5). At the same coverage of surface by two components $i$ and $j$, the following relation will hold:

$$-\mathcal{X}_i(\epsilon_{ci}) = -\mathcal{X}_j(\epsilon_{cj}). \tag{10.7}$$

Now let us assume that the adsorption isotherms of both components obey the LF behaviour originating from the fully symmetrical, Gaussian-like energy distribution (2.3). Then, according to eq. (10.7) we have:

$$\left[1 + \exp\left(\frac{\epsilon_{ci} - \epsilon_{0i}}{c_i}\right)\right]^{-1} = \left[1 + \exp\left(\frac{\epsilon_{cj} - \epsilon_{0j}}{c_j}\right)\right]^{-1}. \tag{10.8}$$

From eq. (10.8) we obtain the following linear relation

$$\epsilon_j = \frac{c_j}{c_i}\epsilon_i + \left(\epsilon_{0j} - \frac{c_j}{c_i}\epsilon_{0i}\right), \tag{10.9}$$

where we have already omitted the superscript 'c' in $\epsilon_c$.

Thus in the case of the Gaussian-like adsorption energy distribution (2.3), $\epsilon_j$ is a linear function of $\epsilon_i$, and its slope is related to the ratio $(c_j/c_i)$ of the variances of that function for the components $j$ and $i$.

Similarly, when the single-gas adsorption isotherms of all the components are described by the DA isotherm (2.4), then, from eqs. (3.11) and (10.7) we obtain the following interrelation

$$\epsilon_j = \epsilon_{lj} + E_j \left(\frac{\epsilon_i - \epsilon_{li}}{E_i}\right)^{r_i/r_j}. \tag{10.10}$$

Figure 20 shows the relation (10.10) for two components the adsorption energy distributions of which have the same variance $E$, but different symmetry, characterized by two values $r_i$ and $r_j$.

As the function $\mathcal{X}(\epsilon)$ is called the 'cumulative adsorption-energy distribution', using eq. (10.7) is called 'seeking for the adsorption energy interrelations through the cumulative distributions'. That idea has been proposed independently by Valenzuela et al.[56] and by Jaroniec and coworkers[57], in the same year (1988).

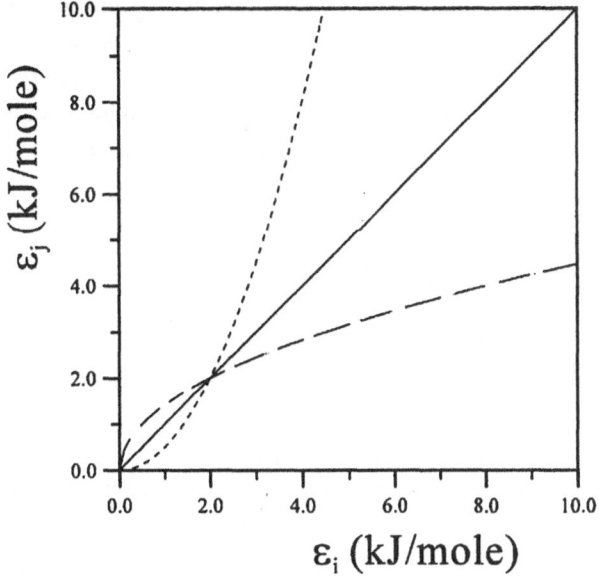

Figure 20. The relation $\epsilon_j = \epsilon_j(\epsilon_i)$ given by eq. (10.10), when $E_j = E_i = 2$ kJ/mole $\epsilon_{li} = \epsilon_{lj} = 0$, for the three values of $(r_i/r_j)$: 0.5 (- - -), 1.0 (—) and 2.0 (- - - -).

## 11. The Case when the Adsorption Energies of Various Components are not Correlated at all.

This is the case of coadsorption of components exhibiting a much different character of interactions with the same solid surface. These differences are usually due to different chemical nature of the coadsorbed components.

There is a certain tendency in literature to apply again eq. (9.1) to such cases, and to consider $\chi_{(n)}(\epsilon)$ to be the following product:

$$\chi_{(n)}(\epsilon) = \prod_{i=1}^{n} \chi_i(\epsilon_i). \tag{11.1}$$

This seems to be an improper strategy. Even if the molecules of different components have similar sizes, their different nature of the gas-solid interactions will result in various depths and positions of the local minima in the gas-solid potential function. It means, we have to consider various non-overlapping and even not correlated lattices of asdorption sites for each of the coadsorbing components.

The case when $\epsilon_i$ and $\epsilon_j$ (with $j \neq i$) are not correlated at all was considered by Wojciechowski et al.[58]. They used the following argument: When the energies $\epsilon_i$ and $\epsilon_{j \neq i}$

are not correlated at all, there should also be no spatial correlations between the local minima - adsorption sites of the components $i$ and $j$. So, when the adsorption energies of the components $i$ and $j$, $\epsilon_i, \epsilon_j$ (with $i, j = 1, 2 \cdots, n, i \neq j$) are not correlated at all, the presence of other components will affect the adsorption of $i$ only by random blocking of adsorption sites, proportional to their total coverages $\theta_{tj}$-s. (The probability that the molecule $j$ will be adsorbed on a certain site does not depend on $\epsilon_i$.)

Thus,

$$\theta_{ti}(\mathbf{p}, T) = -\left(1 - \sum_{j \neq i}^{n} \theta_{tj}(\mathbf{p}, T)\right) \mathcal{X}_i(\epsilon_{ci}), \qquad (11.2)$$

Equation (11.2) is another kind of a master equation from which various expressions for the mixed-gas isotherm can be derived by assuming various adsorption energy distributions $\chi(\epsilon)$.

The equation system (11.2) is linear with respect to $\theta_{ti}(\mathbf{p}, T)$, so, it can be solved easily, to express $\theta_{ti}(\mathbf{p}, T)$ by $\mathcal{X}_i(\epsilon_{ci})$'s. For the purpose of illustration we consider the adsorption from a binary gaseous mixture $(1 + 2)$. Then, the solution takes the following form:

$$\theta_{t1}(p_1, p_2, T) = \frac{-\mathcal{X}_1 + \mathcal{X}_1 \mathcal{X}_2}{1 + \mathcal{X}_1 \mathcal{X}_2}, \qquad (11.3)$$

$$\theta_{t2}(p_1, p_2, T) = \frac{-\mathcal{X}_2 + \mathcal{X}_1 \mathcal{X}_2}{1 + \mathcal{X}_1 \mathcal{X}_2}, \qquad (11.4)$$

Generally if all $\mathcal{X}_i$-s are the LF isotherms for single component adsorption, the solution of the equation system (11.2) yields the generalized LF mixed-gas adsorption isotherm

$$\theta_{ti}(\mathbf{p}, T) = \frac{\left[K_i p_i \exp\left(\frac{\epsilon_{0i}}{kT}\right)\right]^{kT/c_i}}{1 + \sum_{j=1}^{n} \left[K_j p_j \exp\left(\frac{\epsilon_{0j}}{kT}\right)\right]^{kT/c_j}}, \qquad i = 1, 2, \cdots, n. \qquad (11.6)$$

When $\mathcal{X}_i$ is given by equation, eq. (2.4), we arrive at the following explicit expressions:

$$\theta_{t1}(p_1, p_2, T) = \frac{\left[1 - \exp\left(-\left[\frac{kT}{E_2} \ln\left(\frac{p_{12}}{p_2}\right)\right]^{r_2}\right)\right] \exp\left(-\left[\frac{kT}{E_1} \ln\left(\frac{p_{11}}{p_1}\right)\right]^{r_1}\right)}{1 - \exp\left(-\left[\frac{kT}{E_1} \ln\left(\frac{p_{11}}{p_1}\right)\right]^{r_1} - \left[\frac{kT}{E_2} \ln\left(\frac{p_{12}}{p_2}\right)\right]^{r_2}\right)}, \qquad (11.7)$$

and

$$\theta_{2t}(p_1, p_2, T) = \frac{\left[1 - \exp\left(-\left[\frac{kT}{E_1} \ln\left(\frac{p_{11}}{p_1}\right)\right]^{r_1}\right)\right] \exp\left(-\left[\frac{kT}{E_2} \ln\left(\frac{p_{12}}{p_2}\right)\right]^{r_2}\right)}{1 - \exp\left(-\left[\frac{kT}{E_1} \ln\left(\frac{p_{11}}{p_1}\right)\right]^{r_1} - \left[\frac{kT}{E_2} \ln\left(\frac{p_{12}}{p_2}\right)\right]^{r_2}\right)}. \qquad (11.8)$$

This equation is to be compared with eq. (10.6) which is the extension of the DA isotherm for the case of mixed-gas adsorption, when the adsorption energies of various components are strongly correlated.

Now, there still remains the problem of relating the theoretically calculated functions $\theta_{ti}(\mathbf{p}, T)$, to the experimentally monitored quantities $\mathcal{N}_{ti}(\mathbf{p}, T)$.

Most of the authors applying the multicomponent LF equation, eq. (11.6) assumed that $N_{ti} = \mathcal{N}_{si}\theta_{ti}$. This statement is not so obvious and deserves further theoretical studies.

Very frequently the so-called "phase diagrams" are used to represent mixed-gas adsorption equilibria. In the case of a binary adsorbed mixture the phase diagram shows the dependence of the mole fraction of the component 1, $X_1$, on the mole fraction of this component $Y_1 = p_1/(p_1 + p_2)$ in the equilibrium bulk phase, when the total pressure $(p_1 + p_2)$ is constant. Figure 21 shows schematically such phase diagram.

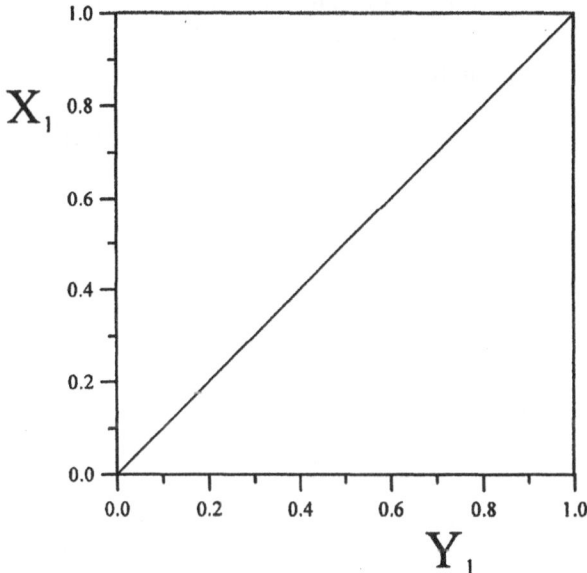

Figure 21. The schematic phase diagram for adsorption from a binary gas mixture. The solid line corresponds to a hypothetical case when none of the components is preferentially adsorbed.

The points lying above the solid line in Figure 21 correspond to preferential adsorption of component 1 whereas the points lying below that solid line correspond to a preferential adsorption of component 2. Sometimes it happens that part of a phase diagram lies above that line or vice versa. It is stated then, that the adsorbed mixture shows an "azeotrope" on the phase diagram.

The term "azeotrope" was taken from the theories of liquid-vapour equilibria, and had there its source in non-ideality of a liquid mixture. There is, therefore, a strong tendency in adsorption literature to look for interactions between the adsorbed molecules as the source of the azeotropes in the phase diagrams. Below, we are going to show that surface energetic heterogeneity may affect strongly the shape of these phase diagrams, and may also be a source of the observed azeotropes.

This fact has not received enough attention in the hitherto literature reporting on mixed–gas adsorption on solids.

For the purpose of illustration we will consider here equations for mixed–gas isotherms, developed by using the integral equation approach and the case of a binary adsorbed mixture. We will define further $X_i$ as $\theta_{ti}/(\theta_{t1} + \theta_{t2})$ and $Y_i = p_i/(p_1 + p_2)$.

First we check how the correlations between adsorption energies of components 1 and 2 affect the phase diagram $X - Y$. For that purpose we compare in Figure 22 the phase diagrams predicted by eq. (10.5) and eq. (11.6) respectively, which are generalizations of the Langmuir–Freundlich isotherm (3.8) for adsorption from a binary gas mixture. When all $kT/c_i$ in eqs. (10.5) and (11.6) are the same, the difference between them is only due to the high correlations between adsorption energies (10.1) leading to eq. (10.5), and a total lack of correlations leading to eq. (11.6).

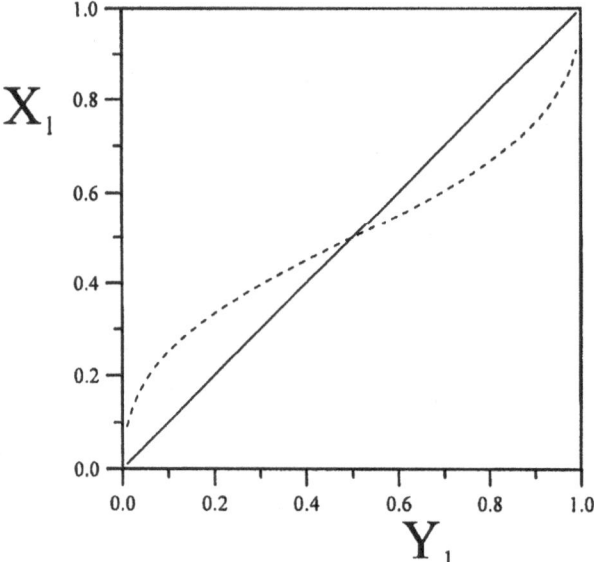

Figure 22. The solid line is the phase diagram calculated from eq. (10.5), whereas the slightly broken line is the phase diagram calculated from eq. (11.6), when $kT/c_i = 0.5$ and $(p_{r1} + p_{r2}) = 1$.

While calculating the phase diagram 22 we introduced the "reduced" pressures $p_{ri} = K_i p_i \exp\left(\frac{c_{0i}}{kT}\right)$.

One can see from Figure 22 that lack of correlations between adsorption energies of the components 1 and 2 leads to azeotropic behaviour of the phase diagram $X - Y$. Figure 23 shows that in the absence of correlations between adsorption energies strong surface heterogeneity for one of the components creates strong tendencies for its preferential adsorption in a large range of the composition in the equilibrium bulk phase $Y$.

It is interesting to note that mixed–gas isotherms developed by assuming that the high correlations (10.1) exist between adsorption energies, never lead to azeotropes in the corresponding $X - Y$ diagrams. Figures 24 and 25 show this for the case of the generalized of Dubinin–Astakhov equation.

Figure 25 shows that in the case of the Dubinin-Astakhov isotherm generalized for the case when no correlations exist between adsorption energies, the growing value of the parameter $r$ causes stronger tendencies to the azeotropic behaviour. Then, Figure 26 shows that strong surface heterogeneity for one of the components creates strong tendencies for its preferential adsorption in a large region of bulk phase compositions. That result observed for the strongly non–symmetrical (right hand widened) adsorption energy distribution, is essentially similar to that observed in Figure 23 corresponding to fully symmetrical adsorption energy distribution.

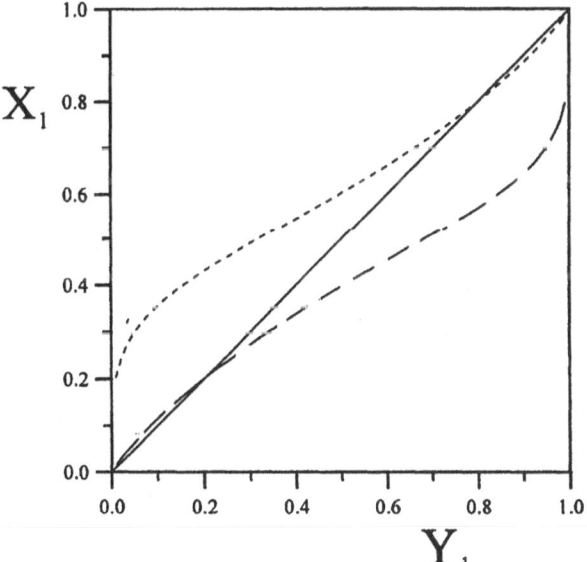

Figure 23. The slightly dashed line is the phase diagram calculated from eq. (11.6) when $kT/c_1 = 0.9$ and $kT/c_2 = 0.3$, whereas the strongly broken line is for $kT/c_1 = 0.3$ and $kT/c_2 = 0.9$. The total equilibrium (reduced) bulk pressure $(p_{r1} + p_{r2}) = 1.0$.

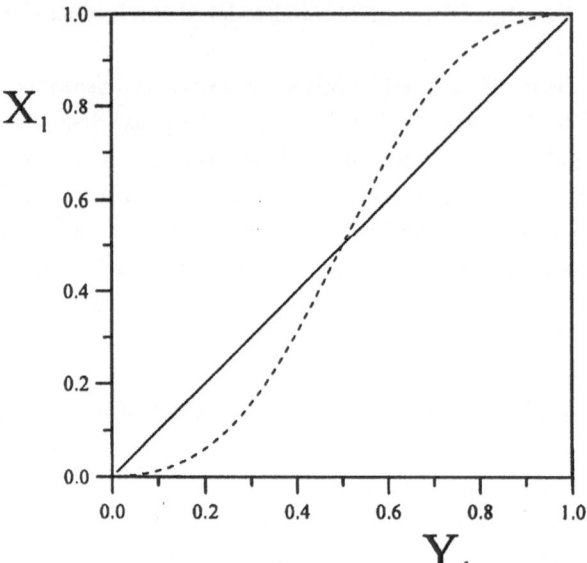

Figure 24. The solid line is the phase diagram calculated from eq. (10.6), whereas the slightly broken line (– – –) is the phase diagram calculated from eqs. (11.7-11.8), when $kT/E_1 = kT/E_2 = 1.0$, $r_1 = r_2 = 1$ and the sum of the relative pressures $p_{ri} = p_i/p_{li}$ in the equilibrium bulk phase is equal to 0.99.

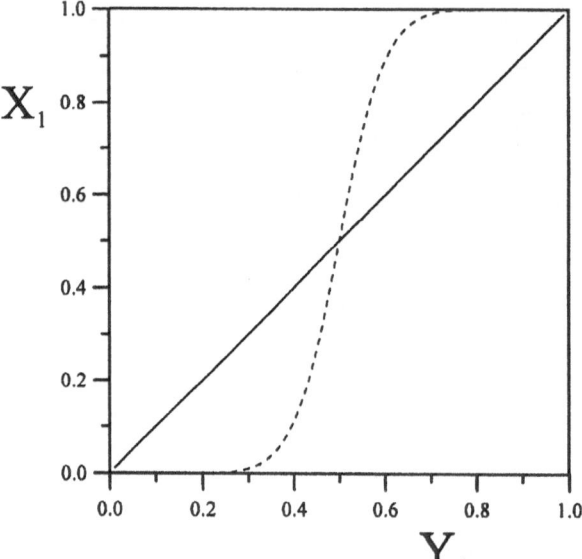

Figure 25. The phase diagrams calculated for the case $r_1 = r_2 = 3$. All other parameters are the same as in Figure 303.

229

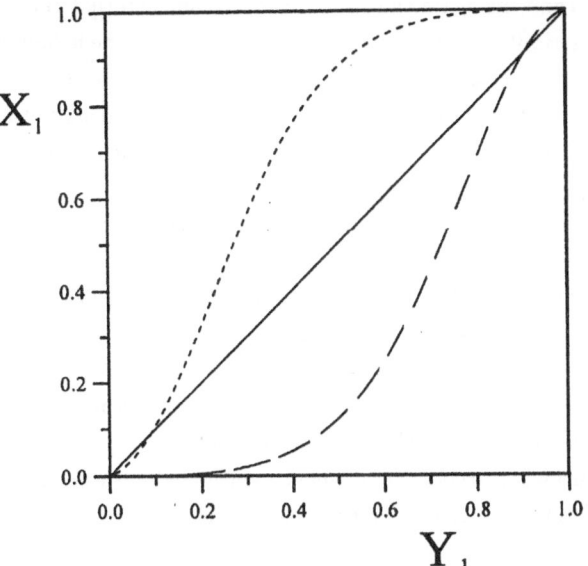

Figure 26. The slightly dashed line (– – –) is the phase diagram calculated for eqs. (11.7-11.8) by assuming that $kT/E_1 = 2.0$, and $kT/E_1 = 0.5$, whereas the strongly broken line (- - - -) is for the case when $kT/E_1 = 0.5$ and $kT/E_2 = 2.0$. Both the phase diagrams were calculated for the case $r_1 = r_2 = 1$, and when $(p/p_{1l} + p_2/p_{2l}) = 0.99$.

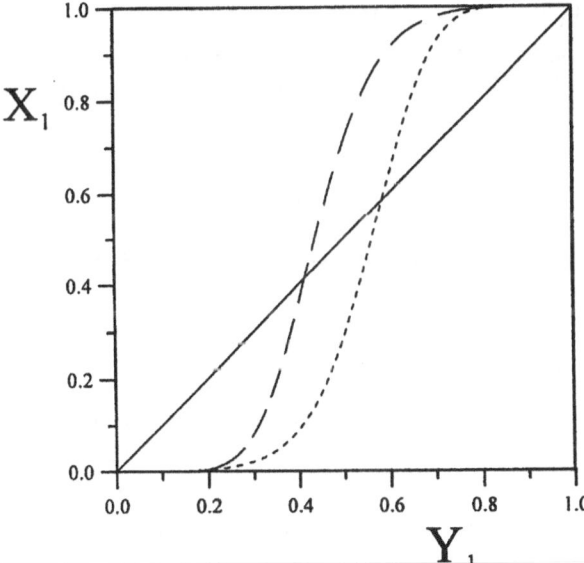

Figure 27. The slightly broken line (– – –) is the phase diagram calculated from eqs. (11.7-11.8) for $r_1 = 3$, $r_2 = 1$, whereas the strongly broken line (- - - -) is for the case when $r_1 = 1$, $r_2 = 3$. The calculations were performed for $kT/E_1 = kT/E_2 = 1.0$ and the total pressure $(p_1/p_{1l} + p_2/p_{2l}) = 0.99$.

Finally Figure 27 confirms again the above drawn conclusion that the symmetry of the adsorption energy distribution is of a secondary importance for the behaviour of the phase diagrams, calculated for the case when no correlations exist between adsorption energies. The differences alone in the symmetry of the adsorption energy distributions cannot be a source of azeotropes. The essential source of azeotropes is the lack of correlations between adsorption energies of co-adsorbed gases.

Preferential adsorption of one of the components has, usually, been ascribed to stronger gas-solid interactions. Now, we can see that preferential adsorption may also be related to stronger surface heterogeneity of one of the components. Next, the appearance of azeotropes on the $X - Y$ diagrams has, usually, been ascribed to strong interactions between the adsorbed molecules. Meanwhile, our model calculations show that the lack of correlations between the adsorption energies of components may also be a source of azeotropic behaviour. Thus, we could conclude the above discussion as follows. It seems, that the effects of surface energetic heterogeneity on mixed-gas adsorption on solids are not understood well yet and deserve further extensive studies.

## 12. The Ideal Adsorbed Solution Approach IAS: Theoretical Principles.

The IAS approach was launched first in the work by Myers and Prausnitz[54] and reexamined recently by Rudisil and Le Van[59].

The IAS theory is based on the assumption that the adsorbed phase can be treated as an ideal solution of the adsorbed components. The reduced spreading pressures $\Pi_i^*$,

$$\Pi_i^* = \frac{\Pi_i A}{RT} = \int_0^{p_i^*} \frac{\mathcal{N}_i}{p_i} dp_i, \tag{12.1}$$

of the components $i = 1, 2, \cdots, n$, in their single-gas standard states, are equal to the reduced spreading pressure of the adsorbed mixture, $\Pi^*$:

$$\Pi_1^* = \Pi_2^* = \ldots = \Pi_n^* = \Pi^*. \tag{12.2}$$

The function $\mathcal{N}_i(p_i, T)$ denotes the adsorbed amount, (moles per unit mass of adsorbent), at the pressure $p_i$ and temperature $T$, and $A$ is the specific surface area. $\mathcal{N}_i(p_i, T) = \mathcal{N}_{si}\theta_i(p_i, T)$, where $\theta_i$ is the fractional coverage by the component $i$ of the surface, and $\mathcal{N}_{si}$ is the maximum adsorbed amount when $\theta_{ti} = 1$. It is assumed in the present consideration that $\mathcal{N}_i$ and $\mathcal{N}_{si}$ are expressed in moles (per unit area or unit weight).

The relation between the mole fraction in the gaseous phase $Y_i$, and the mole fraction in the adsorbed phase $X_i$ is described by Raoult's law for ideal solutions,

$$PY_i = p_i^*(\Pi^*)X_i, \tag{12.3}$$

where $p_i^*$ is the pressure of the single component $i$ in its standard state which is fixed by the spreading pressure of the mixture according to eq. (12.2). In addition, the mole fraction constraints are $\sum_{i=1}^{n} X_i = 1$ and $\sum_{i=1}^{n} Y_i = 1$.

The total amount of the adsorbed gases $\mathcal{N}^0$ depends on the amounts $\mathcal{N}_i^0$ adsorbed in the standard state:

$$\frac{1}{\mathcal{N}^*} = \sum_{i=1}^{n} \frac{X_i}{\mathcal{N}_i^*}, \tag{12.4}$$

where $\mathcal{N}_i^0$ is the value $\mathcal{N}_i$ for the pure component $i$, at $p_i = p_i^*(\Pi^*)$. The actual amount of each component adsorbed is $\mathcal{N}_i = \mathcal{N}^* X_i$.

There is one fundamental problem accompanying the application of IAS to predict the mixed-gas adsorption equilibria on heterogeneous solid surfaces. That problem has never been explicitly discussed in the papers reporting on application of IAS, except for the recent paper by Rudziński et al[50].

The term 'ideal solution' refers to the systems of interacting molecules in which the 'interchange energy' $W$ is equal to zero:

$$W = W_{ii} + W_{jj} - 2W_{ji} = 0, \quad i, j = 1, 2, \cdots, n,$$

were $W_{ij}$ is the interaction energy between two molecules $i$ and $j$ adsorbed on two neighbouring adsorption sites.

As discussed in Sect. 4, when the adsorption of interacting molecules (collective adsorption) on a heterogeneous surface is considered, one must take also into account the surface topography. So far two extreme models of surface topography have been commonly considered:

1. The 'patchwise' model assuming that the surface is composed of large homogeneous surface domains — 'patches', grouping identical adsorption sites. Thus, the adsorption systems have to be considered as composed of macroscopically different subsystems being only in thermal and material contacts. At a certain adsorbate pressure $p$ and temperature $T$ the adsorbed amount $\mathcal{N}_i$ and the corresponding spreading pressure $\Pi_i^*$ are to be referred to a certain patch and will be a function of the adsorption energy $\epsilon$ on this patch. Thus, the correct use of IAS to predict the mixed-gas adsorption equilibria will require the following sequence of operations. First, IAS has to be applied to every homogeneous patch to calculate $\{X_i\}$ values. They are, of course, a function of the energy set $\epsilon$. Next, the calculated $X_i$-s values have to be averaged over all possible sets $\epsilon$ as in the IE approach.

2. The 'random' model assuming that adsorption sites characterized by different adsorption energies are distributed on a heterogeneous surface completely at random. As a result, the local 'microscopic' composition $\{X_i\}$ of the surface phase is the

same across the surface. The adsorbed phase is a thermodynamic entity, the spreading pressure of which is the same across the surface, and has to be calculated by replacing $\mathcal{N}_i$ by $\mathcal{N}_{ti}$, in eq. (12.1).

## 13. Application of IAS to the Surface with Random Topography.

Myers and coworkers were was the first to consider the extension of IAS for the case of heterogeneous solid surfaces did it for the surfaces with random topography first. In his work concerning this possibility, he accepted the rectangular adsorption energy distribution to represent $\chi_i(\epsilon_i)$. Then, the problem was considered more extensively in his second paper coauthored by Richter and Schütz[60].

Readers may appreciate using the analytical approximation for $\Pi^*(p)$, developed by Myers and coworkers for the case when $\mathcal{N}_t(p)$ is the DR isotherm,

$$\Pi^* = \frac{\Pi A}{kT} = \frac{\mathcal{N}_s(\Pi)^{1/2}}{2\,(kT/E)}\mathrm{erfc}\left(\frac{kT}{E}\ln\left(\frac{p_s}{p}\right)\right). \tag{13.1}$$

As the complementary error function $\mathrm{erfc}(\cdot)$ is the standard one in computers now, eq. (13.1) may therefore be considered as the analytical expression for $\Pi^*$.

The problem of the transition to Henry's Law of the equations used to calculate $\Pi^*$ in IAS is of primary importance for the successful application of IAS. But then, also a good fit of the experimental isotherms at mediate coverages is another essential condition. In the mediate coverage region a good fit is obtained usually by using isotherm equations which do not reduce to Henry's Law. (LF, DR, DA, ...). This is a serious difficulty accompanying the use of the IAS approach to predict mixed-gas adsorption equilibria.

Rudziński et al.[50] have proposed a general solution for that problem, by using the RJ approach.

In view of what was said above, the integral (12.1) should be written as follows

$$\Pi^* = \mathcal{N}_s \int_0^{p_m} \theta(\epsilon_1)\frac{dp}{p} - \mathcal{N}_s \int_{p_m}^{p^*} \mathcal{X}(\epsilon_c)\frac{dp}{p}, \tag{13.2}$$

where

$$p_m = 1/K\exp\left(-\epsilon_m/kT\right). \tag{13.3}$$

Equation (13.2) is quite general and any function $\mathcal{X}(\epsilon_c)$ i.e. any $\chi(\epsilon)$ can be accepted. It is convenient to carry out the integration by writing the second integral on the r.h.s. of eq. (13.2) in the following form

$$\int_{p_m}^{p^*} \mathcal{X}(\epsilon_c)\frac{dp}{p} = \frac{1}{kT}\int_{\epsilon_c(p_m)}^{\epsilon_c(p^*)} -\mathcal{X}(\epsilon_c)d\epsilon_c. \tag{13.4}$$

For the particular case of the function (2.5), we have

$$\int_{P_m}^{p^*} \mathcal{X}(\epsilon_c)\frac{dp}{p} = -\frac{c}{F_N kT} \ln \left( \frac{1 + \exp\left(\dfrac{\epsilon_0 - \epsilon_c(p^*)}{c}\right)}{1 + \exp\left(\dfrac{\epsilon_0 - \epsilon_c(p_m)}{c}\right)} \right)$$

$$= -\frac{c}{F_N kT} \ln \left( \frac{1 + \left(Kp^* \exp\left(\dfrac{\epsilon_0}{kT}\right)\right)^{\frac{kT}{c}}}{1 + \exp\left(\dfrac{\epsilon_0 - \epsilon_m}{c}\right)} \right). \tag{13.5}$$

Then, the first integral on the r.h.s. of eq. (13.2) takes the form

$$\int_0^{P_m} \theta(\epsilon_1)\frac{dp}{p} = -\frac{1}{kT} \int_{+\infty}^{\epsilon_c(p_m)} \left[1 + \exp\left(\frac{\epsilon_c - \epsilon_1}{kT}\right)\right]^{-1} d\epsilon_c$$

$$= \ln\left(1 + \exp\left(\frac{\epsilon_1 - \epsilon_c(p_m)}{kT}\right)\right) \tag{13.6}$$

$$= \ln\left(1 + \exp\left(\frac{\epsilon_1 - \epsilon_0}{kT} - \frac{\epsilon_m - \epsilon_0}{kT}\right)\right).$$

Finally

$$\Pi^* = \Pi_m^* + \frac{\mathcal{N}_s c}{F_N kT} \ln\left(1 + \left(Kp^* \exp\left(\frac{\epsilon_0}{kT}\right)\right)^{kT/c}\right), \tag{13.7}$$

where

$$\Pi_m^* = \mathcal{N}_s \ln \left( \frac{1 + \exp\left(\dfrac{\epsilon_1 - \epsilon_0}{kT} - \dfrac{\epsilon_m - \epsilon_0}{kT}\right)}{\left[1 + \exp\left(\dfrac{\epsilon_0 - \epsilon_m}{c}\right)\right]^{c/F_N kT}} \right). \tag{13.8}$$

For the purpose of further considerations we will rewrite eq. (13.7) to the following form

$$\frac{\Pi A}{k'T} = \Pi^* = \Pi_m^* - \mathcal{N}_s \frac{c}{k'T} \ln(1 - \theta_t).$$

After solving eq. (13.7) for $p^*$, we have

$$p^* = \frac{1}{K} \exp\left(-\frac{\epsilon_0}{kT}\right) \left[\exp\left(\frac{F_N kT}{Mc}(\Pi^* - \Pi_m^*)\right) - 1\right]^{c/kT}, \tag{13.9}$$

where $K\exp(\epsilon_0/kT)$ is found from the linear regression (155) for the experimental adsorption isotherm.

Let us consider the simplest case of a binary adsorbed mixture. Then

$$p_1 = p_1^* X_1 \quad \text{and} \quad p_2 = p_2^*(1 - X_1). \tag{13.10}$$

The quantity $p^*$ defined in eq. (13.9) is a function of $\Pi^*$ and of the parameters $K \exp\{\epsilon_0/kT\}$, $kT/c$, $(\epsilon_0 - \epsilon_1)/c$, and $(\epsilon_m - \epsilon_0)/c$. The first two parameters are easily found from the linear regression (3.14) of the adsorption data measured at moderate surface coverages for a single adsorptive. The other two parameters: $(\epsilon_0 - \epsilon_1)/c$ and $(\epsilon_m - \epsilon_0)/c$ can be found by an appropriate numerical analysis of low-coverage data, based on more general equation (4.9). Such low-coverage data are not commonly available, so the parameters $(\epsilon_0 - \epsilon_1)c$ and $(\epsilon_m - \epsilon_0)/c$ are likely to be treated as best-fit ones.

Recently, Rudzinski et al.[61] have shown, that the PT (Potential Theory) approach to mixed gas adsorption is nothing else, but a special edition of IAS approach which is applicable only to strongly heterogeneous surfaces, characterized by random surface topography. The potential theory of adsorption, introduced by Polanyi, has been widely accepted as the basis for correlating the effect of temperature on the adsorption isotherms of pure gases. A number of authors have modified Polanyi's original method of treating pure-gas adsorption isotherms. These modifications attempt both to improve the temperature correlation and to account for the effect of the nature of the adsorbate on the adsorption isotherm for a given adsorbent.

The PT method for predicting mixed-gas adsorption equilibria is based on the assumption that the adsorption data for all pure components fall on the same line when plotted on appropriate coordinates. A number of methods based on the PT have been advocated for correlating pure-gas adsorption isotherms. The most recent extensive experimental and theoretical study by Mehta and Danner[62] showed that the method of Lewis et al. appeared to be the most useful one. It consists in expressing the adsorption isotherm of a pure component $i$ in the following functional form

$$N_{ti} \cdot V_i^s = F\left(\frac{kT}{V_i^s} \ln\left(\frac{f_{si}}{f_i}\right)\right), \tag{13.11}$$

where $F$ is the temperature independent function of $(kT/V_i^s)\ln(f_{si}/f_i)$, $V_i^s$ is the saturated liquid molar volume of the $i$-th pure adsorbate at a pressure equal to the adsorption pressure $p_i$, $f_{si}$ is the saturation fugacity of pure $i$ at the adsorption temperature $T$, $f_i$ is the fugacity at the pressure $p_i$ and the temperature $T$. Moon and Tien[63] have shown that in most cases one may replace the bulk fugacities by pressures.

Rudziński et al.[61] have shown that the PT approach to mixed gas adsorption is nothing else but a special form of IAS approach which can be used only in the case of strongly heterogeneous surfaces, exhibiting random topography. For such surfaces the integral for the spreading pressure $\Pi^*$ takes the form[61],

$$\Pi^* = \int_0^{p_i^*} N_{ti} \frac{dp_i}{p_i} = -\frac{\mathcal{N}_{si}}{kT} \int_{+\infty}^{\epsilon_{ci}(p_i^*)} \mathcal{X}(\epsilon_{ci})d\epsilon_{ci}. \tag{13.12}$$

Assuming $\Pi_i^* = \Pi_{j\neq i}^*$, we have,

$$\mathcal{N}_{si} \int_{+\infty}^{\epsilon_{ci}(p_i^*)} X_i(\epsilon_{ci})d\epsilon_{ci} = \mathcal{N}_{sj} \int_{+\infty}^{\epsilon_{ci}(p_j^*)} X_j(\epsilon_{cj})d\epsilon_{cj}. \tag{13.13}$$

The above equation will be important for our further consideration. Meanwhile we use the following transformation of the variables:

$$t_i = \frac{\epsilon_{ci} + kT\ln(K_i p_{si})}{V_i^s} = \frac{kT}{V_i^s}\ln\left(\frac{p_{si}}{p_i}\right). \tag{13.14}$$

Then, eq. (13.13) takes the form

$$\int_{+\infty}^{(kT/V_i^s)\ln(p_{si}/p_i^*)} V_i^s N_{ti}(t_i)dt_i = \int_{+\infty}^{(kT/V_j^s)\ln(p_{sj}/p_j^*)} V_j^s N_{tj}(t_j)dt_j. \tag{13.15}$$

Thus, if it happens that $V_i^s N_{ti}$ plotted as a function of $t_i$ coalesces with $V_j^s N_{tj}$ plotted as a function of $t_j$, then the upper integration limits must be equal, i.e.

$$\frac{kT}{V_i^s}\ln\left(\frac{p_{si}}{p_i^*}\right) = \frac{kT}{V_j^s}\ln\left(\frac{p_{sj}}{p_j^*}\right). \tag{13.16}$$

Taking into account the Raoult Law (12.3), we can rewrite eq. (13.6) to the following form

$$\frac{kT}{V_i^s}\ln\left(\frac{X_i p_{si}}{p_i}\right) = \frac{kT}{V_j^s}\ln\left(\frac{X_j p_{sj}}{p_j}\right), \tag{13.17}$$

where $p_i$ means the partial pressure of the adsorbate $i$. Usually, $p_i$ well approximates $f_i$, so, eq. (13.7) becomes essentially the Grant–Manes equation. Thus, it should be clearly stated that the PT approach is nothing else, but a special form of IAS theory, valid only for the case of heterogeneous solid surfaces.

Rudziński et. al. have shown that their theoretical development of the PT approach suggests a possibility of two coalescing factors to be introduced. These authors also give certain recommendations as to the way in which one can use the PT approach to predict mixed-gas adsortpion equilibria better than by applying the traditional 'coalescence' operation.

## 14. Application of IAS to the Surfaces with Patchwise Topography.

In their review, Rudzinski et. al.[50] as the first emphasized it clearly, that the authors proposing various extensions of IAS for energetically heterogeneous surfaces did not address explicitly their approaches to a certain topographical model of surface topography.

To illustrate the problem of surface topography Rudzinski et. al. took extensions of the Langmuir-Freundlich equation into consideration. Their results are briefly reported below.

We start with calculating the spreading pressure $\Pi_i^*$ on a homogeneous surface patch characterized by the adsorption energy $\epsilon$,

$$\Pi_i^* = \mathcal{N}_{si} \ln \left( 1 + K_i p_i^* \exp \left( \frac{\epsilon_i}{kT}, \right) \right), \tag{14.1}$$

or, in another form,

$$\frac{\Pi_i A}{kT} = \Pi_i^* = -\mathcal{N}_{si} \ln \left( 1 - \theta_i \right). \tag{14.2}$$

After solving it for $p_i^*$, we have

$$p_i^* = \frac{1}{K_i} \left[ \exp \left( \frac{\Pi^*}{\mathcal{N}_{si}} \right) - 1 \right] \exp \left( -\frac{\epsilon_i}{kT} \right). \tag{14.3}$$

For two-component adsorption, from eqs. (??) and (??) we have,

$$
\begin{aligned}
p_1 &= X_1 \frac{\exp \left( \Pi^*/\mathcal{N}_{s1} \right) - 1}{K_1 \exp \left( \epsilon_1/kT \right)}, \\
p_2 &= (1 - X_1) \frac{\exp \left( \Pi^*/\mathcal{N}_{s2} \right) - 1}{K_2 \exp \left( \epsilon_2/kT \right)}.
\end{aligned}
\tag{14.4}
$$

Valenzuela et al.[56] were first to combine IAS with the concept of seeking the correlations between $\epsilon_i$ and $\epsilon_j$ for $j \neq i$ through their cumulative distributions.

Valenzuela et. al. took for illustration the experimental data reported by Talu and Zwiebel[64] on the adsorption of the binary and ternary mixtures of $H_2S$, $CO_2$, and $C_3H_8$ on H-modernite at 303.15 K.

While predicting mixed-gas adsorption equilibria, they carried out their calculation twice. First they used the single-gas adsorption isotherms to calculate mixed-gas equilibria in terms of the ordinary IAS theory. Next, they calculated the mixed-gas isotherms by using the theoretical approach described above, and called by them the heterogeneous ideal adsorbed solution (HIAS) theory.

Generally, the improvement in agreement obtained by adopting HIAS is only moderate when compared to that achieved by IAS, so, Valenzuela *et al.* concluded that still "something was missing in the theory".

Just at the same time (1987) the concept of the 'site matching' was elaborated independently by Moon and Tien[65]. They made, however, one step forward by generalizing it through the assumption that instead of eq. (10.7) we may have in general a more complicated relation.

# Literature

1. M.Jaroniec, and E.Madey, "Physical Adsorption on Heterogeneous Solids", Elsevier, 1988.

2. W.Rudziński, and D.H.Everett, "The Adsorption of Gases on Heterogeneous Surfaces", Acadaemic Press, 1992.

3. A. Kapoor, J.A. Ritter, and R.T. Yang, Langmuir, (1989).

4. A. Kapoor, and R.T. Yang, A.I.Ch.E.J., 1989; 35, 1735.

5. A. Kapoor, and R.T. Yang, Chem. Engng. Sci., 1990; 45, 3261.

6. V. Pereyra, G. Zgrablich, and V.P. Zhdanov, Langmuir, 1990; 6, 691.

7. V. Pereyra, G. Zgrablich, Langmuir, 1990; 6, 118.

8. V. Mayagoitia, F. Rojas, V. Pereyra, and G. Zgrablich, Surface Sci., 1989; 221, 394.

9. V. Mayagoitia, F. Rojas, J.L. Riccardo, and G. Zgrablich, Phys. Rev., 1990; B41, 7150.

10. J.L. Riccardo, V. Pereyra, G. Zgrablich, F. Royas, V. Mayagoitia, and I. Kornhauser, Langmuir.

11. T.L.Hill, J.Chem.Phys., 1949; 17, 762.

12. S.Ross, and J.P.Olivier, "On Physical Adsorption", Interscience Publishers, Inc., N.Y. 1964.

13. W.A.Steele, J.Phys.Chem., 1963; 67, 2016.

14. R.A.Pierotti, and H.E.Thomas, Trans.Faraday Soc., 1974; 70, 1725.

15. W.Rudziński, and L.Lajtar, J.C.S. Faraday Trans.2, 1981; 77, 153.

16. W.Rudziński, and J.Baszynska, Z.Phys.Chem., Leipzig, 1981; 262, 533.

17. W.Rudziński, and J.Jagiełło, J.Low Temp.Phys., 1981; 45, 1, 1982; 48, 307.

18. W.Rudziński, J,Jagiełło, and Y.Grillet, J.Coll.Interface Sci., 1992; 87, 478.

19. P.Ripa, and G.Zgrablich, J.Phys.Chem., 1979; 79, 2118.

20. L.Riccardo, V.Pereyra, J.L.Rezzano, D.A.Rodriguez Saa, and G.Zgrablich, Surface Sci., 1988; 204, 289.

21. Y.K.Tovbin, Dokl.Akad.Nauk SSSR, 1977; 235, 641.

22. Y.K.Tovbin, Zhur.Fiz.Khim., 1982; 56, 686.

23. R. Rippa, and G. Zgrablich, J. Phys. Chem., 1975; 79, 2118.

24. A. Clark, "The Theory of adsorption and Catalysis", Academic Press, New York, 1970, p.210.

25. P.Kisliuk, J. Phys. Chem. Solids, 3, (1957) 95.

26. L.K. Koopal, and C.H.W. Vos, Langmuir 9, 2593 (1993).

27. T. Nakahara, M. Hirata, and H. Mori, J. Chem., Eng. Data, 27, 317 (1982).

28. J.T. Nolan, T.W. McKeehan, and R. P. Danner, J. Chem. Eng. Data 26, 112 (1981).

29. L.A. Rudnitsky, and A.M. Alexeyev, J.Catal. 37, 232 (1975).

30. G.L. Kington, A.C. McLeod, Trans. Faraday Soc.

31. G. Welder, F.J. Broecker, Surface Sci., 26 454 (1971).

32. O. Beeck, Adv. Catal., 2, 151 (1950).

33. G.C. Schuit, L.L. van Reijen, Adv. Catal. 10, 242 (1988).

34. See the review by M.J.D. Low, Chem. Rev., 60 (1960) 267.

35. S. Roginski, Ya. Zeldovich, URSS 1 (1934), 554.

36. Roginski, S.; Zeldovich, Ya. Acta Physicochim., URSS 1, (1934), 595.

37. Aharoni, C.; Ungarish, M., J.C.S. Faraday Trans. I, 73, (1977), 1943.

38. Aharoni, C.; Ungarish, M., J.C.S. Faraday Trans. I, 74, (1978), 1507.

39. Aharoni, C.; Suzin, Y., J.C.S. Faraday Trans. I, 78, (1982), 2329.

40. Tovbin, Yu., "Lattice–Gas Model in Kinetic Theory of Gas–Solid Interface Processes", Progress in Surface Science, 34, (1991), 1.

41. Cerofolini, G.F., in "Adsorption and Chemisorption on Inorganic Sorbents", Dabrowski, A.; Tertych, V.A., Editors, Elsevier, 1996.

42. For most recent and exhaustive review see, Bhatia, S.; Beltramini, J.; Do, D.D., Catal. Today, 7, (1990), 309.

43. Ward, C.A.; Findlay, R.D., J. Chem. Phys., 76, (1982), 5615. See also proceeding paper by Ward, A., Findlay, R.D., and Rizk, M., J. Chem. Phys., 76 (1982), 5599.

44. W. Rudzinski, "A New Theoretical Approach to Adsorption-Desorption Kinetics on Energetically Flat Solid Surfaces Based on Statistical Rate Theory of Interfacial Transport", in "Equilibria and Dynamics of Gas Adsorption on Heterogeneous Solid Surfaces", (W. Rudzinski, W.A. Steele, and G.Zgrablich Editors) Elsevier (1996).

45. Talbot J.; Jin, X.; Wang, N.-H., Langmuir, 10, (1994), 1663.

46. T.Nitta, M.Kuro-Oka, and T. Katayama, J. Chem. Eng. Japan, 17, 45 (1984).

47. W. Marczewski, A.Derylo-Marczewska, and M. Jaroniec, Colloid Interface Sci. 109, 310 (1986).

48. W. Rudzinski, K. Nieszporek, and A. Dąbrowski, Adsorption Sci. & Technol. 10, 35 (1993).

49. W. Rudzinski, K. Nieszporek, J.M. Cases, L.I. Michot and F. Villieras, Langmuir 12, 170 (1996).

50. W. Rudzinski, K. Nieszporek, H. Moon and H-K. Rhee, Heterogeneous Chemistry Reviews, 1, 275 (1994).

51. G. F. Cerofolini, and W. Rudzinski, "Theoretical Principles of Single- and Mixed-Gas Adsorption Equilibria on Heterogeneous Solid Surfaces", in "Equilibria and Dynamics of Gas Adsorption on Heterogeneous solid Surfaces", (W. Rudzinski, W. A. Steele, and G. Zgrablich, Editors), Elsevier (1996).

52. E. Hoory, and J. M. Prausnitz, Chem. Eng. Sci. 22, 1025 (1967).

53. A. L. Myers, and J. M. Prausnitz, AIChE J. 11, 121 (1965).

54. E. C. Markham, and A. F. Benton, J. Am. Chem. Soc. 53, 497 (1931).

55. D. P. Valenzuela, A. L. Myers, O. Talu, and I. Zwiebel, AIChE J. 35, 959 (1989).

56. A. W.Marczewski, A. Derylo-Marczewska, and M. Jaroniec, Chemica Scripta 28, 173 (1988).

57. B.W. Wojciechowski, C.C. Hsu, and W. Rudzinski, Canad J. Chem. Eng. 63, 789 (1985).

58. E. N. Rudisil, and M. D. LeVan, Chem. Eng. Sci. 47, 1239 (1992).

240

59. E. Richter, W. Schütz, and A. L. Myers, Chem. Eng. Sci. $\underline{44}$, 1609 (1989).

60. W. Rudzinski, K. Nieszporek, H. Moon, and H-K Rhee, Chem. Eng. Sci $\underline{50}$, 2641 (1995).

61. S. D. Mechta, and R.P. Danner, I.& EC Fundamentals, $\underline{24}$, 235 (1985).

62. H. Moon, and C. Tien, Sep. Technol. $\underline{3}$, 161 (1993).

63. O. Talu, and I. Zwiebel, AIChE J. $\underline{32}$, 1263 (1986).

64. H .Moon, and C. Tien, Chem. Eng. Sci. $\underline{43}$, 2967 (1988).

# FUNDAMENTALS OF DIFFUSION IN POROUS AND MICROPOROUS SOLIDS

Douglas M. Ruthven
*Chemical Engineering Department*
*University of Maine*
*Orono, ME 04469*

## Abstract

The major mechanisms of transport within porous and microporous solids and the basic definitions of diffusivity are reviewed and the relationship between the differently defined coefficients (transport diffusivity, self-diffusivity and thermodynamically corrected diffusivity) is discussed in relation to simple models for the diffusion process. The Stefan-Maxwell approach and its extension to multi component diffusion are discussed with reference to recently reported experimental data for uptake kinetics in binary systems. Other topics such as diffusion in non-isotropic systems, diffusion in a uni-dimensional pore system and recent work directed to explaining the kinetic selectivity of carbon molecular sieves are also considered briefly.

## 1. Introduction

The widespread application of zeolites and other microporous solids as catalysts and selective adsorbents has stimulated considerable interest in the study of diffusive transport of guest molecules in small pores[1]. Because of the regularity and uniformity of their pore networks much of this work has been focused on the study of diffusion in zeolites although, from a practical point of view, the random pore networks of amorphous adsorbents such as activated carbon and carbon molecular sieves are of at least equal interest. In general the expectation that the structural regularity of the zeolites would lead to relatively straightforward and easily understandable transport behavior has not been fulfilled. Although a number of general trends have been established, the reported experimental data show many anomalies and inconsistencies as well as a rich variety of different patterns of behavior. It is still not clear to what extent this apparent complexity is real or simply the result of inadequate experimentation coupled with incorrect interpretation of the resulting data.

Largely because of the small size of the available zeolite crystals, the macroscopic determination of intracrystalline diffusivities from macroscopic sorption rate measurements has proved surprisingly difficult. For example, early sorption rate measurements for several hydrocarbons in NaX crystals showed a well defined minimum in the concentration dependence of the apparent diffusivity[2]; this was later shown to be the result of the combined effects of heat transfer and intracrystalline diffusional resistances.[3,4], Similarly, sorption rate measurements carried out in a variant of the

*J. Fraissard (ed.), Physical Adsorption: Experiment, Theory and Applications,* 241–259.

"Carberry Mixer"[5] were shown to have been controlled mainly by external mass transfer resistance.[6] More recently the widely quoted 'window effect', which was suggested by Gorring[7] to explain the unusual trend of uptake rate with carbon number for a series of n-paraffins in zeolite T, has also been shown to be false[8,9] but, ironically, not before a detailed mechanistic theory had been put forward[10,11] to explain the erroneous data!

The practical problems involved in measuring intracrystalline transport and diffusion are discussed in detail in a later paper at this meeting. The aim of the present article is to present a concise summary of the underlying theory on which our understanding of micropore diffusion is based.

## 2. Definitions of Diffusivity

### 2.1. TRANSPORT DIFFUSION

Since the transport of guest molecules within small pores is a stochastic process, driven by differences in local concentration levels (or more strictly by differences in chemical potential) it is logical to correlate rate data in terms of a Fickian diffusivity (D), defined according to Fick's first equation:

$$J = D \frac{\partial q}{\partial z} \tag{1}$$

The diffusivity defined in this way is properly referred to as the 'transport' diffusivity since it measures the rate of transport under the influence of a concentration gradient. In general this quantity can be expected to be a function of concentration although, in the low concentration region, it should approach a constant value.

Since the true driving force for diffusive transport is the gradient of chemical potential an alternative and arguably more logical formulation of the basic expression for diffusive transport is:

$$J = - Bq \frac{\partial \mu}{\partial z} \tag{2}$$

Substitution of the usual expression for the chemical potential with the assumption that the activity can be replaced by the equilibrium vapor pressure then yields:

$$J = -BRT \frac{d \ln p}{d \ln q} . \frac{\partial q}{\partial z} \tag{3}$$

and, by comparison with Eq (1):

$$D = BRT \frac{d \ln p}{d \ln q} = D_o \frac{d \ln p}{d \ln q} \qquad (4)$$

where $D_o$ = BRT is commonly referred to as the (thermodynamically) "corrected" diffusivity. The correction factor dlnp/dlnq is simply the gradient of the equilibrium isotherm in logarithmic coordinates and is thus easily determined provided that the equilibrium relationship is known. Within the Henry's Law region ($q \propto p$) dlnp/dlnc = 1.0 and Eq. 3 Reduces to the simple Fickian expression (Eq.1).

In a manner similar to Eq. 1 one may also define the 'tracer' or 'self' diffusivity ($\mathcal{D}$):

$$J^* = -\mathcal{D} \left. \frac{\partial q^*}{\partial z} \right|_{q-const.} \qquad (5)$$

This quantity measures the rate of migration of marked molecules under equilibrium conditions where there is no net gradient of concentration, whereas the transport diffusivity defined according to Eq. 1 corresponds to the physically different non-equilibrium state in which there is a gradient of concentration. The tracer and transport diffusivities are therefore physically different quantities although, since the basic diffusion mechanisms must be the same in both situations, they are closely related. From the principles of irreversible thermodynamics it may be shown that this relationship must be of the general form:

$$\frac{D(q)}{\mathcal{D}(q)} = \frac{d \ln p}{d \ln q} \left[ 1 + q \frac{\beta(q)}{\alpha(q)} \right] \qquad (6)$$

where the functions $\alpha(q)$ and $\beta(q)$ are related to the straight and cross coefficients of the irreversible thermodynamic formulation. The function $\beta$ is directly proportional to the cross coefficient so it is evident that, when the adsorbed phase concentration is low ($q \to 0$) or when interference between diffusing molecules is small relative to the frictional drag exerted by the pore walls ($\beta \ll \alpha$) Eq. 6 reduces to:

$$D = \mathcal{D} \frac{d \ln p}{d \ln q} \qquad (7)$$

and, by comparison with Eq. 4, we see that $\mathcal{D} \to D_o$

## Self-Diffusion

$$N^*_{l \to r} = q^*(z) \frac{\lambda A}{6} \frac{dt}{\tau(q)}$$

$$N^*_{r \to l} = q^*(z + \lambda) \frac{\lambda A}{6} \frac{dt}{\tau(q)} \approx \left[ q^*(z) + \frac{dq^*}{dz} \lambda \right] \frac{\lambda A}{6} \frac{dt}{\tau(q)}$$

$$J^* = \frac{N^*_{l \to r} - N^*_{r \to l}}{A \, dt} = \frac{-\lambda^2}{6\tau(q)} \frac{dq^*}{dz}$$

$$\mathcal{D} = \frac{\lambda^2}{6\tau(q)}$$

## Transport Diffusion

$$N_{l \to r} = q(z) \frac{\lambda A}{6} \frac{dt}{\tau[q(z)]}$$

$$N_{r \to l} = q(z + \lambda) \frac{\lambda A}{6} \frac{dt}{\tau[q(z + \lambda)]} = \frac{\lambda A}{6} dt \left[ \frac{q(z)}{\tau[q(z)]} + \frac{d}{dq}\left(\frac{q}{\tau}\right) \frac{dq}{dz} \lambda \right]$$

$$J = -\left[ \frac{\lambda^2}{6} \frac{d}{dq}\left(\frac{q}{\tau}\right) \right] \frac{dq}{dz}, \qquad D = \frac{\lambda^2}{6} \frac{d}{dq}\left(\frac{q}{\tau}\right)$$

## Relationship

$$D(q) = \frac{d}{dq}\left(\mathcal{D}(q)q\right)$$

Figure 1.  The relationship between transport and self-diffusion derived from microdynamic considerations.

## 2.2. THE RANDOM WALK MODEL

An alternative approach to the definition of self-diffusivity makes use of the random walk model in which the diffusing molecule is assumed to execute a sequence of random jumps (mean jump length $\lambda$, and mean time interval $\tau$ between jumps). This leads to:

1 dimension:

$$\mathcal{S} = \frac{\lambda^2}{2\tau} = \frac{\langle z^2 \rangle}{2t}$$

(8)

3 dimensions:

$$\mathcal{S} = \frac{\lambda^2}{6\tau} = \frac{\langle r^2 \rangle}{6t}$$

These expressions are commonly referred to as the Einstein equations. By simple algebra it may easily be shown that the self diffusivity defined in this way is formally equivalent to that defined according to Eq. 6. Furthermore, the simple *balance* shown in Figure 1 shows that, according to this model the relationship between self and transport diffusion is:

$$D(q) = \frac{d}{dq} = \left[ q \cdot \mathcal{S}(q) \right]$$

(9)

which shows that if $\mathcal{S}$ is independent of concentration $D = \mathcal{S}$ which is evidently true in the low concentration region. If $\mathcal{S} \propto 1/q$ then $D = 0$ which corresponds to the situation in a binary mixture at the point of incipient phase separation ($\partial\mu/\partial z = 0$ or $d\ln p/d\ln q = 0$).

For an ideal Langmuir system, $d\ln p/d\ln q = (1-q/q_s)^{-1}$ so that:

$$D = \frac{D_o}{\left(1 - q/q_s\right)} = \frac{D_o}{1 - \theta}$$

(10)

$$\frac{D(q)}{D_o(q)} = \frac{1}{\theta} \ln\left(\frac{1}{1-\theta}\right) \approx 1 + \frac{\theta}{2} + \frac{\theta^2}{3} + \dots$$

(11)

The limiting behavior $\mathcal{S} \to D_o$ as $\theta \to 0$ is evident from the series expansion.

## 3. Diffusion Regimes

The diffusion mechanism in a pore (see Figure 2) depends primarily on two factors; the pressure and the ratio of pore diameter/sorbate molecular diameter. When the sorbate

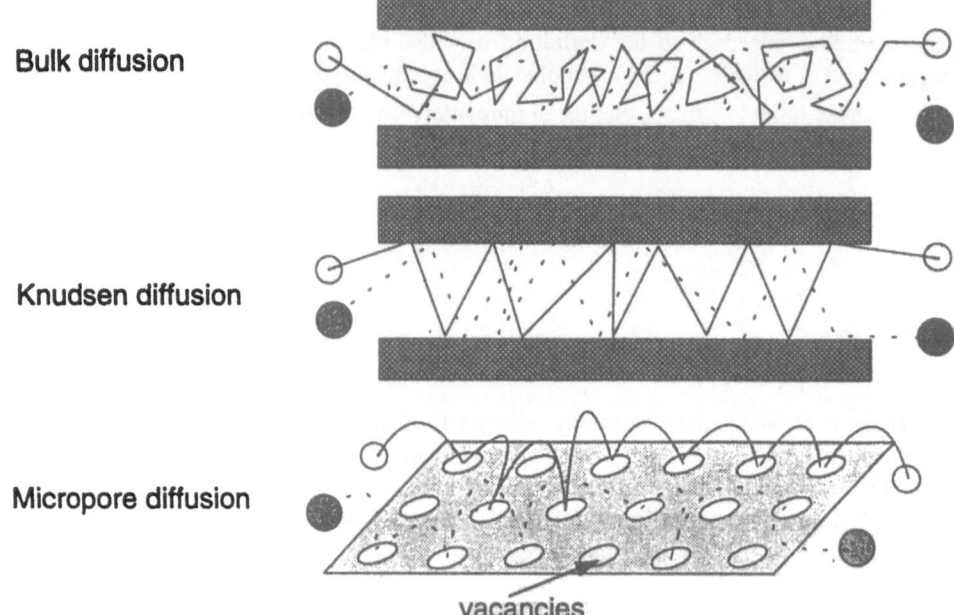

**Bulk diffusion**

**Knudsen diffusion**

**Micropore diffusion**

vacancies

Figure 2.   Three distinct mechanisms of diffusion in a porous particle.
(a) Bulk or molecular diffusion
(b) Knudsen diffusion
(c) Surface diffusion

(from R. Krishna <u>Gas Separation and Purification</u> <u>7</u>, 91 (1993) )

TABLE 1.  Comparison of Experimental Pre-Exponential Factors ($\mathcal{D}_\infty$) with Values from Simple Theoretical Models

| Sorbent | Sorbate | Values of $\mathcal{D}_\infty = \lambda^2 v$ | $\mathcal{D}_\infty \times 10^4$ Exp. | $(cm^2 \cdot s^{-1})$ $\mathcal{D}_\infty = \lambda(E/2M)^{1/2}$ |
|---------|---------|------------|------|------|
| 5A Zeolite | $N_2$ | 0.4 | 0.62 | 53 |
| | $CO_2$ | 0.8 | 1.0 | 40 |
| | $CH_4$ | 0.17 | 0.18 | 55 |
| | $C_2H_6$ | 0.13 | 0.06 | 39 |
| 13X Zeolite | $O_2$ | 0.11 | 130 | 58 |
| | $N_2$ | 0.09 | 77 | 61 |
| | $C_2H_6$ | 0.27 | 160 | 61 |

$\lambda$ is taken as the lattice parameter (12·3Å).
$v$ is estimated from the Henry's Law equilibrium constant.

diameter approaches the pore diameter, the diffusing molecule never escapes from the force field of the pore wall and the effects of steric hindrance become important and often dominant. This regime is referred to variously as micropore diffusion, intracrystalline diffusion, configurational diffusion or surface diffusion. The last of these terms is perhaps the least satisfactory since, under the relevant conditions, the intracrystalline fluid is probably best regarded as a single phase.

## 3.1. ACTIVATED DIFFUSION IN SMALL PORES

The diffusion process involves a sequence of activated jumps between regions or 'sites' of relatively low potential energy separated by barriers of higher energy, thus giving rise to an Arrhenius type of temperature dependence: $\mathcal{D} = \mathcal{D}_{\infty} e^{-E/RT}$. The activation barriers may arise from constrictions in the pore (windows) the passage through which involves a significant repulsive interaction energy or, in somewhat larger pores, simply from the local variation of the attractive potential.

The molecules at their favored sites may be regarded as localized oscillators having a vibration frequency ($\nu$) corresponding to the thermal energy (kT per degree of vibrational freedom), which occasionally acquire sufficient kinetic energy in the appropriate direction, to surmount the energy barrier and reach the adjacent site. Regardless of the dimensionality of the pore system this yields:

$$\mathcal{D}_{\infty} = \lambda^2 \nu \tag{12}$$

This expression or its equivalent, may be derived in a somewhat more rigorous way from transition state theory. For a zeolite system the structural regularity allows the site-site distance to be estimated and the mean vibration frequency can be derived from the Henry's Law equilibrium constant. An approximate *a priori* estimate of the pre-exponential factor can therefore be made. For several systems such estimates have been shown to be in reasonable agreement with experimental data[12,13] (see Table 1). Estimation of the activation energy is also possible, in principle, but such estimates are extremely sensitive to the precise values used for the molecular force constants used to calculate the potential energy profile. This is particularly important when repulsive interactions are significant since a small adjustment of the molecular diameter of the sorbate may then lead to a change of several kilocalories in the estimated energy barrier and may even change the location of the barrier.

An alternative model in which the diffusion process is considered as movement in a three dimensional sinusoidal energy field has also been considered. Such a model, which is more appropriate for diffusion in a large pore zeolite, leads to a very simple relationship between the pre-exponential factor and the diffusional activation energy [$\mathcal{D}_{\infty} = \lambda(E/2M)^{1/2}$] which, for several light molecules in NaX, is in reasonable agreement with the experimental data (Table 1).

In the derivation of Eq. 13 the effects of all sorbate-sorbate interactions are neglected, which is a reasonable approximation at low loadings (within the Henry's Law region). At higher loadings there are three distinct effects that must be accounted for:

248

- the adsorption energy may be affected by the presence of other sorbate molecules;
- the vibration frequency ( $v$ ), which is directly related to the entropy of adsorption; depends on the space available to the adsorbed molecule and therefore on the density of the adsorbed phase;
- the passage of molecules through the windows (the transition state) may be affected by the counter-flow of molecules moving in the opposite direction.

Allowing for these effects, based on simple mechanistic assumptions, yields:

$$\frac{D}{\mathcal{D}} = \frac{d \ln p}{d \ln q} \cdot \frac{1}{1 - \gamma p(q)} \qquad (13)$$

Where p(q) represents the equilibrium pressure at sorbate loading q and $\gamma$ is a constant of proportionality. Comparing with Eq.6 we see that, for $\gamma p \ll 1$:

$$\gamma \frac{p(q)}{q} \cong \frac{\beta(q)}{\alpha(q)} \qquad (14)$$

suggesting that the ratio of the coefficients $\alpha/\beta$ in the irreversible thermodynamic formulation depends directly on the adsorption equilibrium constant.

## 4. Diffusion in Larger Pores - the Stefan-Maxwell Equation

In larger pores ($d_{pore} \gg d_{sorbate}$) diffusion is controlled by collisions between diffusing molecules, as in gas phase diffusion, and by collisions with the pore wall (Knudsen flow). We consider first diffusion in a free binary gas phase. Whenever there is a concentration gradient there is a corresponding gradient of partial pressure and so, a net force exerted by the molecules of type B on type A which, under conditions of no net flow, is balanced by an equal and opposite force exerted on the molecules of type A and those of type B. The gradient of partial pressure is then exactly balanced by the rate of momentum transfer. Using a straightforward momentum balance together with the expressions for collision frequency and mean molecular velocity derived from kinetic theory yields, for hard sphere molecules:

$$-\frac{dp_A}{dz} = \frac{P y_A y_B}{D^1_{AB}} \left( u_A - u_B \right) \qquad (15)$$

Where $u_A$, $u_B$ represent the net flow velocities of the components and:

$$D'_{AB} = \frac{1}{2\sqrt{2}\pi\sigma^2_{AB}} \left(\frac{kT}{P}\right)\left(\frac{kT}{m\pi}\right)^{1/2} \tag{16}$$

which is the familiar kinetic theory expression for the binary diffusivity for hard sphere molecules. (A more sophisticated calculation taking account of intermolecular forces in accordance with the Lennard-Jones potential and using a more accurate averaging procedure to estimate the mean molecular velocities yields the familiar Chapman-Enskog expression for $D_{AB}$.)

Eq. 16 may be conveniently expressed in terms of the fluxes to yield the Stefan-Maxwell equation for a binary system:

$$\frac{dy_A}{dz} = \frac{y_A y_B}{D^1_{AB}}\left(u_A - u_B\right) = \frac{1}{cD'_{AB}}\left(y_B N_A - y_A N_B\right) \tag{17}$$

or in terms of the gradient of chemical potential:

$$-\frac{1}{RT}\cdot\frac{d\mu_A}{dz} = \frac{1}{D'_{AB}}\left(\frac{y_B N_A - y_A N_B}{cy_A}\right) \tag{18}$$

The extension to a multi-component system follows naturally:

$$-\frac{1}{RT}\frac{\partial\mu_i}{\partial z} = \sum_{j=1}^{n}\frac{y_j N_i - y_i N_j}{c_i D'_{ij}} \tag{19}$$

The flux with respect to a fixed frame of reference is given by:

$$J_A = N_4 - y_A\left(N_A + N_B\right) = y_B N_A - y_A N_B = -D'_{ABC}\frac{dy_A}{dz} = -\frac{c_A}{RT}\cdot\frac{d\mu_A}{dz} \tag{20}$$

The Stefan-Maxwell formulation thus leads naturally to the result that, in a binary system, the diffusivity is independent of concentration, whereas classical kinetic theory suggests that the binary diffusivity is concentration dependent, a result which is at variance with the experimental evidence.

To extend this approach to diffusion in a pore we must consider the exchange of momentum with the pore wall. Using the Joule classification, the momentum flux to the wall of a cylindrical pore (radius r) is given by:

250

$$M = \left(\frac{c\bar{v}}{6}\right)\left(\overline{mu_z}\right)2\pi r dz \qquad (21)$$

and, hence, from a force balance:

$$\bar{u}_z = \frac{-3r}{mc\bar{v}}\cdot\frac{dp}{dz} \qquad (22)$$

and with $\bar{v} = \sqrt{8kT/\pi m}$, from kinetic theory:

$$J = c\bar{u}_z = -D_K\frac{dc}{dz} \qquad (23)$$

thus yielding, for the Knudsen diffusivity:

$$D_K = r\sqrt{\frac{kT}{m}}\left(\frac{9}{8}\right)^{1/2} \qquad (24)$$

A more rigorous analysis yields the familiar result:

$$D_K = r\sqrt{\frac{kT}{m}}\left(\frac{32}{9\pi}\right)^{1/2} = 9700r\sqrt{\frac{T}{M}} \quad (cm^2.s^{-1}) \qquad (25)$$

When the transfers of momentum to both the pore wall and other diffusing molecules are significant, assuming simple additivity we have:

$$\frac{N_A}{D} = -\frac{1}{kT}\left[\left(\frac{dp_A}{dz}\right)_{gas} + \left(\frac{dp_A}{dz}\right)_{pore}\right] = \frac{N_A}{D_K} + \frac{N_A - y_A(N_A + N_B)}{D_{AB}} \qquad (26)$$

from which the familiar result obtained by Scott and Dullien[14] and others[15] for pore diffusion in the transitional range is derived:

$$\frac{1}{D} = \frac{1}{D_K} + \frac{1}{D_{AB}}\left[1 - y_A(1 + N_A/N_B)\right] \qquad (27)$$

## 5. Diffusion in a Mixed Adsorbed Phase: Extension of the Stefan-Maxwell Equations

The Stefan-Maxwell approach has been extended by Krishna to diffusion in an adsorbed phase by considering the idealized molecular hopping model and regarding the unoccupied sites (vacancies) as the $(n + 1)$ component in the system.[16,17] The validity of the momentum transfer argument from which the Stefan-Maxwell equation is derived seems questionable in such a situation but the approach may nevertheless be justified on the basis that the results are consistent with experimental observations. With concentrations expressed in terms of fractional occupation of the available sites $(\theta_i = q_i / q_s)$, Eq. 18 becomes:

$$- \frac{1}{RT} \cdot \frac{\partial \mu_i}{\partial z} = \sum_{j=1}^{n} \frac{\theta_j \left( u_i - u_j \right)}{D_{ij}^1} + \frac{\theta_{n+1}}{D^1_{iv}} \cdot u_i \tag{28}$$

The coefficient $D^1_{ij}$ measures the rate at which components i and j can exchange directly while $D^1_{iv}$ measures the rate of migration to vacant sites, the velocity of which $(u_{n+1})$ is zero.

For single component diffusion ($\theta_B = 0, N_B = 0$) this reduces to:

$$N_A = J_A = - \frac{D^1_{AV}}{\left(1 - \theta_A\right)} \cdot \frac{q_A}{RT} \cdot \frac{d\mu_A}{dz} \tag{29}$$

which is equivalent to Eq. 2 with $D_o = D^1_{AV}/(1-\theta)$, as for a Langmuirian system.

For tracer or self-diffusion we have $N_A = -N_B$ and $D^1_{AB} = D^1_A$ which gives:

$$\frac{1}{\beta} = \frac{\theta}{D^1_A} + \frac{1 - \theta}{D^1_{AV}} \tag{30}$$

showing that with increasing loading, the balance between the two mechanisms should shift in favor of direct exchange.

### 5.1. DIFFUSION IN MULTICOMPONENT SYSTEMS

Habgood and co-workers[18,19] were the first to provide a rational analysis of sorption rates in a binary adsorbed phase. Using the principle of the chemical potential driving force (Eq. 2) with due allowance for the effect of each component on the chemical potential of the other, they obtained, for a Langmuir system:

$$\frac{\partial \theta_i}{\partial t} = \nabla \cdot \left[ D_{eff} \nabla \theta_i \right] \tag{31}$$

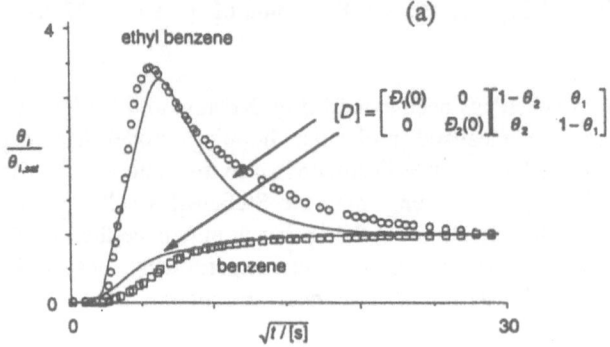

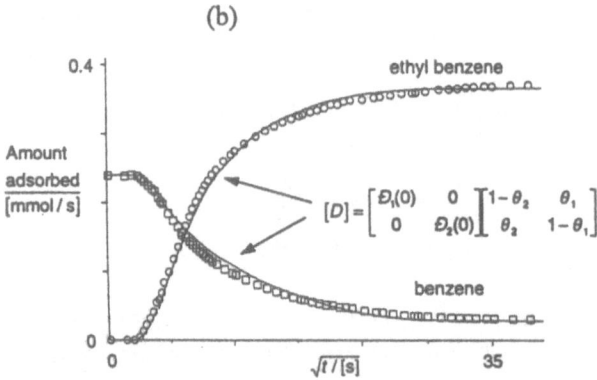

Figure 3. (a) Co-diffusion and (b) Counter-diffusion transient sorption curves for benzene and ethylene in silicalite. Pressure 0-115Pa at 375K. For benzene $D_o = 2.6 \times 10^{-10}$ cm$^2$.s$^{-1}$ and for ethylbenze $D_o = 1.0 \times 10^{-10}$ cm$^2$.s$^{-1}$. These values were derived from the single component measurements. Original experimental data from thesis of Niessen[22]; analysis from thesis of van den Broeke[21].

$$D_{effA} = \frac{D_{OA}}{\left(1 - \theta_A - \theta_B\right)}\left[\left(1 - \theta_B\right) + \theta_A \frac{\nabla\theta_B}{\nabla\theta_A}\right]$$
(32)

with a similar expression for $D_{effB}$. Precisely the same form may be derived from Eq. 28 if the effect of direct exchange of the components is neglected and the second term representing activated jumps to an unoccupied site is assumed dominant. Two variants of this model are plausible; if the jump frequency is independent of surface coverage $D^1_{iv}$ is constant and we obtain Eq. 32 whereas if we assume that the diffusing molecule can successfully execute a jump only when the receiving site is unoccupied we obtain:

$$D_{effA} = D_{OA}\left[\left(1 - \theta_B\right) + \theta_A \frac{\nabla\theta_B}{\nabla\theta_A}\right]$$
(33)

This is in effect the model of Quereshi and Wei.[20]

This model has been used by van den Broeke[12] to model the experimental co and counter-sorption experiments of Niessen[22,24] for $C_8$ aromatic isomers on HZSM-5 crystals - see figure 3. The theoretical curves calculated with independently evaluated equilibrium data and single component diffusivity values evidently provide a reasonably satisfactory representation of the experimentally observed behavior.

## 6. Single File Diffusion

Self-diffusion in a three dimensional pore network can be described by the Einstein expression (Eq8) from which it is evident that the mean square molecular displacement increases linearly with time. This is true also for a one dimensional pore system provided that the pore is large enough that diffusing molecules can pass each other. When this condition is not fulfilled we have an entirely different pattern of behavior known as single file diffusion or 'string of pearls' diffusion. The mean square displacement increases with the square root of time [25,26].

$$z^2(s) = \lambda^2\left(\frac{1-\theta}{\theta}\right)\sqrt{\frac{2t}{\pi\tau}}$$
(34)

which is equivalent to a time varying self-diffusivity given by:

$$\mathcal{D} = \lambda^2\left(\frac{1-\theta}{\theta}\right)\frac{1}{\sqrt{2\pi\theta t}}$$
(35)

254

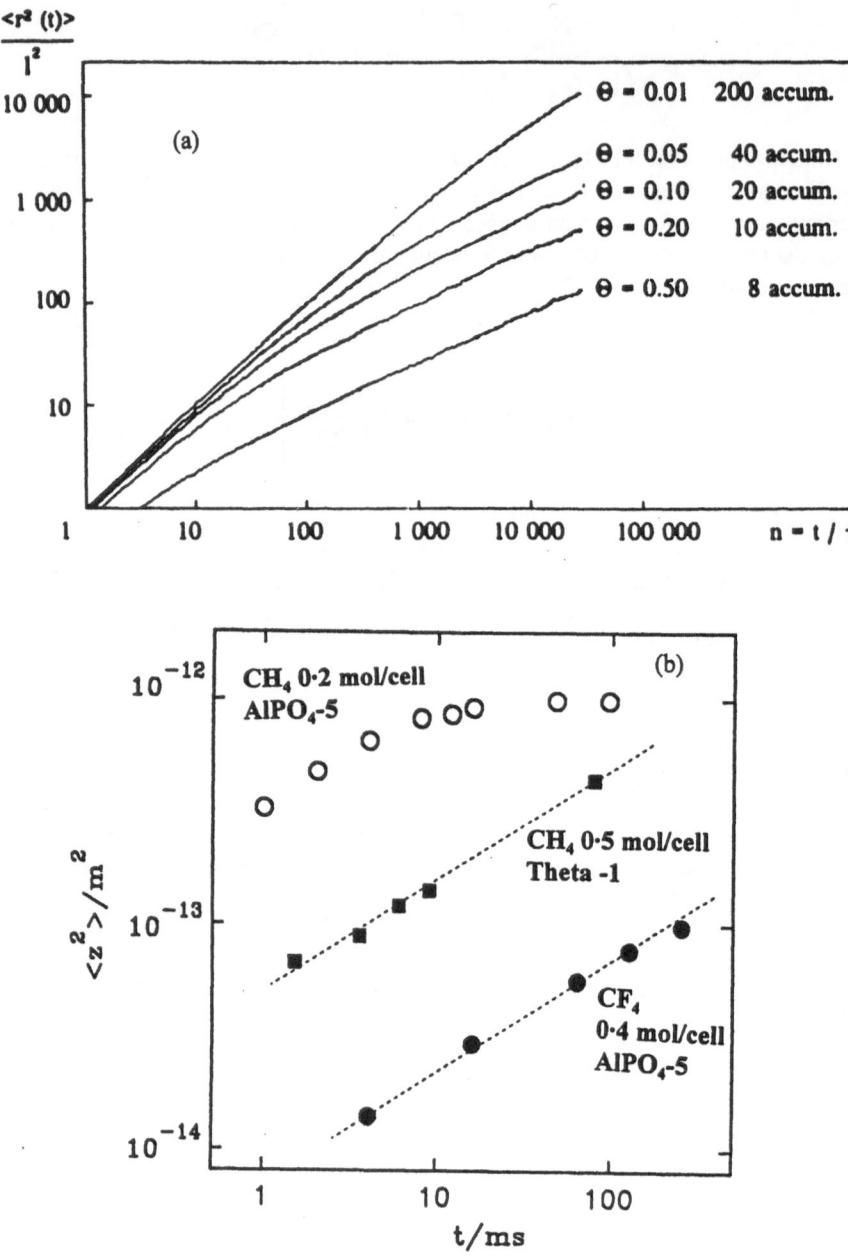

Figure 4. (a) Variations of molecular mean square displacement with time derived from Monte Carlo simulation of $N=10^4$ sites for various degrees of occupancy $\theta$ under conditions of single file diffusion. (b) Experimental data showing time dependence of mean square molecular displacement, as measured by PFGNMR, for $CH_4$ in $AlPO_4$-5 and zeolite theta. From Karger et al.[26] and Kukla et al.[27].

This pattern of behavior has been verified by both Monte Carlo simulation and PFGNMR self-diffusion measurements - see Figure 4[27].

The sorption-desorption rate is not affected by whether or not the diffusing molecules can pass one another so we reach the interesting conclusion that single file diffusion can be observed experimentally only by measuring self-diffusion and, under single file conditions, we may expect a substantial divergence between the (time dependent) self-diffusivity and the transport diffusivity derived from sorption rate measurements.

## 7. Zeolite Membranes

The production of a coherent zeolite membrane has been a long time goal of zeolite researchers. The Delft group has achieved significant progress towards this objective and, although their membranes are still too thick to yield economically viable fluxes some interesting features have been observed.[28,29] Permeation measurements for single component systems provide no great surprises and yield diffusivities which are generally consistent with Eq. 4 - see Figure 5.[29] The behavior of binary and multicomponent systems is more complex (see figure 6)[28] but is quite consistent with the predictions derived from the chemical potential driving force model.[29,30] Clearly predictions of binary system behavior derived directly from single component permeation rate measurements would be seriously in error.

## 8. Non-Isotropic Systems

One of the well known results of diffusion theory is that for a system in which the diffusional properties are non-isotropic it is always possible to define a new coordinate system in which Fick's equations are applicable. This means that except in the extreme case of one-dimensional diffusion, the form of the transient sorption/desorption curves will be essentially the same as for an isotropic system although the apparent diffusivity will then be a complex function of the diffusivities in the three principal directions. Detailed studies of diffusion in non-isotropic zeolite crystals such as silicalite can, however, be achieved using oriented crystals with selected faces coated with an impermeable resin, or embedded in an oriented membrane. Such studies show that for n-hexane in silicalite, the ratio of diffusivities in the transverse and longitudinal directions is about three[3] and this is confirmed by PFGNMR measurements.

## 9. Diffusion in Carbon Molecular Sieves

Although they do not have the uniform pore size characteristic of a zeolite, carbon molecular sieves show very similar kinetic selectivities (large differences in internal diffusivity depending on molecular size). The explanation seems to be that in the pore network of a CMS adsorbent the sorption rate is almost entirely controlled by pores of the critical size.[32]. Pores either smaller or larger have very little influence. This may be

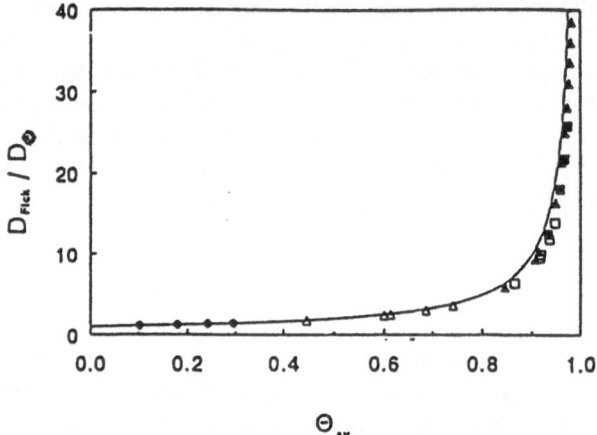

Figure 5.    Concentration dependence of Fickian diffusivities for light alkanes at 292K showing conformity with Eq. 4.  The line represents the function $1/(1-\theta)$ to which the correction factor dlnp/dlnq reduces for an ideal Langmuir system. From Kapteijn et. al.[29].

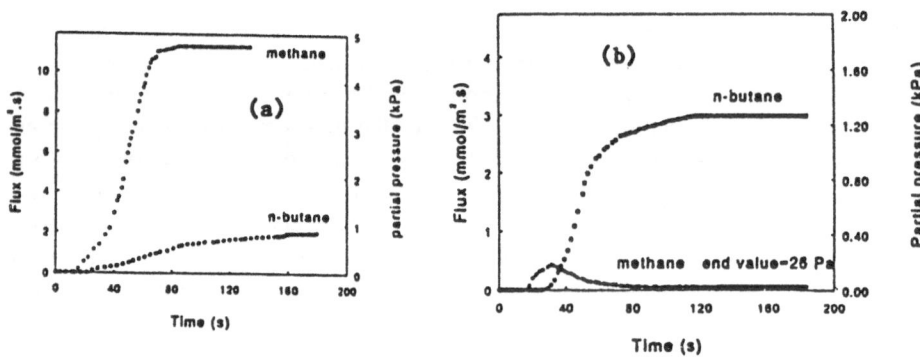

Figure 6.    Transient permeation for a silicalite membrane (a) single component at 300K, 50 k Pa (b) binary mixture at 300K, 50 k Pa for each component. From Bakker et. al.[28].

understood by considering the pore network shown schematically in Figure 7. Consider the pores all to be removed and then replaced in sequence decreasing size according to their critical diameter. Until the percolation threshold is reached there will be no continuous transport path from one side to the other. However, once the critical pore is inserted a continuous path is formed and diffusion will evidently be controlled by this critical pore. Insertion of the remaining (smaller) pores has very little further effect since the molecules cannot effectively pass through them. It is then possible to understand how, provided the sorption capacity of the larger pore is not too great, a high kinetic selectivity can be achieved even in an adsorbent with a relatively broad distribution of pore size.

This explanation implies that the main barriers to molecular transport are distributed throughout the microparticle, rather than concentrated at the surface. Conformity of the transient uptake curves to a diffusion model is consistent with such a mechanism. Some recent experimental studies, however, suggest surface rate control, rather than diffusion control. In an interesting series of experiments Armor et. al.[32] observed that the kinetic selectivity drops sharply when the particle size is reduced to less than 1 mm. This observation is consistent with surface rate control in domains of this order of size. On the other hand, our own studies[33] with the Bergbau Forschung sieve suggest microparticle diffusion control with rates essentially independent of the size of the macroparticle. However the particle size was never reduced much below the 1 mm level at which, according to Amor, dramatic changes in the kinetic behavior occur. Evidently there remains some disagreement concerning the precise nature of the controlling mass transfer resistance in carbon sieves but, considering the way in which such materials are prepared, it would not be surprising if the controlling mechanisms in different CMS preparations are not always the same.

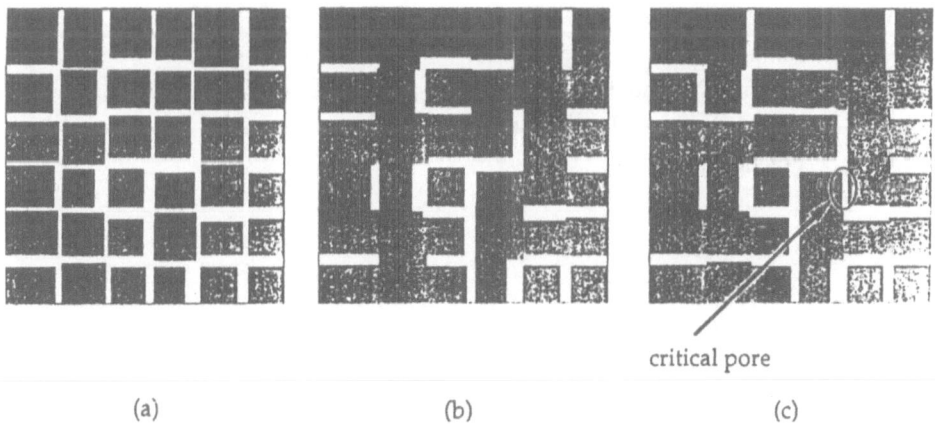

critical pore

(a)                                    (b)                                    (c)

Figure 7.   Idealized pore network showing the loss of connectivity on removal of the critically sized pore. (a) original network (b) network with small pores removed showing no connectivity. (c) network after reinsertion of critically sized pore establishes connectivity. Seaton, et. al.[31].

## NOTATION

| | |
|---|---|
| B | Mobility |
| c | Concentration in gas phase |
| D | Diffusivity $(cm^2.s^{-1)}$ |
| $D_o$ | Corrected diffusivity $(cm^2.s^{-1})$ |
| $D^I_{AB}$ | Stefan-Maxwell diffusivity $(cm^2.s^{-1})$ |
| $\mathcal{D}$ | Self-diffusivity $(cm^2.s^{-1})$ |
| $D_K$ | Knudsen diffusivity $(cm^2.s^{-1})$ |
| E | Diffusional activation energy |
| k | Bolzmann's constant |
| p | Partial pressure |
| P | Total pressure |
| q | Adsorbed phase concentration |
| r | Pore radius |
| R | Gas constant |
| t | Time |
| T | Temperature |
| $J_i$ | Flux of component measured with respect to plans of no net volume flow |
| M | Molecular weight |
| $N_i$ | Flux of component i relative to fixed reference frame |
| y | Mole fraction |
| $u_i$ | Diffusion velocity |
| | |
| $\alpha, \beta$ | Functions related to straight and cross coefficients of irreversible thermodynamics |
| $\gamma$ | Constant of proportionality |
| $\nu$ | vibration frequency |
| $\lambda$ | Jump distance |
| $\mu$ | Chemical potential |
| $\tau$ | Time between molecular jumps |
| $\theta$ | Fractional loading |

REFERENCES

1. Karger, J. and Ruthven, D.M. (1992) *"Diffusion in Zeolites and other Microporous Solids"* John Wiley, New York.
2. Ruthven, D.M. and Doetsch, I.H., (1976) AIChE Jl. 22, 882.
3. Ruthven, D.M., Lee, L.K. and Yucel, H., (1980) AIChE Jl., 26, 16.
4. Ruthven, D.M. and Lee, L-K., (1981) AIChE Jl., 27, 654.
5. Ma, H.A. and Lee, T.Y., (1976) AIChE Journal, 22, 147.
6. Taylor, R.A., (1979) Ph.D. Thesis, University of New Brunswick, Fredericton.
7. Gorring, R.L., (1973) J. Catalysis, 31, 13.
8. Cavalcante, C.L., Eic, M. Ruthven, D.M. and Occrelli, M. (1995) Zeolites 15, 293.
9. Ruthven, D.M., (1995) Proc. Zeolite Symposium., Quebec City, October 1995, published as *Zeolites: A Refined Tool for Catalyst Design* p. 223. L. Bonneviot and S. Kaliaguine eds. Elsevier, Amsterdam.
10. Nitsche, J. and Wei, J. (1991) AIChE Jl. 37, 661.
11. Wei, J., (1994) I and E.C. Research 33, 2467.
12. Kärger, J., Bülow, M. and Haberlandt, K., (1977) J. Colloid Interface Science: 66, 386.
13. Xu, Z., Eic, M. and Ruthven, D.M., (1993); Ninth Internat. Zeolite Conference, Montreal, July 1992, Proceedings, R.von Ballmoos et. al. Eds. p. 147, Butterworth-Heinemann, Stoneham, MA.
14. Scott, D.L. and Dullien, F.A.L., (1962) AIChE, Jl. 8, 113.
15. Evans, R.B., Watson, G.M. and Mason, E.A., (1961) J. Chem. Phys. 33, 2076.
16. Krishna, R., (1990) Chem. Eng. Sci. 45, 1779.
17. Krishna, R., (1993) Chem. Eng. Sci. 48, 845.
18. Habgood, H.W., (1958) Can. J. Chem. 36, 1384.
19. Round, G.F., Habgood, H.W., and Newton, R., (1966) Sep. Sci: 1, 219.
20. Qureshi, W.Q. and Wei, J., (1990) J. Catal. 126, 126 and 147 (1990) and 147.
21. van den Broeke, L.J.P., (1994) *The Maxwell-Stefan Theory for Micropore Diffusion*, Ph.D Thesis, University of Amsterdam, December 1994.
22. Niessen, W., (1991) Ph.D. Thesis, Fritz-Haber Institut der Max Planck Gesellschaft, Berlin, Germany.
23. Niessen, W. and Karge, H.G. (1991) in *"Studies in Surface Science and Catalysis"*, 60, 213 T. Innui, S. Namba and T. Tatsumi eds., Kodansha-Elsevier, Tokyo.
24. Niessen, W., Karge, H.G. and Jozefowicz, L., (1992) Proceeding 4th Int. Conf. on Adsorption, Kyoto, May 1992 p. 475, M. Suzuki, Ed. Kodansha, Tokyo.
25. Fedders, P.A., (1978) Phys. Rev. B 17, 40.
26. Kärger, J., Petzold, M., Pfeifer, H., Ernst, S. and Weitkamp, J., (1992) J. Catalysis 136, 283.
27. Kukla, V., Kärger, J., Pfeifer, A., Kornatowski, J., Caro, J., Girnus, I., Demuth, D., Schink, S., Unger, K. and Rees, L.V.C. Nature - in press.
28. Bakker, W.J.W., Zheng, G., Kapteijn, F., Makkee, M., Moulijn, J.A., Guns, E.R. and van Bekkum, H., (1993) in *Precision Process Technology* p. 425, M.P.C. Weijnen and A.A.H. Drinkenberg eds., Kluwer.
29. Kapteijn, F., Bakker, W.J.W., Zheng, G., Poppe, J. and Moulijn, J.A., (1995) Chem. Eng. J., 57, 145-153.
30. Krishna, R., van den Broeke, L.J.P., (1995) Chem. Eng. J., 57, 155-162.
31. Caro, J., Moack, M., Marlow, F., Petersohn, D., Griepentrog, M. and Kornatowski, J., J Am. Chemi. Soc.
31. Seaton, N.A., Friedman, S.P., MacElroy, J.M.D. and Murphy, B.J. - Private communication.
32. Armor, J.N., (1994) *Carbon Molecular Sieves for Air Separation* in Separation Technology, 163-199 E.F. Vansent - ed. Elsevier, Amsterdam.
33. Ruthven, D.M., (1992) Chem. Eng. Sci 47, 4305-4308.

# MEASUREMENT OF DIFFUSION IN MICROPOROUS SOLIDS BY MACROSCOPIC METHODS

DOUGLAS M. RUTHVEN AND STEFANO BRANDANI*
*Department of Chemical Engineering*
*University of Maine*
*Orono, ME 04469-5737*


*\*Universita del L'Aquila*
*I-67040 Monteluco di Roio*
*L'Aquila Italy*

## Abstract

In Part I of this paper the main macroscopic experimental methods for measuring diffusion in microporous solids are reviewed and the advantages and disadvantages of the various techniques are discussed. For several systems experimental measurements have been made by more than one technique and in Part II of the paper the results of such comparative studies are reviewed. While in many cases the results show a satisfactory consistency there are also several systems for which there are substantial discrepancies. A number of possible explanations have been suggested but none of these is really convincing.

## 1. Introduction

Because of the practical importance of microporous materials, notably zeolites and carbon molecular sieves, as catalysts and selective adsorbents, the problem of measuring micropore diffusivities has attracted considerable attention.[1,2] This task has proved more difficult than might have been anticipated largely because these materials are generally available only as rather small particles (or crystals). A wide range of different experimental techniques have been applied including microscopic and macroscopic methods and both transient and steady state measurements. An historical summary is given in Table 1. In this review we consider only macroscopic methods in which the diffusive flux is measured under well defined experimental conditions; the microscopic approach in which the movement of the molecules is tracked directly is discussed in another paper at this symposium[30].

A simple classification of the main microscopic techniques is shown in Table 2 and this provides a useful framework for our review. Macroscopic measurements

*J. Fraissard (ed.), Physical Adsorption: Experiment, Theory and Applications, 261–296.*
© *1997 Kluwer Academic Publishers.*

## Table1

## Measurement of Diffusion in Zeolite Crystals

### Historical Development

| Method | Author | Reference |
|---|---|---|
| Direct monitoring of transient conc. Profile ($H_2O$ - Heulandite) | Tiselius (1934) | (3,4) |
| Transient uptake rate | Barrer (1938-) | (5,6) |
| NMR Relaxation | Resing (1967) | (7,8) |
| PFG NMR | Pfeifer, Kärger, Lechert (early 1970s) | (9-11) |
| Chromatography | Haynes, Ma, Ruthven (mid 1970s) | (12-14) |
| Membrane Permeation | Hayhurst, Wernick (1983) | (15,16) |
| Effectiveness Factor | Haag, Post (early 1980s) | (17,18) |
| ZLC | Eic (mid 1980s) | (19,20) |
| Frequency Response | Yasuda, Rees (late 1980s) | (21,22) |
| QENS | Cohen de Lara, Jobic (late 1980s) | (23,24) |
| FTIR | Niessen and Karge (1991) | (25,26) |
| IR and IR/Freq. Response | Grenier, Meunier, Bourdin (1993) (CNRS) | (27,28) |
| Tracer ZLC | Hufton et al. (1994) | (29) |

## Table 2

## Classification of Macroscopic Methods for Measurement of Diffusion in Microporous Solids

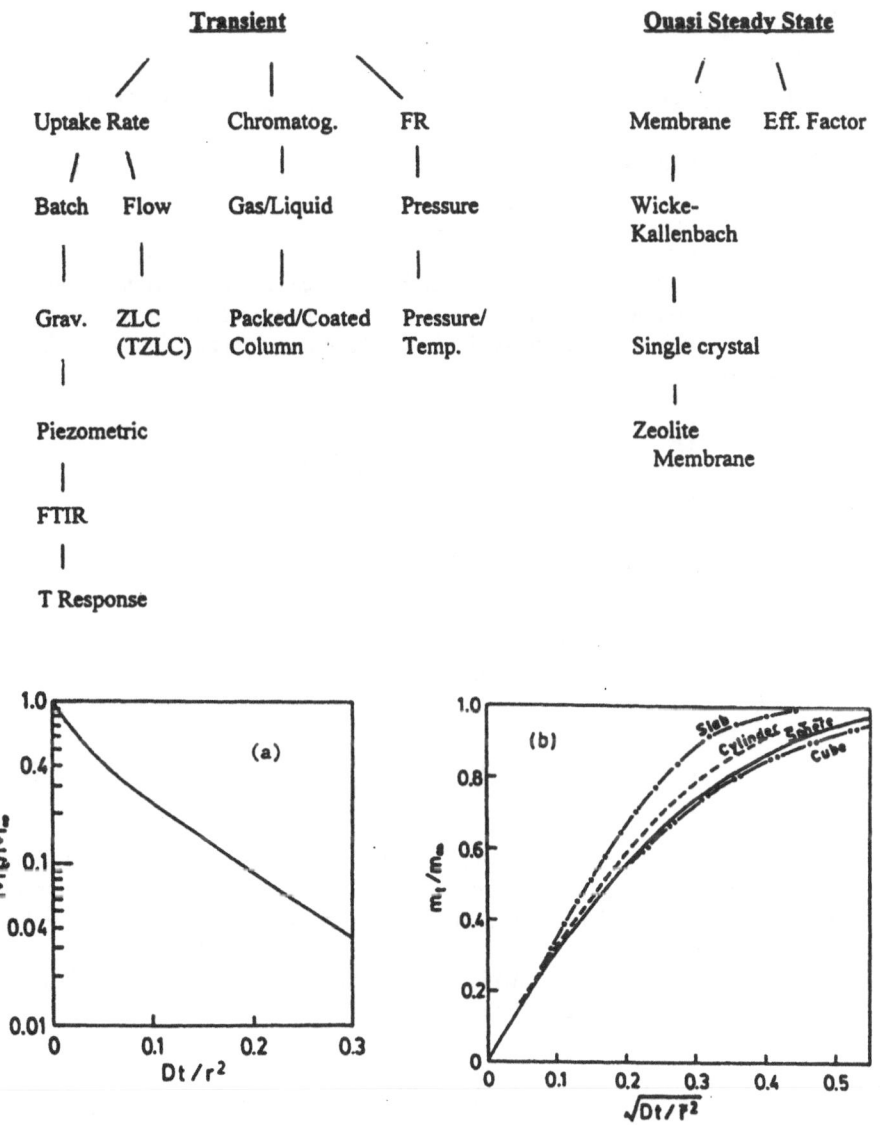

| Transient | | | Quasi Steady State | |
|---|---|---|---|---|
| Uptake Rate | Chromatog. | FR | Membrane | Eff. Factor |
| Batch    Flow | Gas/Liquid | Pressure | Wicke-Kallenbach | |
| Grav.    ZLC (TZLC) | Packed/Coated Column | Pressure/ Temp. | Single crystal | |
| Piezometric | | | Zeolite Membrane | |
| FTIR | | | | |
| T Response | | | | |

Figure 1. Theoretical uptake curves for an isothermal diffusion controlled system showing (a) the long time asymptote according to Eqs 1 or 3 and (b) the form of the short time response plotted in accordance with Eq 2. Note that the initial slope is quite insensitive to the particle shape. (The equivalent radius r is defined as the radius of a sphere with the same surface to volume ratio as the actual particles.)

generally yield 'transport' diffusivities, although variants of the techniques, using isotopically tagged tracers, can be devised to measure self-diffusivities. The large majority of the macroscopic techniques involve transient measurements. Steady state or quasi-study state methods, notably membrane permeation and catalyst effectiveness measurements have been demonstrated but their application has been limited to a few systems.

## 2. PART I: Survey of Macroscopic Methods for Measuring Micropore Diffusion

### 2.1 UPTAKE RATE MEASUREMENTS

Perhaps the most obvious experimental approach to the determination of micropore diffusivities is to measure directly the sorption/desorption rate for an adsorbent particle subjected to a well defined change in the ambient concentration of sorbate. For an isothermal spherical particle subjected to a step change in sorbate concentration at the external surface at time zero, the approach to equilibrium, under conditions of diffusion control is given by[1]:

$$\frac{M_t}{M_\infty} = 1 - \frac{6}{\pi^2} \sum_1^\infty \frac{1}{n^2} \exp\left(- n^2 \pi^2 Dt / R^2\right) \qquad (1)$$

The short and long time asymptotes are given by:

$$\text{Short Time:} \quad \frac{M_t}{M_\infty} = \frac{6}{R} \sqrt{\frac{Dt}{\pi}} \qquad (2)$$

$$\text{Long Time:} \quad \frac{M_t}{M_\infty} = 1 - \frac{6}{\pi^2} \exp\left(- \pi^2 Dt / R^2\right) \qquad (3)$$

The form of the response curves is shown in figure 1. The corresponding expressions for other particle shapes are easily derived but there is little numerical difference from the response for a spherical particle of the same external area to volume ratio (i.e. based on an equivalent radius).

In contrast, for the same situation, a particle with surface resistance control follows a simple exponential approach to equilibrium:

$$\frac{M_t}{M_\infty} = 1 - \exp\left(- 3kt / R\right) \qquad (4)$$

When mass transfer is rapid and the approach to equilibrium is controlled entirely by heat transfer, the uptake curve obeys[31]:

$$\frac{M_t}{M_\infty} = 1 - \left(\frac{\beta}{1+\beta}\right) \exp\left[\frac{-3ht}{\rho C_p R} \cdot \frac{1}{(1+\beta)}\right] \tag{5}$$

where $\beta = (\Delta H/\rho C_p)(\partial q^*/\partial T)_p$. A plot of $\log\left(1 - M_t / M_\infty\right)$ vs t thus provides clear evidence concerning the nature of the rate limiting resistance and a convenient way to extract the time constant ($R^2/D$, $R/3k_s$ or $\rho C_p R/3h$).

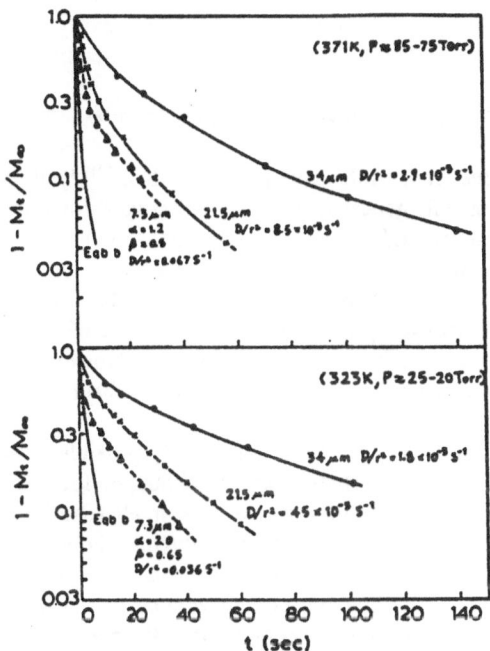

Figure 2. Experimental uptake curves for $CO_2$ in 4A zeolite crystals showing near isothermal behavior in large (34 and 21.5 μm) crystals ($D=9\times10^{-9}$ cm$^2$s$^{-1}$ at 371 K and $5.2\times10^{-9}$ cm$^2$s$^{-1}$ at 323 K). The solid lines are the theoretical curves for isothermal diffusion from Eq. 1 with the appropriate value of $D_c/r^2_c$. The uptake curves for the small (7.3 μm) crystals show considerable deviation from the isothermal curves but conform well to the theoretical nonisothermal curves with the values of $D_c$ estimated from the data for the large crystals, the value of β calculated from the equilibrium data, and the value of α estimated using heat transfer parameters estimated from uptake rate measurements with a similar system under conditions of complete heat transfer control. The limiting isothermal curve is also shown by a continuous line with no points. From Ruthven et al.[31].

The apparent simplicity of this approach is, however, deceptive. For measurement of intracrystalline diffusion the method works well when diffusion is relatively slow (large crystals and/or low diffusivity) but when sorption rates are rapid the uptake rate may be controlled by extracrystalline diffusion (through the interstices of the adsorbent bed) and/or by heat transfer. The intrusion of such effects is not always obvious from the shape of the uptake curve but it may generally be detected by changing the sample quantity and/or the sample configuration. It is in principle possible to allow for such effects in the mathematical model used to interpret the uptake curves (figure 2) and, indeed, the modeling of non-isothermal systems has been studied in considerable detail [31-35]. However, any such intrusion will obviously diminish the accuracy and confidence with which the intracrystalline diffusivities can be determined.

This technique may also be used to measure effective macropore diffusivities in bi-porous adsorbent pellets. For such a system with a linear equilibrium isotherm and assuming rapid intracrystalline diffusion, the governing diffusion equation is of the same form as for micropore control. The solution is identical to Eq 1 except that R now refers to the particle radius and the diffusivity D is replaced by the effective diffusivity $D_e = $
$$\frac{D_P \varepsilon_P}{\varepsilon_P + (1 - \varepsilon_P) K}.$$ Since the equilibrium constant (K) is generally large and varies with temperature according to the vant Hoff equation ($K = K_o e^{-\Delta H / RT}$) it is clear that a macropore controlled system will generally yield an effective diffusivity that is much smaller than the pore diffusivity and shows a strong (Arrhenius) form of temperature dependence but with an apparent activation energy that approximates the heat of adsorption. This may lead the unwary investigator to conclude that the rate is controlled by an activated (micropore) diffusion process whereas in fact the effect results from the temperature dependence of the capacity of the microparticles. However, a change in the gross particle size can often be used to provide an unequivocal diagnosis of the nature of the controlling resistance.

With vapor phase gravimetric systems it is generally possible to achieve a good approximation to a step change in sorbate pressure although, for strongly adsorbed species this requires a large system volume and a very small adsorbent sample. For piezometric measurements and for liquid phase systems, however, the external concentration (or pressure) is the measured quantity and to achieve accuracy it is therefore necessary to minimize the external system volume. In solving the diffusion equation, the time dependence of the boundary condition at the particle surface must then be considered and this leads to a slightly more complex solution for the uptake curve. The piezometric method has been widely applied[36] but a detailed analysis of this technique (see Appendix A) reveals that the restrictions on the conditions required to obtain reliable diffusivity data by this method are in fact more severe than had been generally appreciated.

In both gravimetric and piezometric experiments it is generally desirable to make the measurement over a small differential concentration change in order to ensure that the assumption of system linearity is fulfilled. Under these conditions the transient sorption curve (expressed as fractional approach to equilibrium) should be independent of either the step size or direction (adsorption or desorption). Varying the step size and direction thus provides a simple and sensitive experimental test for system linearity.

Uptake rate measurements with large oriented crystals have been used by Caro[37] to demonstrate the non-isotropy of silicalite. Diffusion coefficients for the longitudinal and transverse directions differed by a factor of about three.

An ingenious alternative to the gravimetric method has been developed by Karge and co-workers[25,26] who tracked the progress of sorption by monitoring the intensity of an IR band characteristic of the adsorbed species. This approach offers the important advantage that, by a judicious choice of the IR wavelength, it may be applied to follow sorption of one (or more) components in a multi-component system.

Regardless of the way in which progress of the sorption is followed, all uptake rate measurements are subject to the intrusion of heat transfer resistance and such effects are more severe for strongly adsorbed and rapidly diffusing species. To circumvent this problem Meunier et al[27] introduced a novel approach in which progress of the uptake is followed indirectly by monitoring the temperature of the adsorbent sample, using a sensitive IR detector. A representative response curve is shown in figure 3. A non-isothermal model continuing time constants for both heat transfer and diffusion is needed to interpret the experimental data. However, it emerges that the initial rising portion of the response curve is determined primarily by diffusion while the tail is determined mainly by heat transfer. Both parameters can therefore be determined with confidence, even from a single experimental response curve.

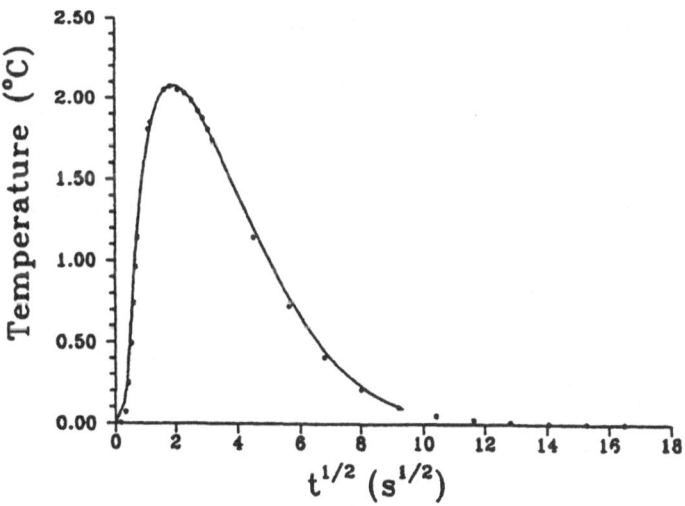

Figure 3. Transient temperature response for $CH_3OH$-NaX, pressure step 48-80 Pa (step A, run 13) showing conformity between experimentally observed temperature and the theoretical curve, calculated from the diffusion model with $D_o = 2.6 \times 10^{-12}$ m$^2$ s$^{-1}$; $h = 2.3$ Wm$^{-2}$K$^{-1}$. From Grenier et al.[27].

## 2.2 FREQUENCY RESPONSE

In the frequency response method, first applied to the study of zeolitic diffusion by Yasuda[21] and further developed by Rees and co-workers,[2,22] the volume of a system containing a widely dispersed sample of adsorbent, under a known pressure of sorbate, is subjected to a periodic (usually sinusoidal) perturbation. If there is no mass transfer or if mass transfer is infinitely rapid so that gas-solid mass transfer equilibrium is always maintained, the pressure in the system should follow the volume perturbation with no phase difference. The effect of a finite resistance to mass transfer is to cause a phase shift so that the pressure response lags behind the volume perturbation. Measuring the 'in phase' and 'out of phase' responses over a range of frequencies yields the characteristic frequency response spectrum which may be matched to the spectrum derived from the theoretical model in order to determine the time constant of the mass transfer process. As with other methods the response may be influenced by heat transfer resistance so, to obtain reliable results, it is essential to carry out sufficient experimental checks to eliminate such effects or to allow for them in the theoretical model. Being a quasi steady state measurement the frequency response technique offers certain advantages over direct measurement of the step response. The form of the frequency response spectrum depends on the nature of the dominant mass transfer resistance and can therefore be helpful in distinguishing between diffusion controlled and surface resistance controlled processes. However, the argument that the cyclic nature of the perturbation eliminates the intrusion of heat effects must be treated with caution. For both p-xylene and 2-butyne in silicalite Shen and Rees[38,39] observed a bimodal response spectrum and they interpreted the two peaks as indicative of two different transport processes corresponding to diffusion through the straight and sinusoidal channels. There is some NMR evidence to support the view that such molecules cannot easily reorient themselves at the channel intersections and for silicalite-2, which contains only straight channels of similar dimensions, only a single response peak is observed; so this hypothesis is certainly plausible. However, Sun and Bourdin[40] have shown that an alternative explanation is also possible. If the heat balance equations are included in the theoretical model the predicted response assumes a bimodal form and the heat transfer parameter required to match the experimental data appear to be quite reasonable.

In a recent development of the frequency response technique, Bourdin[28] has applied the frequency response approach to the IR temperature measurement system. In this experiment the volume of the system is perturbed sinusoidally and both the pressure and temperature responses are measured. It was found that the phase differences between the pressure and temperature were more reliable and reproducible than the phase differences between the pressure and the volume. The explanation seems to be that since the quantity of adsorbent is quite small, a small amount of superficial adsorption on the walls of the apparatus can distort the phase relationship between the pressure and the volume but will not affect the temperature-pressure phase relationship. In the light of this observation there may be a need to re-examine some of the earlier FR results which were derived from the phase difference between the driver piston and the pressure response. The main advantage claimed for this approach is that it provides a clearer

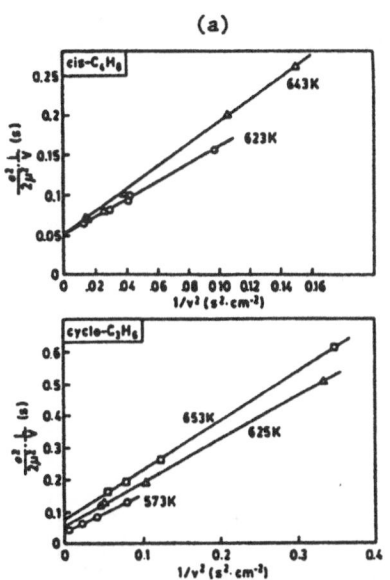

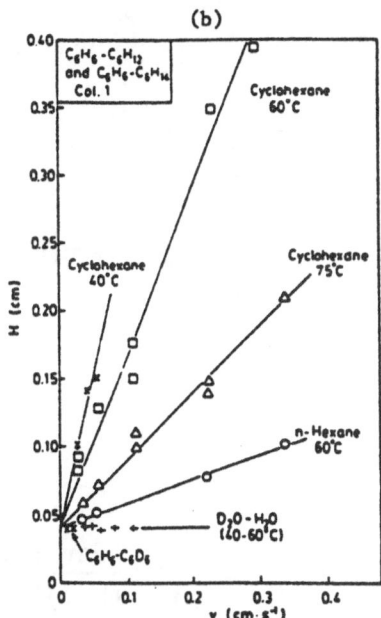

Figure 4. Chromatographic HETP data. (a) Plots of H/2v vs 1/v² for cis-butene and cyclopropane in 5A zeolite. From Haq and Ruthven[43]. (b) Plots of H vs v for benzene/n-hexane in NaX (40 μm crystals). Note that the axial dispersion term (2$D_L$/v ≅ 0.04 cm) is essentially the same for $C_6H_6$-$C_6D_6$ and for $H_2O$/$D_2O$. From Awum et al.[44].

discrimination between different mass transfer resistances (internal diffusion/surface barrier).

## 2.3 CHROMATOGRAPHIC MEASUREMENTS

For fast diffusing systems the limitations imposed by extracrystalline resistances to mass and heat transfer make it impossible to derive reliable intracrystalline diffusivity values from direct sorption rate measurements, regardless of the technique used to follow the uptake. Since both heat and mass transfer are enhanced in a flow system the possibility of deriving reliable diffusion values from measurements of the dynamic response of a packed adsorption column has attracted considerable attention. The early models for mass transfer resistance in a chromatographic column were based on the equilibrium stage concept. The Kubin-Kucera model[41,42] by showing the relationship between the height equivalent to a theoretical plate (HETP) and the diffusional time constant provided the essential theoretical basis for the chromatographic approach to the measurement of intra-particle diffusivities. The generalization to bi-porous particles was provided by Sarma and Haynes[12].

The chromatographic response is conveniently analyzed in terms of the first and second moments of the pulse response:

$$\mu \equiv \frac{\int_0^\infty c.tdt}{\int_0^\infty cdt} = \frac{L}{V}\left[1+\left(\frac{1-\varepsilon}{\varepsilon}\right)K\right] \qquad (6)$$

where, for a composite particle:

$$K = \varepsilon_P + (1-\varepsilon_P)wK_C \qquad (7)$$

$$\sigma^2 = \frac{\int_0^\infty (t-\mu)^2 dt}{\int_0^\infty cdt} \qquad (8)$$

$$HETP \equiv \frac{\sigma^2}{\mu^2}L = 2\frac{D_L}{v} + \frac{2\varepsilon v}{(1-\varepsilon)}\left[\frac{R}{3k_f} + \frac{R^2}{15\varepsilon_P D_P} + \frac{r_c^2}{15KD_c}\right]\left[1+\frac{\varepsilon}{(1-\varepsilon)K}\right]^{-2} \qquad (9)$$

It is evident that the HETP measures only the overall resistance to mass transfer and cannot provide evidence concerning the nature of this resistance. The chromatographic response is indeed remarkably insensitive to differences in the nature of the mass transfer resistance so, regardless of the way in which the data are analyzed, it is not possible to obtain any such information except by varying critical parameters such as the particle size.

The major difficulty in the analysis of chromatographic data is to separate the axial dispersion and mass transfer contributions since, except for gaseous systems at very low flow rates, the axial dispersion coefficient $(D_L)$ is velocity dependent. For liquid systems $D_L$ varies essentially linearly with velocity so a plot of HETP vs $\varepsilon v$ should be linear with the mass transfer resistance directly related to the slope (see figure 4). For gaseous systems at high Reynolds number this same plot can be used but in the low Reynolds number region a plot of H/v vs $1/v^2$ may be more convenient since in this region $D_L$ is essentially constant and the intercept thus yields the mass transfer resistance[43-45].

In the application of the chromatographic method to the measurement of intracrystalline diffusivity it is preferable to pack the column directly with unaggregated crystals rather than with composite (pelleted) material since this eliminates the possible intrusion of macropore resistance. The small crystal size of commercial zeolite samples presents a significant practical problem. Early attempts to utilize a column packed directly with such crystals were not very successful. Erroneously small apparent diffusivities were obtained, probably reflecting the anomalously high axial dispersion which is observed in beds of very small particles due to the tendency of such particles to form agglomerates. Columns packed with large synthetic crystals work well[46] and this approach has the advantage that the use of large crystals increases the relative importance of the diffusional resistance (the second term on the right hand side of Eq 9), thus minimizing the impact of any errors in the estimation of the axial dispersion. However, this approach obviously precludes the use of small commercial crystallites. Small commercial crystallites have however been successfully applied in the form of a wall coated column[47] and this may be the best approach when larger crystals are not available.

## 2.4 THE ZERO LENGTH COLUMN (ZLC) METHOD

The zero length column (ZLC) method is a chromatographic technique that eliminates the uncertainty due to axial dispersion. It depends on following the desorption of sorbate from a previously equilibrated (small) sample of adsorbent into an inert carrier stream. The ZLC method was introduced, for gas phase adsorption systems, in the late '80s[19] and since then it has found widespread application as a simple and relatively inexpensive way of measuring

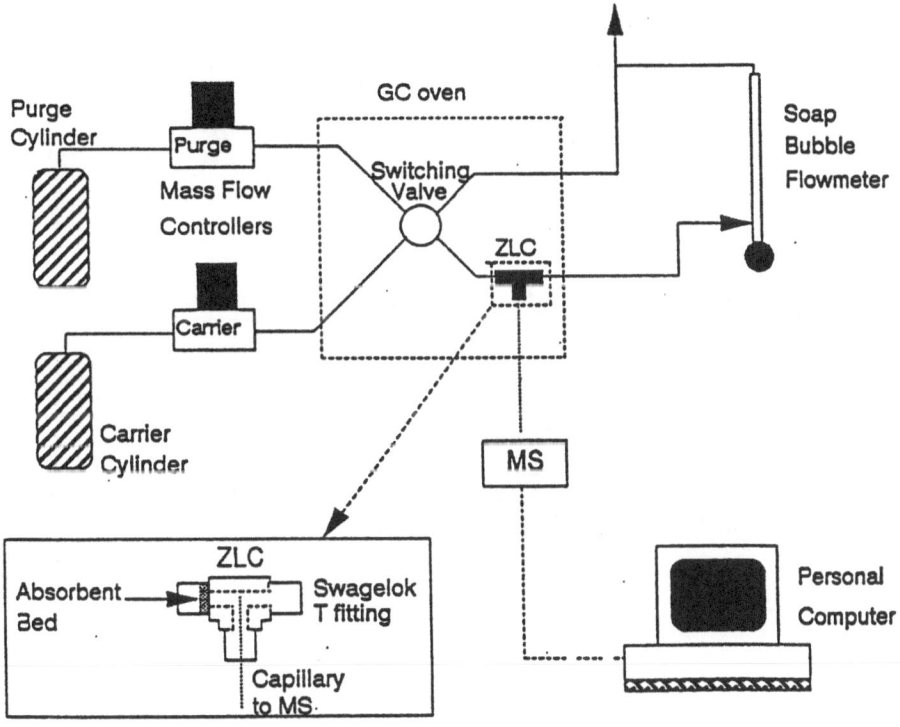

Figure 5. Experimental set–up for gaseous ZLC or TZLC.

limiting diffusivities for hydrocarbons and other simple molecules in zeolites and other microporous adsorbents[1]. The method has been extended to the measurement of counter–diffusion in liquid phase adsorption systems by Ruthven and Stapleton[48]. Recently a new variant of the technique, tracer ZLC, has been presented[29], which allows the measurement of self–diffusivities over the entire range of sorbate loadings. An experimental set–up for gaseous systems is shown in figure 5.

The actual ZLC column consists of a thin layer of adsorbent material placed between two porous sinter discs. The individual particles (or crystals) are dispersed approximately as a monolayer across the area of the sinter. This minimizes the external resistances to heat and mass transfer, so that the adsorption cell can be considered as a perfectly mixed isothermal, continuous flow cell.

Brandani and Ruthven[49] have derived the solution for the general model which accounts for the capacity of the fluid phase as well as that of the adsorbent:

$$\frac{c(t)}{c_o} = \Sigma \frac{2L}{\beta_n^2 + (1 - L + \gamma \beta_n^2)^2 + L - 1 + \gamma \beta_n^2} \exp\left(-\beta_n^2 \frac{D}{R^2} t\right) \qquad (10)$$

where the $\beta_n$ are the positive roots of:

$$\beta_n \cot \beta_n + L - 1 - \gamma \beta_n^2 = 0 \qquad (11)$$

with

$$L = \frac{1}{3} \frac{F}{KV_s} \frac{R^2}{D} \; ; \quad \gamma = \frac{1}{3} \frac{V_f}{KV_s} \qquad (12)$$

It can be clearly seen that the parameter $\gamma$ is of the order of $1/K$. Therefore the liming case $\gamma = 0$ provides a very useful approximation for gaseous systems where the hold up in the fluid phase can generally be neglected in comparison with the adsorbed phase accumulation. In this situation, Eq. 10 reduces to

$$\frac{c}{c_o} = 2L\Sigma \frac{\exp\left(-\beta_n^2 \frac{D}{R^2} t\right)}{\beta_n^2 + L(L-1)} \qquad (13)$$

which is the form obtained from the original ZLC model[19]. It has been shown that eq. 13 can be used with confidence for $\gamma < 0.1$. This model for gas phase adsorption systems has been studied in detail by Brandani and Ruthven[50].

Most of the experimental applications of the ZLC technique have been with gaseous systems[51] and for these systems, the technique may now be regarded as a standard method. Based on our experience it is possible to suggest some guidelines as to how the experiments should be carried out. The key parameter is L, which from its definition, Eq. 12, can be considered the ratio of the diffusional and washout time constants; $R^2/D$ and $KV_s/F$. This parameter is also equal to the dimensionless adsorbed phase concentration gradient* at the surface of the solid at time zero. From either of these definitions it is evident that L gives an indication of how far removed the system is from equilibrium control. This parameter is proportional to the flow rate, so it can be easily varied and in order to extract a reliable time constant, it is necessary to run the experiment at at least two different flow rates.

A major advantage of the ZLC method is that, for any particular system, the validity of the basic assumptions, under the experimental conditions can be verified directly by a series of simple experimental tests. To establish the validity of the zero length limit, measurements are repeated with columns containing different amounts of adsorbent (see figure 6). To exclude external mass transfer resistances, experiments may be run with two different carrier gases, typically argon and helium (see figure 7). Heat effects can be minimized using high flow rates and system linearity can be checked varying the inlet composition.

If L>10 the time constant can be easily extracted from the long time asymptote where Eq. 13 reduces to:

$$\frac{c}{c_0} = 2L \frac{\exp\left(-\beta_i^2 \frac{D}{R^2} t\right)}{\beta_i^2 + L(L-1)} \tag{14}$$

A semilog plot of the dimensionless concentration yields L and $D/R^2$ from the intercept and slope. Figure 8 shows the dependence of the intercept and $\beta_1^2$ on L. It is clear that for L>10 the slope should remain constant. Possible problems can be encountered if the accuracy of the measuring device is limited or if the experimental conditions are beyond the range of the Henry law limit. For hydrocarbons a FI detector is typically used and ensures

---

* from the column mass balance $\dfrac{KV_s}{\overline{q_0}} \dfrac{d\overline{q}}{dt} = -F \dfrac{c}{c_0}$ it is possible to

obtain $\dfrac{3KV_sD}{R^2}\left(\dfrac{\partial Q}{\partial \xi}\right)_{\xi=1} = -F\dfrac{c}{c_0}$ and $\left(\dfrac{\partial Q}{\partial \xi}\right)_{\xi=1}^{t=0} = -L$

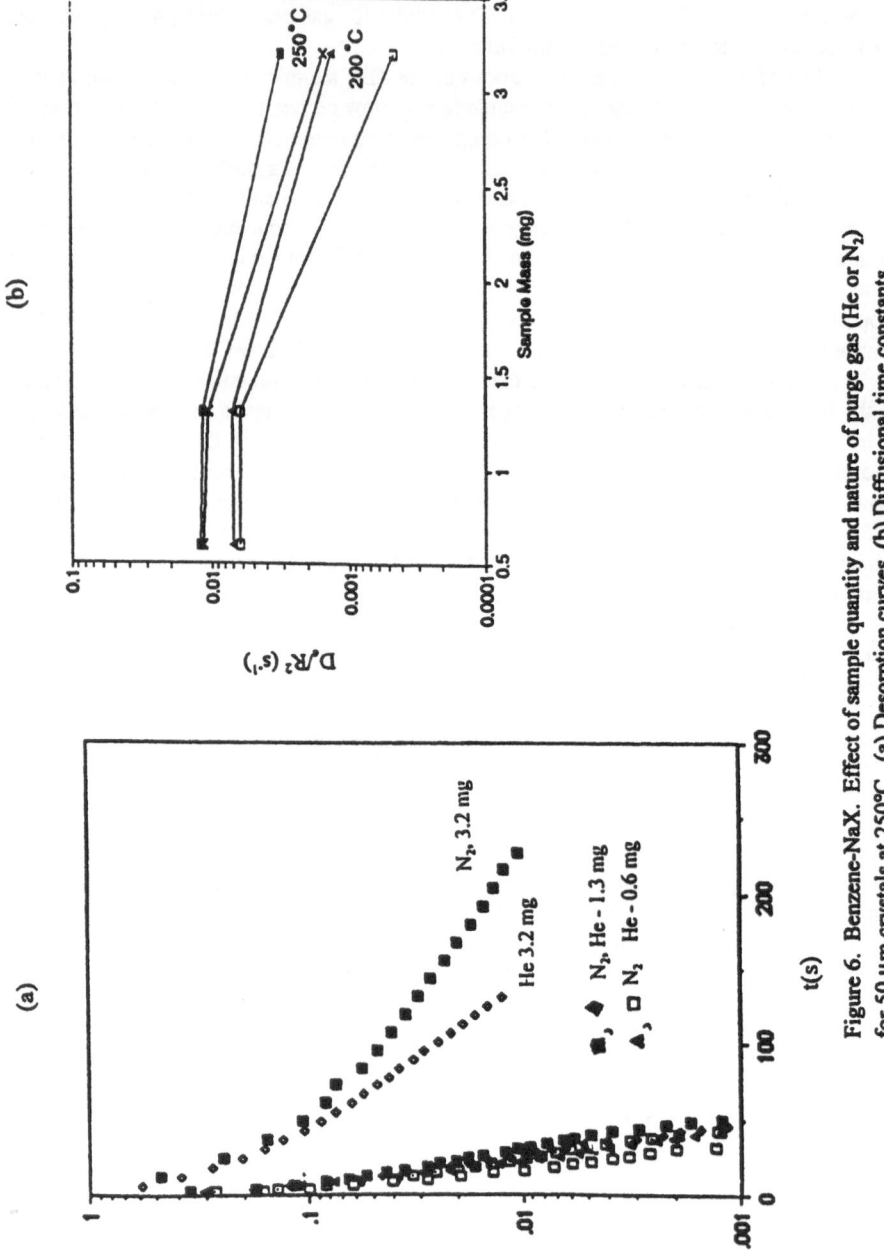

Figure 6. Benzene-NaX. Effect of sample quantity and nature of purge gas (He or $N_2$) for 50 μm crystals at 250°C. (a) Desorption curves, (b) Diffusional time constants.

275

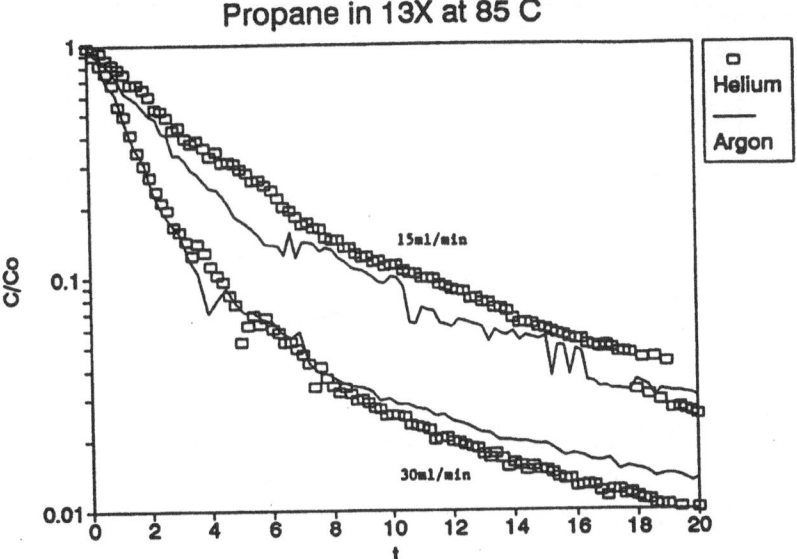

Figure 7. Experimental desorption curves for the system Propane – NaX with two different carrier gasses.

a value in which the linearity assumption is valid and the long time asymptote should therefore still yield reliable information on the diffusional time constant. If this asymptote can be measured, the slope should always increase with the flow rate. Since the ZLC response is determined primarily by the limiting behaviour at low concentration, the method is not well suited to determination of the concentration dependence of diffusivity from integral measurements.

It has been suggested[52,53] that in the application of the ZLC method the diffusional time constant should be extracted from the experimental response curves by a numerical parameter optimization procedure ("total curve fitting") using a program such as ZEUS. A major problem with this approach is easily seen from figure 9 which shows the response curves, at two different flow rates and concentrations, calculated from the parameters derived in this way by Bulow et al.[52]. The curves show that the slope of the long time asymptote decreases with increasing flow rate, which is physically inconsistent, thus raising doubts concerning the validity of the derived time constants. Careful examination reveals that the reported response curves were probably affected by erroneous baselining. However, the total curve fitting approach can be applied to yield optimized rate parameters provided that several response curves are regressed simultaneously[49].

In tracer ZLC (TZLC) the experiment is similar to the standard method but the monitored species is the deuterated form of the sorbate. This introduces an additional cost for the material and the requirement for an on–line mass spectrometer. The advantages are the elimination of all possible heat effects, strict linearity of the equilibrium between the

276

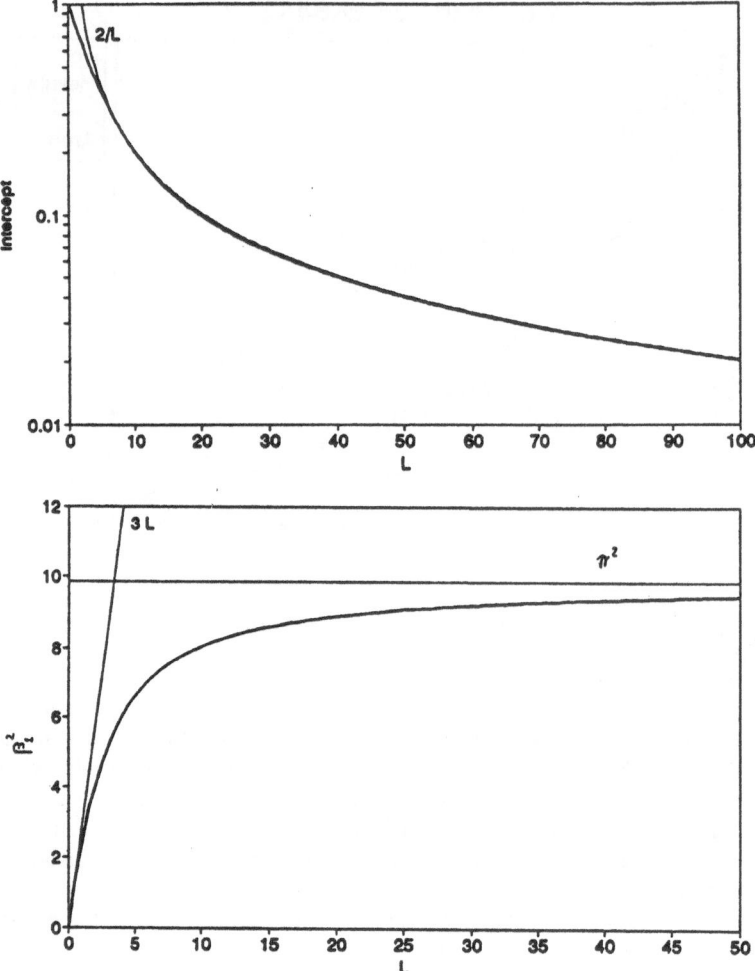

Figure 8. ZLC Desorption curves. (a) intercept of long time asymptote vs L. (b) $\beta_1^2$ vs L

fluid phase and the adsorbed phase, and the possibility of measuring directly the tracer diffusivities (which should be the same as the microscopically measured self–diffusivity) over a wide range of concentration. To reduce the costs the carrier is prepared with a mixture of pure and deuterated hydrocarbons. It has been shown that small imbalances in the concentration of the carrier and the purge streams do not affect the desorption dynamics[54] (see figure 10).

There is still the possibility that the desorption curves may be misinterpreted due to a surface barrier controlled process. The response curve for surface control would be qualitatively similar but it is possible to obtain experimental evidence to exclude this

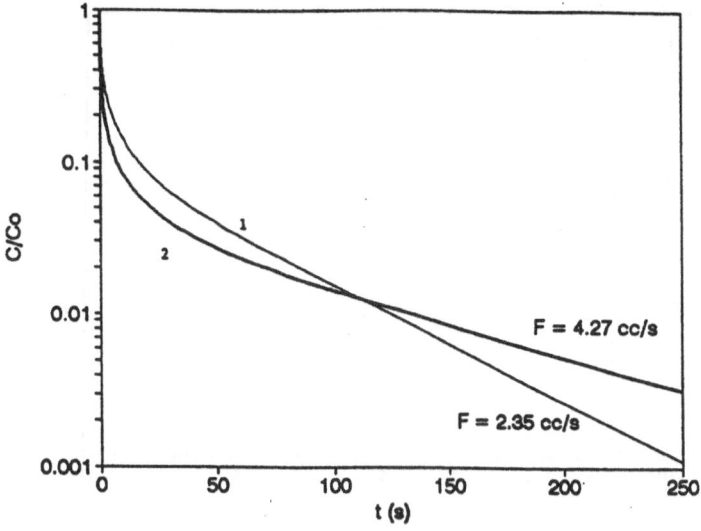

Figure 9. Desorption of Benzene from HZSM–5 obtained from numerical total curve fitting of experimental ZLC desorption curves[52] (The theoretical curves matched perfectly the experimental data). The following values were used: R=4.7 μm, $V_s$=4.18×10⁻³ cm³, $V_f$=2.18×10⁻³ cm³ (1) Initial composition $c_0$=1.25×10⁻⁸ mol/cm³, flow rate F=2.35 cm³/s, K=12600, D=4.3×10⁻¹⁰ cm²/s; (2) Initial composition $c_0$=2.66×10⁻⁹ mol/cm³, flow rate F=4.27 cm³/s, K=17600, D=2.2×10⁻¹⁰ cm²/s;
Note: To reproduce the curves of Micke et al.[52] L had to be calculated as $FR^2/DKV_s$

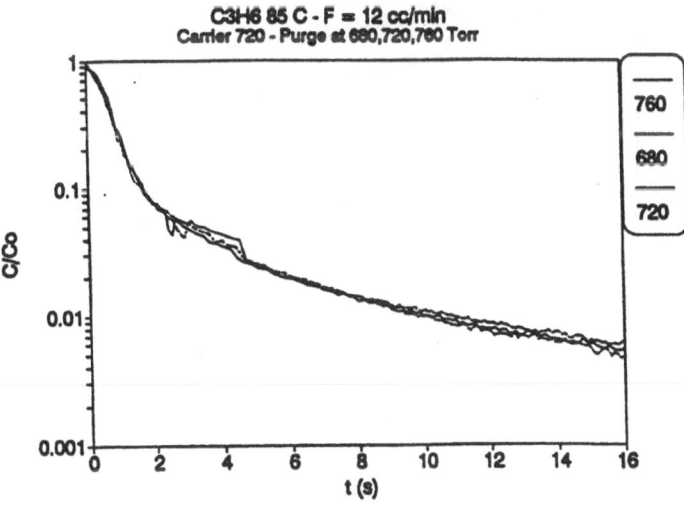

Figure 10. Experimental desorption curves measured with a partial pressure of 720 Torr of Propylene in the carrier gas and 680, 720 and 760 in the purge gas[54].

278

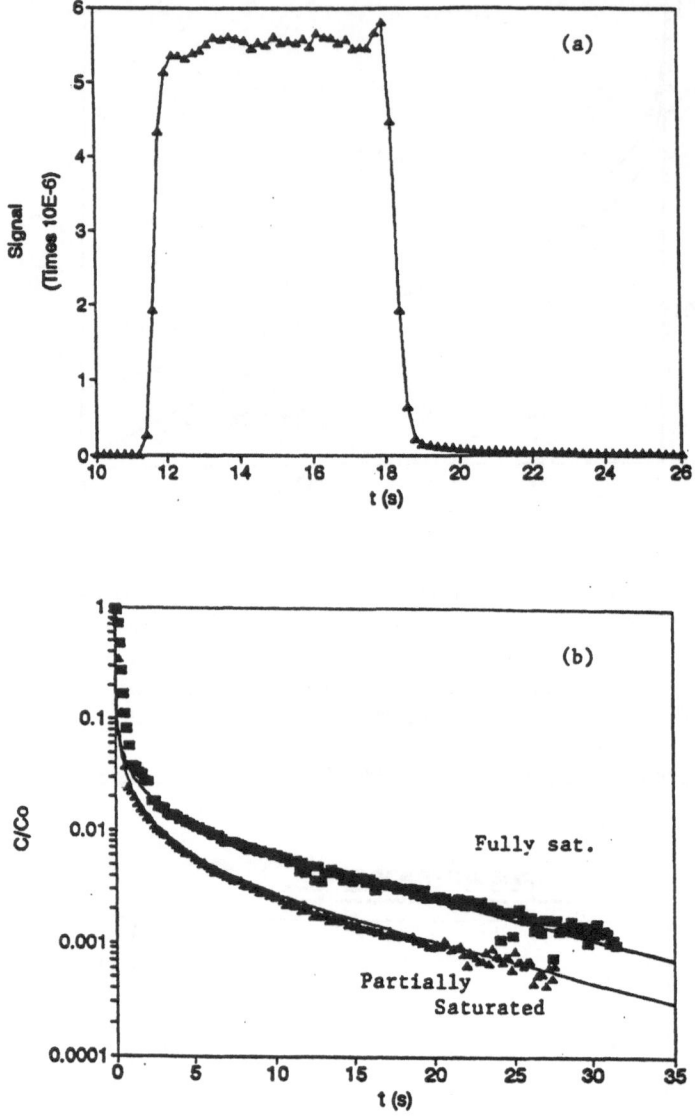

Figure 11. Experimental response to a partial loading experiment[50].
(a) Signal vs time. (b) Semilog ˙plot showing a comparison between full saturation and partial loading response. Lines are calculated from theory.

mechanism through a partial loading experiment. The adsorbent is exposed to the carrier flow for a limited period, not allowing sufficient time to approach equilibrium in the adsorbent. The theory has been discussed by Brandani and Ruthven[50] and an experimental validation has been carried out for the system propane – NaX. A typical response curve is shown in figure 11. There is a shift in the long time asymptote which can be predicted theoretically. The adsorbed phase concentration can be calculated from integration of the experimental response:

$$\frac{\overline{q}}{q_o} = 1 - \frac{\displaystyle\int_0^t \frac{c}{c_o} dt}{\displaystyle\int_0^\infty \frac{c}{c_o} dt} \qquad (15)$$

For the system considered the curve shown in figure 12 was obtained. If the process is controlled by a surface barrier the plot of the adsorbed phase concentration will not vary. In contrast, if the process is controlled by intracrystalline diffusion there will be a shift of the long time asymptote. This can be qualitatively clarified considering figure 13. The partial loading experiment is applicable if the conditions are well removed from equilibrium control, L>20. If L<20 it is possible to obtain an indication by running the system at different flow rates and considering the first and second moments of the response[55]. If there is a surface barrier $\sigma^2/\mu^2$ should be independent of flow rate, while for a diffusion controlled process this quantity increases almost linearly in the range 0<L<15.

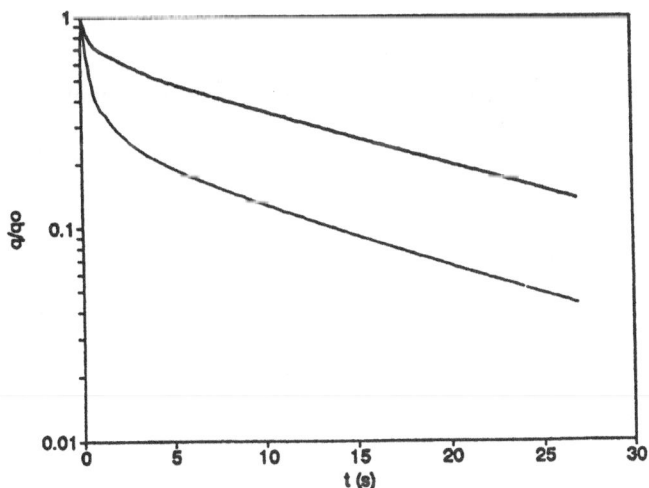

Figure 12. Average adsorbed phase concentration calculated from numerical integration of the experimental desorption curves[54]. Fully equilibrated sample (heavy line). Partially loaded sample (light line).

280

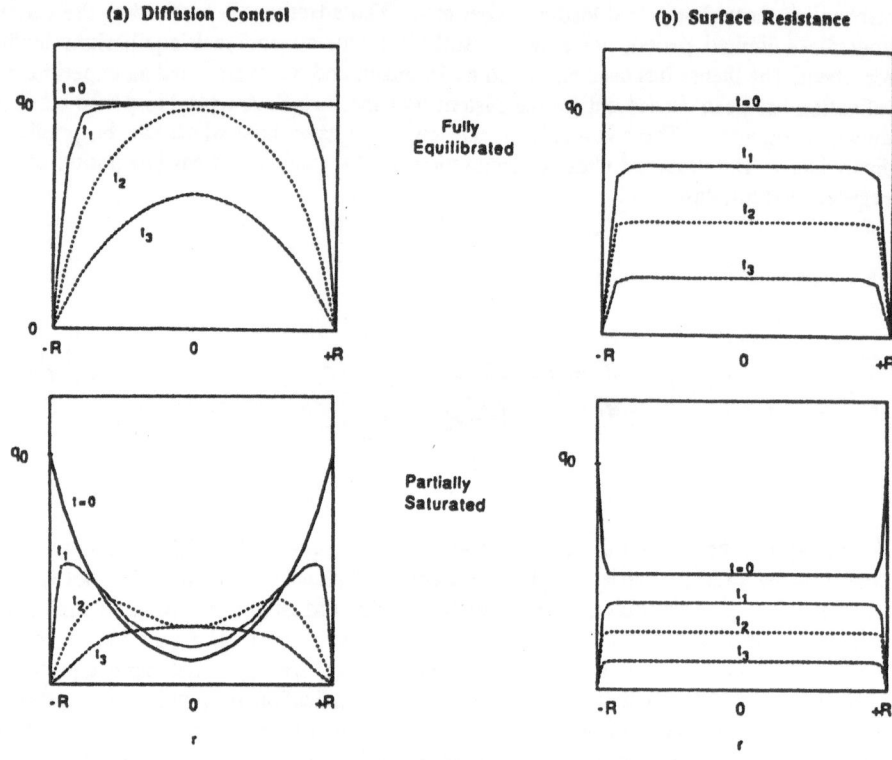

Figure 13. Comparison of the concentration profiles for diffusion control (a) and surface resistance (b).

Further experimental validation is needed to establish the applicability of the technique to liquid systems and to commercial solids characterized by a biporous structure with two controlling mass transfer resistances.

## 2.5 MEMBRANE MEASUREMENTS

Measurement of the flux through a zeolite crystal membrane under well defined conditions of sorbate pressure on the high and low pressure faces provides an intrinsically attractive approach to the measurement of intracrystalline diffusion. However, in practice this approach is not at all straightforward. The earliest attempts were by Wernick and Osterhuber[16] and Paravar and Hayhurst.[15] Both these groups made their membrane from a single large zeolite crystal held in a small hole and sealed with epoxy. Wernick and Osterhuber studied diffusion of butane in NaX while Paravar and Hayhurst studied light paraffins in silicalite. Both used the quasi-steady state mode of operation in which a relatively high sorbate pressure is maintained on one side of the crystal with a vacuum on the other side. The flux is then determined from the rate of pressure increase on the low pressure side. The form of the pressure response is shown in figure 14. The initial transient leads to a time delay while in the long time region a linear asymptote of pressure vs time is approached. Values for the diffusivity can be derived from both the slope and intercept of such plots: and the consistency of these two values provides a check on the validity of the measurements. The main problem with this approach is associated with the difficulty of properly mounting the small crystal and ensuring that the active faces are not blocked by stray epoxy. More recently measurements have been made with coherent layers of silicalite crystals, grown *in situ* on a stainless steel support[56].

## 2.6 EFFECTIVENESS FACTOR

The overall rate of a catalytic reaction, under diffusion limited conditions, depends on the Thiele Modulus $\left( R\sqrt{k/D} \right)$. From measurements of reaction rate in a series of particles of different size it is therefore possible to extract both the intrinsic rate constant (k) and the micropore diffusivity. This approach was first suggested by Haag[17]. A more complete experimental study (for 22 dimethyl butane in HZSM-5) was presented by Post et al[18] - see figure 15.

## 3. PART II: Review of Experimental Diffusivity Data for Selected Systems

The techniques outlined above have been used to study diffusion in a wide range of zeolite systems. In general we find that there is, in most cases, reasonable agreement between the different macroscopic methods and also between the microscopic methods (QENS, PFGNMR). However, although for several systems the macroscopic and

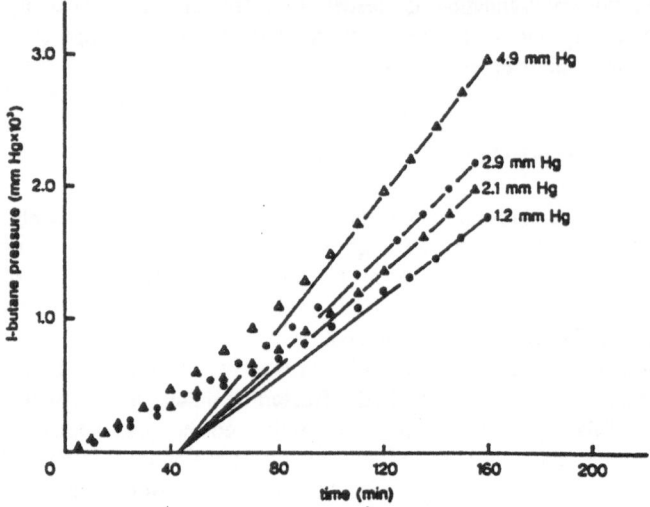

Figure 14. Permeability measurement through a single (100 μm) crystal of silicalite (I-butane at 334 K). Note asymptotic approach to the same intercept regardless of pressure. This implies that the diffusivity is independent of concentration over the relevant range. From Hayhurst and Paravar[15].

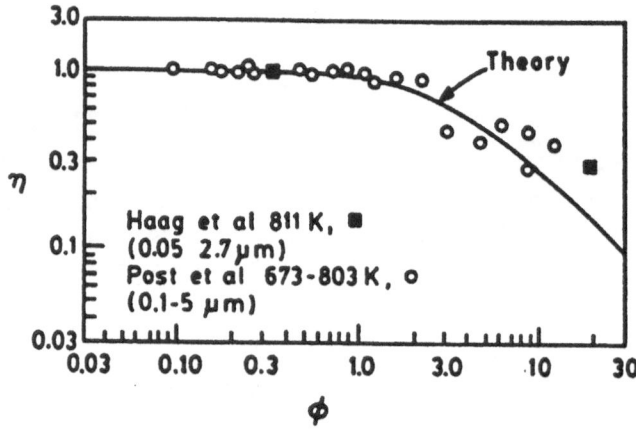

Figure 15. Experimental verification of the Thiele analysis as applied to intracrystalline diffusion control of the catalytic reaction of 2,2-dimethylbutane in H-ZSM-5 catalysts. Open symbols are from Post et al.[18], filled symbols are from Haag et al.[17]. From Post et al.[18].

microscopic measurements are also consistent, there are at least as many systems for which we see significant discrepancies.

## 3.1 SYSTEMS WHICH SHOW CONSISTENT BEHAVIOR

(a) Benzene - Silicalite[57]

Diffusion of benzene in silicalite has been studied by several different macroscopic techniques with broadly consistent results (see figure 16). However, diffusion in this system is too slow to allow reliable PFGNMR measurements so no comparison between micro and macro techniques is possible.

(b) Xe, CO, C, H₁, n C, H₁₀ - 5A

The consistency between sorption rate measurements and PFGNMR measurements in large crystals of 5A zeolite was noted many years ago. In more recent studies a similar pattern of consistency between sorption rate, ZLC and PFGNMR data has been observed for Xe, $CO_2$ and $C_3 H_8$ in 5A zeolite (see figure 17).

(c) CH, OH - NaX

This system has been studied by ZLC, tracer ZLC, and infra red temperature rise measurements as well as by PFGNMR with striking consistency between the reported results (see figure 18). It is notable that diffusion rates in this system are as fast or even faster then in most of the systems for which discrepancies are found.

## 3.2 SYSTEMS SHOWING DISCREPANCIES BETWEEN MICRO AND MACRO MEASUREMENTS

(a) Aromatics - NaX

The striking discrepancy between the result of sorption rate measurements with large (100 μm, 250 μm) crystals and the PFGNMR data for the xylenes in NaX was pointed out in 1988[20]. The behavior of benzene is similar[60]. Diffusion of benzene is about an order of magnitude faster then the xylenes and direct derivation of diffusivities from uptake rate measurements is therefore possible only over a very limited range of conditions. This system has, however, been studied in great detail by FTIR and by the ZLC and tracer ZLC methods, which all yield diffusivities that are quite consistent with lower temperature sorption rate measurements but substantially smaller than the PFGNMR values (see figure 19).

(b) Propane, Propene - NaX

The recently reported tracer ZLC data for propane and propene in NaX show a similar discrepancy and for these systems there is evidently a clear difference in the trend of diffusivity with sorbate loading, as well as in the order of magnitude of the diffusivity values.

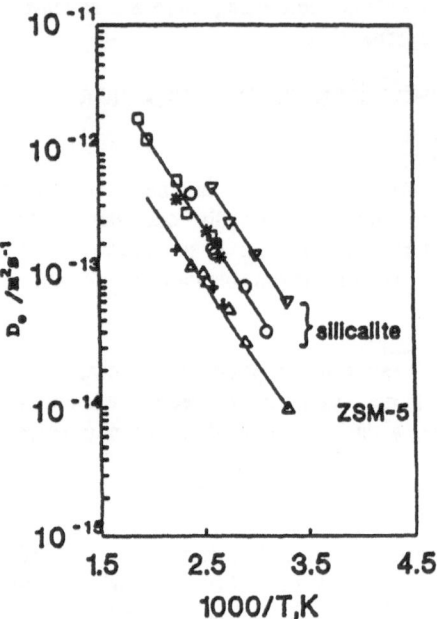

Figure 16. Comparison of corrected diffusivity of benzene in silicalite-1 and HZSM-5 at low sorbate concentrations. (□ Van Den-Begin et al., square wave; (○) Eic and Ruthven, ZLC; (△, ▽) Zikanova et al., piezometric; (*) silicalite-1, (+) HZSM-6, SSFR; (■) n.m.r. tracer exchange. From Shen and Rees[46].

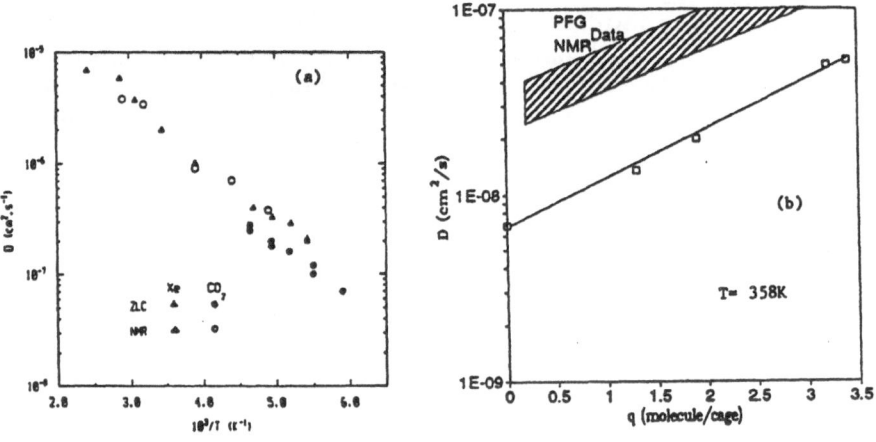

Figure 17. Comparison of PFG NMR and ZLC diffisivities for (a) Xe and $CO_2$ in 5A zeolite[58]; (b) Propane in 5A zeolite[29].

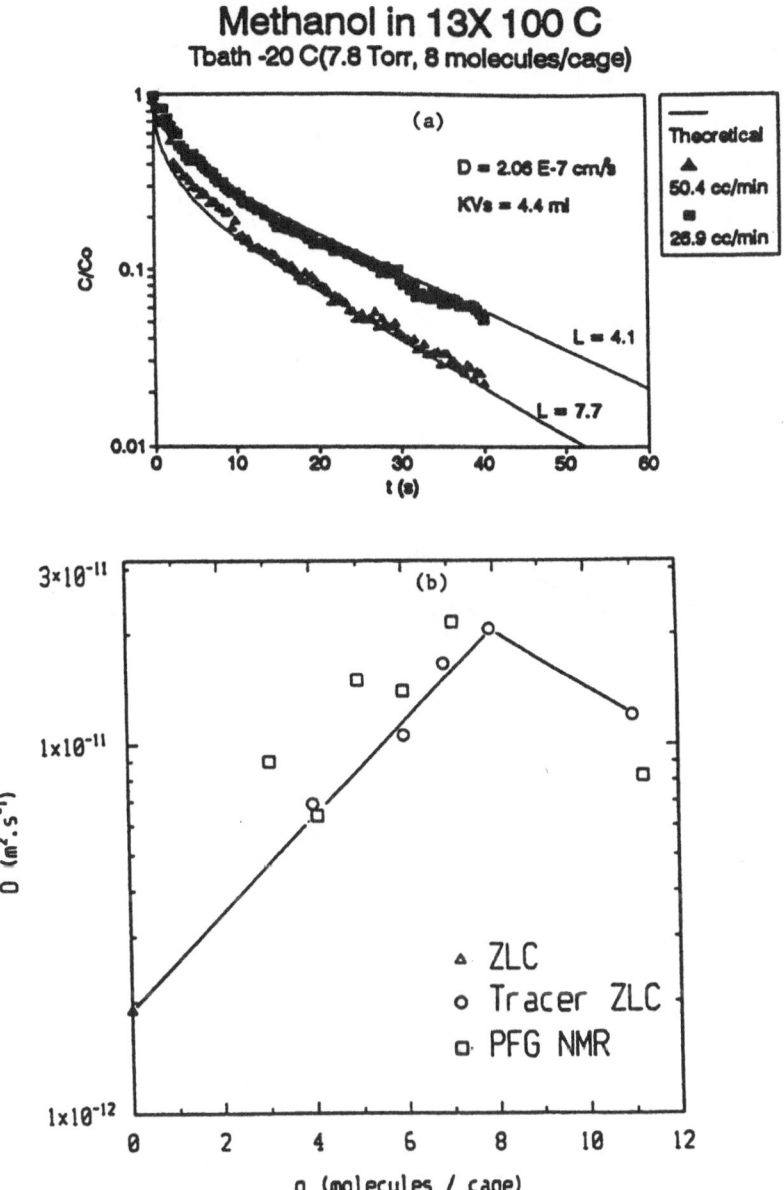

Figure 18. Diffusion of methanol in NaX zeolite crystals at 100°C (a) Tracer ZLC response curves; (b) variation of self-diffusivity with loading showing comparison of ZLC and PFG NMR data. From Brandani et al.[59].

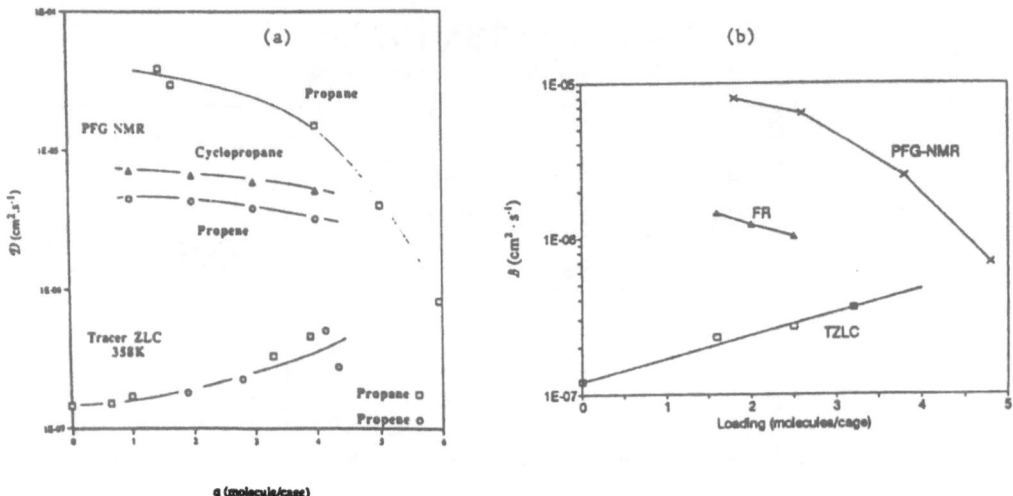

Figure 19. (a) Comparison of PFG NMR diffusifity data (325 K) and Tracer ZLC data (358 K) for propane and propene in NaX zeolite crystals. From Brandani et al.[54]. (b) Comparison of PFG NMR frequency response and tracer ZLC diffusivity data for benzene-NaX at 468 K. From Brandani et al.[60].

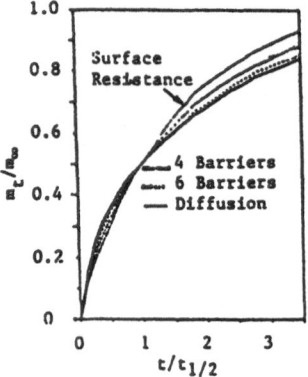

Figure 20. Theoretical uptake curves for a one dimensional system showing the effect of internal (and surface) barriers on the form of the uptake curve.

## TABLE 3

### Diffusion in AlPO$_4$-5 (~300K)

| Sorbate | Technique | Time Scale (sec) | Mechanism | D(cm$^2$.s$^{-1}$) |
|---------|-----------|------------------|-----------|---------|
| CH$_4$ | PFG-NMR | 0.01 - 0.1 | Single File [D $\propto$ 1/$\sqrt{t}$] | 5x10$^{-9}$ - 5x10$^{-8}$ |
| C$_3$H$_8$ | TZLC | 1-10 | Fickian [D const.] | 10$^{-6}$ - 5x10$^{-5}$ |

* Undimensional pore system diam. ~ 7.3Å

(c) Propane - AlPO$_4$5

There has recently been much interest in the phenomenon of single file diffusion, which occurs in a undimensional pane system when the diffusing molecules are too large to pass one another. In this situation the mean square displacement increases in proportion to the square root of elapsed time, rather then linearly with time, as in a Fickian system. Such behavior was recently observed in an experimental PFGNMR study of diffusion of methane in large crystals of AlPO$_4$-5. However, when diffusion in the same crystals was studied by the tracer ZLC method, the results were entirely consistent with normal one-dimensional Fickian diffusion. Remarkably, in this case the macroscopically observed diffusivities were larger than the values estimated (via the Einstein equation) from the PFGNMR data. This appears to be the first reported instance in which the micro measurements suggest slower diffusion than the macroscopic measurement.

The time scales of the two sets of measurements are, however, quite different (see Table 3). This raises the possibility that the difference in the observed pattern of behavior may reflect a lack of long range coherence in the channel system which could give rise to single file behavior over a short range (of distance and time) but, over distances of the order of the crystal length the behavior may be dominated by the passing places arising from the defects in the structure.

A wide range of explanations have been suggested to account for the observed discrepancies between macro and micro measurements, including:

Heat effects
External diffusional resistance
Errors in analysis of macro rate data
Surface barrier resistance
Mobile/Immobile phase behavior
Structural defects

None of these explanations seems capable of explaining the observations in a totally convincing way and the role played by structural defects may be much more significant than has generally been thought. In this context it is worth noting that a relatively small number of internal diffusion barriers can easily replicate the dynamic behavior of a relatively slow macroscopic diffusion controlled process, while the true micropore diffusion rate between barriers (measured by microscopic techniques) may be much higher (figure 20). In the absence of real evidence for such structural defects, however, any such explanation remains very tentative.

## Notation

| | |
|---|---|
| a | surface area/volume ratio for adsorbent *sample* |
| c | sorbate concentration in fluid phase |
| $c_o$ | initial steady volume of c |
| $C_p$ | heat capacity of adsorbent |
| D | diffusivity |
| $D_c$ | intracrystalline diffusivity |
| $D_e$ | effective diffusivity for macroporous pellet $\{D_e = \varepsilon_p D_p/[\varepsilon_p+(1-\varepsilon_p)K]\}$ |
| $D_p$ | macropore diffusivity |
| h | overall heat transfer coefficient |
| k | solid "film" mass transfer coefficient |
| $k_f$ | external fluid film mass transfer coefficient |
| K | dimensionless Henry constant |
| $K_o$ | pre-exponential factor in temperature dependence of K |
| L | column length; parameter defined by Eq 12 |
| q | sorbate concentration in adsorbed phase |
| $q_o$ | value of q at equilibrium with $c_o$ |
| q̄ | value of q averaged through a particle |
| R | particle radius; gas constant |
| $r_c$ | crystal radius |
| t | time |
| T | temperature (K) |
| v | interstitial fluid velocity |
| w | volume fraction of binder in adsorbent pellet |
| α | heat transfer parameter $(ha/\rho C_p)/(D_c/r_c^2)$ |
| β | parameter defined below Eq 5 |
| $\beta_a$ | defined by Eq 11 |
| ε | voidage of adsorbent bed |
| $\varepsilon_p$ | porosity of adsorbent particle |
| -ΔH | heat of adsorption |
| μ | mean retention time |
| $\sigma^2$ | variance of chromatographic response peak |
| η | effectiveness factor |
| φ | Thiele modulus |

# References

1. Kärger, J. and Ruthven, D.M. (1992) *Diffusion in Zeolites and Other Microporous Solids*, John Wiley, New York.
2. Rees, L.V.C. (1994) "Exciting New Advances in Diffusion of Sorbates in Zeolites and Microporous Solids" in J. Weitkamp et al, eds., *Zeolites and Related Materials: State of the Art 1994*, pp. 1133-1150. Studies in Surface Science and Catalysis Vol. 84, Elsevier, Amsterdam (1994).
3. Tiselius, A. (1934) *Z. Phys. Chem.* A 169, 425.
4. Tiselius, A. (1935) *Z. Phys. Chem.* A 174, 401.
5. Barrer, R.M. (1941) *Trans. Faraday Soc.* 37, 590.
6. Barrer, R.M. (1978) *Zeolites and Clay Minerals*, Academic Press, London.
7. Resing, H.A. and Thompson, J.K. (1967) *J. Chem. Phys.* 46, 2876.
8. Resing, H.A. and Murday, J.S. (1973) "NMR Relaxation and Molecular Motion in Zeolites". Proc. 3$^{rd}$ Internat. Zeolite Conf., Zurich (1974), Published as *Molecular Sieves*, Adv. Chem. *121*, 414-440, ACS, Washington (1973).
9. Kärger, J. (1971) *Z. Phys. Chem. Leipzig 248*, 27.
10. Pfeifer, H. (1972) "Nuclear Magnetic Resonance of Molecules Adsorbed on Solids" in *NMR - Basic Principles and Progress*, 7, 53-153, Springer, Berlin.
11. Lechert, H., Wittern, K.P. and Schweitzer, W., (1978) "NMR Studies of Mobility of Aromatics in Faujasite", *Proc. Zeolite Symp. Szeged, Hungary*, Sept. 1978. Published as *Acta Physica et Chemica, 24*, 201-206.
12. Haynes, H.W. and Sarma, P.N. (1974) *Adv. Chem. 133*, 205.
13. Ma, Y.H. and Mancel, C. (1973) *Adv. Chem. 121*, 392.
14. Shah, D.B. and Ruthven, D.M. (1977) *AIChE Jl 23*, 804.
15. Paravar, A. and Hayhurst, D.T. (1983) "Direct Measurement of Diffusivity for Butane across a Single Large Silicalite Crystal" *Proc. 6$^{th}$ Internat. Zeolite Conf.*, Reno, pp 217-224, D. Olson and A. Bisio eds, Butterworth, Guildford (1984).
16. Wernick, D.L. and Osterhuber, E.J. (1983) *Ibid*, pp 122-130.
17. Haag, W.O., Lago, R.M. and Weisz, P.B. (1982) *Faraday Discussions* (Chem. Soc. London) 72, 317.
18. Post, M.F.M., van Amstel, J. and Kouwenhoeven (1983) *Proc. 6$^{th}$ Internat. Zeolite Conf*, Reno pp 517-525, D. Olsen and A. Bisio eds., Butterworth, Guildford (1984).
19. Eic, M. and Ruthven, D.M. (1988) *Zeolites 8*, 40-45.
20. Ruthven, D.M. and Eic, M. (1988) *Am. Chem. Soc. Symp. Series 368*, 362.
21. Yasuda, Y. (1976) *J. Phys. Chem. 80*, 1867.
22. van den Begin, N.G. and Rees, L.V.C. (1989) *Proc. 8$^{th}$ Internat. Zeolite Conf.* Amsterdam, p. 915-924, P.A. Jacobs and R.A. van Santen eds., Elsevier, Amsterdam (1989).
23. Jobic, H., Bee, M. and Kearley, G.J. (1989) *Zeolites 9*, 312.
24. Cohen de Lora, E., Kahn, R. and Mezei, F. (1983) *J. Chem. Soc. Faraday Trans 79*, 1911.
25. Niessen, W. and Karge, H. (1991) *Studies in Surface Sci. and Catalysis 60*, 213-221.
26. Niessen, W., Karge, H. (1993) *Microporous Materials 1*, 1-8.
27. Grenier, Ph., Meunier, F., Gray, P.G., Kärger, J., Xu, Z. and Ruthven, D.M. (1994) *Zeolites* 14, 242-249.
28. Bourdin, V. (1996) Contribution a *L'Etude Experimentale de la Cinetique d'Adsorption das les Materiaux Poreux Ph.D. Thesis*, Univ. Paris XI.
29. Hufton, J.R., Brandani, S. and Ruthven, D.M. (1994) "Measurement of Intracrystalline Diffusion by Zero Length Column Tracer Exchange " *Proc. 10$^{th}$ Internat. Zeolite Conf., Garmisch*, pp 1323-1330. J. Weitkamp, H.G. Karge, H. Pfeifer and W. Holderich eds. Elsevier, Amsterdam.
30. Kärger, J. (1996) - paper presented at NATO ASI - Physical Adsorption.
31. Ruthven, D.M., Lee, L-K. and Yucel, H. (1980) *AIChE Jl*. 26, 16.
32. Ruthven, D.M. and Lee, L-K. (1981) *AIChE Jl.* 27, 654.
33. Lee, L-K and Ruthven, D.M. (1979) *J. Chem. Soc. Faraday Trans I 75*, 2406.
34. Haul, R. and Stremming, H. (1984) *J. Colloid Interface Sci 97*, 348.
35. Sun. L.M. and Meunier, F. (1987) *Chem. Eng. Sci 42*, 1585.
36. Bulow, M., Mietk, W., Struve, P., Schirmer, W., Kocirik, M. and Kärger, J. (1983) *Proc. Sixth Internat. Zeolite Conf., Reno* pp 242-250. D. Olson and A. Bisio eds., Butterworth, Guildford (1984).

290

37.  Caro, J., Noack, M., Marlow, F., Petersohn, D, Griepentrog, M. and Kornatowski, K., (1993) *J. Am. Chem. Soc.*
38.  Shen, D.M. and Rees, L.V.C., (1991), Zeolites *11*, 684-689.
39.  Shen, D.M. and Rees, L.V.C., J. Chem. Soc. Faraday Trans *89*, 1063-1065.
40.  Sun, L.M. and Bourdin, V., (1993) Chem. Eng. Sci *48*, 3783-3793.
41.  Kubin, M., (1965) *Coll. Czech. Chem. Commun. 30*, 1104 and 2900.
42.  Kucera, E.J. (1965), *J. Chromatog. 19*, 237.
43.  Haq, N. and Ruthven, D.M., (1986), *J. Colloid Interface Sci. 112*, 164.
44.   Awum, F., Narayan, S. and Ruthven, D.M., (1988), *I and EC Research 27*, 1510.
45.  Ching, C.B. and Ruthven, D.M. (1988) *Zeolites 8*, 68-73.
46.  Hufton, J.R., Ruthven, D.M. and Danner, R.P., (1995) *Microporous Materials 5*, 39-52.
47.  Delmas, M., Cornu, C. and Ruthven, D.M., (1995), *Zeolites 15*, 45-50.
48.  Ruthven, D.M. and Stapleton, P., (1993), *Chem. Engng Sci.*, *48*, 89-98.
49.  Brandani, S. and Ruthven, D.M., (1995), *Chem. Engng Sci.*, *50*, 2055-2059.
50.  Brandani, S. and Ruthven, D.M., Adsorption , 1996, 2, should appear in No. 2.
51.  Brandani, S., (1995), *AIDIC Conference Series*, *1*, 19-28.
52.  Micke, A., Kocirik M. and Bulow M., (1993), *Microporous Materials*, *1*, 363-371.
53.  Micke, A., Kocirik M. and Bulow M., (1994), *Ber. Bunsenges. Phys. Chem.*, *98*, 242-248.
54.  Brandani, S., Hufton, J.R. and Ruthven, D.M., (1995), *Zeolites*, *15*, 624-631.
55.  Brandani, S. and Ruthven, D.M., (1996), *Ind. Eng. Chem. Res.*, *35*, 315-319.
56.  Kapteijn, F., Bakker, W.J.W., Zheng, G., Pappe, J. and Moulijn, J.A., (1995), *Chem. Eng. J. 57*, 145-153.
57.  Shen, D. and Rees, L.V.C., (1991), *Zeolites 11*, 666-671.
58.  Ruthven, D.M., (1993), *Zeolites 13*, 594.
59.  Brandani, S., Ruthven, D.M. and Kärger, J.(1995), *Zeolites 15*, 494-496.
60.  Brandani, S., Xu, Z. and Ruthven, D.M., Microporous Materials - in press.

# Appendix A - The Piezometric Method

The piezometric method involves following the pressure response in a dosing cell connected to an uptake cell which contains the adsorbent. A schematic representation is given in figure A1. According to the results reported in the literature, the piezometric method can be used to accurately measure intracrystalline diffusivities for fast diffusing and strongly adsorbed species such as benzene on NaX[A1]. Furthermore it is also claimed to provide the required accuracy needed to study combined intracrystalline processes such as diffusion and first order reaction[A2].

The analysis of the response curves is not as direct as for other methods, since we have to consider the flow through the valve. Furthermore the time required to fully open the valve is typically 0.5–0.7 s. Qualitatively it is evident that the initial part of the response is influenced by the flow through the valve, the opening of the valve and the mass transfer to the solid. It is therefore obvious that the initial portion of the response cannot be used to obtain reliable diffusional time constants. The final stages of the response are not affected by the initial opening of the valve, and the pressure difference across the valve becomes very small. Qualitatively it can be argued useful diffusion information can be obtained from the long time region of the response curve. However, it has been clearly shown that this portion of the response curve is strongly affected by heat effects[A3,A4]. It follows from this observation that the ability of the piezometric method to yield reliable intracrystalline diffusivity values is costrained, especially for fast diffusing and strongly adsorbed species. All these considerations have not been considered by Bulow and Micke[A5], who simply suggest a total curve fitting procedure using a numerical solution of the model equations, assuming isothermal behaviour.

Even if isothermal behaviour is assumed, there is still a limitation to the range of applicability of the method. Considering an instantaneous opening of the valve, and linearizing the model equations it is possible to derive[A6] a simple analytic solution for the dimensionless pressure in the dosing cell.

$$\frac{p_d}{p_d^0} = \frac{3\delta}{1+3\delta+3\gamma} + \sum a_i \exp\left(-\beta_i^2 \tau\right) \tag{A1}$$

where

$$a_i = \frac{2w^2\delta\beta_i^2}{2w^2\delta\beta_i^2 + \left(w-\beta_i^2\right)^2\left(\beta_i^2+z_i^2-z_i+2\gamma\beta_i^2\right)} \tag{A2}$$

$$z_i = 1+\gamma\beta_i^2+\frac{w\delta\beta_i^2}{w-\beta_i^2} \ ; \ \gamma = \frac{1}{3}\frac{V_u}{KV_s} \ ; \ \delta = \frac{1}{3}\frac{V_d}{KV_s} \ ; \ w = \frac{\Re T}{V_d}\frac{\chi\left(P_d^0+P_u^0\right)D}{R^2}$$

$$\tag{A3}$$

and $b_i$ are the positive nonzero roots of

$$\beta_i \cot \beta_i - z_i = 0 \tag{A4}$$

The dimensionless parameters defined in eq. (A3) clearly show that the response is influenced by the ratios of the accumulation in the fluid and solid phase, g and d, and the ratio of the characteristic time constants of the valve and of the diffusion process, w. The ratio d/g depends on the geometric configuration of the system and has a typical value of 1.5[A1].

The actual response will be confined between the response of the empty cell and the response corresponding to an infinitely rapid mass transfer, i.e. equilibrium control, which are given by:

$$\frac{p_d}{p_d^0} = \frac{\delta}{\gamma + \delta} + \frac{\gamma}{\gamma + \delta} \exp\left(-\frac{\gamma + \delta}{\gamma} w\tau\right) \tag{A5}$$

and

$$\frac{p_d}{p_d^0} = \frac{3\delta}{1 + 3\delta + 3\gamma} + \frac{1 + 3\gamma}{1 + 3\delta + 3\gamma} \exp\left(-\frac{1 + 3\delta + 3\gamma}{1 + 3\gamma} w\tau\right) \tag{A6}$$

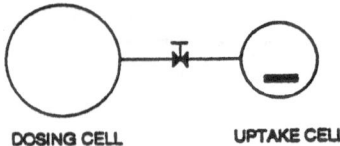

DOSING CELL          UPTAKE CELL

Figure A1. Schematic representation of the experimental set-up.

The system response is strongly influenced by the adsorption equilibrium as can be seen from figure A2. If the system is characterized by a high value of w, the pressures in the two cells will equilibrate rapidly and there will be a slow decay to the final equilibrium. In this case the mass transfer in the solid is controlling the process. If w is relatively low, the system is closer to equilibrium control and the controlling process becomes the flow through the valve. In this case, the form of pressure response in the dosing cell does not change, since it is always monotonic. It is therefore not possible to distinguish the two regimes from an observation of the experimental curves. However, if the monitored

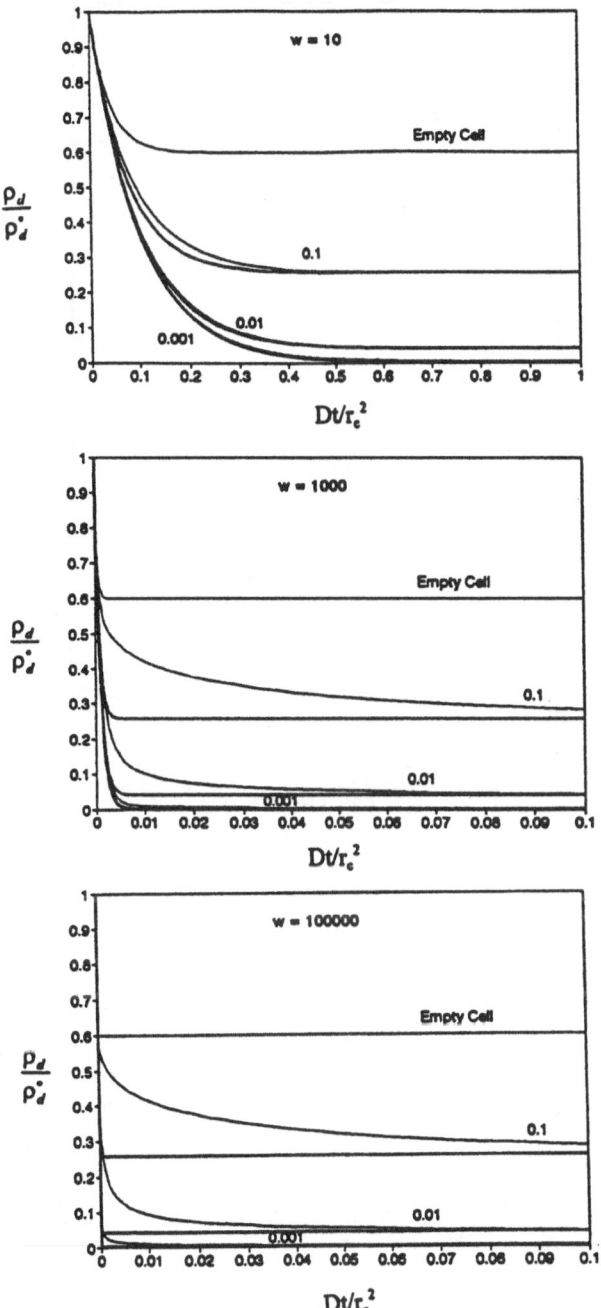

Figure A2. Dimensionless pressure response in the dosing cell vs dimensioless time as a function of the parameter γ (δ/γ=1.5), showing the effect of the adsorption equilibrium. Heavy solid lines represent the corresponding equilibrium controlled processes. (a) w = 10; (b) w = 1000; (c) w = 100000.

294

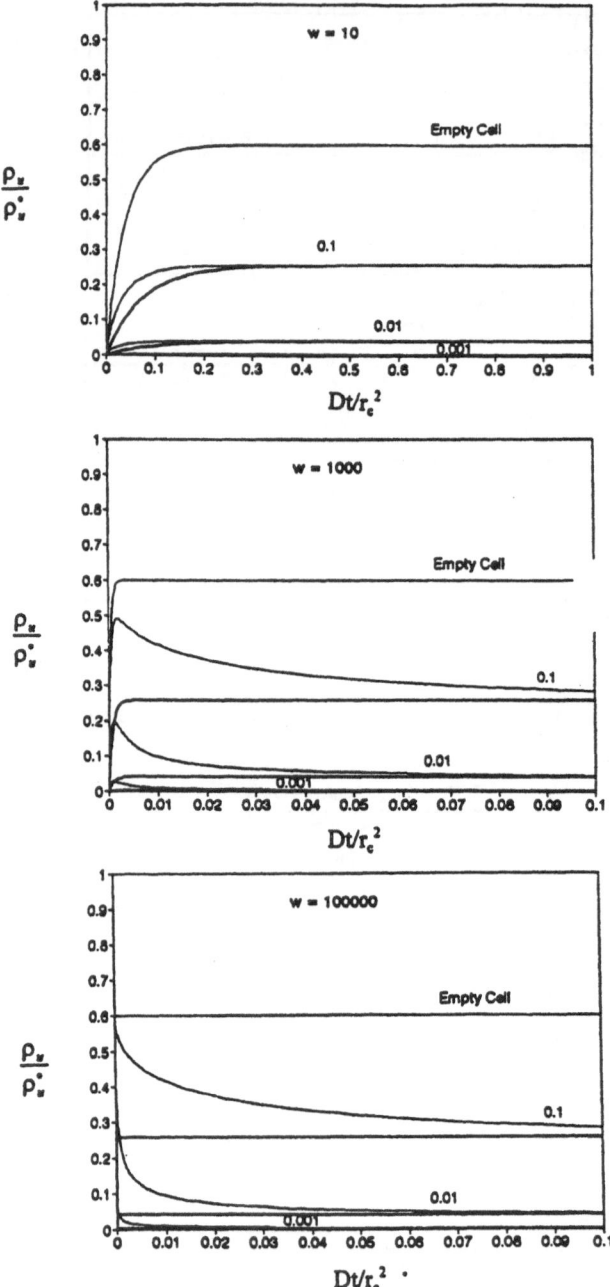

Figure A3. Dimensionless pressure response in the uptake cell vs dimensioless time as a function of the parameter γ (δ/γ=1.5), showing the effect of the adsorption equilibrium. Heavy solid lines represent the corresponding equilibrium controlled processes. (a) w = 10; (b) w = 1000; (c) w = 100000.

TABLE A1. Parameter values used for the system Benzene on NaX.

| | |
|---|---|
| Temperature[A1] | 353 K |
| Volume of Dosing Cell[A1] | 120 cm$^3$ |
| Volume of Uptake Cell[A1] | 80 cm$^3$ |
| H | $10^7$ |
| Intracrystalline Diffusivity[A1], $D_0$ | $2 \cdot 10^{-7}$ cm$^2$ s$^{-1}$ |
| Mass of Solid[A1] | 8 mg |
| Radius of Crystals[A1] | 60 $\mu$m |
| $\chi_2^{(A2)}$ | $1.3\ 10^{-8}$ mol Pa$^{-2}$ s$^{-1}$ |
| $P_d^0$ | 30 Pa |
| $P_u^0$ | 20 Pa |
| $\gamma$ | 0.0005 |
| $\delta$ | 0.00075 |
| w | 2860 |

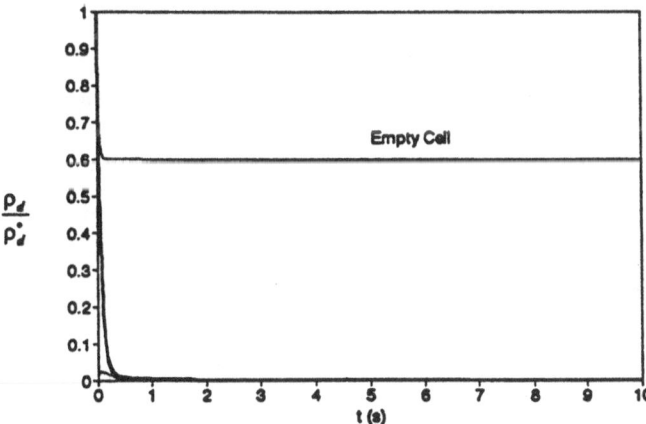

Figure A4. Dimensionless pressure response vs time for the system Benzene on NaX, for the parameter values reported in Table A1.

quantity is the pressure in the uptake cell, it should be possible to clearly distinguish the controlling process. Figure A3 shows the dimensionless pressure response in the uptake cell for the same parameter values as in figure A2. If w is relatively large, the pressure responses in the two cells give essentially the same information. On the other hand, if the process is closer to equilibrium control the pressure in the uptake cell can indicate the controlling process. When this curve is monotonic the system is in equilibrium control and the overall process is limited by the flow through the valve. If the pressure in the uptake cell exhibits a maximum, then it is evident that the response can be used to obtain information on the mass transfer in the solid.

Unfortunately in many of the reported applications of this technique only the pressure in the closer cell was monitored. This greatly reduces the sensitivity of the method and the reliability of the reported results. By way of an example let us consider the system benzene-NaX. With the parameter values in table A1 the response curve shown in figure A4 is obtained. There is almost no difference between the curves for equilibrium control and for the full model. It is therefore doubtful whether any reliable diffusivity values may be extracted for this system, regardless of the method employed for data analysis, especially since heat effects will certainly modify the behaviour in the long time region.

The applicability of the piezometric method appears to be limited to relatively slow diffusing systems especially for strongly adsorbed species. The best results can be obtained by monitoring the uptake cell, as well as the dosing cell and a match of both pressure responses may yield a more reliable diffusivity value. Based on our analysis it is possible to state that all measurements reported in the literature which used the pressure in the dosing cell as the monitored quantity should be critically reviewed, since these measurements cannot provide definitive evidence concerning the nature of the rate controlling process.

**References**

A1. M. Bulow, W. Mietk, P. Struve and P. Lorenz , (1983), *J. Chem. Soc. Faraday Trans. 1*, *79*, 2457–2466.

A2. A. Micke and M. Bulow, (1994), *Microporous Materials*, *3*, 185–193.

A3. Lee and Ruthven, (1980), *AIChE J.*

A4. Grenier et al. 1994

A5. M. Bulow and A. Micke, (1995), *Adsorption*, *1*, 29–48.

A6. Brandani 1996 in preparation.

# NMR STUDIES OF DIFFUSION IN POROUS SOLIDS

J. KÄRGER
Leipzig University, Physics Department
Linnéstr. 5, D-04103 Leipzig, Germany

## 1. Introduction

Porous media are of considerable interest in a very diverse range of activities. They comprise a wide variety of substances, including ceramics, rocks, carbons, porous glasses, silica gels and zeolites. Depending on the nature of these substances, their pore diameters may range from the subnanometer region (like in the case of zeolites) up to macroscopic dimensions. In many cases, their properties and particularly the efficiency of their technical use are strongly influenced by the rate of molecular transportation. Owing to their complex structure, molecular transport within these materials generally proceeds in a sequence of processes of quite different nature. It may therefore be rather complicated to deduce unambiguous messages about the diverse processes of molecular transport from an observation of overall transport phenomena, such as e. g. the uptake curve of a conventional adsorption-desorption experiment. In this case, data analysis is additionally complicated by the fact that the experiments are carried out under non-equilibrium conditions. Heat effects during molecular adsorption/desorption may significantly influence the sorption rates [1, 2]. Extensive accounts on this topic highlighting the relevance of a rather close inspection of the sorption data for their unambiguous interpretation are given in this volume in the contributions by D. M. Ruthven [3, 4] as well as in the literature [5-7].

Among the methods of studying molecular diffusion in porous media, the pulsed field gradient (PFG) NMR technique has attained particular relevance. As a non-invasive technique, PFG NMR is able to trace molecular displacements within the system under study over a space scale from about 100 nm up to 100 μm. Since the modes of molecular migration depend on the space scale over which they occur, in this way a discrimination between different modes of molecular propagation becomes possible. Moreover, since NMR spectroscopy is sensitive to a particular nucleus and moreover - in high-resolution experiments - even to a particular chemical surrounding, PFG NMR is able simultaneously to determine the diffusivities of the individual species in multicomponent systems.

J. Fraissard (ed.), Physical Adsorption: Experiment, Theory and Applications, 297–329.
© 1997 Kluwer Academic Publishers.

From a historical point of view PFG NMR deserves particular interest. By revealing large discrepancies with the results of uptake measurements it was this technique which initiated a reconsideration of the conditions under which conventional adsorption-desorption measurements are able to provide unambiguous information about intracrystalline diffusion in zeolites [8].

The present contribution is devoted to the application of PFG NMR to diffusion studies in porous solids, with particular emphasis on zeolites. Chapter 2 provides a survey about the different situations under which molecular diffusion may be observed, correlating the different measuring techniques with the conditions under which they work and with the corresponding diffusivities. After the presentation of the fundamentals of the PFG NMR method in chapter 3, it is demonstrated in chapter 4 that the wealth of information provided by PFG NMR is best reflected by the propagator representation. If over the distances covered in the PFG NMR studies the medium may be considered to be quasi-homogeneous, molecular transport is satisfactorily well described by the laws of normal diffusion. Chapter 5 provides different examples for the measurement of normal diffusion in porous media. Heterogeneities and hierarchies in the pore structure, however, may lead to substantial deviations from normal diffusion. Chapter 6 describes a series of possibilities of transport patterns generated by deviations from normal diffusion.

## 2. The „various" diffusivities

Diffusion is the process of molecular transport associated with the stochastic movement of the individual molecules. Figure 1 illustrates 3 different situations of diffusion measurement. Any heterogeneity in the distribution of the molecules under study over the sample will give rise to molecular fluxes into the direction of decreasing concentration (fig. 1a). Flux density and concentration gradient are related to each other by Fick's first law

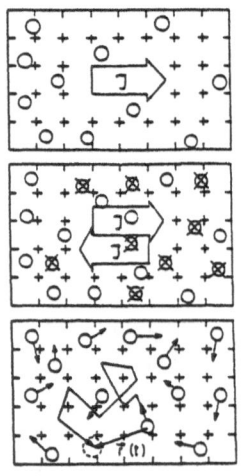

**Figure 1.** Microscopic situation corresponding to the measurement of transport diffusivity (eq. (1); a) and self-diffusivity (eq. (5); b and eq. (11); c). The flux of the labelled molecules ($\otimes$) in (b) is counterbalanced by that of the unlabelled molecules (O). If the mobility of the unlabelled molecules is unaffected by the presence of the labelled molecules, the fluxes in (a) and (b) are equal and the transport and self-diffusion coincide. From ref. [5] with permission.

$$j_z = -D_\mathrm{d} \frac{dc}{dz}. \tag{1}$$

$D_\mathrm{d}$ denotes the so-called transport diffusivity. The subscript d has been added to distinguish this quantity from the coefficient of self-diffusion ($D$) which is the main subject of this contribution. With the conservation equation

$$\dot{c} = -\frac{\partial j_z}{\partial z}, \tag{2}$$

eq. (1) becomes

$$\dot{c} = \frac{\partial}{\partial z}\left(D_\mathrm{d}\frac{\partial c}{\partial z}\right) \tag{3}$$

which is referred to as the diffusion equation or Fick's second law. If the transport diffusivity does not depend on the concentration, eq. (3) becomes

$$\dot{c} = D_\mathrm{d}\frac{\partial^2 c}{\partial z^2}. \tag{4}$$

Eq. (1) is the starting point for diffusion measurements by permeation techniques [9, 10]. Transient adsorption-desorption measurements are based on the application of eqs. (3) or (4) with the relevant initial and boundary conditions [3-7]. For the observation of the adsorption/desorption process a multitude of techniques including gravimetry, piezometry and, more recently, also spectroscopic methods like IR microscopy [11, 12] or [129]Xe NMR [13, 14] may be applied.

Figure 1b illustrates the situation of a tracer exchange experiment. It is assumed that the molecular species under study is available in two modifications (e.g. as different isotopes) which are identical with respect to their transport properties. Again, as in the case of transport diffusion, a relation of the type of Fick's first equation combines the flux density and the concentration gradient of either of these species. One has

$$j_z^* = -D\frac{dc^*}{dz} \tag{5}$$

where $c^*$ refers to the concentration of the labelled (or unlabelled) species with the understanding, that the flux of this species $\left(j_z^*\right)$ is exactly compensated by the flux of the other species so that the overall concentration, i.e. the sum of the concentrations of the two species, remains unchanged. D denotes the coefficient of tracer diffusion or self-diffusion. It depends exclusively on the overall concentration, and not on the concentration of the labelled (or unlabelled) molecules. Hence, Fick's second law of tracer exchange may quite generally be written in the form of eq. (4), viz.

$$\dot{c}^* = D\frac{\partial^2 c^*}{\partial z^2}. \tag{6}$$

Equation (6) is the basis of self-diffusion studies by tracer techniques [15, 16].

Equation (6) may be used to determine the probability that a molecule, initially at position 0 has migrated to position $z$ after a time interval $t$. For this purpose one has to solve eq. (6) with the initial condition

$$c^*(z, t = 0) = \delta(z) \tag{7}$$

and with the boundary condition

$$c^*(z = \pm\infty, t) = 0. \tag{8}$$

$\delta(z)$ denotes Dirac's delta function, which is equal to zero for any $z \neq 0$, and whose integral $\int_{-\infty}^{\infty} \delta(z)dz$ is equal to one. By insertion into eq. (6) one may easily see that the solution is a Gaussian of the form

$$c^*(z, t) \equiv P(z, t) = (4\pi Dt)^{-\frac{1}{2}} \exp\left(-\frac{z^2}{4Dt}\right). \tag{9}$$

P $(z,t)$ contains the maximum information attainable about the stochastic process of molecular diffusion and has been termed propagator. The particular shape of the propagator as given by eq. (9) results for molecular diffusion in an unconfined space as expressed by the boundary condition (8). Molecular confinement and/or the existence of ranges of different mobility have to be taken into account by a corresponding change of the boundary conditions and/or the introduction of matching conditions between the different regions.

With eq. (9), the mean square value of molecular displacement in $z$ direction during the observation time $t$ results to be

$$\langle z^2(t) \rangle \equiv \int_{-\infty}^{\infty} P(z,t) z^2 dz = 2Dt, \qquad (10)$$

which is commonly referred to as Einstein's equation of diffusion [17]. Fig. 1c schematically shows the vector of displacement for an arbitrarily chosen molecule. Self-diffusion studies by quasielastic neutron scattering (QENS) [18, 19] and by PFG NMR [5, 20] are based on the measurement of the mean square displacement or - more generally - of the propagator. In an isotropic system all coordinate directions are equivalent. The mean square displacement therefore becomes

$$\langle r^2(t) \rangle = \langle x^2(t) \rangle + \langle y^2(t) \rangle + \langle z^2(t) \rangle = 6Dt \qquad (11)$$

Using the basic relation between the space vector $\vec{r}$ and the velocity vector $\vec{v}$

$$\vec{r}(t) = \int_0^t \vec{v}(t') dt', \qquad (12)$$

from eq. (11) one may obtain

$$D = \frac{1}{3N} \int_0^{\infty} \sum_i \langle \vec{v}_i(t) \vec{v}_i(0) \rangle dt \qquad (13)$$

where the sum has to be extended over all (N) molecules of the system.

By comparing figs. 1a and b it becomes obvious that in the limit of small concentrations, i.e. in the case of negligibly small interaction between the diffusants the coefficients of transport diffusion and self-diffusion should coincide. With increasing concentration, however, the mutual interaction of the diffusants is expected to lead to differences in the flux densities in cases a) and b).

Transport diffusivities may be more easily compared with the self-diffusivities by introducing a „corrected" transport diffusivity

$$D_{d0} = D_d \frac{d \ln c}{d \ln p} \tag{14}$$

with c denoting the adsorbate concentration in equilibrium with the adsorbate pressure p. One may show that the corrected transport diffusivity may be represented in the form [21, 22]

$$D_{d0} = \frac{1}{3N}\left[\int_0^\infty \sum_i \langle \vec{v}_i(t)\vec{v}_i(0)\rangle dt + \int_0^\infty \sum_{i \neq j} \langle \vec{v}_i(t)\vec{v}_j(0)\rangle dt\right]. \tag{15}$$

The first term on the right hand side of eq. (15) coincides with the right hand side of eq. (13), while the second term takes account of the correlation between the velocities of different molecules.

If this correlation is negligibly small - as to be expected for sufficiently small concentrations - the self-diffusivity therefore coincides with the corrected transport diffusivity. However, also in the case of larger concentrations, the second term on the right hand side of eq. (15) should still be smaller than the first one, so that quite generally the self-diffusivities and corrected transport diffusivities should be comparable with each other. It is one of the most intriguing problems of current zeolite research that there is a number of systems where this assumption is reasonably well fulfilled while for other systems large discrepancies between the self-diffusivities and the corrected transport diffusivities are observed [3 - 6].

The proportionality between the mean square displacement and the observation time as expressed by eqs. (10) and (11) may be intuitively understood by considering the total displacement $\vec{r}$ as the vector sum of displacements $\vec{r}_k$ during equal time intervals, which are large enough to that the displacements in subsequent intervals are uncorrelated. In this case one has

$$\langle r^2 \rangle = \left\langle \left( \sum_k r_k \right)^2 \right\rangle = \sum_k \langle r_k^2 \rangle \tag{16}$$

since any cross term ($\langle \vec{r}_k \vec{r}_l \rangle$ with k ≠ l), which otherwise also would occur on the right hand side of eq. (16), becomes zero. For a given length of the time interval, the mean square displacement $\langle r^2 \rangle$ is thus found to be proportional to the total number of time intervals considered, i.e. to the total observation time. Molecular propagation which proceeds under such conditions is called normal diffusion. Only under such conditions, Fick's laws as expressed by eqs. (5) and (6) are applicable. PFG NMR studies dealing with different aspects of normal zeolitic diffusion are reviewed in chapter 5.

Molecular confinement generally leads to a correlation of subsequent displacements. Depending on the nature of the confinement, quite different patterns of deviation from normal diffusion, i.e. from the proportionality between the mean square displacement and the observation time, may be observed. Examples of phenomena of anomalous diffusion and of their investigation by PFG NMR are presented in chapter 6.

## 3. Fundamentals of PFG NMR

Diffusion measurement by NMR is based on the fact that in an inhomogeneous magnetic field the Larmor frequency (i.e. the precessional frequency of the spins about the magnetic field in the classical model of NMR)

$$\omega \left( \vec{r} \right) = \gamma B \left( \vec{r} \right) \tag{17}$$

becomes space dependent. In the PFG NMR method, the constant magnetic field $B_0$ is superimposed by an inhomogeneous field

$$B_{add} = gz \tag{18}$$

over two short time intervals $\delta$ at distance $\Delta$, where the field gradient $g$ is assumed to point into the direction of the $z$ coordinate. Hence, the precessional phase accumulated during the i-th field gradient pulse by a spin at position $z$ is

$$\varphi = \delta\gamma\left(B_0 + gz\right). \tag{19}$$

By applying suitable sequences of RF pulses, an NMR signal ("spin echo") may be generated. The intensity of this signal is proportional to the magnetization in the plane prependicular to the direction of the constant magnetic field. This magnetization results as the vector sum of the individual spins and is therefore determined by the distribution of their precessional phases. By applying a pair of field gradient pulses either with opposite signs or with identical signs but separated by a so-called $\pi$ pulse, the net precessional phase of a particular spin at the instant of the spin echo results to be

$$\Delta\varphi = \delta\gamma\, g\left(z_2 - z_1\right) \tag{20}$$

where $z_1$ and $z_2$ denote the $z$ components of the positions of the considered spin at the instants of the first and second field gradient pulses. The total signal intensity $S$ results as the vector sum of the individual spins, i.e. as the sum of their projection on the direction with no net phase shift, yielding

$$\psi \equiv S/S_0 = \int\cos\left(\delta\gamma\, gz\right)P\left(z,\Delta\right)dz \tag{21}$$

where $S_0$ is the signal intensity without field gradient pulses applied. $P(z, \Delta)$ represents the probability density that during the time interval $\Delta$ between the field gradient pulses within the sample a molecule (strictly speaking, a nuclear spin of the species under study) has been shifted over a distance $z$ with respect to the $z$ coordinate. In a homogeneous system, this quantity clearly coincides with the propagator as given by eq. (9). In a heterogeneous system, the quantity $P(z, \Delta)$ is an "average" propagator [23], resulting by taking the weighted mean over all possible starting positions within the sample.

Inserting eq. (9) into eq. (21) yields

$$\psi = \exp\left(-\gamma^2\delta^2 g^2 D\Delta\right) \tag{22}$$

as the PFG NMR spin echo attenuation in the case of normal diffusion. Using eq. (10), eq. (22) may be likewise expressed as

$$\psi = \exp\left(-\gamma^2\delta^2 g^2 \left\langle z^2\right\rangle / 2\right) \tag{23}$$

which more properly reflects that it is essentially the mean square displacement in the direction of the applied field gradient ($z$ direction) which is recorded by PFG NMR. Equation (23) allows an order-of-magnitude estimate of the lower limit of molecular displacements $\langle z^2 \rangle^{1/2}$ still accessible by PFG NMR. If we assume that for an unambiguous attribution of the echo attenuation to molecular diffusion, under the influence of the field gradient pulses the spin echo must be damped to a value $\psi = e^{-1}$, with typical maximum values for the field gradient amplitude ($g = 25\ T\ m^{-1}$) and the pulse width ($\delta = 2\ ms$), for hydrogen containing molecules ($\gamma_H = 2.67 \cdot 10^8\ T^{-1}s^{-1}$) one obtains $\langle z^2 \rangle^{1/2}_{min} \approx 100$ nm .With eq. (10), the lower limit of the diffusivity accessible by PFG NMR is $\langle z^2 \rangle_{min}/2t_{max}$, where $t_{max}$ denotes the maximum possible observation time. If the spin echo is generated by a $\pi/2-\pi$ pulse sequence („primary spin echo"), $t_{max}$ is controlled by the transverse nuclear magnetic relaxation time. For zeolites its value is typically of the order of milliseconds. The range of observation times may be significantly enhanced by applying the pulse sequence of the stimulated echo ($\pi/2-\pi/2-\pi/2$). In this case the observation time is controlled by the longitudinal nuclear magnetic relaxation time, which may attain hundreds of milliseconds.

As a crucial condition for the correctness of diffusion measurements by PFG NMR, the product $\delta g$ of the width and amplitude of the field gradient pulses of one pair must be required to be identical. This necessitates both high mechanical and high electrical stability. Experimental procedures to meet these requirements may be found in the literature [20, 24, 25].

## 4. The Propagator

The propagator of molecular diffusion within a sample is given by the Fourier transform of the spin echo attenuation $\psi$. With eq. (21) one obtains

$$P(z,\Delta) = \frac{1}{2\pi} \int \psi(\delta\gamma g, \Delta) \cos(\delta\gamma gz) d(\delta\gamma g), \qquad (24)$$

where the spin echo attenuation $\psi \equiv S/S_0$ is considered as a function of the „generalized scattering vector" $\delta\gamma g$ [26 - 29]. In ref. 23 for the first time relations of this type were used to monitor molecular propagation in heterogeneous systems. Since the heterogeneity of a sample, in particular in the case of structural confinement, becomes also apparent in the propagation patterns of the diffusants, this type of investigation has been termed dynamic imaging [20, 26, 27].

Figure 2 shows the propagators of methane in beds of zeolite crystallites of type NaCaA with two different crystallite radii as obtained by Fourier transformation of the spin echo attenuation [23].

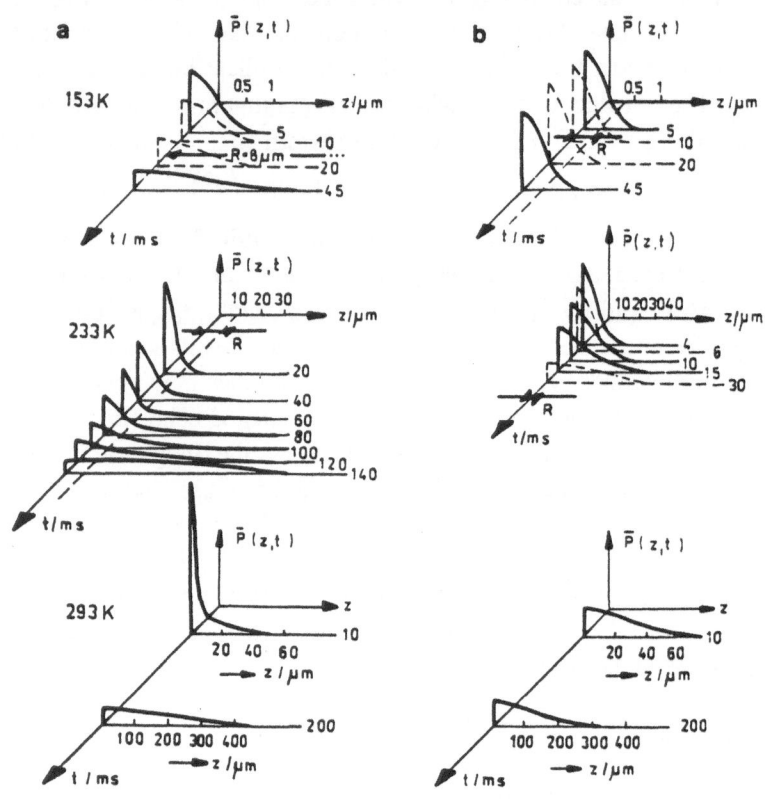

**Figure 2.** Propagator representation of the self-diffusion of ethane in zeolite NaCaA with mean crystallite radii of R = 8 μm (a) and 0.4 μm (b). Due to symmetry, the propagator P (z, t) is only represented for z > 0. From ref. [23] with permission.

At the lowest temperature, in the large zeolite crystallites during the considered observation times molecular propagation proceeds like in an infinitely extended medium. This is the regime under which one is able unambiguously to trace intracrystalline diffusion. In sects. 5.2 - 4 different aspects of intracrystalline diffusion shall be considered. In the smaller crystallites, however, molecular diffusion remains restricted with root mean square displacements comparable with the radii of the crystallites. The thermal energy is evidently not yet sufficient to allow a substantial number of molecules to leave the individual crystallites and to attain the higher level of potential energy in the gas phase in the intercrystalline space. Due to this fact, the crystallite size may be directly determined by measuring the molecular displacement.

With further increasing temperature more and more molecules are able to leave the crystallites due to the enhancement of their thermal energy. Hence, in the small crystallites, during the observation times considered in the experiments at 233 K essentially all molecules are able to leave their crystallites. Leaving and entering the crystallites, the molecules behave as in a quasi-continuum. This type of molecular transport in beds of zeolites has been termed long-range diffusion. It will be discussed in more detail in section 5.1. During the covered range of observation times, in the bed of large crystallites, however, only a fraction of molecules is able to leave the individual crystallites, while for another fraction the diffusion paths are not long enough to take the molecules out of the crystallites. In this case, the propagator consists of a superposition of a narrow constituent which corresponds to those molecules, which remain in the interior of the crystallites, and a broad constituent corresponding to those molecules which have left the crystallites. The relative intensity of both constituents coincides with the relative amount of molecules belonging to these two groups. As to be expected, with increasing observation time the intensity of the broad line increases at the expense of the narrow one. Plotting the intensity of the broad line as a function of the observation time yields the same information as a tracer exchange experiment. This way of analysing the PFG NMR data has therefore been called NMR tracer desorption technique [30, 31]. In sect. 6.3 it will be applied to investigate surface barriers on zeolite crystallites after different coking procedures.

Comparing the distribution curves for the two types of crystallites one finds that in the low and high temperature cases (for sufficiently large observation times) the distribution widths observed with the large crystallites were larger, while the situation is reversed at medium temperatures. At low temperatures the restriction in the small crystallites clearly leads to the smaller distribution widths, while with increasing temperature the propagation in the bed of small crystallites proceeds faster. This is due to the fact that in the small crystallites the diffusants get faster into the range of high mobility, i.e., into the intercrystalline space. Finally, at the highest temperature and for the largest observation times, again the rate of propagation is fastest for the large crystallites, probably as a consequence of the larger free paths in the intercrystalline space and/or a reduced tortuosity.

## 5. Normal Diffusion

### 5.1. LONG-RANGE DIFFUSION

For root mean square displacements much larger than the diameters of the individual particles, according to the central limit theorem of statistics the propagator of molecular transportation in a bed of adsorbent particles (crystallites) is given by a Gaussian (eq. (9)). In this approach it is assumed that the mean life time of the molecules in the interior of the individual crystallites is much smaller than the observation time. Under this condition it may be intuitively understood that the effective diffusivity defined on the basis of eq. (10) for root mean square displacements much larger than the crystallite diameters (the long-range diffusivity $D_{l.r.}$) is determined by the relation [32]

$$D_{l.r.} = p_{inter} D_{inter} \qquad (25)$$

with $p_{inter}$ and $D_{inter}$ denoting respectively the relative amount of molecules in the intercrystalline space and their diffusivity. Fig. 3 shows the long-range diffusivity of cyclohexane in a bed of zeolite crystallites of type NaX [32].

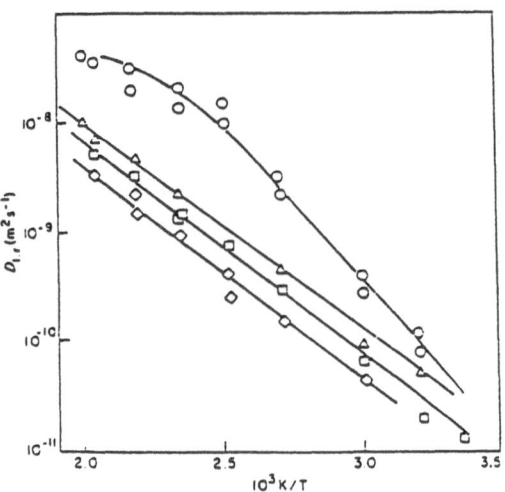

**Figure 3.** Coefficients of long-range diffusion of cyclohexane in zeolite NaX at a sorbate concentration of 1.9 molecules per cavity in the pure adsorbent-adsorbate system (O), and under the influence of an argon atmosphere of $\leq 0.06$ MPa ($\Delta$), 0.13 MPa ($\square$), and 0.2 MPa ($\diamond$). From ref. [32] with permission.

The zeolite crystallites are contained in fused glass tubes. The total amount of sorbate molecules within the sample tube, which have been introduced after sample activation and before the sample was fused, corresponds to a total loading of 1.9 molecules per cavity. Fig. 3 shows the data for cyclohexane self-diffusion both in the pure adsorbate-adsorbent system and under the influence of argon atmospheres of different pressures which were introduced into the sample tube in addition to the cyclohexane molecules.

The pure adsorbate-adsorbent system shows a marked deviation from the Arrhenius dependency. This may be referred to a change in the dominating mechanism of intercrystalline diffusion. At sufficiently low temperatures the gas phase concentration in equilibrium with the amount adsorbed is so small that the mean free path $\lambda$ of the molecules in the intercrystalline space is determined by the collisions with the outer surface of the crystallites. Representing the intercrystalline diffusivity in a simple gas-kinetic approach by the relation

$$D_{inter} \approx \frac{1}{3} \lambda \overline{v}, \tag{26}$$

the second term in eq. (25) is thus found to increase with increasing temperature only in proportion with $\overline{v}$, i.e. with $\sqrt{T}$. In this temperature regime the temperature dependence of $D_{l.r.}$ is therefore predominantly given by that of $p_{inter}$. Taking into account, that for gas phase concentrations negligibly small in comparison with the sorbate concentration, $p_{inter}$ is proportional to the gas phase pressure of sorbate, one has

$$p_{inter} \propto \exp(-E_{iso} / RT) \tag{27}$$

with $E_{iso}$ denoting the isosteric heat of adsorption. The activation energy of $D_{l.r.}$ in this temperature range should therefore be close to the isosteric heat of adsorption. From the slope of $D_{l.r.}$ in this temperature range one does in fact obtain an activation energy $(58 \pm 6)$ kJ mol$^{-1}$ which is in good agreement with the heats of adsorption for the same system (55 kJ mol$^{-1}$).

With further increasing temperature, the mean free path is more and more controlled by mutual collisions of the molecules in the gas phase, whose concentration is continuously increasing. The increase of $D_{l.r.}$ due to the increase of $p_{inter}$ is therefore partially compensated by the decrease of $D_{inter}$ in consequence of the decrease of $\lambda$. This behaviour leads in the Arrhenius plot to the observed deviation from the straight line to smaller diffusivities.

Under the influence of an argon atmosphere, the long-range diffusivity is reduced as a consequence of the reduced mean free path. This effect increases with increasing argon pressure, which leads to a decrease in the mean free path. Over the whole temperature range considered, there is no deviation from a linear Arrhenius plot. This indicates that intercrystalline diffusion of the cyclohexane molecules is controlled over the whole temperature range by essentially one process, viz. the collision with the argon atoms. The activation energy of long-range diffusion results as the difference between the heats of adsorption of cyclohexane and of argon, since the former quantity effects an increase of $D_{l.r.}$ via the increase of $p_{inter}$, while the latter quantity reduces $D_{l.r.}$ due to a decrease of $D_{inter}$, brought about by the reduction of $\lambda$ as a consequence of the increasing argon concentration in the gas phase with increasing temperature. With an adsorption heat of 12 kJ mol$^{-1}$ for argon on NaX [33], one obtains a theoretical value of 43 kJ mol$^{-1}$, which is in satisfactory agreement with the experimental result of $(39 \pm 5)$ kJ mol$^{-1}$.

## 5.2. INTRACRYSTALLINE DIFFUSION

For molecular root mean square displacements much less than the crystallite radii, the propagation of the vast majority of the adsorbed molecues does not interfere with the external boundary of the crystallites. Molecular diffusion is monitored therefore as if proceeding in an infinitely extended medium, yielding intracrystalline diffusivities. If the root mean square displacements get into the order of the crystallite diameters, they are in general reduced in comparison with the case of unrestricted diffusion. In this case formal application of Einstein's equation (10) would lead to an „effective" diffusivity, which is smaller than the genuine intracrystalline diffusivities. However, by quantitatively taking into account the effect of molecular confinement, the effective diffusivities may be transferred into the real values of intracrystalline diffusion [34]. Such a procedure is of particular use for the investigation of the intracrystalline diffusion in commercial zeolites, which are generally available as small crystallites.

Systematic studies of intracrystalline diffusion have mainly been carried out with laboratory-synthesized large crystallites. In general, as to be required in the case of normal diffusion, the molecular mean square displacement is found to increase in proportion with the observation time [35]. In some cases, however, a deviation to lower mean square displacements corresponding to effective diffusivities decreasing with increasing observation times is observed [36, 37]. Such a behaviour can be explained by the existence of transport resistances within the crystallites. The existence of such resistances has been suggested as one possibility to explain the difference between the results of macroscopic and microscopic measurements of intracrystalline diffusion [3, 4, 6].

The variety of the mechanisms controlling intracrystalline diffusion is reflected by the different patterns of concentration dependence. So far, 5 different types of concentration dependence have been observed, which are illustrated in fig. 4 by particular examples. For n-alkanes in zeolites NaX [38] and ZSM-5

[39, 41], e.g., the coefficient of self-diffusion is found to decay monotonically with increasing sorbate concentration over up to two orders of magnitude (pattern I).

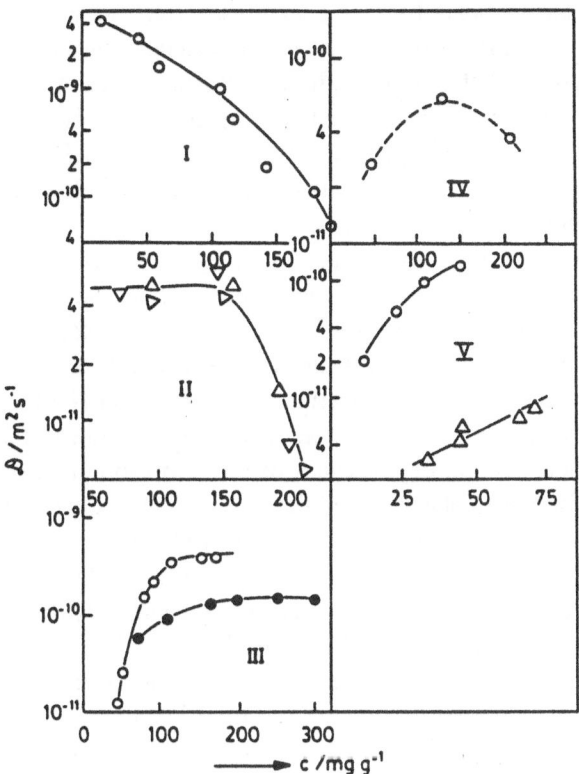

**Figure 4.** Patterns of concentration dependence of intracrystalline self-diffusivities as determined from PFG NMR measurements. (I) $n$-hexane in NaX at 358 K; (II) ortho ($\Delta$), meta ($\triangleright$) and para ($\nabla$) xylenes in NaX at 393 K; (III) ammonia (O) and water ($\bullet$) in NaX at 298 K; (IV) acetonitrile in NaX at 393 K; (V) ethane (O) at 173 K and propane ($\nabla$) at 413 K in NaCaA. From ref. [5] with permission.

Unsaturated hydrocarbons in NaX follow the same pattern from medium concentration, while yielding diffusivities which are essentially constant for smaller concentrations (pattern II). This constancy should be referred to the specific interaction between the cations and the double bonds of the unsaturated hydrocarbons, while the decrease of the diffusivity with increasing concentration is a simple consequence of the increasing mutual hindrance of the molecules. For small molecules undergoing a particularly strong interaction with the adsorption sites, after an initial fast increase, the diffusivity is essentially found to remain constant with further increasing concentration (pattern III). Pattern IV results as a superposition of patterns I or II with pattern III, and is most likely to be observed for larger molecules undergoing specific interaction with adsorption sites. Such a pattern has been observed for methanol in NaX by both PFG NMR and tracer ZLC measurements [42]. In some cases, over the whole concentration range an increase of the diffusivity with increasing concentration has been observed. This type V behaviour might be understood as a limiting case of pattern III if up to highest loadings molecular propagation is controlled by the interaction with active sites [43] or by the corresponding concentration dependence of the escape rate from cavities with „small" windows on the basis of the theory of absolute reaction rates [44].

The discrepancy observed for a number of systems between the results of transport diffusion and self-diffusion measurement does not only concern the absolute values of the respective diffusivities. In most cases, the corrected diffusivities as following from the transport diffusion measurements tend to remain constant with varying concentration, while the PFG NMR measurements provide a wealth of different concentration dependences as illustrated by fig. 4.

The various concentration dependences of self-diffusion might imply a rather remarkable hint for the explanation of the discrepancy between transport and self-diffusion measurement. So far for systems following patterns I or II, the corrected transport diffusivities were in striking contrast with the results of PFG NMR, while in all other cases satisfactory agreement was observed. It is interesting to note that the second term on the right hand side of eq. (15), representing the difference between the corrected transport diffusivities and self-diffusivites, must be expected to be most essential in exactly these two cases of pattern I and II where the interaction between different molecules has been considered to be more pronounced than in all other cases. So far it is not possible to decide whether these correlations are only incidental or whether they may be considered as an indication of the real origin of this discrepancy.

## 5.3. DIFFUSION ANISOTROPY

Only in isotropic systems, molecular diffusion is completely described by a single quantity. Any anisotropy will generate an orientation dependence of the rate of molecular transportation, which therefore must be described by a diffusion tensor rather than by a diffusion coefficient as a scalar quantity. Though already in the very first diffusion studies with zeolites, carried out with crystals of natural heulandite [45, 46], the direction dependence of molecular transportation was investigated, over many years in most diffusion studies with zeolites it was implicitly assumed that the transport properties are sufficiently well represented by one diffusivity. In fact, due to the small size of the zeolite crystallites,

313

orientation dependent diffusion measurements with synthetic zeolites have only recently become possible [47 - 49]. PFG NMR has been able to trace diffusion anisotropy in essentially two different ways. As one possibility, large crystallites of type ZSM-5 have been introduced in an array of parallel capillaries, so that their crystallographic z axes (cf. Fig. 5) were essentially aligned in parallel with the capillary axes [47].

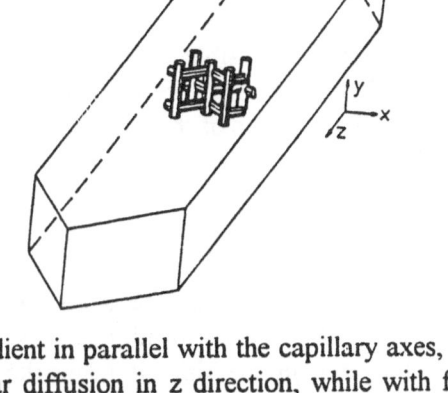

**Figure 5.** Schematic representation of the orientation of the internal channel system within zeolite crystallites of type ZSM-5. From ref. [48] with permission.

In this way, by applying the field gradient in parallel with the capillary axes, one is predominantly measuring molecular diffusion in z direction, while with field gradients applied perpendicular to the z direction one is mainly observing molecular diffusion in the crystallographic xy plane. Figure 6 shows the diffusivities measured for methane within a sample of oriented zeolite crystallites of type ZSM-5 with field gradients applied perpendicular and parallel to the direction of the capillaries.

**Figure 6.** Arrhenius plot of the diffusivities of methane in ZSM-5 at a sorbate concentration of 12 molecules per unit cell with field gradients applied parallel (●) and perpendicular (○) to the capillaries of the container system at 293 K. The broken (dotted) lines indicate the expected range of $D_z$ and ½ $(D_x + D_y)$. From ref. [47] with permission.

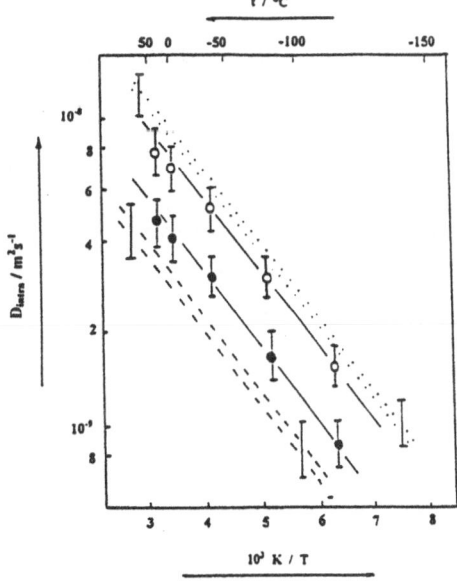

Since the alignment of the crystallites in the capillaries is not perfect, for the determination of the real values for the diffusivities in z direction ($D_z$) and in the xy plane ($D_{xy}$) additional assumptions about the degree of alignment are unevitable. Fig. 6 represents the intervals for the values of $D_z$ and $D_{xy}$ which have been estimated by assuming a mean deviation of 25° to 30° from perfect alignment [47]. As to be expected, molecular diffusion in the xy plane is much faster than in z direction. In complete agreement with the schematic representation of fig. 5, which shows that the two channel systems are aligned in parallel with the x and y directions, the diffusivity in z direction is found to be much slower than in the xy plane. By this procedure, a separate determination of the diffusivities in x and y direction is excluded since an alignment of the crystallites is only possible with respect to their z axes.

In an alternative way, information about diffusion anisotropy may be deduced by analysing the shape of the spin echo attenuation with increasing field gradient intensity [48]. In a powder sample, the rate of molecular propagation in the direction of the field gradient covers the whole spectrum of orientation dependent diffusivities within the crystallites. For a quantitative analysis of the spin echo attenuation curves it is of great importance that in ZSM-5 the diffusivities in the x, y and z directions may be shown to be interrelated [50, 51]. Under the assumption that the time to migrate from one channel intersection to another is much longer than the time it takes the diffusants to lose memory of their configuration, one has

$$c^2 / D_z = a^2 / D_x + b^2 / D_y \qquad (28)$$

where $a$, $b$ and $z$ are the unit cell edge lengths. This correlation has been found to work well for small diffusants as well as for larger rigid molecules such as benzene [52]. For long-chain molecules the assumption that the diffusants may lose their memory sufficiently fast may not hold anymore since the chain acts as a „tracer" of the diffusion path [53]. Figure 7 shows the results of the PFG NMR studies of diffusion anisotropy by an analysis of the spin-echo attenuation curve for methane in ZSM-5, which may clearly be expected to follow eq. (28). It turns

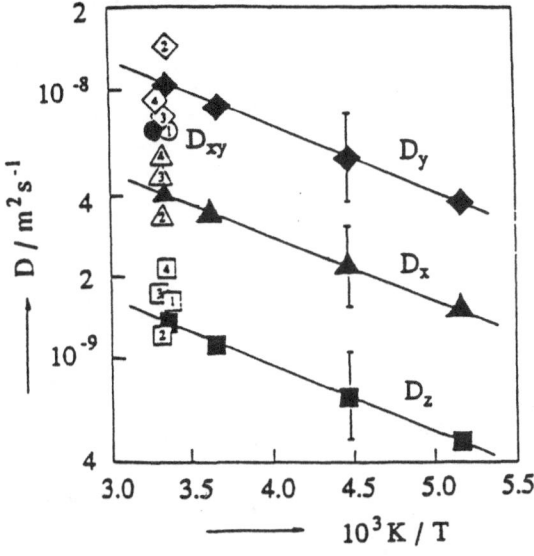

**Figure 7.** Arrhenius plot of the principal values of the diffusion tensor for methane adsorbed in ZSM-5, as determined on the basis of Fig. 3 (full symbols), and comparison with the results of the measurement with oriented samples (1) and of MD simulations presented in Ref. [54] (2), [55] (3) and [56] (4) (open symbols with inserted numbers). From ref. [48] with permission.

out that the experimental data are in satisfactory agreement with both MD simulations [54 - 56] and the measurements with oriented crystallites.

## 5.4. DIFFUSION DURING CHEMICAL REACTION

As a non-invasive method, PFG NMR is able to monitor the diffusivity of the reactant and product molecules during chemical reactions. As an example, fig. 8 shows the diffusivities of isopropanol and propene at 200 °C during the conversion of isopropanol to propene and water in zeolite NaX, as a standard reaction for acid catalysis [57]. The measurements have been carried out using $^{13}C$ PFG NMR with $^{13}C$ labelled isopropanol. It is interesting to note that over the investigated conversion range the mobility of isopropanol remains essentially unchanged while the propene diffusivity is increasing. Such a behaviour might be explained by the fact that with a decreasing number of the less mobile isopropanol molecules their retarding influence on the propagation of the propene molecules as well decreases.

316

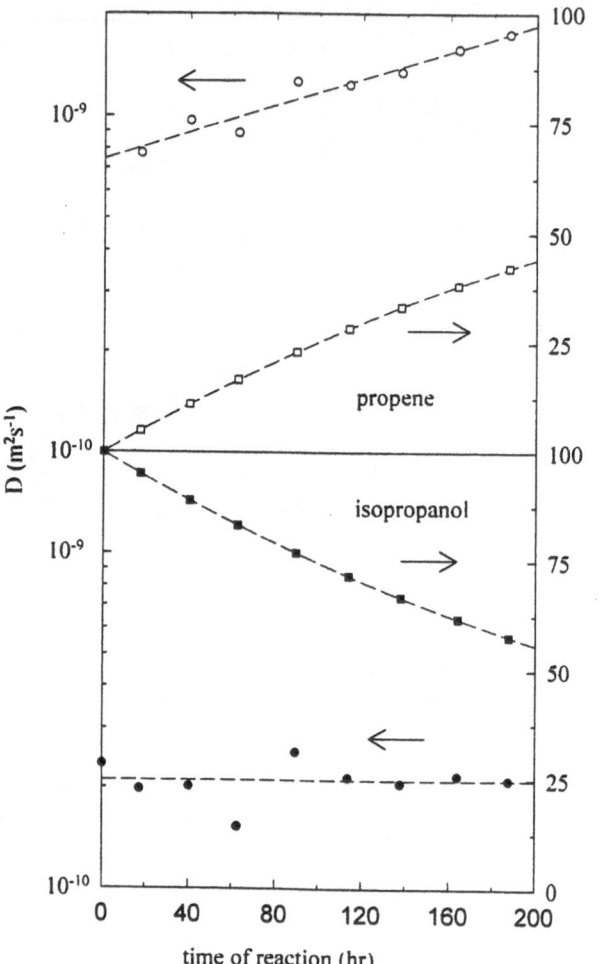

**Figure 8.** Time dependence of the relative amount of isopropanol and propene during the conversion of isopropanol in NaX and their self-diffusivities at 473 K (open symbols: propene, closed symbols: isopropanol). From ref. [57] with permission.

## 6. Diffusion under Confinement

### 6.1. ANOMALOUS DIFFUSION

Under the influence of confining geometries subsequent displacements $\vec{r}_k$ during the process of molecular propagation have most likely opposite signs, so that the cross terms $\langle \vec{r}_k \vec{r}_l \rangle$ become negative. This leads to a mean square displacement which increases less than linearly with the observation time. If in hierarchical structures the influence of confinement increases with increasing propagation - such as, e.g., in fractal networks [58] - the mean square displacement may again be found to follow a power law

$$\langle r^2(t)\rangle \propto t^\kappa. \tag{29}$$

Now, however, in contrast to Einstein's equation (10), the time exponent $\kappa$ is smaller than 1. This leads to an effective diffusivity

$$D_{eff} = \frac{\langle r^2(t)\rangle}{6t} \propto t^{\kappa-1}, \tag{30}$$

decreasing with increasing observation time.

The investigation of diffusion phenomena in porous media deviating from Einsstein's equation is a current topic of diffusion studies by NMR [59, 60]. In order to trace subtle structural peculiarities of the pore system, short diffusion paths must be covered. As an example, fig. 9 shows the time dependence of the effective diffusivities of polydimethylsiloxane (PDMS) with a molecular weight of 22,530 g/mol in a polypropylene host matrix at different pore filling factors [61].

**Figure 9.** Effective diffusivities of PDMS (22,530 g/mol) in a polypropylene host matrix at pore filling factors as indicated. From ref. [61] with permission.

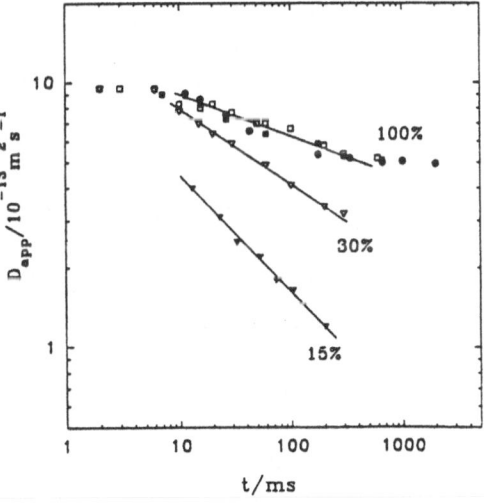

318

From the representation of the effective diffusivities as a function of the displacement (fig. 10) it becomes obvious that anomalous diffusion as reflected by a dependence of the diffusivity on the diffusion time and hence on the diffusion path length occurs for displacements between about 100 and 600 nm.

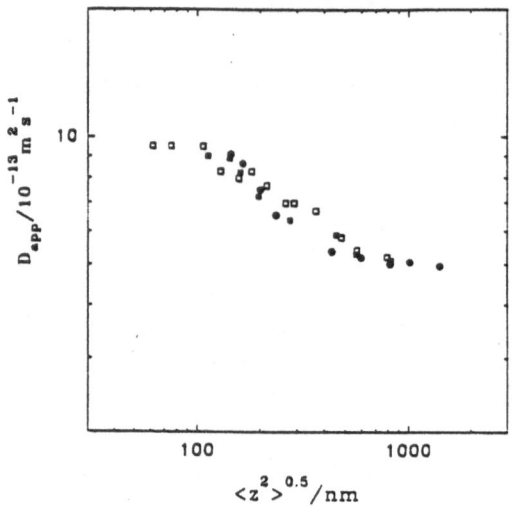

**Figure 10.** Effective diffusivity of PDMS ($M_w$ = 22,530 g/mol) in a polypropylene host matrix at a pore filling factor of 100 % at 293 K as a function of the root-mean-square displacement of PDMS during the NMR experiment. The filled circles refer to effective diffusivities extrapolated from measurements at 343 K to room temperature by applying the well-established temperature-time shift principle of polymer dynamics. From ref. [61] with permission.

It is interesting to note that these lower and upper cut-offs are of the order of the diameters of the smallest and largest pores of the host matrix. Deviation from ordinary diffusion appears to be of increasing significance with decreasing pore filling factor. This might be explained by the fact that the interaction between the pore surface and the diffusants becomes closer with decreasing loading.

## 6.2. RESTRICTED DIFFUSION

If the molecules under study are confined within closed regions so that during the observation time of PFG NMR only a negligibly small amount of diffusants is able to leave these regions, with increasing observation time the mean propagator as resulting from PFG NMR measurements is more and more determined by the extension of the confining regions rather than by molecular mobility. In fact, the probability that a diffusant occupies a certain position within a confining region

319

at the instant of the second gradient pulse eventually becomes independent of its position during the first gradient pulse and coincides with the (a priori) probability distribution of the diffusants. In this limit, the formalism of PFG NMR leads to relations which are well-known from scattering experiments [26 - 29], and by covering a sufficiently large range of gradient intensities (and hence of the „generalized scattering vector"), the spin echo attenuation is in fact found to exhibit oscillations instead of decaying monotonically to zero [26, 29]. For an estimate of the size of the confining regions, however, it is sufficient to only consider the initial part of the spin echo attenuation. This part is sufficiently well represented by eq. (22) with $D$ being replaced by an effective quantity $D_{eff}$. By molecular confinement within spheres of radius $R$, in the limit of sufficiently large observation times one has

$$D_{eff} = R^2 / 5\Delta . \tag{31}$$

In fact, the low-temperature data of fig. 2b provide an example, where PFG NMR measurements allow an estimate of the size of the confining regions, i.e. in this case of the individual crystallites.

As another example, fig. 11 shows the results of PFG NMR self-diffusion measurements of PDMS (polydimethylsiloxane) within the spherical cavities of a PS (polystyrene) matrix [62].

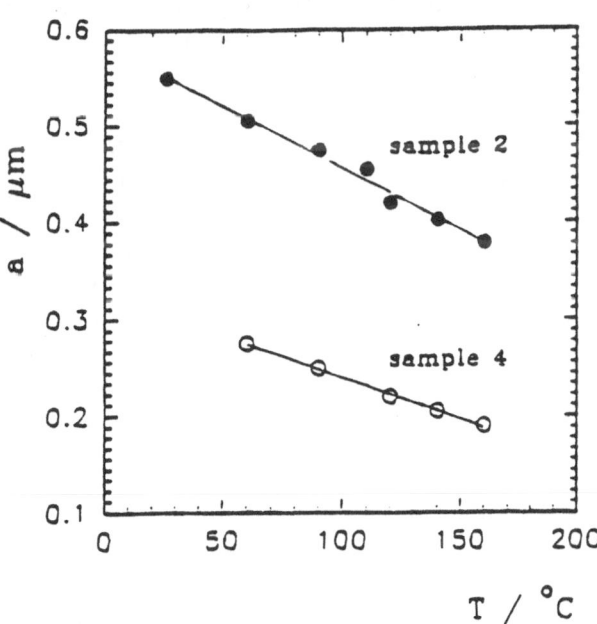

**Figure 11.** The temperature dependence of the droplet radii derived from the spin echo attenuation using eq. 31 for two samples, both having a 20 % oil content but differing in their amount of PS-PDMS copolymer. The straight lines are only guides for the eye. From ref. [62] with permission.

Using eq. (31), the effective diffusivities of PFG NMR were transferred into the corresponding mean radii of the cavities. The order of magnitude of the pore sizes as observed for the two samples under study agree very well with the information provided by scanning electron microscopy. In addition it is found that the size of the cavities decreases with increasing temperature. This effect may be explained by an increase of the solubility of PDMS by the PS matrix effecting a swelling of the PS matrix. This process could be traced quantitatively by monitoring the uptake of PDMS by the PS matrix after temperature enhancement by both solid state NMR and by measuring the time dependence of the decrease of the size of the cavities using the PFG NMR method [63].

## 6.3. SURFACE BARRIERS

Over many years the existence of transport resistances on the outer surface of the zeolite crystallites has been considered as one of the most probable explanations of the discrepancy between the messages of uptake measurements and PFG NMR studies on intracrystalline diffusion [8, 64]. In principle, analysing the time dependence of molecular uptake should allow to discriminate, whether the observed process is controlled by intracrystalline diffusion or by permeation through some surface resistance. Due to the influence of further processes such as the release of the heat of adsorption and molecular transport into the adsorption vessel („valve effect") and through the bed of crystallites [5 - 7], however, the time dependence of the uptake curve may be additionally modified. A decision about the rate controlling mechanism, which is exclusively based on an analysis of the time dependence, is therefore not without some ambiguity. There is a particular need, therefore, for the direct proof and for the quantification of transport resistances on the outer surface of zeolite crystallites. PFG NMR studies of intracrystalline diffusion in combination with NMR tracer desorption measurement of the intracrystalline mean life time [30] do in fact provide such a possibility. Under the assumption that intercrystalline molecular exchange is exclusively controlled by intracrystalline diffusion, the intracrystalline mean life time results to be

$$\tau_{intra} = \tau_{intra}^{Diff} = R^2 / 15D, \qquad (32)$$

where for simplicity the crystallites are assumed to be of spherical shape with $R^2$ denoting the mean square radius. For crystal shapes deviating from this assumption the numerical factor on the right hand side of eq. (32) must be correspondingly modified [65]. If $\tau_{intra}^{Diff}$ coincides with the intracrystalline mean life time, any essential influence of surface barriers on the rate of molecular exchange may be excluded. So far, after a correspondingly careful activation, in all PFG NMR studies with as-synthesized zeolite crystallites this case was observed. The existence of structural surface barriers has to be excluded, therfore, as a general explanation of the discrepancy between PFG NMR and uptake measurements. This conclusion coincides with the results of ZLC measurements carried out after partial and total equilibration of the crystallite bulk phase with the diffusant [4, 6, 66].

Figure 12 shows the results of a PFG NMR study of the formation of transport resistances in H-ZSM-5 by coking with n-hexane and mesitylene at elevated temperature over different reaction times [40, 67].

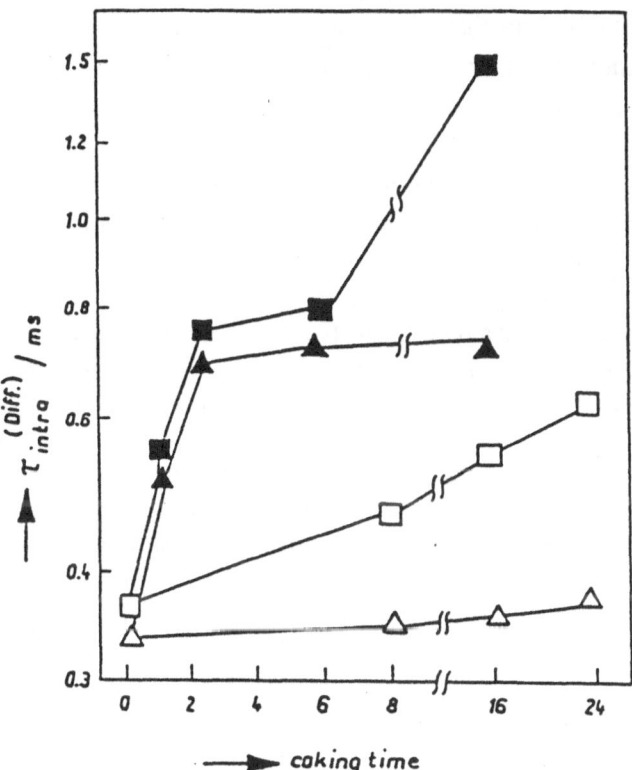

**Figure 12.** Values for the intracrystalline mean lifetime $\tau_{intra}$ ($\square$, $\blacksquare$) and the quantity $\tau_{intra}^{Diff}$ ($\triangle$, $\blacktriangle$) versus time on stream for methane at 296 K and a sorbate concentration of 12 molecules per unit cell, in H-ZSM-5 coked by $n$-hexane (filled symbols) and mesitylene (open symbols). From ref. [67] with permission.

The coking was followed by the usual activation procedure (temperature enhancement up to 400 °C under continuous evacuation), and afterwards the samples were loaded with methane (12 molecules per unit cell) as a probe molecule. Figure 12 once again shows that in the starting material the values for $\tau_{intra}$ and $\tau_{intra}^{Diff}$ coincide, indicating that in this case there is no surface barrier. Using mesitylene as a coking chemical, the methane mobility within the zeolite crystallites remains essentially unaffected by the coking procedure, so that the values for $\tau_{intra}^{Diff}$ remain constant, while the real intracrystalline mean life time increases with increasing coking time. One has to conclude, therefore, that the coke is predominantly deposited in a layer close to the outer surface of the zeolite crystallites, forming a surface barrier. With n-hexane as a coking chemical, the intracrystalline diffusivity of methane is found to decrease to the same extent as the intracrystalline mean life time increases. Hence, $\tau_{intra}$ and $\tau_{intra}^{Diff}$ continue to agree so that the existence of surface barriers must be excluded. It may be concluded, therefore, that the coke deposition occurs more or less homogeneously over the total bulk phase of the zeolite crystallites. Only in a second stage (for coking times > 6 h) $\tau_{intra}$ starts to become larger than $\tau_{intra}^{Diff}$ indicating from now on a preferential coke deposition on the outer surface. Since in contrast to n-hexane the mesitylene molecules are too large to penetrate the ZSM-5 pore network the different coking behaviour of the two chemicals is not unexpected.

While in the considered case of barrier formation by coking only a moderate reduction of the rate of intercrystalline molecular exchange is observed, hydrothermal treatment of zeolite NaCaA has been found to lead to a reduction of the rate of intercrystalline exchange by up to two orders of magnitude [68]. This effect can be referred to a collapse of the zeolite lattice in the vicinity of the crystallite surface while the internal lattice structure has been found to be nearly unchanged [69, 70]. In such cases, PFG NMR studies of intracrystalline diffusion may essentially be carried out under the conditions of total restriction as described in section 6.2.

## 6.4. SINGLE FILE DIFFUSION

A particular case of molecular confinement occurs in zeolites with one-dimensional channels such as ZSM-12, -22, -23, -48, $AlPO_4$ -5, -8, -11, L, theta 1 and omega [71], if the adsorbed molecules are too large to pass each other. This situation may be illustrated by a file of strung pearls, where a given pearl may only be shifted if the adjacent ones are shifted as well, in order to leave the necessary free space for displacement. Molecular transport under such conditions has been termed single-file diffusion [72,73]. It is obvious that under such conditions subsequent molecular displacements are highly correlated. Any displacement of a molecule into a given direction is followed with a higher probability by a displacement into the opposite direction. This is a simple consequence of the requirement that under equilibrium condition the centre of gravity of the system tends to keep its position. Molecular displacement to the right, e.g., necessitates a condensation of the molecules on the right in parallel with a dilution on the left. In a subsequent period of time, therefore, the enhanced concentration of the molecules on the right will effect an „expansion", so that a backward shift of the molecule under study becomes more probable. It should be emphasized that in contrast to the correlation effect well-known from the atomic transport in solids [74], this correlation becomes the more stringent the larger are the considered displacements. This is exactly the situation necessary for the occurence of anomalous diffusion following the time dependence of eq. (29). In the present case of single-file diffusion the time exponent is found to be $\kappa= 0.5$, and for single-file diffusion by activated jumps the mean square displacement in channel direction is found to be correlated to the elementary steps of diffusion by the expression [73, 75, 76]

$$\langle z^2 \rangle = 2F\sqrt{t} \tag{33}$$

with the „mobility factor"

$$F = \lambda^2 \frac{1-\theta}{\theta} \frac{1}{\sqrt{2\pi\tau}}. \tag{34}$$

$\lambda$ and $\tau$ denote the jump length and the mean time between subsequent jump attempts. A jump attempt is only successful if the site where it is directed to, is empty. In the other case, the diffusant remains at its position. 0 stands for the site occupancy. Recently, the validity of eq. (33) could be confirmed quite generally by MD simulations [77].

Equations (33) and (34) may be considered as a special case of a more general relation correlating the translational mobility of an element of the single-file system at infinite dilution with its mobility under single-file conditions. With l denoting the mean free distance (clearance) between adjacent molecules in the single-file system, and with $\langle |s(t)| \rangle$ denoting the mean value for the displacement of the diffusants at infinite dilution, the mean square displacement in the single-file system may be shown to be [78]

$$\langle z^2(t)\rangle = l\langle|s(t)|\rangle. \qquad (35)$$

Figure 13 shows the results of PFG NMR measurements of $CF_4$ in $AlPO_4$-5 at a temperature of 180 K [79].

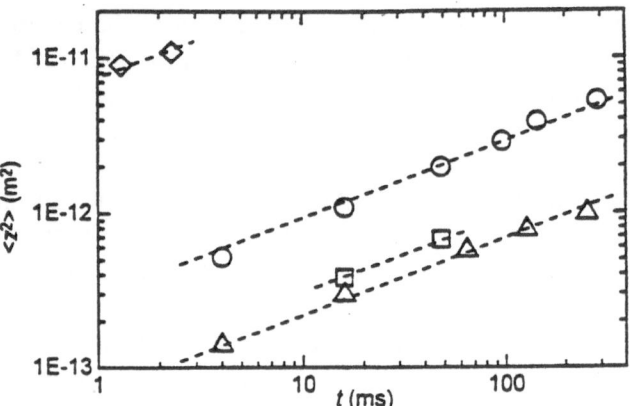

Figure 13. Molecular mean square displacement of $CF_4$ in $AlPO_4$-5 at 180 K as a function of the observation time at a sorbate concentration of 0.005 ($\Diamond$), 0.05 (O), 0.2 ($\square$), and 0.4 ($\triangle$) molecules per unit cell. From ref. [79] with permission.

Over the whole range of observation times, the mean square displacement is found to follow the $\sqrt{t}$ dependence as predicted by eq. (33) for single-file systems. With the respective diameters of 0.47 nm for tetrafluoromethane [80] and of 0.73 nm for the channels of the $AlPO_4$-5 structure [71], the structural prerequisites of single-file behaviour are most likely to be given, so that the observed time dependence is not unexpected. Moreover, also the absolute values of the mobility factors are in reasonable agreement with the concentration dependence given by eq. (34). On the basis of eq. (35), the measured mean square displacements may be expressed in terms of the diffusivity which a sole molecule would have in the channel of the $AlPO_4$-5 structure. With a slight scattering in the magnitude, depending on the given sorbate concentration, the diffusivity of a single $CF_4$ molecule in the channel would be of the order of $10^{-6}$ $m^2$ $s^{-1}$. This value is by two orders of magnitude larger than the largest zeolitic diffusivities for $CF_4$ so far measured under comparable conditions [81, 82]. It was illustrated by MD simulations [79] that this increase in the translational mobility of molecules in straight channels in comparison with three-dimensional pore networks may be associated with the fact that the dissipation of the momentum of the diffusants proceeds much faster in three-dimensional pore networks than in straight cylindrical pores.

The measurement of single-file diffusion in zeolites with straight, non-intersecting channels by the PFG NMR method confirms the great influence of the mutual correlation of the diffusants on their propagation patterns. Similar time dependences have been observed in PFG NMR studies with ethane in $AlPO_4$-5 [83] and with methane in theta-1 [84].

## 7. Conclusion

The PFG NMR method has proved to be a rather versatile technique for studying molecular transport in porous materials. Being able directly to trace molecular displacements from hundreds of nanometers up to hundreds of micrometers, PFG NMR is able to establish direct correlations between the structure of the adsorbent and the molecular mobility. This correlation is illustrated in a most informative way by the study of molecular diffusion in beds of zeolite crystallites. Depending on the relation between the particle size and the observed diffusion paths, PFG NMR is able to measure both the intracrystalline and long-range diffusivities as well as the intercrystalline molecular exchange rate.

The interrelation between structure and molecular displacements allows the estimate of the extension of confining regions within the adsorbate-adsorbent system (method of „dynamic imaging"). A particular case of molecular confinement is observed in zeolites with a one-dimensional channel structure. As soon as the adsorbed molecules are too large to pass each other within a channel, the molecular mean square displacement increases in proportion with the square root of the observation time rather than with the observation time itself. This phenomenon of „single-file" diffusion which has been postulated theoretically since a couple of years, has recently been confirmed experimentally by the PFG NMR method. In materials with hierarchical pore structure, molecular mean square displacements may also be found to increase less than linearly with increasing observation times. It is a current topic of research to correlate such deviations from normal diffusion with the structural properties of the samples under study.

## 8. Acknowledgement

I am obliged to Professor Harry Pfeifer for his continuous support and encouragement over many years. Special thanks are due to Professor Douglas M. Ruthven for many stimulating discussion and due to the Alexander von Humboldt Foundation and the Max-Planck-Society for financial support of this collaboration.

326

## 9. References

1. Chihara, K., Suzuki, M., and Kawazoe, K., Chem. Eng. Sci. 31 (1976) 505
2. Ruthven, D.M. and Lee, L.-K., AIChE Journal 27 (1981) 654
3. Ruthven, D.M., this volume
4. Ruthven, D.M. and Brandani, S., this volume
5. Kärger, J. and Ruthven, D.M. *Diffusion in Zeolites and other Microporous Solids*, Wiley, New York, 1992
6. Ruthven, D.M. In: Bonneviot L., Kaliaguine, S. (eds) Zeolites: a refined tool for designing catalytic sites, Elsevier, Amsterdam (1996), p. 223
7. Bülow, M. and Micke, A., Adsorption 1 (1995) 29
8. Kärger, J. and Caro, J., J. Chem. Soc. Faraday Trans. I, 73 (1977) 1363
9. Wernick, D.L. and Osterhuber, E.J., Diffusional transition in zeolite NaX: 1. Single Crystal gas permeation studies. In: Olson, D., Bisio, A. (eds) Proceedings of the sixth international zeolite conference, Butterworths, Guildford (1984) 122
10. Paravar, A. and Hayhurst, D.T., Direct Measurement of diffusivity for butane across a single large silicalite crystal. In: Olson, D., Bisio, A. (eds) Proceedings of the sixth international zeolite conference, Butterwoths, Guildford (1994) p. 217
11. Karge, H.G. and Nießen, W., Catal. Today 8 (1991) 451
12. Niessen, W. and Karge, H.G., Adsorption 1 (1993) 1
13. Kärger, J., Pfeifer, H., Wutscherk, T., Ernst, S., Weitkamp, J., and Fraissard, J., J. Phys. Chem. 96 (1992) 5059
14. Springuel-Huet, M.A., Nosov, A., Kärger, J., and Fraissard, J., J. Phys. Chem. 100 (1996) 7200
15. Dyer, A. and Yusof, A.M., Zeolites 7 (1987) 191
16. Dyer, A. and Townsend, R.T., J. Inorg. Nuclear Chem. 35 (1973) 3001
17. Einstein, A., Ann. Phys. 17 (1905) 349
18. Jobic, H., Application of Neutron Scattering to Catalysis. In: Vedrine, J.C., (ed) Catalyst characterization, Plenum Press, (1994), New York, 347
19. Jobic, H., Bée, M., and Kearly, G.J., J. Phys. Chem. 98 (1994) 4660
20. Callaghan, P.T., *Principles of Nuclear Magnetic Resonance Microscopy*, Clarendon Press, Oxford, 1991
21. Evans, D.J. and Morris, G.P., *Statistical Mechanics of Nonequilibrium Liquids*, Academic Press, London, 1990
22. Magin, E.J., Bell, A.T., and Theodorou, D.N., J. Phys. Chem. 97 (1993) 4173

23. Kärger, J. and Heink, W., J. Magn. Reson. 51 (1983) 1
24. Callaghan, P.D., J. Magn. Reson. 90 (1990) 177
25. Heink, W., Kärger, J., Seiffert, G., Fleischer, G., and Rauchfuß, J., J. Magn. Reson. 114 (1995) 101
26. Callaghan, P.T. and Coy, A., Phys. Rev. Lett. 68 (1992) 3176
27. Cotts, R.M., Nature 351 (1991) 443
28. Fleischer, G. and Fujara, F., NMR-Basic Principles and Progress 30 (1994) 159
29. Bär, N.K., Kärger, J., Krause, C., Schmitz, W., and Seiffert, G., J. Magn. Reson. A113 (1995) 278
30. Kärger, J., AIChE-Journal 28 (1982) 417
31. Kärger, J., Heink, W., Pfeifer, H., Rauscher, M., and Hoffmann, J., Zeolites 2 (1982) 275
32. Kärger, J., Zikanova, A., and Kocirik, M., Z. Phys. Chem., Leipzig 265 (1984) 587
33. Barrer, R.M., *Zeolites and clay minerals as sorbents and molecular sieves.* Academic Press, London (1978) 197
34. Bär, N.K., Kärger, J., Schwarz, H., Ernst, S., and Weitkamp, J., Micropor. Materials, 6 (1996) 355
35. Heink, W., Kärger, J., Pfeifer, H., and Stallmach, F., J. Am. Chem. Soc. 112 (1990) 2175
36. Völter, J., Caro, J., Bülow, M., Fahlke, B., Kärger, J., and Hunger, M., Appl. Catal. 42 (1988) 15
37. Bär, N.K., thesis, Leipzig University, in preparation
38. Kärger, J., Pfeifer, H., Rauscher, M., and Walter, A., J. Chem. Soc. Faraday I 76 (1980) 717
39. Caro, J., Bülow, M., Schirmer, W., Kärger, J., Heink, W., and Pfeifer, H., J. Chem. Soc. Faraday Trans. I 81 (1985) 2541
40. Caro, J., Bülow, M., Jobic, H., Kärger, J., Zibrowius, B., Adv. Cat. 39 (1993) 351
41. Pfeifer, H., Kärger, J., Germanus, A., Schirmer, W., Bülow, M., and Caro, J., Ads. Sci. Techn. 2 (1985) 229
42. Brandani, S., Ruthven, D.M., and Kärger, J., Zeolites 15 (1995) 494
43. Heink, W., Kärger, J., Ernst, S., and Weitkamp, J., Zeolites 14 (1994) 320
44. Kärger, J., Surf. Sci. 57 (1976) 749
45. Tiselius, A., Z. Phys. Chem. A 169 (1934) 425
46. Tiselius, A., Z. Phys. Chem. A 174 (1935) 401
47. Hong, U., Kärger, J., Kramer, R., Pfeifer, H., Seiffert, G., Müller, U., Unger, K.K., Lück, H.B., and Ito, T., Zeolites 11 (1991) 816

48. Hong, U., Kärger, J., Pfeifer, H., Müller, U., and Unger, K.K., Z. phys. Chem. 173 (1991) 225

49. Caro, J., Noack, M., Richter-Mendau, J., Marlow, F., Petersohn, D., Griepentrog, M., and Kornatowski, J., J. Phys. Chem. 97 (1993) 13685

50. Kärger, J., J. Phys. Chem. 95 (1991) 5558

51. Kärger, J. and Pfeifer, H., Zeolites 12 (1992) 872

52. Snurr, R.Q., Bell, A.T., and Theodorou, D.N., J. Phys. Chem. 98 (1994) 11948

53. Maginn, E.J., Bell, A.T., and Theodorou, D.N., J. Phys. Chem. 100 (1996) 7155

54. Demontis, P. Fois, E.S., Suffritti, G.B., and Quartieri, S., J. Phys. Chem. 94 (1990) 4329

55. Goodbody, S.J., Watanabe, K., McGowan, D., Walton, J.P.B., and Quirke, N., J. Chem. Soc. Faraday Trans. I 87 (1991) 1951

56. June, R.L., Bell, A.T., and Theodorou, D.N., Phys. Chem. 94 (1990) 8232

57. Schwarz, H.B., Ernst, S., Kärger, J., Knorr, B., Seiffert, G., Snurr, R.Q., and Weitkamp, J., J. Catal., submitted

58. Avnir, D., *The Fractal Approach of Heterogeneous Chemistry*, Wiley, Chichester, 1992

59. Stapf, S., Kimmich, R., and Seitter, R.O., Phys. Rev. Lett. 75 (1995) 2855

60. Stapf, S. and Kimmich, R., J. Chem. Phys. 103 (1995) 2247

61. Appel, M., Fleischer, G., Kärger, J., Fujara, F., and Siegel, S., Europhys. Lett. 34 (1996) 438

62. Appel, M., Fleischer, G., Dieng, A.C., and Riess, G., Macromol. 28 (1995) 2345

63. Appel, M., Fleischer, G., Geschke, D., Kärger, J., Winkler, M., Dieng, A.C., and Riess, G., Macromol. Rapid Comm. 17 (1996) 81

64. Bülow, M., Z. Chem. 25 (1985) 81

65. Förste, C., Kärger, J., Pfeifer, H., Riekert, L., Bülow, M., and Zikanova, A., J. Chem. Faraday Trans. I 86 (1990) 881

66. Brandani, S. and Ruthven, D.M., Adsorption 2 (1996) 133

67. Kärger, J., Pfeifer, H., Caro, J., Bülow, M., Schlodder, H., Mostowicz, and R., Völter, J., Appl. Cat. 29 (1987) 21

68. Kärger, J., Pfeifer, H., Richter, R., Fürtig, H., Roscher, W., and Seidel, R., AIChE-Journal 34 (1988) 1185

69. Kärger, J., Bülow, M., Millward, B.R., and Thomas, J.H., Zeolites 6 (1986) 146

70. Seidel, R., and Staudte, B., Zeolites 13 (1993) 348

71. Meier, W.M., and Olson, D.H., *Atlas of Zeolite Structure Types*, Butterworth-Heinemann, London, 1992
72. Riekert, L., Adv. Catal. 21 (1970) 281
73. Kärger, J., Petzold, M., Pfeifer, H., Ernst, S., and Weitkamp, J., 136 (1992) 283
74. Allnatt, A.R. and A.B. Sidiard, *Atomic Transport in Solids*, Cambridge University Press, Cambridge, 1993
75. Fedders, P.A., Phys. Rev. B 17 (1978) 40
76. Kärger, J., Phys. Rev. A 45 (1992) 4173 and E 47 (1993) 1427
77. Hahn, K. and Kärger, J., J. Phys. Chem. 100 (1996) 316
78. Hahn, K. and Kärger, J., J. Phys. A 28 (1995) 3061
79. Hahn, K., Kärger, J., and Kukla, V., Phys. Rev. Lett. 76 (1996) 2762
80. Breck, D.W., *Zeolite molecular sieves*, Wiley, New York, (1974) 588
81. Kärger, J., Pfeifer, H., Rudtsch, S., and Heink, W., J. Fluorine Chem. 39 (1988) 349
82. Fenzke, D. and Kärger, J., J. Z. Phys. D 25 (1993) 345
83. Gupta, V., Nivarthi, S.S., McCormick, and A.V., Davis, H.T., Chem. Phys. Lett. 247 (1995) 596
84. Kukla, V., Kornatowski, J. Demuth, D., Girnus, I., Pfeifer, H., Rees, L.V.C., Schunk, S., Unger, K.K., and Kärger, J., Science 272 (1996) 702

# $^{129}$Xe NMR OF PHYSISORBED XENON
*Principles*

M. A. SPRINGUEL-HUET

*Laboratoire de Chimie des Surfaces, URA-CNRS 1428,*
*Université P. et M. Curie, 4 place Jussieu, 75252 Paris Cedex 05*
*France*

## 1. Introduction

The use of $^{129}$Xe NMR to study the properties of porous solids began at the end of the 70s [1,2]. Since that time a great number of papers (more than 200) and several reviews [3-7] have been published.

The central idea of the pioneers of this technique was to find a molecule which was unreactive and particularly sensitive to its environment, to physical interactions with other chemical species and to the nature of adsorption sites, and which could, therefore, be used as a probe to determine in a new way solid properties difficult to detect by classical physicochemical techniques. In addition, this probe needed to be detectable by NMR, since this technique is particularly suitable for investigating electron perturbations in rapidly moving molecules.

Xenon is an ideal probe because it is an inert gas, with a large spherical electron cloud. From the NMR point of view, the $^{129}$Xe isotope has a spin of one-half, its natural abundance is 26 % and its sensitivity of detection relative to the proton is $10^{-2}$. The xenon atom diameter is 4.4 Å which allows it to access sites of catalytic interest.

The high polarizability of the electron cloud leads to a large chemical shift range and makes its chemical shift an especially sensitive measure of local atomic interactions. This range goes from -40 ppm for Xe adsorbed in AgX zeolite to about 7500 ppm in the compound $XeO_6^{4-}$ [8] (Figure 1). The shift of gaseous Xe extrapolated to zero pressure has been chosen as the chemical shift reference

*J. Fraissard (ed.), Physical Adsorption: Experiment, Theory and Applications, 331–348.*
© *1997 Kluwer Academic Publishers.*

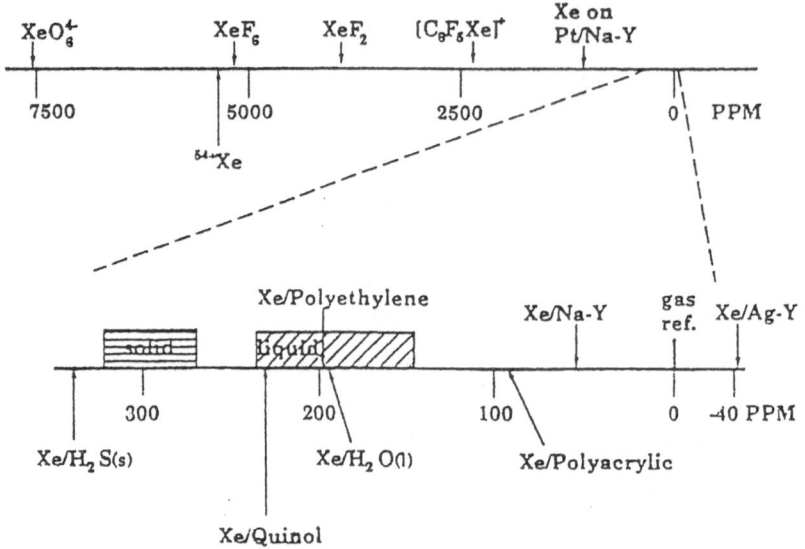

*Figure 1.* $^{129}$Xe NMR chemical shift scale for various xenon compounds, phases, and environments. The low frequency region has been expended in the lower part. (from ref. 6, with permission )

The chemical shift consists of two parts [9]:

- diamagnetic contribution, $\sigma_d$, which arises from the interaction of the electron orbitals with applied fields: according to the Lenz law the electrons precess in a way that opposes the field. This "shielding" of the nucleus from the applied magnetic field reduces the resonance frequency. This contribution depends only on the fundamental state. It has been calculated for isolated xenon atoms: 5638 ppm to lower frequency relative to the bare nucleus [10].

- a paramagnetic contribution, $\sigma_p$, which depends on the excited states and on the symmetry of the valence orbitals. It expresses the departure from spherical symmetry, so is zero for the isolated Xe atom. It generally increases the resonance frequency of the xenon nucleus. In most xenon compounds the electron distribution is not spherically symmetrical, and the $\sigma_p$ values show a large variation, which depends on the nature of the atom to which the xenon is bonded and the bond type. Shifts larger than that of the bare nucleus may be obtained. Except for a very small number of cases, the chemical shifts, $\delta$, measured relative to the isolated atom, are positive (downfield).

Pure monoatomic xenon is expected to have a long relaxation time $T_1$, but relaxation is enhanced by interaction with small amount of paramagnetic species. For example, $^{129}$Xe is easily detectable in xenon gas at 5 atm and at 27 MHz if 1% of $O_2$ is added. The $T_1$ of $^{129}$Xe adsorbed in microporous solids is rather short, typically in the range of 10 ms to a few seconds, even at low temperature.

# 2. $^{129}$Xe NMR Chemical Shift in Xenon Gas

## 2.1 PURE GAS PHASE

After the first detection of gaseous xenon [11] subsequent studies by Hunt and Carr [12] and by Brinkmann [13] showed that the shift can be expressed quite well by a virial expansion of the xenon density $\rho$ over a wide range:

$$\delta(T,\rho) = \delta_0 + \delta_1(T).\rho + \delta_2(T).\rho^2 + \delta_3(T).\rho^3 \qquad (1)$$

where $\delta(T, \rho)$ is the resonance shift at a temperature T and a density $\rho$. $\delta_0$ is the position for the isolated atom. Experimentally this corresponds to virtually zero pressure. $\delta_i(T)$ are the virial coefficients of the shift in density. The most exact values were obtained by Jameson et al. [14]. At 298.15 K

$$\delta_1 = 0.548 \pm 0.004 \ ppm/amagat* \qquad (2\text{-}1)$$
$$\delta_2 = (0.169 \pm 0.02) \times 10^{-3} \ ppm/amagat^2 \qquad (2\text{-}2)$$
$$\delta_3 = (-0.163 \pm 0.01) \times 10^{-5} \ ppm/amagat^3 \qquad (2\text{-}3)$$

For densities below about 100 amagats, terms of order higher than one are negligible. In this case the linear variation yields a value of 0.55 ppm chemical shift per amagat. $\delta_1$ results from binary collisions. Higher-order terms are due to the effect on the NMR frequency of three or more body collisions which become important at higher density. This effect is particularly important in the case of xenon, compared to other nuclei, because of the high polarizability of the electronic cloud. Theoretical calculations of $\delta_1$ from the interaction of two xenon atoms at the moment of collision were first made by Adrian [15]. This author showed that, during the lifetime of a collision, short-range electron exchange forces were responsible for the large density-dependent shift. According to Adrian the van der Waals interactions were not so important. Later Jameson [16] showed that Adrian overestimated the role played by repulsive interactions in the pressure dependence of the chemical shift.

The chemical shift is also very sensitive to temperature [17]. Jameson *et al.* have incorporated temperature into the coefficients $\delta_i$ of equation (1). Fitting $\delta_1(T)$ to a 4$^{th}$ degree polynomial, they obtained [18]:

$$\delta_1(T) = 0.536 - 0.135 \times 10^{-2}\tau + 0.132 \times 10^{-4}\tau^2 - 0.598 \times 10^{-7}\tau^3 + 0.663 \times 10^{-10}\tau^4 \quad (3)$$
in ppm/amagat, where $\tau = T\text{-}300$ K.

---

* One amagat is the density (2.689 x $10^{19}$ atoms/cm$^3$) of an ideal gas at standard temperature (273.15 K) and pressure (1 atmosphere).

It must be remembered that the shift $\delta_0$ of an isolated xenon atom is strictly temperature-independent [18].

Finally, Cowgill and Norberg [19] showed that the effect of the xenon pressure on the chemical shift is of the same type for all three phases: gas, liquid and solid. This result is particularly interesting for the interpretation of the $\delta$ values of adsorbed xenon.

## 2.2. MIXTURES OF XENON AND OTHER GASES

To elucidate the interaction potentiel of two different molecules Jameson *et al.* have studied the density dependence of the $^{129}$Xe chemical shift in mixture of Xe and other gases such as:
- rare gases (Kr, Ar) [14],
- molecules with tetrathedral symmetry ($CH_4$, $CF_4$, $SiF_4$.etc...) [14, 20],
- nonspherical molecules ($N_2O$, $C_2H_4$, $C_2H_6$, and $BF_3$) [21],
- linear molecules (CO, $N_2$, HBr, HCl, $CO_2$ and $C_2H_2$) [14, 20, 21, 22],
- paramagnetic species ($O_2$ and NO) [23-26].

In a mixture of xenon with another gas, A, the shift of xenon depends on the nature of A and the collision frequency with this latter. When gas densities are so low (<100 amagats) that only terms for binary collisions need to be considered, one may generalize equation (1) to a multi-dimensional virial expansion of the shift [14]:

$$\delta(T) = \delta_0 + \delta_{1.Xe}(T) \cdot \rho_{Xe} + \delta_{1.A}(T) \cdot \rho_A \qquad (4)$$

where $\rho_{Xe}$ and $\rho_A$ are the densities of Xe and A, respectively. $\delta_{1,A}$ has been determined for many gases and is also temperature-dependent (Table 1).

In general the effect on the xenon shift of xenon-other gas interactions is smaller than that for interactions between xenon atoms, except for the paramagnetic molecules $O_2$ and NO. The large shift induced by gaseous $O_2$ and NO has been interpreted in terms of a "contact overlap" mechanism. Overlap of the Xe (5s) and the $O_2$ ($\pi_g$) or NO ($\pi^*$) orbitals is the major contributor to the spin density at the Xe nucleus [23, 24]. It should be remarked that this is the first example of a contact shift involving two molecules.

Jameson *et al.* [25] studied the temperature dependence of the $\delta_{1,A}$ terms for $O_2$ and NO and confirmed that this Fermi contact interaction is responsible of the high chemical shift value at least at low temperature (T < 350 K). In this low temperature region $\delta_{1,A}$ depends on $1/T$, this Curie-type behaviour being expected for a Fermi contact interaction.

Previously contact-shift was only observed when the unpaired electron was on the same molecule as the nucleus under observation. This result will also be important in the case of xenon adsorbed on paramagnetic or metallic solids.

TABLE 1. Values of the first order terms $\delta_{1A}$ for various gases A

| Nature of the gas A | $\delta_{1A}$ (ppm/Mole.cm$^{-3}$) | $\delta_{1A}$ (ppm/amagat) |
|---|---|---|
| Ar | 3119 | 0.137 |
| $CO_2$ | 3823 | 0.171 |
| $CF_4$ | 4327 | 0.193 |
| $CHF_3$ | 4287 | 0.191 |
| $CH_2F_2$ | 4965 | 0.222 |
| $CH_3F$ | 4314 | 0.192 |
| $CH_4$ | 6201 | 0.277 |
| HCl | 7678 | 0.343 |
| Kr | 6070 | 0.271 |
| Xe | 12283 | 0.548 |
| NO | 17769 | 0.924 |
| $O_2$ | 21037 | 1.250 |

## 3. $^{129}$Xe NMR of Adsorbed Xenon : Generalities

By analogy with the results obtained in the gas phase, Fraissard *et al.* have shown that the chemical shift of adsorbed xenon is the sum of several terms corresponding to the various perturbations it suffers [27]. In the case of porous solids with homogeneous surfaces, the chemical shift can be written:

$$\delta = \delta_{ref} + \delta_S + \delta_{Xe} \qquad (5)$$

$\delta_{ref}$ is the chemical shift reference (gaseous xenon at zero pressure, i.e. $\delta_{ref} = \delta_0$ = 0). $\delta_S$ arises from interactions between xenon and the surface of the zeolite pores, provided that the solid does not contain any electrical charges. In this case it depends only on the mean free path, $\ell$, of a Xe atom in the porous structure. $\ell$ is the average distance travelled by a Xe atom between two successive collisions with the pore wall. It depends, of course, on the dimensions and the shape of the pores. $\delta_{Xe} = \delta_{Xe\text{-}Xe}\, \rho_{Xe}$ corresponds to Xe-Xe interactions; it increases with the local density of adsorbed xenon and becomes predominant at high xenon pressure.

To explain the hyperbolic relation obtained between $\delta_S$ and the mean free path, $\ell$, for various zeolites (Figure 2 and Table 2), we used a simple model of fast site exchange [28], which was then used also by Cheung [29]. Other models have also been proposed, such as "the surface curvature effect" by Derouane [30], or models using the Lennard-Jones potentiel between Xe and O atoms (representing the solid surface) by Ripmeester [31] and by Cheung [32].

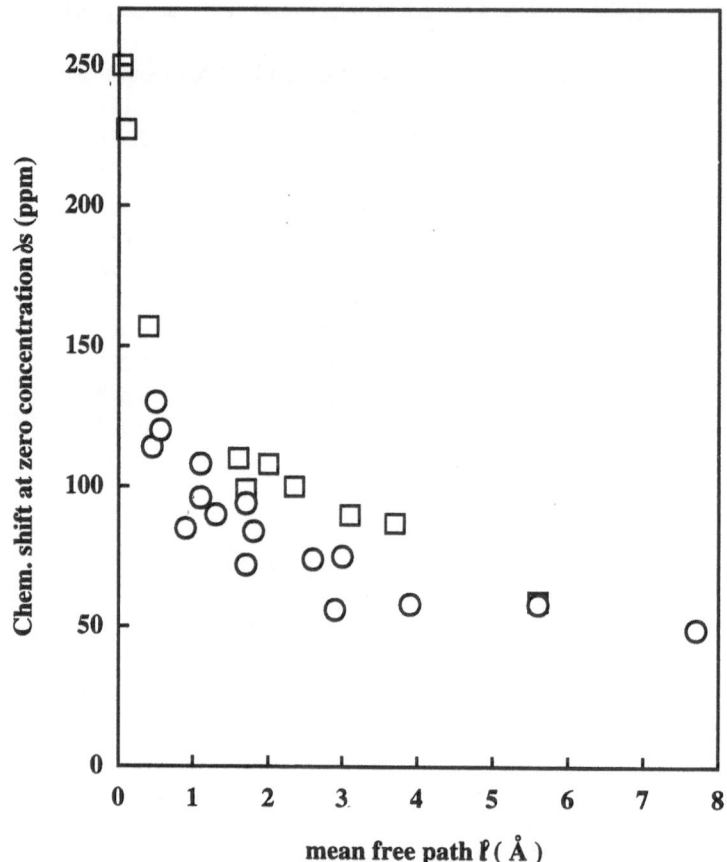

*Figure 2.* Variation of $\delta_S$ with the mean free path, $\ell$, for Xe adsorbed in classical zeolites ( □ ) : A, ferrierite, L, offretite, rho, Y, mordenite, erionite; and in new molecular sieves with low (or zero) cation content ( O ): ZSM-5, ZSM-12, SZM-20, ZSM-23, ZSM-48, EU-1, omega, theta-1, AlPO$_4$-5, AlPO$_4$-11, AlPO$_4$-17, SAPO-31, SAPO-34, MAPO-36, SAPO-37, SAPO-41, VPI-5 (from ref. 7, with permission)

## 3.1. FAST SITE EXCHANGE

For simplicity, exchange between adsorbed and gaseous phase is assumed to be negligible, which is reasonable at room temperature, at least for zeolites.

At 26°C, which is the usual recording temperature, the experimental chemical shift is the average value of the shift of xenon in rapid exchange between a position on the pore surface $A$ (defined by $\delta_a$) and a position in the volume of the cavity or channel (defined by $\delta_v$) (Figure 3):

TABLE 2. Values of $\delta_S$ and mean free parth $\ell$ for various molecular sieves.

| Molecular sieves | $\delta_s$ (ppm) | $\ell$ (Å) | Characteristics of the void spaces accessible to xenon atom |
|---|---|---|---|
| A, ZK4 | 87 | 3.7 | sphere, diameter 11.4 Å with six 8-ring openings of 4-5 Å, depending on the cation |
| L | 90 | 3.1 | onedimensional barrel-shaped channels: 12-ring openings of 7.1 Å, max. diameter 7.4 Å |
| Ferrierite | 227 | 0.1 | c-channel, 10 ring: 5.4x4.2Å |
| | 157 | 0.4 | b-channel, pseudo-sphere, diameter: ≈7 Å with 2 8-ring openings: 4.8x3.5Å |
| Mordenite | 115 | 2.45 | onedimensional channel 12 ring : 6.7 x 7.0 Å |
| | 250 | 0.05 | side-pocket :5.7x2.6x4.8 Å |
| Offretite | 108 | 2 | onedimensional channel 12 ring : 6.7x6.8 Å |
| ZSM-12 | 90 | 1.3 | onedimensional channel, non planar 12 ring 5.5 x 5.9 Å |
| ZSM-20 | 58 | 5.6 | intergrowth of cubic and hexagonal faujasite: normal supercage with four 12 ring apertures of 7.4 Å diameter, "maxi" supercage with five 12 ring openings, 7.1x7.1 Å (for two) and 7.4x6.5Å (for three) and "mini" supercage with three 12 ring openings 7.4x6.5Å) |
| ZSM-23 | 114 | 0.45 | onedimensional channel, 10 ring: 5.2x4.5 Å |
| ZSM-48 | 96 | 1.1 | onedimensional channel, 10 ring: 5.3x5.6 Å |
| EU-1 | 94 | 1.7 | onedimensional channel, 10 ring: 4.1x5.7 Å "side pockets": 6.8x5.8 Å, 8.1 Å deep |
| Omega | 75 | 3.0 | onedimensional channel, 12 ring, 7.4 Å |
| Theta-1, Nu-10, ZSM-22 | 130 | 0.5 | onedimensional channel, 10 ring: 4.4x5.5 Å |
| Erionite | 99 | 1.7 | cavity 6.3x6.3x15.1Å with six openings: 3.6x5.2Å |
| AlPO$_4$-5, SAPO-5 | 56 | 2.9 | onedimensional channel, 12 ring: 7.3 Å |
| AlPO$_4$-8 | 58 | 3.9 | onedimensional channel, 14 ring: 7,9x8,7 Å |
| AlPO4-11, SAPO11 | 120 | 0.56 | onedimensional channel, 10 ring: 3.9x6.3 Å |
| AlPO$_4$-17 | 72 | 1.7 | crionite structure |
| SAPO-31 | 85 | 0.9 | onedimensional channel, non planar 12 ring : 5.3 Å |
| SAPO-34 | 84 | 1.8 | chabazite structure: cavity 6.7x6.7x10 Å with six 8 ring openings 3.8 Å |
| MAPO-36 | 74 | 2.6 | onedimensional channel, 12 ring: 6.5x7.5 Å |
| Y, SAPO-37 | 58 | 5.6 | faujasite structure |
| SAPO-41 | 108 | 1.1 | onedimensional channel, 10 ring: 7.0x4.3 Å |
| VPI-5 | 49 | 7.7 | onedimensional channel, 18 ring: 12.1 Å |

The case of infinite cylinder and sphere, very often encountered in the study of zeolites can be rigourously solved by calculation: Infinite cylinder: $\ell = D_C - D_{Xe}$; sphere $\ell = 0.5 ( D_S - D_{Xe})$, where $D_C$, $D_S$ and $D_{Xe}$ are the diameters of the cylinder, the sphere and the xenon atom respectively. The values of pore size given above are from reference [61].

338

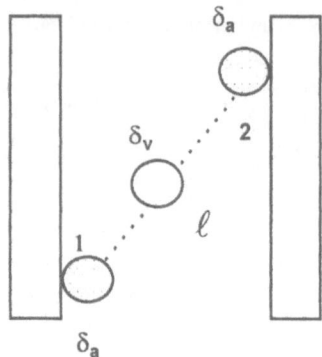

*Figure 3.* Scheme of fast site exchange. (from ref. 7, with permission)

$$\delta = \frac{N_a \delta_a + N_v \delta_v}{N_a + N_v} \tag{6}$$

where $N_a$ and $N_v$ are the numbers of xenon atoms in each state. This equation is valid whatever the xenon concentration; in particular for [Xe] = 0, i.e. when $\delta = \delta_s$. In this case $N_a$ and $N_v$ are the probabilities of there being a Xe atom at the surface ($A$) or in the volume ($V$).

$\delta_a$ depends on xenon-surface interactions. $\delta_v$ is a function of $\delta_a$ and the distance, $\ell$. In fact, $\delta_v = \delta_a$ when the xenon atom leaves the surface; then $\delta_v$ decreases during the journey between sites 1 and 2, whence the need to determine a mean value $<\delta_v> = f(\delta_a, \ell)$.

It is assumed that the Xe atoms are randomly distributed in the pores, which means there are no preferential states. Then $N_a$ and $N_v$ can be expressed in terms of the xenon diameter $D$, the surface area $A$, the void space $V$-$AD$ and the density of xenon either adsorbed on the surface ($\rho_a$) or in the void volume ($\rho_v$), with $\rho_a$ proportional to $\rho_v$ at low concentration ($\rho_a = k \rho_v$). When $\ell$ is not too small, $<\delta_v> \approx 0$ [28].

$$\delta_s = \frac{AD\rho_a}{AD\rho_a + (V\text{-}AD)\rho_v} \quad \delta_a = \frac{\delta_a}{(k\text{-}1)\dfrac{V}{DkA}} = \frac{Dk}{Dk(k\text{-}1) + \dfrac{V}{A}} \delta_a = \frac{a}{b + \dfrac{V}{A}} \delta_a \tag{7}$$

Obviously $\dfrac{V}{A}$ is proportional to $\ell$.

More simply Johnson and Griffiths wrote:

$$\delta_s = \delta_a P + \delta_v (1\text{-}P) \tag{8}$$

with the probability $P$ of a Xe atom being at the cavity wall proportional to the surface area to volume ratio of the cavity. They found a fairly good linear relationship between $\delta_s$ and $A / V$ for a few zeolite structures [33].

Interestingly, Cheung et al. [29] also used the fast site exchange model, even at 144 K, to interpret their [129]Xe NMR data on Xe adsorbed in Y zeolites. They expressed the chemical shift at temperature $T$ and concentration $\rho$ as :

$$\delta(T,\rho) = \frac{b n_a \delta_a + (1 + b \rho_g) \delta_l \rho_g}{1 + b n_a + b \rho_g} \tag{9}$$

where $b = k_a / k_d$ is the ratio of the rate constants for adsorption and desorption of a xenon atom on a site; $n_a = N_a / V$ is the number of adsorption sites in a given volume; $\rho_g$ is the concentration of xenon located in the free volume of faujasite supercages; $\delta_a$ and $\delta_l$ have the same meanings as above.

At low Xe concentration (small $\rho_g$) or for weak adsorption ($b\rho_g \ll 1$), the equation (9) reduces to equation (10):

$$\delta(T,\rho) = \delta_s + \rho_g \frac{(\delta_l - b\delta_s)}{(1 + b n_a)^2} \tag{10}$$

with: $$\delta_s = (b n_a \delta_a)/1 + b n_a = \delta_{(T,\rho = 0)} \tag{11}$$

It can be seen that equation (11) is similar to equation 7.

Equation (10) is of the form of:

$$\delta(T,\rho) = \delta_s + K\rho \tag{12}$$

Indeed the chemical shift variations with of Xe concentration are linear for large-pore zeolites or molecular sieves (Y, beta, ZK4 zeolites; MAPO-36...). One should notice that the slope $K$ is smaller than $\delta_l$, at least by a factor of of $1/(1 + b n_s)^2$, which is consistent with the observation that $K$ is about 0.40 ppm/amagat which can be compared to the slope for the gas phase ($\delta_l = 0.548$ ppm/amagat [27]).

## 3.2. INFLUENCE OF SURFACE CURVATURE.

This model, proposed by Derouane et al., factorizes and scales the physical interaction of a sorbed atom or molecule with the curved surface environment of the micropores, the atom or molecule being represented by its polarizability and the solid (adsorbent) by a frequency-dependent dielectric function. It can be shown that the van der Waals interactions between the trapped species and the curved wall of a sphere can be expressed as the product of two terms accounting for the intrinsic molecular property (polarizability) and extrinsic surface curvature effects [30, 34]:

$$W = -\frac{C}{4d^3}(1-\frac{d}{2a})^{-3}$$

(13)

with $C$ a molecular constant, $d$ the distance from the point molecule to the pore wall (i.e. its van der Waals radius, $r_m$, in the physisorbed state) and $a$ the radius of a pore which is assumed to be spherical. A correction coefficient, $(3/2)^{1/3}$, for a cylindrical pore relative to a sphere with the same curvature is applied.

The first factor in equation 13 is the van der Waals energy for the flat surface ($a = \infty$) and a distance $d$. If we write $W(0)$ for the flat surface ($d/a = 0$), the ratio:

$$W_r = W(d/2a) / W(0) = (1-d/2a)^{-3}$$

(14)

quantifies the surface curvature effect.

Setting $d = m\,r_m$ and $a = n\,r_m$ (with $1< m < n$ and $1 < n < \infty$), the van der Waals energy can also be expressed as:

$$W(m,n) = W(F)\,[m(1-m/2n)]^{-3}$$

(15)

in which $W(F) = -c/4r_m^3$ is the physisorption energy of the adsorbed molecule ($m = 1$ and $d = 2m$) on the flat surface ($n = a = 0$).

Using equation (15) the energy separation, $\Delta W_s$, between the pore center ($m = n$) and the pore wall ($m = 1$) as a function of the cavity size (in units of $r_m$) can be derived:

$$\Delta W_s = W(F)\{(2/n)^3 - [2n/(2n-1)]^3\}$$

(16)

It is seen that the activation energy required to promote a molecule from the pore wall ($m = 1$) to the center of the cavity increases rapidly to reach a maximum equal to about 1.44 $W(F)$ for $n \approx 2.5$ and then decreases asymptotically towards the physisorption value for the flat surface.

An important consequence of this model is that, at temperatures such that $kT < W$, spherical molecules or atoms stick and oscillate with respect to the pore wall and hop as a whole over its corrugation barriers, rather than collide with the wall and be reflected through the cavity or pore, as is generally assumed. The pore or cage center never corresponds to an equilibrium position.

Therefore Derouane and B'Nagy wrote that the chemical shift, $\delta_s$, induced by interaction with the pore wall is directly proportional to the molecular interaction energy with the surface $W_r$ ($A$ being a molecular constant):

$$\delta_s = A\, W_r \qquad (17)$$

They calculated $W_r$ values for xenon adsorbed in a certain number of zeolites and obtained a linear correlation between $\delta_s$ and $W_r$ [34].

These two models, fast site exchange and surface curvature effect, are limiting cases. The real situation should depend on the temperature and the adsorption energy, which depends itself on the pore size (the surface being chemically the same). Everything depends on how values of the xenon-surface interaction are calculated and on the temperature. In the opinion of Derouane *et al.* the xenon atom sticks to the surface and never goes through the cage center, whereas Fraissard *et al.* and Cheung *et al.* consider that the xenon atom samples the void space freely. The energy $W$, calculated by Derouane, corresponds to the chemical shift, $\delta_a$ (equation 6), of the fast site exchange. It is therefore not surprising to obtain a relationship between $W$ and $_{ds}$, since in Fraissard's model $\delta_s$ depends on $\delta_a$ and $<\delta_v>$, $<\delta_v>$ itself being a function of $\delta_a$.

Some other models used a Lennard-Jones potential to describe the physical interaction between a xenon atom and a surface consisting of oxygen atoms.

## 3.3. LENNARD-JONES POTENTIAL CURVES

Ripmeester considered the potential energy between a xenon atom and a spherical shell representing the cage atoms. Oxygen atoms are smeared over this shell with a density similar to that of the A zeolite cage. The Xe-O interaction is described by a Lennard-Jones potential function, $U(r) = -4\varepsilon\,(\alpha^{\,6}/r^6 - \alpha^{\,12}/r^{12})$, with $\varepsilon = 0.8975$ kJ/mole and $\alpha = 3.49$ Å, which can be integrated over the whole sphere [31].

The results are qualitatively the same as those obtained before by Anderson and Horlock for argon adsorbed in slit-shaped pores of active magnesium oxide [35] or by Everett and Powl on a variety of models (infinite cylinder or infinite slit) [36]. It appeared from this previous work that the critical parameter is not the pore size itself but rather the ratio between the size of the pore and that of the adsorbate.

The energy curves are given in terms of the distance $r$, expressed as a fraction of the radius, in Figure 4. Several points emerge:

- the potential energy goes through an absolute minimum for $R = 4.07$ Å, $R$ being the spherical shell radius (which corresponds to $\ell = 0.47$ Å). Therefore, the strength of the binding site is intimately linked to its ability to make efficient contact with xenon atoms. For sphere radii both larger and smaller than 4.07 Å, the Xe-surface interaction is weaker;

- on the other hand, when $R \approx 4.5$ Å ($\ell \approx 0.9$ Å) the potential profile becomes quite flat, and for larger R values a central hump appears. This means that for $R > 4.5$ Å the Xe atom no longer samples the cage uniformly but spends more time near the cage wall than in the center of the cage.

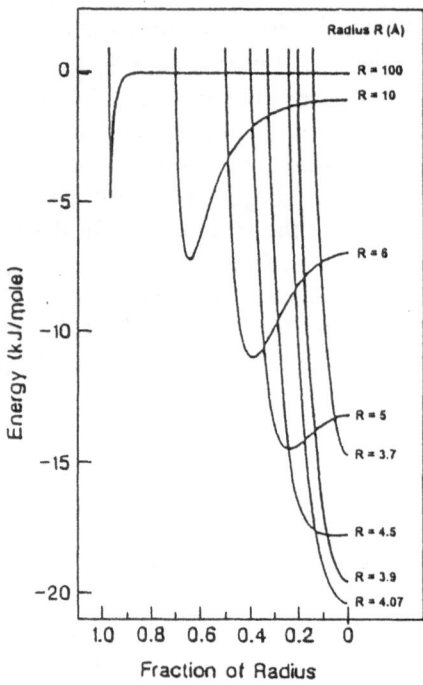

*Figure 4.* Calculated Lennard-Jones potential curves for a xenon atom in a spherical cage shown as a function of cage radius. It is assumed that oxygen atoms are smeared out over the surface at a constant density similar to that of zeolite A (from ref. 31, with permission)

They concluded that for small cages, $\delta_S$ values can be expected to reflect the true void space. However, for large cages, the $\delta_S$ value is a complicated function of sorption energy, void space and temperature, the $\delta_S$ variations with temperature being a means of distinguishing the two types of behaviour.

To determine the chemical shift of Xe adsorbed in such a cage, Ripmeester and Ratcliffe chose a potential curve with a hump which they approximated by a rectangular potential. In this simple picture Xe atoms are either in the potential well (with the chemical shift $\delta_a$) or in the central part of the cage (with the chemical shift $\delta_g$). Writing the chemical shift as the average of $\delta_a$ and $\delta_g$ weighted by the fractional population above ($P_a$) and below ($p_b$) the energy barrier they obtained:

$$\delta = [1 - P_a(r^3/R^3)]\ \delta_a + P_a\ (r^3/R^3)\ \delta_g \tag{18}$$

If it is assumed that $\delta_g = 0$ and at the zero concentration limit, equation (18) reduces to:

$$\delta_s = \{1 - [r^3 exp(-\Delta E/RT)]/[R^3 - r^3 + R^3 exp(-\Delta E/RT)]\}\delta_a \tag{19}$$

When $T$ tends to zero $\delta_s$ tends to $\delta_a$, i.e. the Xe atom sticks to the surface; while when $T$ tends to infinite, $\delta_s$ tends to $2/[1 + R^3/(R^3 - r^3)]\ \delta_a$. Consequently, only the limiting high-temperature $\delta_s$ value should reflects the true void space, as it depends only on $R/r$ and $\delta_a$. At any $T$ value the chemical shift takes an intermediate value which depends on the void space as well as the adsorption energy of the Xe atom on the surface, whence the need to perform variable temperature experiments.

Cheung made calculations similar to those of Ripmeester for a xenon atom trapped between two infinite parallel layers, and obtained similar results: the appearance of a central hump when the distance between the layers increases [32]. He considered that the chemical shift of a xenon atom is directly proportional to the van der Waals interaction energy $(W_r)$, being itself represented by the first term (attractive part) in the Lennard-Jones potentiel, $U$.

When $U$ has a single minimum at the center (micropore of about the same size as the xenon atom), and assuming that the potential near the minimum can be described by a parabola, Cheung showed that the chemical shift increases linearly with increasing temperature. Conversely he obtained a decrease in the chemical shift with increasing temperature when the pore is much larger than the xenon atom. This result is completely consistent with experimental observations [37, 38]

Modelling the potential by a rectangular potential, Cheung also calculated the chemical shift and obtained a general equation:

$$\frac{1}{\delta_S(T)} = A(T) + B(T)\left(\frac{L - 2a_{Xe}}{m}\right) \tag{20}$$

where $A(T)$ and $B(T)$ depend on the well width, the well depth and the temperature; $L$ and $a_{Xe}$ are the pore width and the Xe atom radius, respectively; $m$ is equal to 1, 2 or 3 if the pore is layer-like (1), cylindrical (2) or spherical (3). The expression $(L-2a_{Xe})/m$ is exactly that used by Fraissard's group to calculate the mean free path, $\ell$, of a Xe atom in pores, which can be considered to these simple pore models. Once again, the equation (20) is quite consistent with the empirical relationship obtained by Springuel-Huet et al. [28].

Molecular dynamics simulations, in which the interaction energy is calculated with a Lennard-Jones potential, was used to correlate the chemical shift with the number of binary collisions [39].

## 4. Chemical Shift Anisotropy

Not only the chemical shift but also the lineshape may give interesting information about the local symmetry of the adsorption sites. For example Springuel-Huet *et al.* have shown that, for Xe adsorbed in $AlPO_4$-11 molecular sieves, the ellipsoid-shaped channels of this structure induce a chemical shift anisotropy easily observed in the [129]Xe NMR spectrum (Figure 5) [40]. At low Xe loadings, the anisotropy is axially symmetric and positive, with two of the components of the chemical shift tensor equal ($\sigma_{xx} > \sigma_{yy} = \sigma_{zz}$). When Xe-Xe collisions become more important, $\sigma_{zz}$ increases, corresponding to non-symmetric anisotropy. At high loading the anisotropy is axial but negative, corresponding to the situation: $\sigma_{xx} = \sigma_{zz} > \sigma_{yy}$.

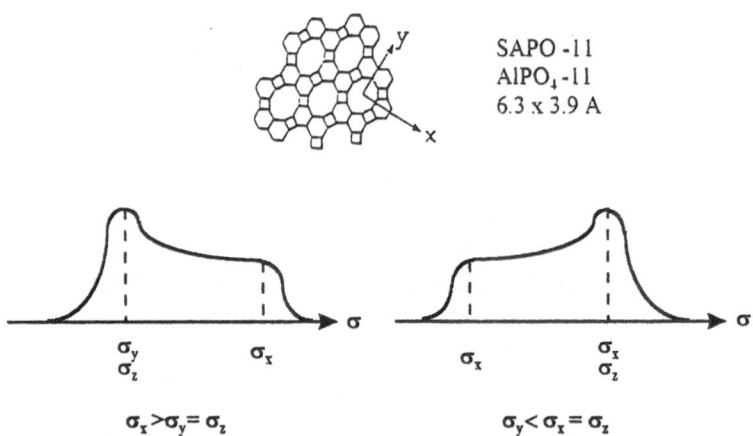

*Figure 5.* Schematic representation of the $AlPO_4$-11 (or SAPO-11) structure and [129]Xe NMR lineshape at low (right) and high (left) Xe loadings, (from ref. 40, with permission)

Other cases of chemical shift anisotropy have been reported, in particular by Ripmeester for Xe trapped in clathrasil compounds [41-43] and Pellegrino *et al.* for Xe adsorbed on Nu-10 zeolite [44].

Finally, Ripmeester and Ratcliffe, studying again the Xe/ $AlPO_4$-11 system, found that at very high Xe loading the anisotropy becomes again non-symmetric as it is for intermediate loadings and, on the other hand, that the two principal components of the shielding tensor vary linearly with Xe loading, the third one being invariant [45]. To explain these data they used a « cell theory »: they consider the $AlPO_4$-11 channels as a succession of small cells which can fit only one Xe atom. They could distinguish three types of sites, in which Xe have one, two or three neighbouring sites occupied by another Xe. Each cell has its own characteristic shielding tensor. Due to rapid exchange between these three types of Xe, the observed tensor is then a dynamic average of the shielding tensor of the three types of site weighted by their statistical probabilities. Assuming a binomial distribution of Xe, they could calculate the three averaged components of the shielding tensor and simulate $^{129}$Xe NMR spectra in terms of Xe loading. The simulated spectra agree well with the experimental ones.

## 5. Influence of Xenon Diffusion

The mobility of Xe in porous systems is usually very high. This is an important parameter which must be taken into account when interpreting the data. Pulsed-field gradient NMR measurements give Xe diffusion rate constants in zeolites (A, Y, ZSM-5) at room temperature of the order of $10^{-8}$ to $10^{-9}$ m$^2$/s [46-49]. In these materials the movement the xenon averages its interactions over a great number (thousands) of zeolites cages and channels on the typical NMR timescale (ca. $10^{-3}$ s). Therefore, depending on experimental conditions, such as the pore size, the particle size or the temperature, exchange with the gas phase may be important and may have a great influence on the NMR spectrum: number of lines, chemical shift, linewidths and relaxation times. For example Figure 6 shows the effect of the particle size on the spectrum for Xe adsorbed in 40 Å

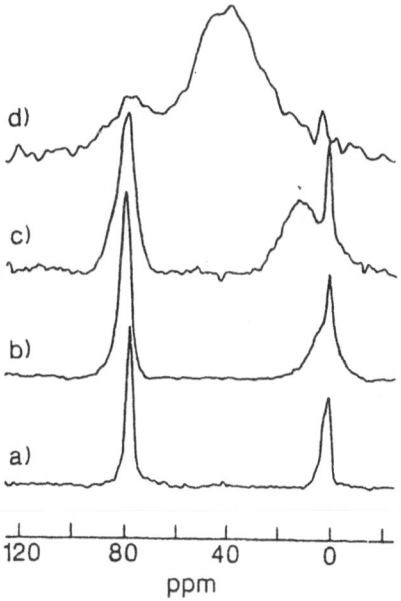

Figure 6. $^{129}$Xe NMR spectra for xenon adsorbed in 40Å Vycor glass of different particle sizes at loading of $9.10^{19}$ Xe atoms/g, (a) D > 250 ; (b) 250 > D > 75 ; (c) 53 > D >38; (d) 38 >D, diameter D in μm (from ref. 49, with permission).

Vycor glass [50]. As the particle decreases, the two lines corresponding respectively to Xe gas (0 ppm) and to Xe adsorbed on the surface (80 ppm) broadens due to exchange between this two kinds of Xe populations. For the smallest particles, an exchange signal appears between the two previous ones.

The high xenon mobility has been used to study macroscopic heterogeneities in assemblages of molecular sieve particles [51-58]. On the other hand it is possible to eliminate intercrystalline diffusion by coating the external surface of the molecular sieves with large molecules. For instance, octamethylcyclotetrasiloxane has been used in the case of faujasite [59-60]

347

## References

1. Ito, T. and Fraissard, J. (1980) in L.V.C. Rees (ed.), *Proceedings of the Fifth International Conference on Zeolites*, Heyden, London, pp 510-515
2. Ripmeester, J. A. and Davidson, D. W. (1980) *Bull. Magn. Reson.* **2**, 139
3. Fraissard, J., Ito, T. and Springuel-Huet, M.-A. (1988) *J. Chim. Phys.* **85**, 747-757
4. Dybowski, C., Bansal, N. and Duncan, T. M. (1991) *Ann. Rev. Phys. Chem.* **42**, 433-464
5. Barrie, P. J. and Klinowski, J. (1992) *Progr. NMR Spectr.* **24**, 91-108
6. Raftery, D. and Chmelka, B. F. (1994) *NMR Basic Principles and Progress* **30**, 111-158
7. Springuel-Huet, M.-A., Bonardet, J.-L. and Fraissard, J. (1995) *Appl. Magn. Reson.* **8**, 427-456
8. Schrobilgen, G. J. (1978) in R.K. Harris and B. E. Mann (eds.), *NMR and the Periodic Table*, Academic Press, London
9. Ramsey, N. F. (1950) *Phys. Rev.* **78**, 699
10. Malli, G. and Fraga, S. (1966) *Theor. Chim. Acta*, **5**, 275
11. Proctor, W. G. and Yu, F. C. (1951) *Phys. Rev.* **81**, 20
12. Hunt, E. R. and Carr, H. Y. (1963) *Phys. Rev.* **130**, 2302
13. Brinkmann, D. and Carr, H. Y. (1966) *Phys. Rev.* **150**, 174
14. Jameson, A. K., Jameson, C. J. and Gutowski, H. S. (1970) *J. Chem. Phys.* **53**, 2310-2321
15. Adrian F. J. (1964) *Phys. Rev.* **136**, A980-A987
16. Jameson, C. J. (1975) *J. Chem. Phys.*, **63**, 5296-5301
17. Kanegsberg, E., Pass, B. and Carr, H. Y. (1969) *Phys. Rev. Lett.* **23**, 572
18. Jameson, C. J., Jameson, A. K. and Cohen, S. M. (1973) *J. Chem. Phys.* **59**, 4540-4546
19. Cowgill, D. F. and Norberg, R. E. (1972) *Phys. Rev.* **B6**, 1636
20. Jameson, C. J., Jameson, A. K. and Cohen, S. M. (1976) *J. Chem. Phys.* **65**, 3401-
21. Jameson, C. J., Jameson, A. K. and Cohen, S. M. (1977) *J. Chem. Phys.* **66**, 5226-5230
22. Jameson, C. J., Jameson, A. K. and Parker, H. (1978) *J. Chem. Phys.* **68**, 3943-3944
23. Jameson, C. J. and Jameson, A. K. (1971) *Molec. Phys.* **20**, 957-959
24. Buckingham, D. and Kollman, P. A. (1972) *Molec. Phys.* **23**, 65-74
25. Jameson, C. J., Jameson, A. K. and Cohen, S. M. (1975) *Mol. Phys.* **29**, 1919-1927
26. Jameson, C. J., Jameson, A. K. and Cohen, S. M. (1976) *J. Chem. Phys.* **65-9**, 3397-3400
27. De Ménorval, L. C., Fraissard, J. and Ito, T. (1982) *J. Chem. Soc. Faraday Trans I* **78**, 403
28. Springuel-Huet, M. A., Demarquay, J., Ito, T. and Fraissard, J. (1988) *Stud. Surf. Sc. Catal.* **37**, 183-189
29. Cheung, T. T. P., Fu, C. N. and Wharry, S. (1988) *J. Phys. Chem.* **92**, 5170-5180
30. Derouane, E. G., André, J. M. and Lucas, A. A. (1987) *Chem. Phys. Lett.* **137**, 336-340
31. Ripmeester, J. A. (1990) *J. Phys. Chem.* **94**, 7652-7656
32. Cheung, T. T. P. (1995) *J. Phys. Chem.* **99**, 7089-7095
33. Johnson, D. W. and Griffiths, L. (1987) *Zeolites* **7**, 484-487
34. Derouane, E. G. and B'Nagy, J. (1987) *Chem. Phys. Lett.* **137**, 341-344
35. Anderson, P. J. and Horlock, R. F. (1969) *Trans. Faraday Soc.* **65**, 251-
36. Everett, D. H. and Powl, J. C. (1976) *J. Chem. Soc. , Faraday Trans. I* **72**, 619
37. Chen, Q. J. and Fraissard, J. (1992) *J. Phys. Chem.* **96**, 1809-1814
38. Springuel-Huet, M. A. and Fraissard, J. (1992) *Zeolites* **12**, 841-845
39. Vigné-Maeder, F. (1994) *J. Phys. Chem.* **98**, 4666-4672
40. Springuel-Huet, M. A. and Fraissard, J.(1989) *Chem. Phys. Lett.* **154**, 299-302
41. Ripmeester, J. A. (1982) *J. Am. Chem. Soc.* **104**, 209-210
42. Davidson, D. W. and Ripmeester, J. A. (1984) in J. L. Atwood and J. E. D. Davies (eds.), *Inclusion Compounds*, Academic Press, New York, vol. 3
43. Ripmeester, J. A., Ratcliffe, C. I. and Tse, J. S. (1988) *J. Chem. Soc., Faraday Trans. I* **84**, 3731-3745

348

44. Pellegrino, C., Ito, T., Gabelica, Z., B'Nagy, J. and Derouane, E. G. (1990) *Appl. Catal.* **61**, L1-L4
45. Ripmeester, J. A., Ratcliffe, C. I. (1995) *J. Phys. Chem.* **99**, 619-623
46. Heink, W., Kärger, J., Pfeifer, H. and Stallmach, F. (1990) *J. Am. Chem. Soc.* **112**, 2175-2179
47. Pfeifer, H., Freude, D. and Kärger, J. (1991) *Stud. Surf. Sci. Catal.* **65**, 397-404
48. Kärger, J. and Pfeifer, H. (1991) *J. Chem. Soc., Faraday Trans.* I **87**, 1989-1992
49. Kärger, J., Pfeifer, H., Stallmach, F. and Spindler, H. (1990) *Zeolites* **10**, 288-292
50. Ripmeester, J. A., Ratcliffe, C. I. (1993) *Anal. Chim. Acta.* **283**, 1103-1112
51. Shoemaker, R. and Apple, T. (1987) *J. Phys. Chem.* **91**, 4024-4029
52. Chen, Q. J. and Fraissard, J. (1992) *Chem. Phys. Lett..* **169**, 595-598
53. Chen, Q. J. and Fraissard, J. (1992) *J. Phys. Chem.* **96**, 1815-1819
54. Ryoo, R., Pak, C., Chmelka, B. F. (1990) *Zeolites* **10**, 790-793
55. Bansal, N. and Dybowski, C. (1990) *J. Magn. Reson.* **89**, 21-27
56. Chmelka, B. F., Pearson, J. G., Liu, S. B., Ryoo, R., de Ménorval, L. C. and Pines, A. (1991) *J. Phys. Chem.* **95**, 303-310
57. Tway, C. and Apple, T. (1992) *J. Catal.* **133**, 42-54
58. Ryoo, R., Pak, C., Ahn, D. H., de Ménorval, L. C. and Figueras, F. (1990) *Catal. Lett.* 7,417-422
59. Ryoo, R., Kwak, J. H. and de Ménorval, L. C. (1994) *J. Phys. Chem.* **98-29**, 7101-
60. Kim, J. G., de Ménorval, L. C., Ryoo, R. and Figueras, F. (1995) *Stud. Surf. Sci. Catal.* **94**, 226
61. Meier, W. H. and Olson, D. H. (1992) *Atlas of Zeolite Structure Types*, 3rd Revised Edition, Butterworth-Heinemann, London

# 129Xe NMR OF PHYSISORBED XENON
*Applications*

M. A. SPRINGUEL-HUET

*Laboratoire de Chimie des Surfaces, URA-CNRS 1428,*
*Université P. et M. Curie, 4 place Jussieu, 75252 Paris Cedex 05,*
*France*

## 1. Introduction

As in the gas phase, most information has been obtained by analysis of the variation of the chemical shift with the xenon concentration, generally at 26°C.

If xenon undergoes additional interactions besides Xe-Xe and Xe-surface, the equation (5) in part I, "Principles" (preceding paper), contains additional terms corresponding to each interaction. For adsorbed xenon Fraissard *et al.* have written the following general equation [1]:

$$\delta = \delta_{ref} + \delta_S + \delta_{Xe} + \delta_{SAS} + \delta_E + \delta_M \qquad (1)$$

$\delta_S$ and $\delta_{Xe}$ have been previously defined (part I). The term $\delta_{Xe}$, which is responsible of the slope of the $\delta = f(N)$ curves, $N$ being the xenon concentration, is proportional to the local Xe density and, therefore, inversely proportional to the internal free volume. When the distribution of Xe-Xe collisions is isotropic, *i. e.* in the case of large pores (Y, ZK4, beta zeolites, etc.), the $\delta = f(N)$ plots are straight lines; if this distribution is anisotropic, as it is for narrow channels with diameters between approximately 4 and 7.5 Å, the slope of this plot increases with $N$.

When there are strong adsorption sites (SAS) in the void space interacting with xenon much more than the cage or channels walls, each xenon spends a relatively long time on these SAS, particularly at low xenon concentration. The corresponding chemical shift will be greater than in the case of a non-charged structure.

When $N$ increases, $\delta$ must decrease if there is fast exchange of the atoms adsorbed on strong adsorption sites with those adsorbed on the other sites. When $N$ is high enough the effect of Xe-Xe interactions becomes again the most important and $\delta$ increases. In this case $\delta_{N \to 0}$, the chemical shift extrapolated to

349

*J. Fraissard (ed.), Physical Adsorption: Experiment, Theory and Applications, 349–368.*
© *1997 Kluwer Academic Publishers.*

350

zero concentration, depends on the nature and number of these strong adsorption sites. Often these latter are cations, more or less charged and sometimes paramagnetic. The theoretical curve is then displaced downfield (Figure 1). The chemical shift difference expresses the effect, $\delta_E$, of the electrical field and, if it exists, $\delta_M$ due to the magnetic field created by these cations.

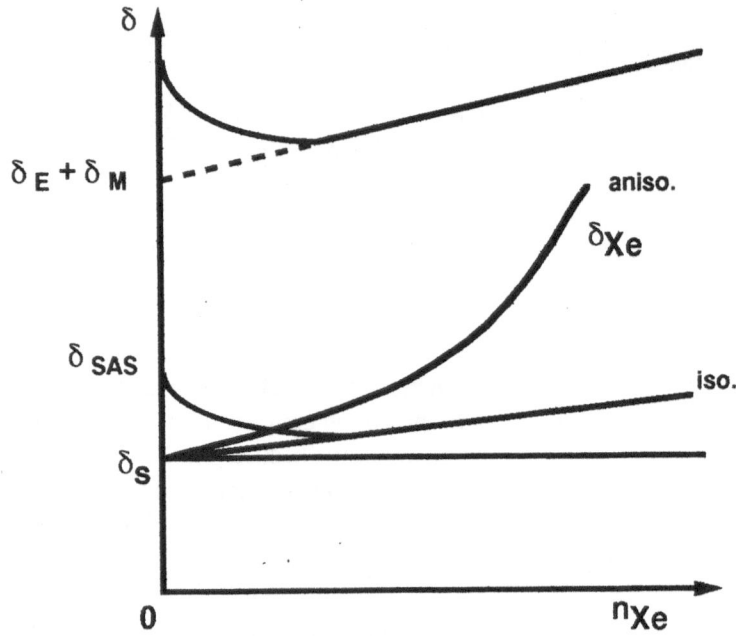

*Figure 1.* Schematic chemical shift variations as a function of xenon concentration associated with the different terms, $\delta_i$.

If strong adsorption sites of different types (for example, various highly charged cations or metal particles) are distributed in solid pores (for example, zeolite supercages), Xe interacts with them, and the chemical shift of xenon is different depending on the presence or the absence of these targets. If the exchange between these various sites is fast with respect to the NMR time-scale, an average signal will be observed, while a multiple signal spectrum corresponds to a slow exchange situation.

Finally, reactive gas (for example, hydrogen on Pt-zeolite, carbon monoxide) may also be adsorbed on the previous samples. In this complex case too, the xenon NMR spectrum gives information about the distribution of the gas.

Study of the different terms of equation (6) leads to a wide range of applications. Several reviews have been already published [2-6]. Some examples are given here.

## 2. Zeolite Structure

### 2.1. ZEOLITES WITH SINGLE TYPES OF VOID VOLUME

When the zeolite structure contains only one type of pore (channels or cavities) accessible to xenon, the spectrum consists of a single line. Since the terms $\delta_S$ and $\delta_{Xe}$ depend on the pore structure, many studies are concerned with the determination of the dimensions and the shape of the pores. It is, of course, very interesting to evaluate these parameters for newly synthesized molecular sieves of unknown structure [7] or to follow the variation of the pore size after specific treatments such as dealumination [8-10]. Not only Y zeolite [11-16] but other less common zeolites or related materials have been investigated: ZSM [17-19], L [19, 20], Omega [21], Theta [22], VPI-5 [23, 24], SAPO-34 [25], SAPO-37 [26, 27], SAPO and AlPO$_4$-5 [27], SAPO and AlPO$_4$-11 [28, 29], etc.

On the other hand, the line intensity is proportional to the number of Xe atoms resonating at the line frequency and corresponding to a given environment. Consequently comparison with the intensity for a reference sample, considered as perfectly crystallized, is a means of measuring the crystallinity of the sample under study. This measurement is interesting especially when the well crystallized domains are too small to give rise to an X-ray diffraction pattern [30].

The case of NaA zeolite has received special attention because the diffusion of xenon between the $\alpha$–cages is very slow due to the blockage of the openings by Na$^+$ cations. The sensitivity of the $^{129}$Xe chemical shift means that signals corresponding to different Xe population densities in the $\alpha$–cages can be resolved (Figure 2) [31].

The distribution of the xenon atoms in the cages is obtained from the intensities of the lines. Various studies of xenon distribution statistics, assuming different models with binomial or hypergeometric distribution, have been carried out [32, 33], sometimes in the presence of other adsorbed molecules [34-36]. Jameson et al. found experimental deviations from these statistical models, explained by attractive Xe-Xe interactions which favour clustering at low to medium loadings [36]. The observed shifts and their temperature dependence are interpreted by using the results of a Grand Canonical Monte Carlo simulation [37].

A cell theory of an interacting gas has been derived by Cheung. This theory is based on the statistical mechanical principle that the equilibrium distribution of $N$ particles among $M$ cells, when both $N$ and $M$ are very large, is given by the most probable distribution, with the constraints that the total number of particles and the average energy remain constant [38]. To describe $^{129}$Xe shifts for Xe adsorbed in NaA zeolites, it was necessary to include, in addition to the attractive interactions between the particles, the repulsive interactions when the Xe atoms begin to fill the α-cages.

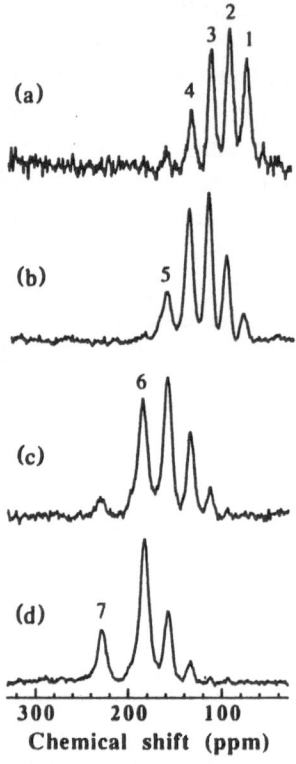

*Figure 2.* Room temperature $^{129}$Xe NMR spectrum of Xe occluded in NaA zeolite at 523 K at different equilibrium pressures: (a) 8 bars, (b) 40 bars, (c) 150 bars and (d) 210 bars. The numbers above the peaks corresponds to the number of Xe atoms per α-cage (from ref. 32, with permission).

With more sophisticated NMR experiments [35, 39, 40] (inversion-recovery or multisite magnetization transfer experiments, two-dimensional exchange NMR) and simulations [37, 41, 42], the dynamics of xenon movement between α−cages can be investigated. For example, Figure 3 shows the changes in the 2D NMR spectra with the increase in the mixing time, during which there is a magnetization exchange between the resolved resonances, from which the intercage xenon exchange dynamics can be characterized. The microscopic rates of intercage motion are related to the adsorption energy of the Xe atoms and the activation energy for their transfer. Larsen *et al.* showed that the dependence of the adsorption energies on Xe cage occupancy reflects the importance of the intracage interactions and is directly related to the cage occupancy distribution. From variable temperature measurements these authors estimated an activation energy of about 60 kJ/mole for the transfer of a Xe atom from one cage to another [35].

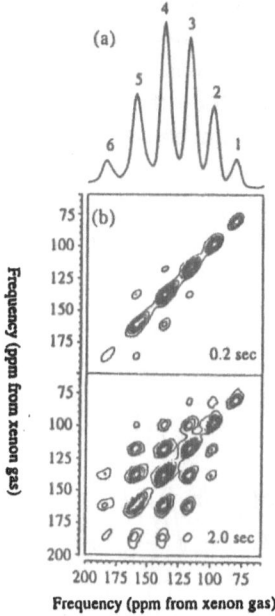

*Figure 3.* $^{129}$Xe 1D NMR spectrum (a) and 2D exchange NMR spectra (b) of Xe adsorbed on NaA zeolite. The latter were recorded using mixing times of 0.2 and 2.0 s. The diagonal peaks correspond to the 1D spectrum. The crosspeaks are a result of intercage motion of Xe during the mixing time of the 2D experiment (adapted from ref. 35, with permission).

## 2.2. ZEOLITES WITH SEVERAL TYPES OF VOID VOLUME

Except in the case of NaA, $^{129}$Xe NMR spectra have as many components as there are different types of void volume in the zeolites, at least if exchange between xenon adsorbed in the different zones is slow on the NMR time-scale. This is the case of ferrierite whose spectrum has two lines corresponding to the two types of channels (*c* and *b*-channels). The *b*-channels of the structure can only contain one Xe atom; therefore, the corresponding chemical shift is independent of the xenon pressure and only the line intensity increases with the filling rate [43].

Very often the number of lines depends on the nature of the cations and the temperature. For example, at room temperature, there are two lines for Na mordenite, corresponding to the channels and the side-pockets, while there is only one for H mordenite, due to rapid exchange between the two sites when the cation is very small (Figure 4) [44, 45]. The coalescence temperature is about

273 K for H⁺ and 370 K for Na⁺. If the cation is Cs⁺, the side-pockets are no longer accessible to xenon, and there is only one line [44]. Analogous results have been obtained with Rho zeolite in which Cs⁺ cations located in the prisms, between cavities, prevent Xe atoms being there [46]

In general, if there are several signals corresponding to various types of void, the line intensities can be used to study the distribution of adsorbed xenon in the different voids for a given total concentration as a function of the temperature and the location of cations. In this way one can obtain interesting information about the structures of zeolites, *e.g.* zeolite intergrowths [43] and crystallinity. The evolution of crystals defects after various treatments can in some cases be followed [23].

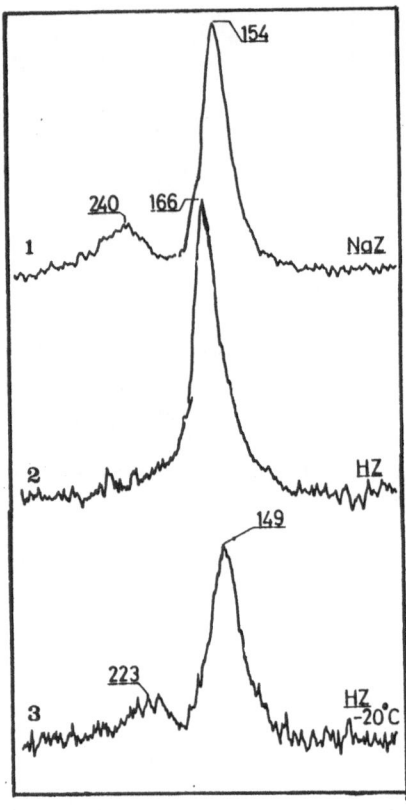

Figure 4. $^{129}$Xe NMR spectra of Xe adsorbed in and Na⁺-mordenite (1) and H⁺-mordenite, at room temperature (2) and -20°C (3) (adapted from ref. 45, with permission).

## 2.3. CATION STUDIES

For most zeolites, the influence of cations on the $^{129}$Xe NMR is negligible at room temperature when the cation is H⁺ or Na⁺. The chemical shift is given by the two terms $\delta_S$ and $\delta_{Xe}$ of equation [6]. However when the cations are more highly charged, the $\delta = f([Xe])$ curves have a characteristic shape due to the $\delta_{SAS}$ term.

The interaction between cations and xenon has been described by simple models ranging from high polarization of the electron cloud [47] to an electron transfer from the xenon atom to the cation [14].

Many studies of X and Y zeolites containing cations like $Mg^{2+}$, $Ca^{2+}$, $Zn^{2+}$, $Cd^{2+}$ and even rare earth cations ($Y^{3+}$, $La^{3+}$, $Ce^{3+}$) have been carried out with different degrees of cation exchange and temperature of thermal treatment (*i.e.* the extent of dehydration of the zeolite). Since Xe atoms can only interact with cations located in the supercages and not with those in the sodalite cages or prisms, the location of cations can be deduced. The migration of cations inside the crystals, between different sites, have been studied in terms of their hydration state. Generally speaking, cations tend to migrate towards the sodalites and prisms when they lose their hydration shell. This migration depends also on the nature and amount of the cation [48-51]. For example, it has been shown that $La^{3+}$ and $Ce^{3+}$ migrate during dehydration while $Y^{3+}$ cations do not. The migration of $La^{3+}$ and $Ce^{3+}$ leads to a charge deficit in the supercages, the surface becoming then hydrophobic, which increases the thermal stability of the zeolites.

The migration of cations between crystals during thermal treatment has also been observed in the case of a mixture of previously dehydrated RbNaX and NaY zeolites [52].

The problem is more difficult for paramagnetic cations ($Ni^{2+}$, $Co^{2+}$). One must consider the $\delta_M$ term, which may be very large, leading to $\delta$ values of several thousand ppm [53, 54].

Another interesting example is the unusual behaviour of $Ag^+$ and $Cu^+$. The chemical shifts observed for CuX and AgX zeolites are small (compared to the $Na^+$ form) and even negative (AgX) [55-58]. The adsorption of Xe in dehydrated AgX is much greater than in NaX. Most remarkably, the shifts decrease with concentration down to negative values, -40 ppm for $[Xe]\rightarrow 0$ (Figure 5) These results have been attributed to specific interactions of xenon with $Ag^+$ cations in the supercages, especially $Ag^+$ in SIII sites [59]. This location of $Ag^+$ allows close contact with Xe which favours electron donation involving the Ag $4d^{10}$ and Xe $5d^0$ orbitals. This process is considered responsible for the observed low frequency.

For CuX and CuY zeolites, the parabolic form of the $\delta = f([Xe])$ curves, which is expected because of the presence of paramagnetic $Cu^{2+}$ ions, is not observed. During dehydration there is at the same time migration of $Cu^{2+}$ from the supercages and partial autoreduction of $Cu^{2+}$ to $Cu^+$ inside the supercages. These very accessible $Cu^+$ cations are able to behave like $Ag^+$ and participate in $3d_\pi$-$5d_\pi$ electron donation [58]. This behaviour is not general for cations of the $nd^{10}$ configuration. Indeed, for $Zn^{2+}$ and $Cd^{2+}$ the $\delta = f([Xe])$ variations are similar to that of $Mg^{2+}$ or $Ca^{2+}$. In this case the higher charge of the cation, responsible for a greater polarization of the electronic cloud and high chemical shift, may compete with electron donation [60].

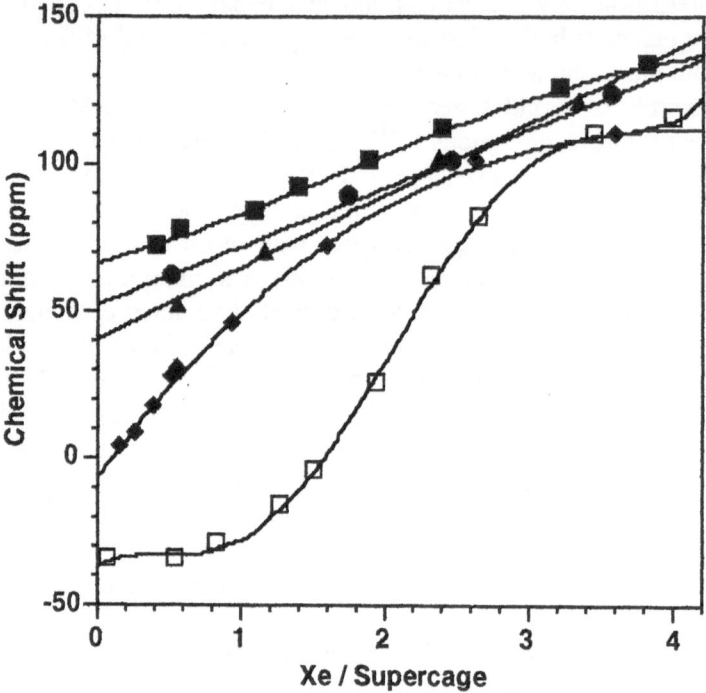

*Figure 5.* $^{129}$Xe NMR chemical shift versus Xe concentration for Ag$_\lambda$X zeolites pretreated at 623 K. λ (%): (■) 0; (●) 20; (▲) 40; (◆) 80 ; (□) 100 (from ref. 56, with permission).

A common and crucial problem when dealuminating zeolites for catalytic purposes is that there are non-framework aluminum species (Al$_{NF}$) in the pores, often difficult to extract. The catalytic properties depend greatly on the chemical nature of these species. A study of several dealuminated ZSM-5 samples has shown that the Xe interactions with Al$_{NF}$ are strongly dependent on the mean free charge of the Al$_{NF}$ rather than on their quantity [61].

## 2.4. COKING AND COADSORBED SPECIES

### 2.4.1. *Coking*

These Al$_{NF}$ play a major role in coke formation during hydrocarbon cracking reactions. After a small coke deposit, it is seen that the Al$_{NF}$ no longer interact with Xe, showing that coke is first deposited on these species [62]. Depending on the location of the coke in the zeolite structure, in the openings of the supercages or in the supercages themselves of a Y zeolite, for example, $\delta_S$ or $\delta_{Xe}$ increases in comparison with a coke-free sample.

When coke is deposited in the openings, it only prevents xenon from diffusing between the cages, the supercage volume being roughly constant (only $\delta_S$ increases).

However, when coke is deposited in the supercages, the reduction of the pore volume results in the increase of both $\delta_S$ and $\delta_{Xe}$ (Figure 6). It is possible to follow the progressive blocking of the pores by the increase in these terms. At high coke loading, a signal at low chemical shift may appear; this has been attributed to Xe adsorbed in mesopores formed by the coke between the zeolite crystallites [62]. Similar studies have been made on HY [63, 64] and on ZSM-5 zeolite [64-67]

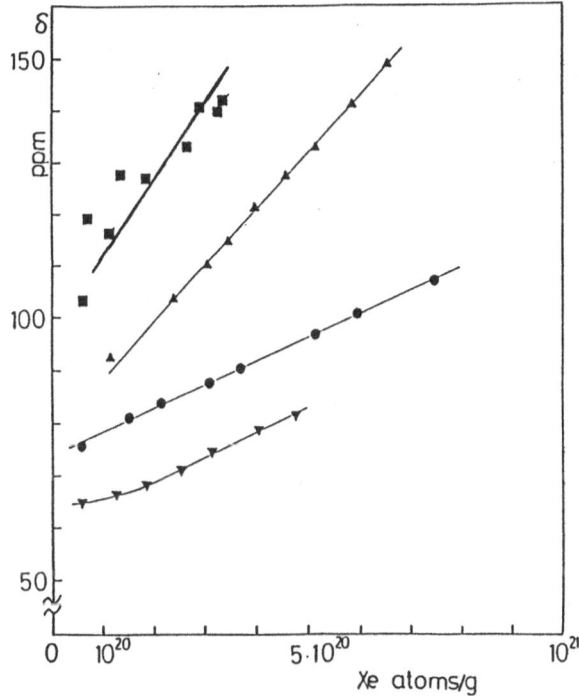

*Figure 6.* $^{129}$Xe NMR chemical shift variations against Xe concentration for H-ZSM-5 at different coke loadings: 0 % (▼), 3 % (●), 10.5 % (▲), 15 % (■) (from ref. 62, with permission)

### 2.4.2. *Water Adsorption*

In the H$_2$O/NaY system, it is easy to differentiate between water adsorbed in regions accessible or not to xenon, to measure the volume of water in pores and to determine the blockage of windows by water molecules. Gédéon *et al.* analysed the variations, in terms of the water content (denoted $C$, the fraction of the saturation amount), of the slopes of the $\delta = f([Xe])$ curves (related to the pore volume) and of $\delta_{N \to 0}$ (related to the mean free path ). They showed that during zeolite dehydration water molecules first leave the supercages (for $1 < C < 0.40$) then the windows, where there are more strongly adsorbed (for $0.4 < C < 0.15$) and finally the sodalite cages and the prisms ($C < 0.15$) (Figure 7) [68].

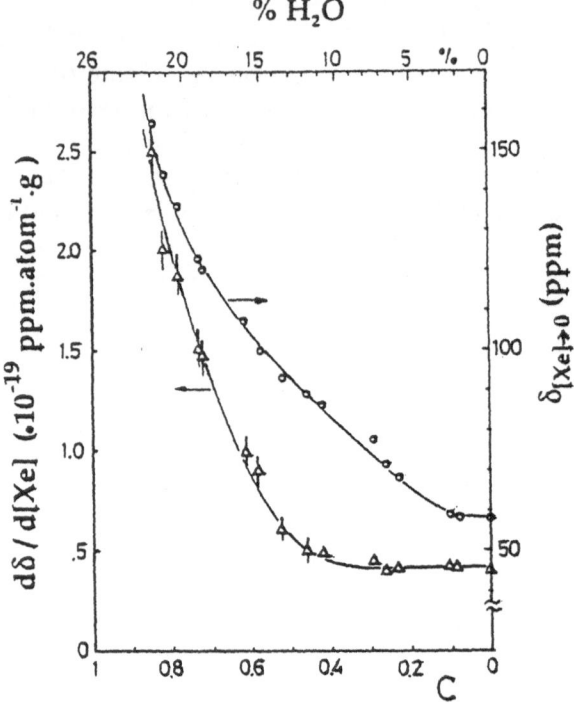

*Figure 7.* Effect of water concentration C on shift $\delta_S$ and on the slope of the $\delta$-N plot. The water concentration is expressed as a percentage (w/w) of anhydrous solid (upper) or as the relative concentration (lower) (from Ref. 68, with permission).

### 2.4.3. *Other Molecules*

In the same way one can study the distribution of organic molecules like benzene [69-72], trimethylbenzene [69, 70], hexamethybenzene [70, 73], n-hexane [69], etc. in NaX or NaY zeolites and observe the aggregation of molecules in the cages. In some cases multiple quantum NMR has been useful [70, 73]. For example, in Figure 8, the presence of several signals in the spectrum after benzene adsorption proves that benzene molecules are distributed heterogeneously in a NaY zeolite sample. Only by thermal treatment at 250°C for 10 hours it is possible to homogenize the distribution [72].

Under certain experimental conditions, it is even possible to study the diffusion of such organic molecules inside crystallites by studying the evolution of spectra with time during their adsorption. This has been done for benzene and n-hexane on ZSM-5 zeolites [74-76]. In the case of a process limited by intracrystalline diffusion, a diffusion coefficient may be determined. Since the chemical shift and the linewidth depend on the local concentration of adsorbate molecules, the simulation of spectra from theoretical intracrystalline concentration profiles allows the determination of diffusion coefficients.

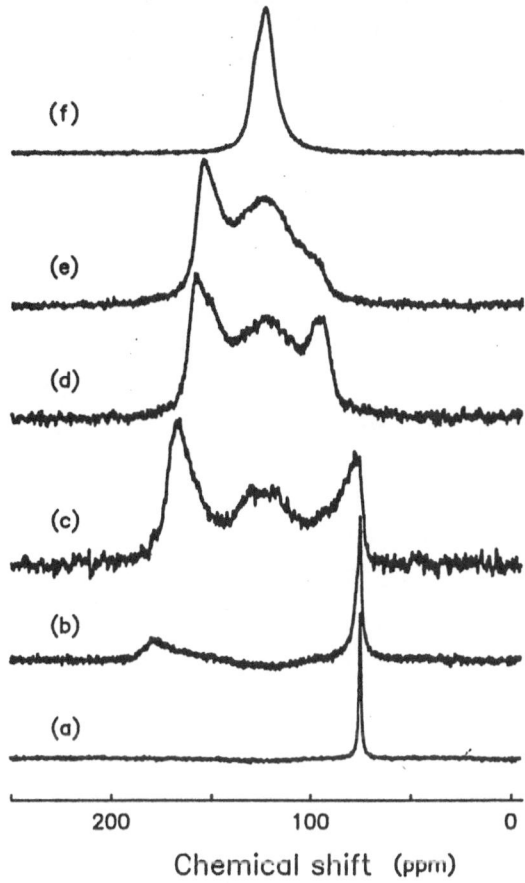

*Figure 8.* $^{129}$Xe NMR spectra of Xe adsorbed in NaY zeolites ($P_{Xe}$ = 300 Torr) with benzene coverage $\theta = 3.0$: (a) immediately after benzene adsorption and after (b) 10 days, (c) 30 days, (d) 90 days, (e) 120 days and, finally, after being heated to 250 °C for 10 hours (f) (from ref. 72, with permission).

## 3. Other microporous solids

$^{129}$Xe NMR spectroscopy has proved to be a valuable tool for studying the microporosity and the symmetry of systems other than zeolites. Some examples follow.

360

## 3.1. CLATHRATES AND CLATHRASILS.

These crystalline compounds are guest-host systems where small molecules are trapped in well defined cages in an ice lattice formed by hydrogen-bonded water molecules (clathrates) or organic molecules (clathrasils). Ripmeester and co-workers were pioneers in the study of such systems. They showed that, depending on the guest molecules, different structures with cages of various geometries exit [77-80].

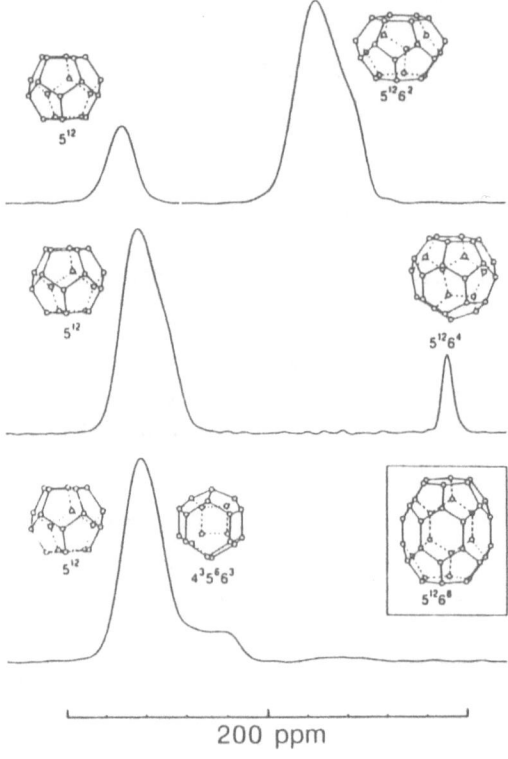

200 ppm

*Figure 9.* Characteristic $^{129}$Xe NMR spectra of Structure I, II and H hydrates recorded at 77 K: Xe structure I hydrate (top), double hydrate of n-butane and Xe (middle), double hydrate of bicyclo-[2.2.2.]oct-2-ene and Xe (bottom). The large cage of structure H hydrate (inset) seems not to be occupied by Xe (from Ref. 80, with permission).

Confinement of xenon atoms in the cavities of these solids of different symmetry has been shown to produce chemical shift anisotropy patterns in the $^{129}$Xe NMR spectra (Figure 9). For instance, subtle structural transformations of clathrate and clathrasils, such as phase transition, can be followed by monitoring changes in the $^{129}$Xe resonance signals. In those systems where $^{129}$Xe atoms are close to the framework protons, proton-xenon cross-polarization allows a signal enhancement and, of course, a discrimination between bulk and occluded xenon.

## 3.2. AMORPHOUS SILICAS.

The rather large space available to Xe between the silica particles appears to suggest that the chemical shifts would be low. On the other hand, the broad distribution of pores sizes should lead to a broad signal. Instead of that, surprisingly, the chemical shift of xenon adsorbed in such systems is very high and the signals are narrow [81, 82], even at low temperature [83]. The chemical shift varies little with the Xe pressure. Low-temperature NMR experiments have shown a marked increase in the chemical shift when the temperature decreases.

The narrowness of the signal, the increasing shift with decreasing temperature and the pressure insensitivity of the shifts have been interpreted as the result of a fast exchange occurring between xenon atoms adsorbed on the silica surface and xenon atoms in the free inter-particle volume. The adsorbed xenon atoms may be located in defects of the aerosil surface or in the small space at the contact point between particles.

Nevertheless, in a series of compressed silicas, the chemical shift follows the general variation observed with zeolites, *i.e.* it increases as the pore sizes decreases with increasing compression.

Terskikh *et al.* obtained analogous data with silica gel [84]. Since the $\delta_S = f(\ell)$ relation cannot be used for these solids, these authors proposed another one to fit their data.

Finally, these NMR features are also observed for the mesoporous M 41S silicas [85]. The surface roughness of these amorphous solids, leading to a stronger Xe-surface interaction, is suspected to be responsible for the high chemical shift. Water adsorption induces a decrease in the chemical shift, as opposed to what it is observed with zeolites; it is assumed that the water molecules mask the roughness to Xe [85]. As before, the same general trend of chemical shift variation with pore size is obtained. The relation of Terskikh *et al* cannot be used with these mesoporous silicas.

## 3.3. AMORPHOUS POLYMERS.

The first $^{129}$Xe NMR spectrum of Xe adsorbed in a polymer was reported by Sefcik *et al.* for polyvinylchloride (PVC) in 1983 [86]. The technique is now widely used and reviews have already been published [87, 88]. The solubility of Xe in most polymers is sufficiently high for signals to be detected. Although Xe is relatively soluble in polymers, generally only the amorphous phase is permeable to the gas.

The relatively low mobility of xenon in those solids allows the investigation of structural heterogeneities which may exist in such systems. Both homopolymers and copolymer blends have been investigated.

[129]Xe NMR signals can be broad and are highly shifted (250 ppm for a PVC polymer, for example). Broad lines reflect a continuous distribution of sites showing the microscopic heterogeneity of the polymer matrix. However, due to high Xe mobility, Stengle and Williamson obtained sharp signals (about 1 ppm wide) for Xe dissolved in low-density polyethylenes [89]. The corresponding chemical shift is about 200 ppm. This value is completely consistent with their measurements of [129]Xe shifts in n-alkanes solutions, which could be described by a "reaction field model". This model assumes that spontaneous electric moments arising in the solute induce a reaction field in the solvent which in turn affects the magnetic shielding of the solute [90]. In a very closely related group of solvents, such as the n-alkanes, the reaction field and hence the shifts vary in a simple way with solvent density.

At temperatures above the glass transition $T_g$, mobile polymer segments allow the xenon atoms to migrate easily between different free volume environments, favouring a motionally narrowed signal [89, 91-94].

[129]Xe NMR is able to probe individual microphases in block polymers and then to follow the densification of polymers after cross-linking [95]. Phase transitions may be also monitored (Figure 10) [96].

The [129]Xe 2D exchange experiment is a particularly powerful tool for measuring xenon diffusivities and exchange between heterogeneous domains of copolymer blends [97, 98]. For example, Figure 11 shows the 2D exchange spectra of [129]Xe adsorbed in a model blend system of polystyrene and polyvinylmethyl ether for different mixing times. The effective xenon diffusion coefficients can be determined by monitoring the relative amplitudes of the cross peaks, which appear when xenon atoms exchange between different phases, with respect to the corresponding diagonal peaks over a range of mixing times.

More recently, xenon optical-pumping techniques have been developed which permit a dramatic enhancement in sensitivity (depending on the Xe pressure and the laser power) with corresponding applications to low-surface area materials [97]. In particular, Raftery et al. obtained [129]Xe NMR spectra of optically pumped Xe adsorbed on polyacrylic acid [98]. From the temperature dependence of the spectrum they deduced the $\delta_a$ term, characteristic of the Xe-surface interaction (95 ppm), finding it to be comparable with that of NaY zeolite and consistent with similar adsorption energies. Experiments on polarization transfer from optically polarized Xe towards other nuclei have also been performed [99, 100].

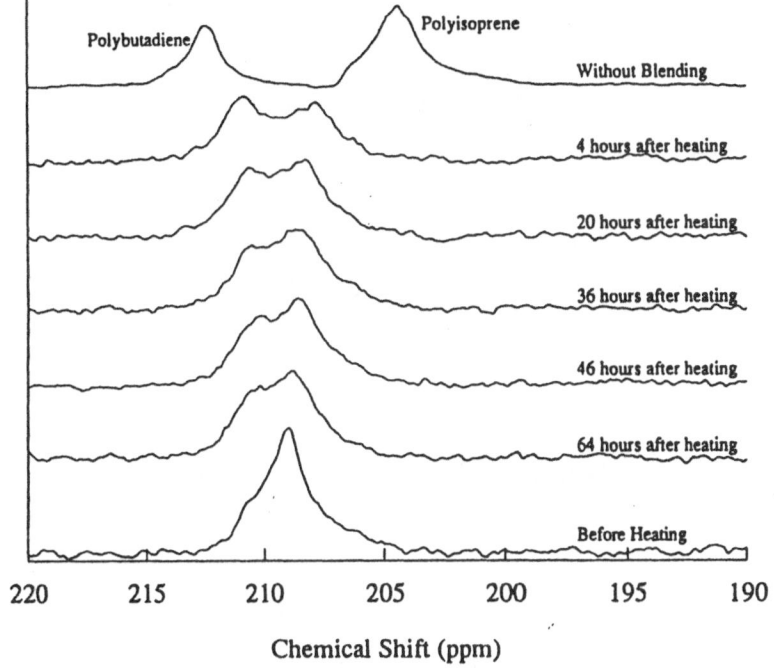

*Figure 10.* $^{129}$Xe NMR spectra of Xe dissolved in a miscible blend (polybutadiene-polyisoprene) (bottom), at various times after phase separation by heating. The top spectrum corresponds to complete phase separation and was obtained by placing physically distinct pieces of the pure polymers in the NMR tube. Redissolution of the components is not complete even after 64 h. (from ref. 94, with permission).

## 3.4. PILLARED CLAYS.

These complex solids are less ordered than zeolites and present a distribution of local environments which leads to broad Xe signals. Moreover, the presence of many impurities favours broad signals. Despite this, there have been a few studies [101-103].

As for xenon adsorbed in amorphous silicas, there is no dependence of the chemical shift upon xenon pressure at room temperature, suggesting no Xe-Xe interaction effect and rapid exchange with the gas phase. No influence of the pillaring of a gelwhite L on the shift has been detected, showing that the interaction of xenon with the alumina pillars is weak; but the chemical shift seems to be related to the interlayer spacing [101, 102]. The situation may, however, be more complex in other cases [103]

To study xenon adsorption on specific sites it is necessary to perform low temperature $^{129}$Xe NMR experiments. Barrie *et al.* have interpreted the low-temperature data in terms of Xe phase transitions [102].

364

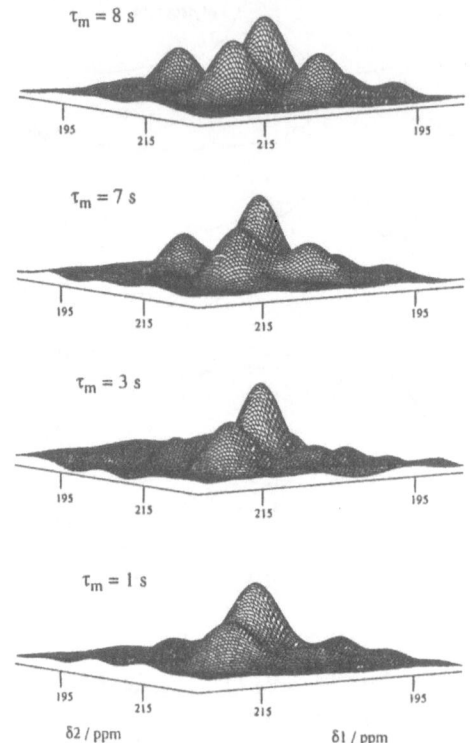

$\tau_m = 8$ s

195    215    215    195

$\tau_m = 7$ s

195    215    215    195

$\tau_m = 3$ s

195    215    215    195

$\tau_m = 1$ s

195    215    215    195

$\delta 2$ / ppm        $\delta 1$ / ppm

*Figure 11.* 2D exchange NMR spectra of $^{129}$Xe adsorbed in a lamellar blend system of polystyrene and polyvinylmethyl ether for different mixing times, indicated on the figure (from ref. 95, with permission).

## 4. Conclusion

$^{129}$ Xe NMR spectroscopy has become an important and widely used technique for characterizing a complex chemical systems and host phases. The sensitivity of the NMR parameters of Xe to local environments makes it an excellent probe, thanks to the different types of interaction that Xe atoms can undergo including the symmetry or geometry of these environments. We have mentioned a few materials which have been studies, but there are many others, such as coals and amorphous carbons, liquid crystals, proteins, or even heteropolyoxometallates. This list will probably come to include more and more types of materials owing to the development of more sophisticated techniques, such as optical pumping, cross-polarisation, multidimensional NMR, etc...

References

1. Springuel-Huet, M.-A., Demarquay, J., Ito, T. and Fraissard, J. (1988) *Stud. Surf. Sc. Catal.* 37, 183-189
2. Reisse, J. (1986) *New Journal of Chemistry* 10, 665-672
3. Dybowski, C., Bansal., N. and Duncan, T. M. (1991) *Ann. Rev. Phys. Chem.* 42, 433-464
4. Barrie, P. J. and Klinowski, J. (1992) *Progr. NMR Spectr.* 24, 91-108
5. Raftery, D. and Chmelka, B. F. (1994) *NMR Basic Principles and Progress* 30, 111-158
6. Springuel- Huet, M.-A., Bonardet, J.-L. and Fraissard, J. (1995) *Appl. Magn. Reson.* 8, 427-456
7. Benslama, R., Fraissard, J., Albizane, A., Fajula, F. and Figueras, F. (1988) *Zeolites* 8, 196-198
8. Shertukde, P. V., Hall, W. K.and Marcelin, G. (1992) *Catalysis Today*, 15, 491
9. Cotterman, R. L., Hickson, D. A., Cartlidge, S., Dybowski, C., Tsiao, C. and Venero, A. F. (1991) *Zeolites* 11, 27-34
10. Chao, K. J. and Shy, D. S. (1993) *J. Chem. Soc. Faraday Trans. I* 89, 3841-3845
11. Ito, T. and Fraissard, J. (1982) *J. Chem. Phys.* 76, 5225-5229
12. Ito, T. and Fraissard, J. (1986) *J. Chim. Phys.* 83, 441-445
13. Ito, T. and Fraissard, J. (1987) *J. Chem. Soc., Faraday Trans. I* 83, 451-462
14. Cheung, T. T. P., Fu, C. M. and Wharry, S. (1988) *J. Phys. Chem.* 92, 5170-5180
15. Cheung, T. T. P. and Fu, C. M. (1989) *J. Phys. Chem.* 93, 3740-3747
16. Pires, J., Brotas de Carvalho, M., Ribeiro, F. R., B'Nagy, J. and Derouane, E. G. (1993) *Appl. Catal.* A95, 75-86
17. Alexander, S. M., Coddington, J. M. and Howe, R.F. (1991) *Zeolites* 11, 368-370
18. Chen, Q. J., Springuel-Huet, M.-A., Fraissard, J., Smith, M. L., Corbin, D. R. and Dybowski, C. (1992) *J. Phys. Chem.* 96, 10914-10917
19. Cheung, T. T. P. (1990) *J. Phys. Chem.* 94, 376-380
20. Ito, T., de Ménorval, L. C., Guerrier, E. and Fraissard, J. (1984) *Chem. Phys. Lett.* 111, 271-274
21. Ito, T., Springuel-Huet, M. A. and Fraissard, J. (1989) *Zeolites* 9, 68-74
22. Pellegrino, C., Ito, T., Gabelica, Z., B'Nagy, J. and Derouane, E. G. (1990) *Appl. Catal.* 61, L1-L7
23. Chen, Q. J., Fraissard, J., Cauffriez, H., and Guth, J.-L. (1991) *Zeolites* 11, 534-538
24. Martens, J. A., Feijen, E., Lievens, J. L., Grobet, P. J. and Jacobs, P. A. (1991) *J. Phys. Chem.* 95, 10025-10031
25. Chen Q. J., Ph. D. thesis, University Paris 6, France (1990)
26. Dumont, N., Ito, T. and Derouane, E.G. (1989) *Appl. Catal.* 54, L1-L6
27. Chen, Q. J., Springuel-Huet, M. A. and Fraissard, J. (1989) *Chem. Phys. Lett.* 159, 117-121
28. Springuel-Huet, M. A., Fraissard, J. (1989) *Chem. Phys. Lett.*, 154, 299-302
29. Ripmeester, J. A. and Ratcliffe, C.I. (1995) *J. Phys. Chem.* 99, 619-623
30. Springuel-Huet, M. A., Ito, T. and Fraissard, J. P. (1984) *Stud. Surf. Sc. Catal.* 18, 13-21
31. Samant, M. G., de Ménorval, L.-C., Dalla Betta, R. A. and Boudart, M. (1988) *J. Phys. Chem.* 92, 3937-3938
32. Chmelka, B. F, Raftery, D., McCormick, A. V., de Ménorval, L.-C., Levine, R.D. and Pines, A. (1991) *Phys. Rev. Lett.* 66, 580-583 and 67, 931
33. Jameson, C. J., Jameson, A. K. Gerald II, R. and de Dios, A. C. (1992) *J. Chem. Phys.* 96, 1676-1689 and 1690-1697
34. Ryoo, R., de Ménorval, L.-C., Kwak, J. H. and Figueras, F. (1993) *J. Phys. Chem.* 97, 4124-4127

366

35. Larsen, R. G., Shore, J., Schmidt-Rohr, K., Emsley, L., Long, H., Pines, A., Janicke, M. and Chmelka, B. F. (1993) *Chem. Phys. Lett.* **214**, 220-226
36. Jameson, A. K., Jameson, C. J., de Dios, A. C., Oldfield, E., Gerald II, R. E. and Turner, G. L. (1995) *Solid State Nuclear Magnetic Resonance* **4**, 1-12
37. Jameson, C. J., Jameson, A. K. Lim, H. M. and Baello, B. I. (1994) *J. Chem. Phys.* **100**, 5965-5976
38. Cheung, T. T. P. (1993) *J. Phys. Chem.* **97**, 8993-9001
39. Jameson, A. K., Jameson, C. J. and Gerald II, R. E. (1994) *J. Phys. Chem.* **101**, 1775-1786
40. Nivarthi, S. S. and McCormick, A. V. (1994) *Microporous Mater.* **3**, 47-53
41. van Tassel, P. R., Davis, H. T. and McCormick, A. V. (1992) *Mol. Phys.* **76**, 411-432
42. van Tassel, P. R., Davis, H. T. and McCormick, A. V. (1993) *J. Chem. Phys.* **98**, 1-10
43. Ito, T., Springuel-Huet, M. A. and Fraissard, J. (1989) *Zeolites* **9**, 68-74
44. Ripmeester, J. (1984) *J. Magn. Reson.* **56**, 247-253
45. Ito, T., de Ménorval, L.-C., Guerrier, E. and Fraissard, J. (1984) *Chem. Phys. Lett.* **3**, 271-274
46. Tsiao, C.-J., Kauffman, J. S., Corbin, D. R., Abrams, L., Carrol Jr, E. E. and Dybowski, C. (1991) *J. Phys. Chem.* **95**, 5586-5591
47. Fraissard, J. and Ito, T. (1988) *Zeolites* **8**, 350-361
48. Ito, T. and Fraissard, J. (1987) *J. Chem. Soc., Faraday Trans. I* **83**, 451-
49. Gédéon, A. and Fraissard, J. (1994) *Chem. Phys. Lett.* **219**, 440-444
50. Chen, Q. J., Ito, T. and Fraissard, J. (1991) *Zeolites* **11**, 239-243
51. Kim, J.-G., Kompany, T., Ryoo, R. Ito, T. and Fraissard, J. (1994) *Zeolites* **14**, 427-432
52. Chen, Q. J. and Fraissard, J. (1990) *Chem. Phys. Lett.* **169**, 595-598
53. Gédéon, A., Bonardet, J.-L. and Fraissard, J. (1989) *J. Phys. Chem.* **93**, 2563-2569
54. Bonardet, J.-L., Gédéon, A. and Fraissard, J. (1995) *Stud. Surf. Sci. Catal.* **94**, 139-146
55. Gédéon, A., Burmeister, R., Grosse, R., Boddenberg, B. and Fraissard, J. (1991) *Chem. Phys. Lett.* **79**, 191-195
56. Grosse, R., Burmeister, R., Boddenberg, B., Gédéon, A. and Fraissard, J. (1991) *J. Phys. Chem.* **95**, 2443-2447
57. Gédéon, A., Bonardet, J.-L. and Fraissard, J. (1993) *J. Phys. Chem.* **97**, 4254-4255
58. Gédéon, A., Bonardet, J.-L., Lepetit, C. and Fraissard, J. (1995) *Solid State NMR* **5**, 201-212
59. Grosse, R., Gédéon, A., Watermann, J., Fraissard, J. and Boddenberg, B. (1992) *Zeolites* **12**, 909-915
60. Springuel-Huet, M.-A., Bonardet, J.-L., Gédéon, A. and Fraissard, J. (1996) *Langmuir* to be published
61. Chen, Q. J., Guth, L., Serve, A., Caullet, P. and Fraissard, J. (1991) *Zeolites* **11**, 798-803
62. Barrage, M.-C., Bonardet, J.-L. and Fraissard, J. (1990) *Catal. Lett.* **5**, 143-154
63. Miller, J. T., Meyers, B. L. and Ray, G. J. (1991) *J. Catal.* **128**, 436-446
64. Liu, S.-B., Prasad, S., Wu, J.-F., Ma, L.-J., Yang, T.-C; Chiou, J.-T., Chang, J.-Y. and Tsai, T.-C. (1993) *J. Catal.* **142**, 664-671
65. Barrage, M.-C., Bauer, F., Ernst, H., Fraissard, J., Freude, D. and Pfeifer, H. (1990) *Catal. Lett.* **6**, 201-208
66. Tsiao, C., Dybowski, C., Gaffney, A. M. and Sofranko, J. A. (1991) *J. Catal.* **128**, 520-525
67. Bonardet, J.-L., Barrage, M.-C., Fraissard, J., Kubelkova, L., Novakova, J., Ernst, H. and Freude, D. (1992) *Collect. Czech. Chem. Commun.* **57**, 733-738
68. Gédéon, A., Ito, T. and Fraissard, J. (1988) *Zeolites* **8**, 376-380
69. de Ménorval, L.-C., Raftery, D., Liu, S.-B., Takegoshi, K., Ryoo, R. and Pines, A. (1990) *J. Phys. Chem.* **94**, 27-31
70. Chmelka, B. F., Pearson, J. G., Liu, S.-B., Ryoo, R., de Ménorval, L.-C. and Pines, A. (1991) *J. Phys. Chem.* **95**, 303-310

71. Liu, S.B., Ma, L.-J., Lin, M.-W., Wu, J.-F. and Chen, T.-L. (1992) *J. Phys. Chem.* **96**, 8120-8125

72. Wu, J.-F., Chen, T.-L., Ma, L.-J., Lin, M.-W. and Liu, S.B. (1992) *Zeolites* **12**, 86-94

73. Ryoo, R., Liu, S.-B., de Ménorval, L.-C., Takegoshi, K., Chmelka, B., Trecoske, M. and Pines, A. (1987) *J. Phys. Chem.* **91**, 6575-6577

74. Kärger, J., Pfeifer, H., Wutscherk, T., Weitkamp, S. E. J. and Fraissard, J. (1992) *J. Phys. Chem.* **96**, 5059-5063

75. Springuel- Huet, M.-A., Nosov, A., Kärger, J. and Fraissard, J. (1996) *J. Phys. Chem.* **100**, 7200-7203

76. Springuel- Huet, M.-A., Nosov, A., Ngokoli-Kekele, ,P., Kärger, J., Dereppe, J.-M. and Fraissard, J. in M. L. Occelli (ed.), Marcel Dekker, Inc., New York , to be published

77. Ripmeester, J. A. and Davidson, D. W. (1981) *J. Molec. Struc.* **75**, 67-72

78. Ripmeester, J. A. (1982) *J. Am. Chem. Soc.* **104**, 289-290

79. Ripmeester, J. A., Ratcliffe, C. I. and Tse, J. S. (1988) *J. Chem. Soc., Faraday Trans. I* **84**, 3731-3745

80. Ripmeester, J. A. and Ratcliffe, C. I. (1990) *J. Phys. Chem.* **94**, 8773-8776

81. Conner, W. C., Weist, E. L., Ito, T. and Fraissard, J. (1989) *J. Phys. Chem.* **93**, 4138-4142

82. Conner, W. C., Weist, E. L., Fraissard, J., Ito, T., Chen, Q. J. and Springuel-Huet, M.-A. in A. B. Mersmann and S. E. Scholl (eds.), *Proc. 3rd Int. Conf. on Fundamentals of Adsorption*, United Engineering Trustees Inc., New York, pp. 977-988

83. Cheung, T. T. P. (1989) *J. Phys. Chem.* **93**, 7549-7552

84. Terskikh, V. V., Mudrakovski, I. L. and Mastikhin, V. M. (1993) *J. Chem. Soc., Faraday Trans. I* **89**, 4239-4243

85. Springuel-Huet, M.-A., Fraissard, J., Schmidt, R, Stöcker, M. and Conner, W. C. to be published by the Royal Chemical Society in the Proceedings of the COPS IV meeting held in Bath (UK) Sept. 1996

86. Sefcik, M. D., Schaefer, J; Desa, J. A. E. and Yelon, W. B. (1983) *Polym. Prepr., Am. Chem. Soc., Div. Polym. Chem.* **24**, 85-86

87. Miller, J. B. (1993) *Rubber Chemistry and Technology* **66**, 455-461

88. Walton, J. H. (1994) *Polymers & Polymer Composites* **2**, 35-41

89. Stengle, T. R. and Williamson, K. L. (1987) *Macromolecules* **20**, 1430-1431

90. Rummens, F. H. A. (1975) *Chem. Phys. Lett.* **31**, 596

91. Miller, J. B., Walton, J. H. and Roland, C. M. (1993) *Macromolecules* **26**, 5602-5610

92. Kentgens, A. P. M., van Boxtel, H. A., Verweel, R.-J. and Veeman, W. S. (1991) *Macromolecules* **24**, 3712-3714

93. Kennedy, G. J. (1990) *Polymer Bulletin* **23**, 605-608

94. Walton, J. H., Miller, J. B., Roland, C. M. and Nagode J. B. (1993) *Macromolecules* **26**, 4052-4054

95. Tomaselli, M., Meier, B. H., Robyr, P., Suter, U. W. and Ernst, R. R. (1993) *Chem. Phys. Lett.* **205**, 145-152

96. Mansfeld, M. and Veeman, W. S. (1993) *Chem. Phys. Lett.* **213**, 153-157

97. Raftery, D., Long, H., Meersmann, T., Grandinetti, P. J., Reven, L. and Pines, A. (1991) *Phys. Rev. Lett.* **66**, 584-587

98. Raftery, D., Reven, L., Long, H., Pines, A., Tang, P. and Reimer, J. A. (1993) *J. Phys. Chem.* **97**, 1649-1655

99. Bowers, C. R., Long, H. W., Pietrass, T., Gaede, H. C. and Pines, A. (1993) *Chem. Phys. Lett.* **205**, 168-170

100. Long, H. W., Gaede, H. C., Shore, J., Reven, L., Bowers, C. R., Kritzenberger, J., Pietrass, T. and Pines, A. (1993) *J. Am. Chem. Soc.* **115**, 8491-8492

101. Fetter, G., Tichit, D., de Ménorval, L.-C. and Figueras, F. (1990) *Appl. Catal.* **65**, L1-L4

102. Barrie, P. J., McCann, G. F., Gameson, I., Rayment, T. and Klinowski, J. (1991) *J. Phys. Chem.* **95**, 9416-9419

103. Yamanaka, S., Takahama, K. I., Kunii, K. and Shiotani, M. (1994) *Clay Sc.* **9**, 149-158

# SOLID/LIQUID INTERACTION ON HYDROPHILIC/HYDROPHOBIC ADSORBENTS:

# SORPTION, MICROCALORIMETRIC AND SAXS EXPERIMENTS

IMRE DÉKÁNY

*Department of Colloid Chemistry Attila József University*

*H-6720 Szeged, Hungary.*

## 1. Introduction

In pure liquids the solid-liquid interfacial interactions are quantitatively characterized by heats of immersion. The energy of the surface may be adequately characterized, however, by immersing surfaces of different hydrophobicities into the same liquid and measuring heats of immersion. Likewise, when applying the same liquid pair (e.g. a polar-apolar binary liquid) to surfaces of various polarities, different adsorption selectivities are observed as a function of surface modification. Hydrophilic and hydrophobic silica gel, clay minerals of lamellar structure and their derivatives hydrophobized by cationic surfactants are eminently suitable for this purpose. An enormous advantage of lamellar adsorbents as opposed to compact or porous ones is that in the given liquid or mixture the lamellae move apart, the mineral swells, since the adsorbing liquid molecules penetrate among the lamellae (interlamellar adsorption or intercalation). Interlamellar expansion is easily monitored by X-ray diffraction and from the Bragg reflexions the interlamellar liquid volume can be calculated may be compared with adsorption capacities determined in liquid sorption experiments. The advantage of applying parallel techniques is that the simple X-ray diffraction experiment offers a control for liquid sorption.

However, S/L interfacial free energy functions may also be calculated from adsorption excess isotherms on the basis of Gibbs' isotherm equation. These data are combined with calorimetric data and, based on the second law of thermodynamics, the entropy functions characterizing the adsorption exchange process are also given. Surface thermodynamic potential functions are listed for adsorbents with different surface energies (clay minerals and hydrophobic clays). These data quantitatively characterize the themodynamical properties of the solid/liquid interfacial layer. Simple Bragg reflexions, however, cannot be measured in dispersions of spherical particles. In this case interparticle interactions and states of aggregation in pure liquids and in binary mixtures were determined by the SAXS technique. The value of correlation length, the Porod

369

*J. Fraissard (ed.), Physical Adsorption: Experiment, Theory and Applications, 369–406.*

constant, the specific surface area of the particles and the distance distribution function were therefore calculated for the system of the given composition. These SAXS parameters are in a direct correlation with the adsorption layer composition and the rheological and optical properties of suspensions, characterizing the solid/liquid interaction.

The introduction of the SAXS technique for studying suspensions of pure liquids and mixtures also yields new information on the structural parameters of disperse systems and helps determine how these data can be influenced by modifying solid/liquid interfacial interactions and the composition of the bulk phase.

## 2. Adsorption on binary liquid mixtures and enthalpy of displacement on solids

The sorption exchange process taking place at the solid/liquid interface may be described in thermodynamically exact terms when the activities of the interfacial layer and of the bulk phase are known. In accordance with the exchange equilibrium at the solid/liquid interface, the liquid sorption equilibrium constant is given by the following formula:

$$(1) + r(2)^s \rightleftharpoons (s)^s + r(2)$$

$$K = \frac{x_1^s f_1^s (x_2 f_2)^r}{(x_2^s f_2^s)^r x_1 f_1} \tag{1}$$

Assuming that $f_1^s/f_2^s \cong 1$, i.e. the activity coefficients of the interphacial layer compensate for each other [1,2]:

$$K' = \frac{x_1^s (a_2)^r}{(x_2^s)^r a_1} \tag{1/a}$$

If the activity data of the bulk phase are known, the value of K' can be calculated at a given value of $r = V_{m,2}/V_{m,1}$ by means of computer iteration. The Redlich-Kister equation has proven perfectly reliable for calculating the activities: its application allows the calculation of the activity coefficients of components 1 and 2 and, on this basis, functions $a_1 = f(x_1)$ and $a_2 = f(x_2)$ can be given [3,4].

The applicability of equation 1/a is demonstrated in Figure 1. It is revealed by the adsorption equilibrium diagrams that when considering the adsorption of a liquid pair made up of components significantly different in polarity, the value of the equilibrium constant K' decreases with increasing hydrophobicity of the surface [5,6].

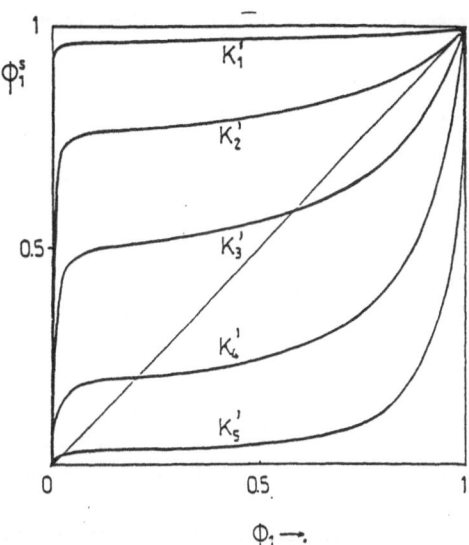

*Figure 1.* Adsorption equilibrium diagramms on S/L interface in methanol-benzene mixtures at different equilibrium consants. $K_1 = 10^3$, $K_2 = 10^2$, $K_3 = 10$, $K_4 = 1$, $K_5 = 10^{-1}$, Calc. after Eq. (1/a)

Calculations regarding the composition of the adsorption layer and the determination of adsorption capacity and the heterogeneity of the surface have been discussed in several publications by Schay and Nagy [1,2,7], Kipling and Everett [8-10] and later by the Polish adsorption school [11-15]. The energetics of the adsorption displacement process taking place in the adsorption layer have been the subject of considerably fewer publications and these mostly deal with the adsorption of dilute solutions. The change in enthalpy accompanying the adsorption exchange process in organic liquids on apolar surfaces has been studied in detail by Findenegg et al. [16-17], Groszek [17] and Denoyel and Rouquerol [18]. Studies published by Allen and Patel [20] and by Woodbury and Noll [21, 22] already include a simultaneous analysis of adsorption excesses and calorimetric data. Billett, Everett and Wright [23] determined heats of immersion wetting in a system of benzene-cyclohexane/activated carbon in the entire range of mixing, parallel with the determination of adsorption excess isotherms.

Our present work gives a parallel analysis of the liquid sorption excess isotherms and adsorption exchange enthalpy isotherms of benzene-n-heptane and methanol-benzene mixtures on adsorbents with polar and apolar surfaces. Systems with U- or S-shaped excess isotherms were selected. Our aim is to examine the connection between adsorption excess isotherms and enthalpy isotherms for the various isotherm types.

The determination of enthalpy changes occurring in the course of flow microcalorimetric measurements absolutely necessitate a simultaneous analysis of the material balance and enthalpy balance of adsorption. These relationships, expounded according to either the adsorption layer model or the Gibbs model of adsorption excess amounts are suitable for the exact determination of changes in exchange enthalpy observable in flowing systems and for the correct interpretation of the sorption exchange process.

## 2.1. THE ADSORPTION LAYER MODEL

When binary mixtures are adsorbed, the composition of the adsorption layer formed by the effect of surfacial forces is usually different from the composition of the bulk phase equilibrated with it. The liquid phase fully wetting a solid surface may therefore be divided into an adsorption layer directly contiguous with the adsorbent and a bulk phase equilibrated with the adsorption layer, in stationary as well as in flowing systems. It has to be noted, however, that there is no way for the exact demarcation of the adsorption layer from the bulk phase, and that no reservations are made as to whether the surfacial layer consists of one or several layers. Thus, by adsorption layer we mean the totality of the molecules located within the range of surfacial forces. The range of surfacial forces, i.e. the thickness of the adsorption layer may also be a function of the composition of the layer.

Knowing molar enthalpies and material amounts, the total enthalpy change occurring in the course of the adsorption exchange process ($\Delta_d H$) may be calculated by formulating the difference between enthalpy terms effective at the beginning and at the end of the displacement process. The driving force of the displacement is that the flowing liquid enforces a new equilibrium, corresponding to its own composition, on the interfacial layer and the flowing bulk phase. For slow flows and ideal binary mixtures:

$$\Delta_d H = (n_o - \bar{n}_o)\, h + n\,^c h + n_1^s h_1^s + n_2^s h_2^s = \bar{n}\,^c h - n_1^s h_1^s - n_2^s h_2^s \qquad (2)$$

The equation above may be extended for non-ideal mixtures in the following way:

$$\Delta_d H = (n_o - \vec{n}_o)\, h + n^c h + h^e(x_1) + n_1^s h_1^s + \tfrac{s}{2} h_2^s + h^{se}(x_1^s) - \vec{n}^c h - h^e(x_1) - n_1^s h_1^s - n_2^s h_2^s - h^{se}(x_1^s)$$

$$(3)$$

In equations (2) and (3) $\bar{n}_o$ and $n_o$ are the moles of liquid mixture befor and after the displacement in the reservoir, $\bar{n}_c$ and $n_c$ are the moles of liquids in the sorption cell, $h_1^s$ and $h_2^s$ are the molar enthalpies of components 1 and 2, respectively, in the adsorption layer; h is the molar enthalpy of the bulk phase; the function $h^e(x_1)$ is the mixing excess enthalpy of the bulk phase; and $h^{se}(x_1^s)$ is the excess enthalpy of the interfacial layer. The latter term results from the non-ideal behaviour of the molecules present in the adsorption layer and is therefore dependent upon composition.

Thus the exchange enthalpy measurable by calorimetry is the total of the enthalpy changes due to changes in the composition of the interfacial layer and the bulk phase, plus the excess enthalpies of these same phases. The enthalpy terms relative to the bulk phase in the equation above may be determined in separate experiments without adsorbent but in the same free cell volume by calorimetry, because — in addition to the heat effect of displacement in the interfacial layer — an enthalpy of mixing ($\Delta H_{mix}$) caused by the exchange process itself must be taken into consideration in the bulk phase as well, since, strictly speaking, flowing of a liquid through a cell cannot be accomplished without mixing [24-26].

Equation (3) may then be formulated in the following way:

$$\Delta_d H = \Delta n_1^s h_1^s + \Delta n_2^s h_2^s + \Delta H^{se} + \Delta H_{mix} \qquad (4)$$

If the behaviour of the bulk phase is ideal, $\Delta H_{mix} = 0$. In non-ideal mixtures, however, it is advisable to correct the gross change in enthalpy (measurable by calorimetry) by the values of $\Delta H_{mix}$:

$$\Delta_d H' = \Delta_d H - \Delta H_{mix} = \Delta n_1^s h_1^s + \Delta n_2^s h_2^s + \Delta H^{se} \qquad (5)$$

The so-called corrected exchange enthalpy, obtained after substracting the heat effect of mixing, gives information only on the change in enthalpy associated with the adsorption exchange process occurring within the interfacial layer. It is revealed by equations (5) that it is the difference of the molar adsorption enthalpies of the pure components that will decisively determine the change in enthalpy produced by the displacement process.

Addition of values of $\Delta_d H'$ calculated according to equation (5) starting from $x_1 = 0$ and proceeding towards $x_1 \rightarrow 1$ yields the integral exchange enthalpy isotherm:

$$\sum_{x_1=0}^{x_1} \Delta_d H' = \Delta_{12} H = f(x_1) \qquad (6)$$

In the case of reversible exchange processes described by equation (6), it can therefore be formulated also for integral exchange enthalpies: $\Delta_{21} H = -\Delta_{12} H$.

By introducing summation according to equation (6), i.e. using equation (5) describing integral exchange enthalpy covering a range of compositions from $x_1 = 0$ to an arbitrary value of $x_1$:

$$\Delta_{21}H = n_1^s \left( h_1^s - \frac{1}{r} h_2^s \right) + \Delta_{21}H^{se}(\phi_1^s) \qquad (7)$$

and the change the enthalpy of displacement process involving compositions from $x_1 = 0$ to $x_1 = 1$ [24-26]:

$$\Delta_{21}H_t = n_{1,0}^s \left( h_1^s - \frac{1}{r} h_2^s \right) + \Delta_{21}H^{se}(\phi_1^s) \qquad (8)$$

As the composition of the bulk phase changes, that of the interfacial layer $(x_1, \Phi_1^s)$ will also change according to the adsorption equilibrium distribution. Knowing the adsorption excess isotherm the composition of the adsorption layer can be calculated by the following equation. If the size of the molecules present in the mixture is not uniform, i.e. $r \neq 1$, the volume fraction of the interfacial layer is [24-26]:

$$\phi_1^s = \frac{n_1^s}{n_{1,0}^s} = \phi_1 + \frac{r n_1^{\sigma(n)}}{n_{1,0}^s (x_1 + r x_2)} \qquad (9)$$

then, from equations (7)-(9):

$$\Delta_{21}H = \Delta_{21}H_t \phi_1^s + \Delta_{21}H^{se} \qquad (10)$$

The equation above allows the determination of the excess enthalpy of the interfacial layer, provided that the adsorption excess isotherm of the given system is known.

## 2.2. THE CLASSIFICATION OF ENTHALY ISOTHERMS

In the case of adsorption of ideal or quasi-ideal mixtures on adsorbents with polar or apolar surfaces, usually U-shaped excess isotherms are obtained. On adsorbents with polar surfaces, in the case of liquid pairs made up of components with a large difference in polarity (e.g. alcohol-benzene), the polar component is preferentially adsorbed on the surface and again U-shaped excess isotherms are obtained. Thus in these systems the composition of the interfacial layer $(\Phi_1^s$ or $x_1^s)$ increases monotonously as a function of equilibrium composition, consequently — according to equation (7) — the enthalpy isotherm $\Delta_{21}H$ will also be a monotonously increasing function (Fig. 2.A).

When alcohol-benzene mixtures are adsorbed on adsorbents with low surface energies [25], S-shaped excess isotherms are measured. When changes in enthalpy accompanying the adsorption displacement process are examined in these systems, the integral exchange enthalpy isotherms usually do not increase monotonously but possess a backward section. After having reached adsorption azeotropic composition $\phi_1^a$, changes in the bulk concentration ($\Delta x_1$) are accompanied by endothermic heat effects and these changes do not follow changes in the composition of the surfacial layer (Figs. 2.B).

For S-shaped excess isotherms $\Phi_1^s$ is nearly constant in a relatively wide concentration range and at $n_1^{\sigma(n)}=0$, $\Phi_1^s=\Phi_1^a=\Phi_1$, i.e. adsorption azeotropic composition appears. In this case the value of $\Delta_{21}H$ changes very little in the middle section of the enthalpy isotherm, where $\Phi_1^s\approx$constant; then at the azeotropic composition a maximum is observed ($\Delta_{21}H^a$). Finally, after $n_1^{\sigma(n)}$ has reversed signs, the function $\Delta_{21}H=f(x_1)$ decreases monotonously, and if this endothermic section is not very long, $\Delta_{21}H_t$ is still exothermic at $x_1=1$ [25].

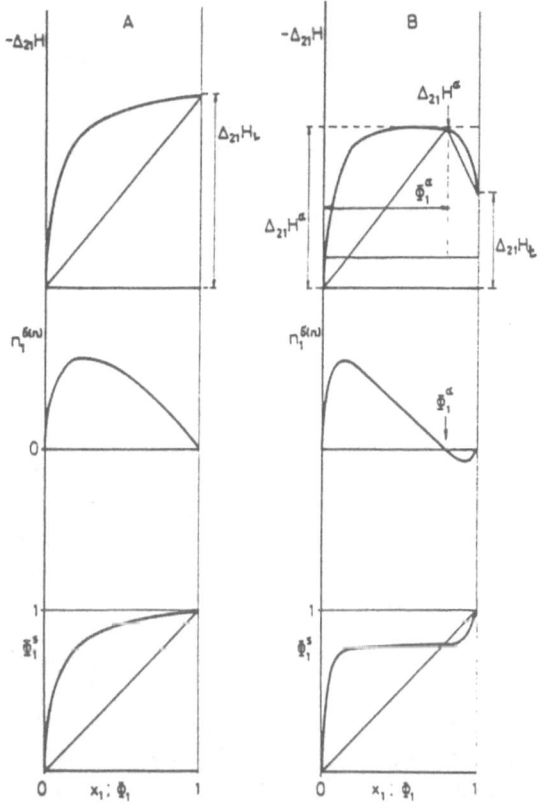

*Figure 2.* Classification of the enthalpy of displacement isotherms. A: U-shaped excess isotherm with corresponding enthalpy of displacement isotherm, B: S-shaped excess isotherm with corresponding enthalpy of displacement isotherm.$\phi_1^s=f(\phi_1)$ the adsorption equilibrium diagram

If adsorption azeotropic composition appears as early as at the beginning of the mixture series, then the endothermic section is more and more extensive and $\Delta_{21}H_t$ is already endothermic at $x_1=1$ [25]. Since the isotherms $\Delta_{21}H=f(x_1)$ do not increase or decrease monotonously in the case of S-shaped excess isotherms, it is not possible to calculate the surface composition $\phi_1^s$ from the ratio $\Delta_{21}H/\Delta_{21}H_t$. In this case the function has to be transformed so as to yield another function with a course identical with $\Phi_1^s$ [24-26].

## 2.3. COMBINATION OF ADSORPTION EXCESS ISOTHERMS AND ENTHALY ISOTHERMS

### 2.3.1. Determination of adsorption capacity

The adsorption of binary systems is described by the Ostwald-de Izaguirre equation which establishes a relationship between the specific reduced adsorption excess amount $(n_1^{\sigma(n)})$ and the material amount in the interfacial layer $(n^s=n_1^s+n_2^s)$ [1,2,8]:

$$n_1^{\sigma(n)} = n_1^s x_2 - n_2^s x_1 = n_1^s - n^s x_1 = n^s(x_1^s - x_1) \qquad (11)$$

The adsorption volume filled by the components of the mixture being adsorbed on a solid surface is: $v^s=n_1^s V_{m,1}+n_2^s V_{m,2}$, in general:

$$\Sigma n_i^s V_{m,i} = V^s \qquad (12)$$

where $V_{m,i}$ is the partial molar volume of the components in the adsorption layer. This equation is formally identical with the description of the so-called pore filling model, still it is also applicable for planar non-porous adsorbents, since the thickness of the adsorption layer is determined solely by the adsorption forces. Thus the adsorption volume $v^s$ designates the volume falling within the range of adsorption forces; at certain regions of the isotherm its value may be nearly constant, but it may also be a function of equilibrium composition.

If molecular sizes within the adsorption space are not identical, i.e. $r=V_{m,2}/V_{m,1}\neq1$ and $v^s=n_{1,0}^s V_{m,1}$, equation (12) may also be formulated for $n_{1,0}^s=n_1^s+rn_2^s$. Knowing the values of the integral exchange enthalpy, the adsorption capacity $(n_1^s)$ and the molar adsorption entalpy of the layer $(h_1^s, h_2^s)$, a combination of equations (7) and (9) and the substitution $n_1^s=n_1^{\sigma(n)}+n^s x_1$ (for an ideal adsorption layer and for the case $\Delta_{21}H^{se}=0$) yield:

$$\frac{\Delta_{21}H}{n_1^{\sigma(n)}} = h_1^s - \frac{1}{r}h_2^s + n^s(h_1^s - \frac{1}{r}h_2^s)\frac{x_1}{n_1^{\sigma(n)}} \tag{13}$$

The ideal behaviour of the adsorption layer also means that $h_1^s$ and $h_2^s$ are independent of the concentration of the mixture, a condition that is rarely met.

Equation (13) is also a linear function, the intersection of which is $b = h_1^s - h_2^s/r$, its slope is $S = n^s(h_1^s - h_2^s/r)$, i.e. adsorption capacity is $n^s = S/b$. Equation (13) was first applied for the interpretation of flow microcalorimetric measurements (on dilute solutions only) by Woodbury and Noll [21].

In the case of S-shaped excess isotherms, the value of the adsorption azeotropic composition $x_1^a$ is known. At the azeotropic point $n_1^{\sigma(n)} = 0$, therefore $x_1 = x_1^a$, according to equation (9). Since $x_1^s = n_1^s/n^s$, adsorption capacity $n^s$ can be divided into two terms with the help of the following equations: $n_1^s = n^s x_1^a$ and $n_2^s = n^s(1-x_1^s)$

Taking into consideration this equations and $(n_{1,0}^s = n_L^s + rn_2^s)$, adsorption capacity relative to the pure component is: $n_{1,0}^s = n^s x_1^a + rn^s(1+x_1^a)$

The free energy of displacement can be calculated in analogy to Eq. (7):

$$\Delta_{21}G = \phi_1^s n_{1,0}^s(g_1^s - g_2^s/r) + G^{se}(\phi_1^s) \tag{14}$$

Free energy of wetting is the isothermic reversible work done when an S/L interface forms and is calculated from the Gibbs equation [2,16-18]:

$$\Delta_{21}G = \sigma - \sigma_2^0 = -RT\int_{a_1=0}^{a_1}\frac{n_1^{\sigma(n)}}{x_2 a_1}da_1 \tag{15}$$

where $\sigma_2^0$ is the S/L interfacial energy in pure component 2 and $\sigma$ is at activity $a_1 = x_1 f_1$ in the mixture, respectively.

Knowing the liquid-vapour equilibrium data, the activities can be calculated ($a_i = f_i x_i$). Combining Eqs. (14) and (9) gives (16) if, in a certain concentration range of the adsorption layer, $G^{se} \approx 0$. This assumption can be justified with Eq. (14) because $\Delta_{21}G = f(\phi_1^s)$ gives a linear function in a wide concentration range:

$$\frac{\Delta_{21}Gx_1}{n_1^{\sigma(n)}} = (g_1^s - g_2^s/r) + n^s(g_1^s - g_2^s/r)\frac{x_1}{n_1^{\sigma(n)}} \tag{16}$$

## 2.4.    MATERIALS AND METHODS

Adsorbent used for adsorption and microcalorimetric measurements are as follows:
1.   Reanal silicagel (R-silicagel, Hungary), its specific surface area is $a^s=537$ $m^2g^{-1}$.
2.   Kieselguhr (K-60, Merck, Germany), $a^s=439$ $m^2g^{-1}$.
3.   Hydrophobic kieselguhr (K-60-$C_{18}$), its surface is modified with $CH_3$-$(CH_2)_{17}$-$SiCl_3$, $a^s=435$ $m^2g^{-1}$.
4.   Charcoal (Chemviron F-400, N.J., USA), $a^s=835$ $m^2g^{-1}$.
5.   Na-illite, natural layer silicate (Füzérradvány, Hungary), $d\leq2$ $\mu m$, $a^s=48$ $m^2g^{-1}$.
6.   HDP-illite organocomplexes: purified illite made organophilic through treating with hexadecylpyridinium-chloride to different extent [28].
     The above adsorbent were used for adsorption measurement in binary liquids of methanol, benzene and n-heptane.
7.   Dodecylammonium-vermiculite ($C_{12}$-$NH_3^+$-vermiculite) and dodecyldiammonium-vermiculite ($C_{12}$-$(NH_3^+)_2$-vermiculite were prepared from Na-vermiculite [30].
     The hydrophobic vermiculites were used for adsorption from n-butanol-water solution.
8.   Hydrophobic $SiO_2$ (R-972, Degussa, Germany) hydrophobized by trimethylchlorosilane ($a^s=123$ $m^2g^{-1}$) was used for adsorption, rheological optical and SAXS experiment in benzene-n-heptane mixtures.

Excess adsorption isotherms for the mixtures $n_1^{\sigma(n)}=f(x_1)$ were determined at $298\pm0.05$ K, putting 0.2-0.5 g of adsorbent and 10 $cm^3$ of mixture into a test tube. The change in concentration was determined by liquid interferometry. Standard deviation of measurements was between $\pm4.4$-$\pm8.8$ %, depending on the composition.

Enthalpy change due to the change in the composition of the surface layer was determined by isothermic flow microcalorimetry (LKB 2107 instrument) at $298\pm0.01$ K. Flow rate of the mixture was 15 $cm^3$/h. The integral displacement enthalpy isotherms ($\Delta_{21}H=f(x_1)$) were determined with the so-called cumulative method. The mixing term ($\Delta H_{mix}$) was measured separately and the necessary corrections were made [24-26].

The immersion wetting enthalpies ($\Delta_w H=f(x_1)$) in pure liquids and mixtures were also measured by the LKB 2107 microcalorimeter in an immersion cell.

Small angle X-ray scattering was measured by $CuK_\alpha$ X-radiation ($\lambda=0.154$ nm) in a PW 1830 generator. The primary beam was directed through a Ni-filter into a compact Kratky camera, type KCEC/3, in which the suspension samples in 1 mm capillary tube were used, and measurements were done in He athmosphere for 4-6 hours. The intensity of the scattered radiation  was measured by a proportional detector (slit width 100 $\mu m$) controlled by a PW 1710 microprocessor, and SDC (Scattering Data Controlling) programm, in an angle range of $2\theta=0.05$-$7°$.
The viscosity of the suspensions were studied with HAAKE Rotovisco RV-20/CV 100 system in low shear ($D=0$-100 $s^{-1}$) range.

## 2.5. ENTHALPY OF DISPLACEMENT ISOTHERMS ON RIGID, POROUS ADSORBENTS

### 2.5.1. *U-shaped excess isotherms and the enthalpy isotherms*

Adsorption excess isotherms of benzene(1)-n-heptane(2) mixtures are shown in Figure (3). The isotherm is U-shaped. Adsorption capacities relative to the pure components $(n^s_{1,0})$ were determined from the Schay-Nagy extrapolation method [1,2,5-7] (Table 1). These data were needed for the comparative evaluation of enthalpy isotherms.

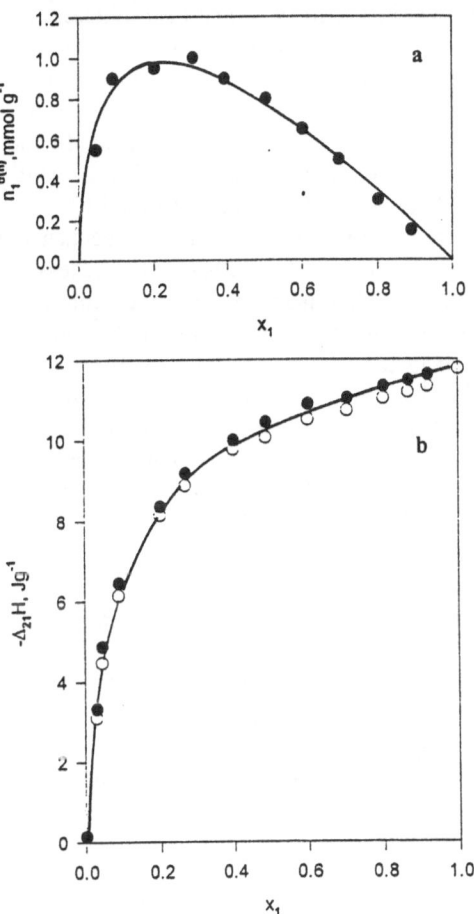

*Figure 3.* Adsorption excess(a) and enthalphy of displacement(b) isotherms in benzene(1)- n-heptane(2) mixture on R-silica gel.
On (b)the (o) adsorption,(●)desorption branche.

Enthalpy of displacement isotherms is parallel presented in Figure 3. Calorimetric data as well as enthalpy isotherms corrected for the enthalpy effect of mixing according to equation (5) are given. It can be established that after correction by $H_{mix}$, the adsorption exchange process is reversible within the limits of experimental error. The relationship between the shape of adsorption excess isotherm and that of enthalpy isotherm is immediately apparent from the Figure 3.

In the case of benzene-n-heptane mixtures the excess isotherm has no linear region: parallel with increasing the concentration of component 1 in the bulk phase, the composition of the interfacial layer changes continuously. This is well reflected by the enthaply isotherm shown in Figure 3, which also indicates a gradual, step-by-step heat exchange and, correspondingly, a gradual process of displacement. Enthalpy changes accompanying the full exchange of the components are listed in Table 1. These data reveal that the full exchange of n-heptane for benzene on the surface of silica gel is an exothermic process and results of -11.3 J $g^{-1}$ of heat. In the case of an exchange of molecules in the reverse direction, on the other hand, identical but endothermic heat effects are obtained (● experimental points of Fig. 3.b).

Adsorption excess isotherm of methanol(1)-benzene(2) mixtures are shown on Figure 4. Pretreatment of silica gel by methanol was necessary in order to eliminate the effect associated with the chemisorption of methanol, so that data reflecting only physical adsorption be measured. In the case of highly preferential adsorption, about 89-90% of the total exchangeable amount of heat ($\Delta_{21}H_t$) is produced at the initial section of the isotherms. In that range of composition where the excess isotherm of the methanol(1)-benzene(2) liquid pair is linear (and here the composition of the interfacial layer is $x_1^s$=constant), some heat effect of exchange is still observable. This means that the composition of the interfacial layer is still changing; these changes are too small to be adequately monitored by our analytical methods but are readily detected by calorimetry. This observation is a direct experimental proof for a theoretical statement by Rusanov [27]: acccording to this author, $(\delta x_1^s/\delta x_1)_{T,P}>0$, i.e. strictly speaking, the composition of the interfacial layer may not be constant within the linear section of the isotherm.

The excess isotherm determined on the surface of graphitized carbon has an inverse U-shape, i.e. the adsorption excess for polar methanol is negative within the entire concentration range (Figure 5). Accordingly, when the amount of methanol in the mixture is increased from $x_1=0$ to $x_1=1$, the effects measured are always endothermic, since the displacement of benzene from the surface is an energy consuming process.

381

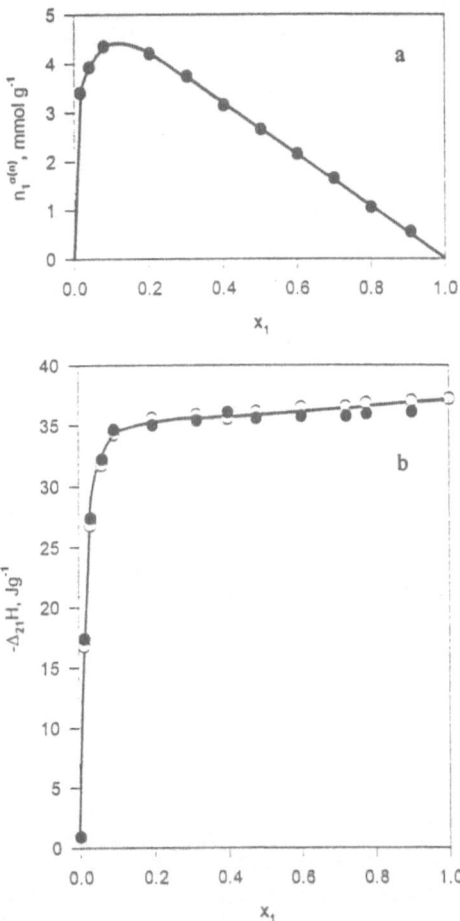

*Figure 4*. Adsorption excess(a) and enthalphy of displacement(b) isotherms in metanol(1)-benzene(2) mixture on R-silica gel.On (b)the (o) adsorption, (●) desorption branche.

## 2.5.2. *S-shaped excess isotherms and enthalpy isotherms*

An excess isotherm determined in a methanol-benzene mixture on silica gel is U-shaped. If, however, the surface is modified by octadecyldimethyl chlorosilane, in methanol-benzene mixtures an S-shaped excess isotherm is obtained, the azeotropic point of which is $x_1^a$=0.615 (Figure 6). When the displacement process is started from benzene, it can be seen that up to a molar fraction of $x_1$=0.6, the incorporation of methanol into the adsorption layer results in a change in enthalpy of ca. 1.35 J g$^{-1}$.

382

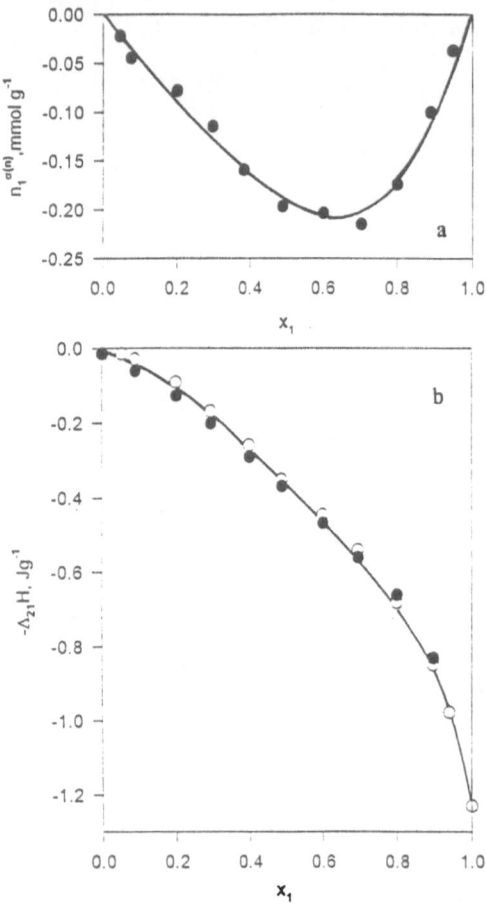

*Figure 5. Ads*orption excess(a) and enthalphy of displacement(b) isotherms in methanol(1)- benzene(2) mixture on Degussa Printex 300. On (b)the (o) adsorption, (●)desorption branche.

However, the displacement of benzene is not yet complete: up to the azeotropic point, the adsorption layer will also contain benzene. A common characteristic of the two isotherms presented in Figure 6 is that the isotherm $n_1^{\sigma(n)}=f(x_1)$ is linear in a very wide range of composition ($x_1=0.1 \ldots 0.7$), while the enthalpy isotherm $\Delta_{21}H=f(x_1)$ is strictly speaking it is not a constant, i.e. $(\delta\Delta_{21}H/\delta x_1)_{T,P}\neq0$. In other words, within this range of

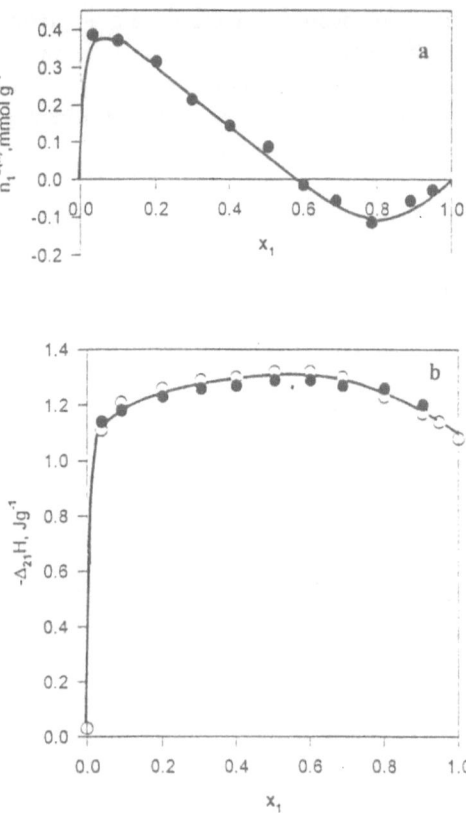

*Figure 6.* Adsorption excess(a) and enthalphy of displacement(b) isotherms in methanol(1)- benzene(2) mixture on hydrophobized CPC-silica. On (b)the (o) adsorption, (●) desorption branche.

composition a very small exchange heat effect is produced, consequently the composition of the layer is not constant in the strict sense of the word, i.e. $(\delta x_1^s / \delta x_1)_{T,P} \neq 0$. The maximum (extreme) value of the enthalpy isotherm is at the azeotropic composition; further heat effects are endothermic and the integral enthalpy isotherm therefore exhibits a decreasing tendency. Endothermic effects occurring from $x_1=0.6$ to $x_1=1$ are associated with the displacement of benzene from the adsorption layer.

384

The azeotropic composition of the excess isotherm determined on Chemviron activated carbon, an adsorbent with a large specific surface area is $x_1^a=0.095$, indicating that the adsorption layer still contains a little methanol (Figure 7.). This methanol is bound to the polar regions of the surface of the adsorbent, as shown by the exothermic heat effect detectable from $x_1=0$ to $x_1=0.1$.

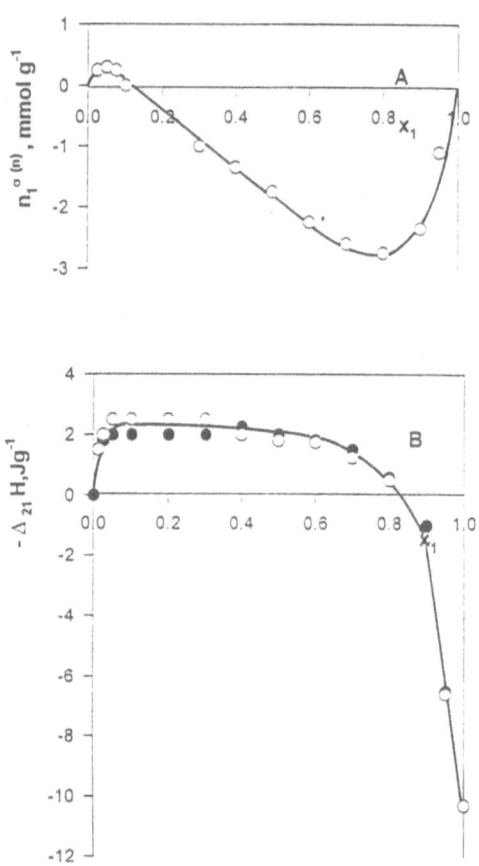

*Figure 7.* Adsorption excess(a) and enthalphy of displacement(b) isotherms in methanol(1)-benzene(2) mixture on Chemviron F-400 charcoal. On (b)the (o) adsorption, (●)desorption branche.

If $x_1 > x_1^*$ — the excess isotherm is only nearly linear between $x_1 = 0.1$ and $0.7$ — the enthalpies of exchange measured are moderately endothermic. If the molar fraction of the bulk phase is larger than $x_1 = 0.8$, the displacement of benzene by methanol will preferentially occur, resulting in a considerable endothermic effect on the porous apolar surface. This process is demonstrated by the parallel changes in the values of $n_1^{\sigma(n)}$, $\Phi_1^s$ and $\Delta_{21}H$ in Figure 7.

When adsorption displacement proceeds from $x_1 = 1$ towards $x_1 = 0$, the heat effects detected are of a reverse sign in the case of each isotherm. This means that the processes of displacement are reversible to a close approximation. A very little irreversibility was detected only in the case of Chemviron F 400 which is probably caused by an incomplete displacement of methanol from the micropores by benzene.

### 2.5.3. Functions of $\Delta_{21}H/n_1^{\sigma(n)} = f(x_1/n_1^{\sigma(n)})$

The applicability of equation (13) for U- and S-shaped excess isotherms is next discussed. Functions $\Delta_{21}H/n_1^{\sigma(n)} = f(x_1/n_1^{\sigma(n)})$ for the case of U-shaped excess isotherms are shown in Figure 8. Adsorption capacities — obtained in this representation as the ratio of the slope and the intersection — are identical with the values mentioned above. The advantage of equation (13) is that the intersection yields the value of $h_1^s - (1/r)h_2^s$.

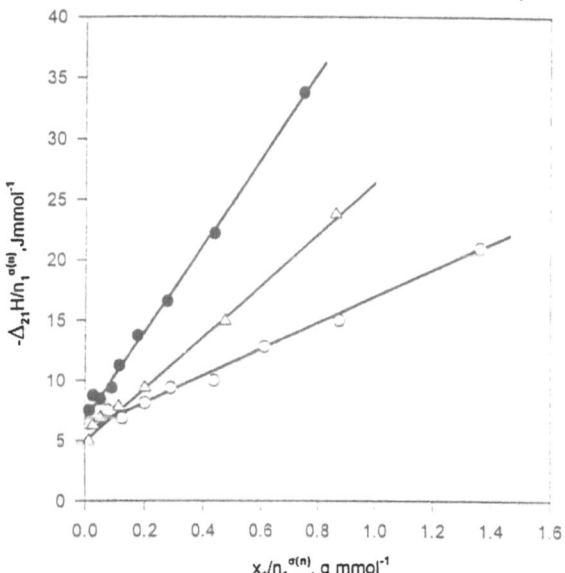

*Figure 8.* Combination of the adsorption excess amounts and calorimetric data (at U-shaped istherms)in benzene(1)- n-heptane(2) mixture (o)on R-silica gel,in methanol(1)-benzene(2) mixtures (Δ)on K-60 silica gel and(●) on R-silica gel.

In Figure 9    S-shaped excess isotherms and calorimetric data combined acccording to equation (13) are shown. The data may be utilized only up to the azeotropic composition $(x_1^a)$, still they are suitable for the determination of $n^s_{1,0}$. In this case the parameters of the straight lines allow only the determination of $n^s$, which can be used for the calculation of the adsorption capacity of the pure component according to equation, if the value of $x_1^a$ is known.

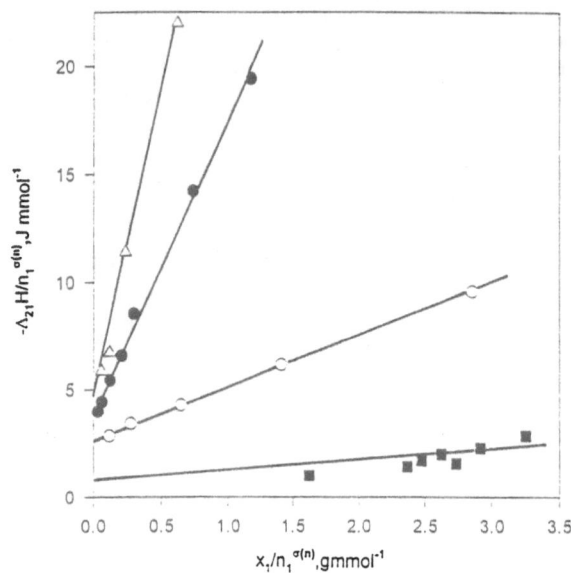

*Figure 9.* Combination of the adsorption   excess amounts and calorimetric data (at S-shaped isotherms) (Δ)on Chemviron F 400, (●)on K-60-C-18 silica gel, (o)CPC-C-18 silica, and (■)on Printex 300 in methanol(1)-benzene(2) mixtures.

However, the analysis of the data in Table 1 reveals that the combination of adsorption and calorimetric data, i.e. the application of equation (13) open a universal possibility for the determination of adsorption capacity in the case of the adsorption of binary liquids in systems with U and S-shaped excess isotherms.

The interaction of adsorbents with various surface energies with the liquid components studied are adequately characterized by the differences in molar adsorption enthalpies between components 1 and 2, $[h_1^s-(1/r)h_2^s]$ listed in Table 1. The largest difference is obtained for the methanol-n-heptane liquid pair, followed by methanol-benzene mixtures on the surface of silica gel. In the case of the adsorption of the methanol-benzene liquid pair, these enthalpy differences in the adsorption layer are decreased by the effect of hydrophobization.

TABLE  1. Results of analysis of adsorption isotherms and calorimetric data

| Adsorbent | Liquid mixture * | $n^s_{1,0}$ (mmol/g) | | $-\Delta_{21}H_t$ (J g-1) | $-(h_1^s-h_2^s/r)$ (kJ/mol) |
|---|---|---|---|---|---|
| | | S.N. | (Eq.13) | | |
| R-silicagel | 1-3 | - | 5.61 | - | 11.20 |
| R-silicagel | 1-2 | 5.31 | 5.14 | 36.71 | 7.10 |
| R-silicagel | 2-3 | - | 2.02 | 11.30 | 5.60 |
| K-60 silicagel | 1-2 | 4.00 | 4.02 | 19.10 | 5.15 |
| K-60-$C_{18}$ silicagel | 1-2 | 4.80 | 4.10 | 9.50 | 4.00 |
| CPC-$C_{18}$-silicagel | 1-2 | 1.13 | 1.48 | 1.10 | 2.51 |
| Chemviron F 400 | 1-2 | 9.10 | 10.41 | -10.52 | 5.31 |
| Printex 300 | 1-2 | 1.01 | 1.00 | - 1.23 | -0.50 |
| Na-illite | 1-2 | 0.84 | 0.85 | 11.10 | 13.65 |
| HDP-illite 1. | 1-2 | 1.25 | 0.95 | 1.70 | 4.32 |
| HDP-illite 2. | 1-2 | 1.30 | 1.18 | 1.35 | 3.21 |
| HDP-illite 3. | 1-2 | 1.42 | 1.38 | 0.85 | 2.27 |

* liquids: (1) methanol, (2) benzene, (3) n-heptane, S.N.: Schay-Nagy extrapolation method [5,6].

## 2.6. Adsorption and microcalorimetric experiments on hydrophobic clays

Silicate surface are usually hydrophilic against water because inorganic counter ions compensate the negative charges created by isomorphous substitutions in the silicate layer structure. By exchanging these counter ions by cationic surfactants the surface can be made hydrophobic. Suitable cations are, for instance, alkylammonium ions or alkylpyridinium ions.

Surface excess adsorption isotherms from binary liquid mixtures respond sensitively to the degree of hydrophobicity of the adsorbent [1-6]. The composition of the adsorption phase changes with the character of the surface. Preceding studies with montmorillonite as adsorbent revealed the mutual influence of the liquid molecules and the long chain organic cations onto the surface structure of the adsorption layer (4-6).

## 2.6.1. *Non-swelling illite and HDP-illites*

The adsorption capacities for U- and S-shaped excess isotherms were calculated by the Schay-Nagy extrapolation and the adsorption space filling model. Figures 10 and 11 show the excess isotherms and surface layer compositions determined for illite and for three gradually organophilized HDP-illites in methanol(1)-benzene(2) mixtures. These adsorption equilibrium functions (Eq. (1/a) indicate the composition of the adsorption layers $\phi_1^s$ as a function of the composition of the bulk phase $\phi_1$ in volume fractions. The amount of methanol in the adsorption layer decreases with increasing coverage by hexadecylpyridinium cations.

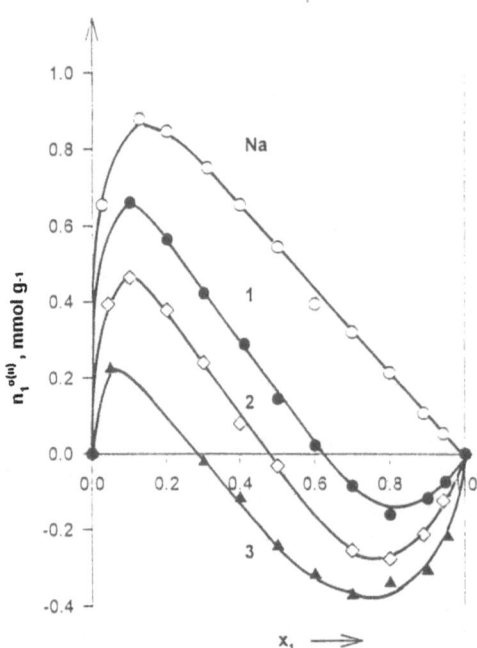

*Figure 10.* Adsorption excess isotherms on Na-illite and on HDP-illite derivatives in methanol(1)-benzene(2) mixtures.Na:sodium- illite, 1,2,3,: hexadecylpyridinium-illites.

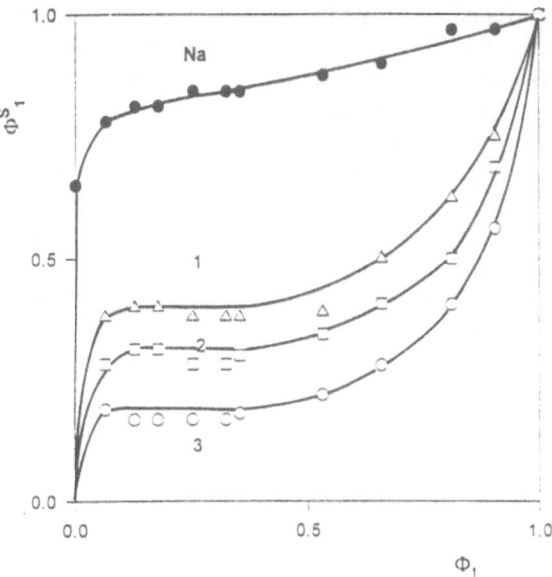

*Figure 11.* Adsorption equilibrium digrammes on Na-illite and on HDP-illite derivatives in methanol(1)-benzene(2) mixtures.Na:sodium-illite, 1,2,3,: hexadecylpyridinium-illites.

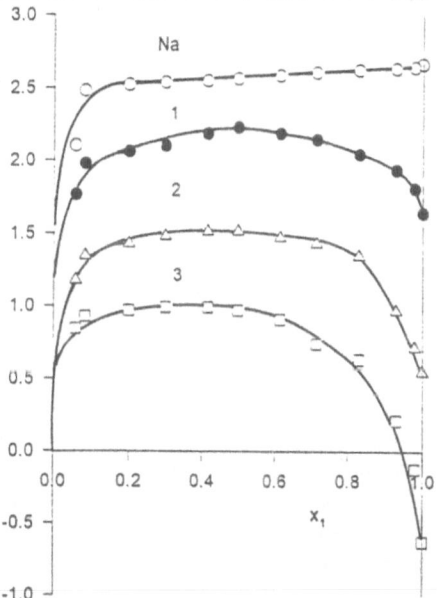

*Figure 12.* Free enthapy of adsorption on Na-illite and on HDP-illite derivatives in methanol(1)-benzene(2) mixtures.Na:sodium illite, 1,2,3,: hexadecylpyridinium-illites.

The excess free energy functions, given by integration of the excess isotherms, Eq (15) reflect the extent hydrophobization (Fig. 12.). Methanol displaces benzene with a maximum change in free energy on Na-illite. The displacement process results in smaller free energy changes on HDP-treated surfaces: the functions show plateau-like maxima at the azeotropic compositions. The free energy function for the sample with maximum hydrophobicity changes sign, which means that the displacement of benzene by methanol is not favoured. Illites und their organophilic derivatives can be dispersed well in methanol-benzene mixtures, therefore their wetting properties can be studied with batch microcalorimetry. Thus, the solid-liquid interaction can be studied only with immersion techniques. The displacement enthalpy can be given as $\Delta_{21}H = \Delta_w H - \Delta_w H_2^{\circ}$; its integral isotherm is plotted in Figure 13.

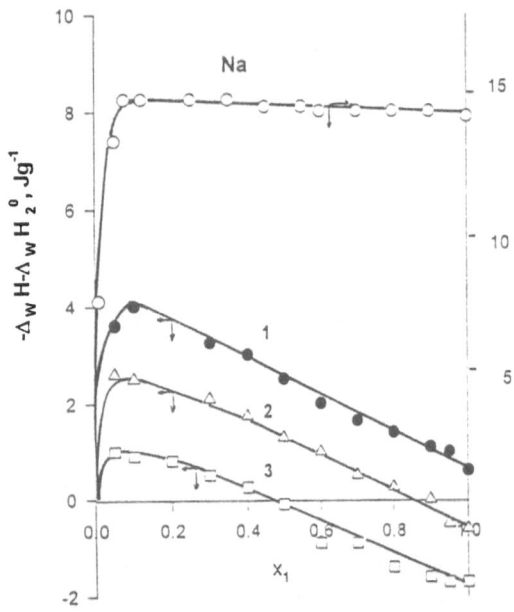

*Figure 13.* Immersional wetting enthalpy isotherms on Na-illite and on HDP-illite derivatives in methanol(1)-benzene(2) mixtures. Na:sodium illite, 1,2,3,: hexadecylpyridinium-illites.

The immersion wetting enthalpy is appreciable on Na-illite. The majority of heat evolution is due to preferential adsorption of methanol. The enthalpy change decreases upon hydrophobization and it becomes endothermic even in $x_1 > 0.5$ compositions.
The application of Eq (13) is more favourable, since it enlightens the difference between the molar adsorption enthalpies of the components (Fig. 14).

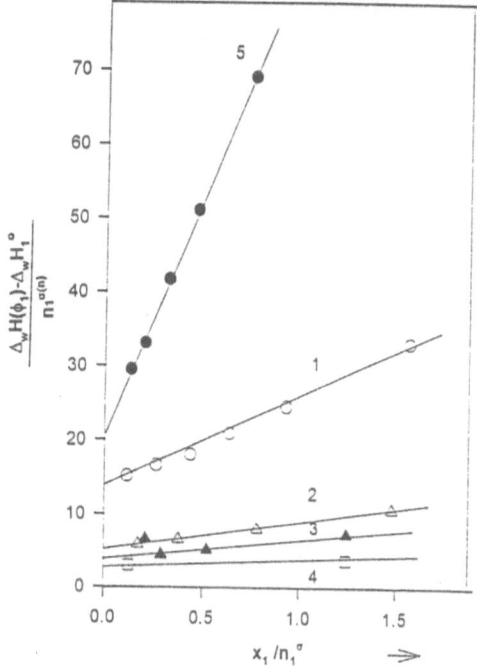

*Figure 14*. Determination of the adsorption capacities from Eq.(13). 1:Na-illite,2,3,4:HDP-illites, 5:Na-montmorillonite in methanol(1)-benzene(2) mixtures.

The linearized free energy and enthalpy functions for HDP-illites can be seen in Figs. 15 and 16. respectively. The parameters of Eqs. (13,14) give the adsorption capacities, the molar wetting energy changes $(g_1^s-g_2^s/r)$, and the molar wetting enthalpy changes $(h_1^s-h_2^s/r)$. These data show that the change in molar wetting data decrease with increasing hydrophobicity.

### 2.6.2. *Adsorption from diluted solution on hydrophobic vermiculites*

During the flow experiment, the adsorption layer of the preferentially adsorbing solute builds up successively by displacement processes at the surface, where solvent molecules at solid surface are displaced by solute molecules from the flowing bulk liquid. Thus, the term displacement involves surface and bulk phenomena, both of which are included in the integral enthalpy $(\Delta_{21}H)$, free enthalpy $(\Delta_{21}G)$ and entropy $(\Delta_{21}S)$ of displacement. It is desirable to find thermodynamic quantities which yield primary information about the adsorbed layer itself. To do this, we define the integral enthalpy

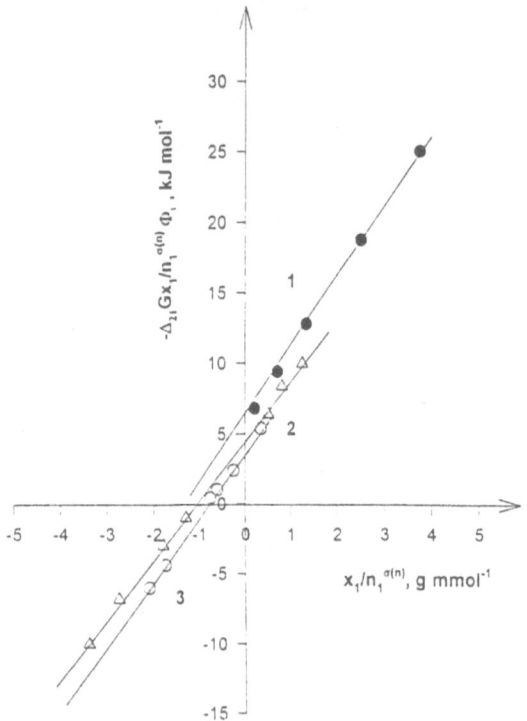

*Figure 15.* Linear free enthalpy functions on illite derivatives. Notation as in Fig.10.

of adsorption ($\Delta_{21}H^s$) as the enthalpy of the adsorbed layer at particular composition minus the enthalpy of the surface layer in the pure solvent. Clearly, the term 'adsorption' is reserved here for the adsorbed layer only, and all the bulk effects, which accompany the adsorption during the displacement process, are eliminated by this definition. The integral enthaly ($\Delta_{21}H^l$) and free enthalpy ($\Delta_{21}G^l$) of 'bulk dilution' yield link between the corresponding displacement and adsorption quantities [ 29]

$$\Delta_{21}H = \Delta_{21}H^s + \Delta_{21}H^l \qquad (17)$$

$$\Delta_{21}G = \Delta_{21}G^s + \Delta_{21}G^l \qquad (18)$$

where $\Delta_{21}H^l = -n_1^{\sigma(n)}\Delta_{sol}h_1^\infty$ and $\Delta_{21}G^l = -n_1^{\sigma(n)}(\Delta_{sol}\mu_1^\infty + RT\ln x_1)$ and $\Delta_{sol}h_1^\infty$ is the standard enthalpy and $\Delta_{sol}\mu_1^\infty$ the free enthalpy of solution of component 1 in 2 at infinitive dilution. In n-butanol-water at infinitive dilution are $\Delta_{sol}h_1^\infty$=-9.30, $\Delta_{sol}\mu_1$=+9.84 kJ mol⁻¹, respectively [29].

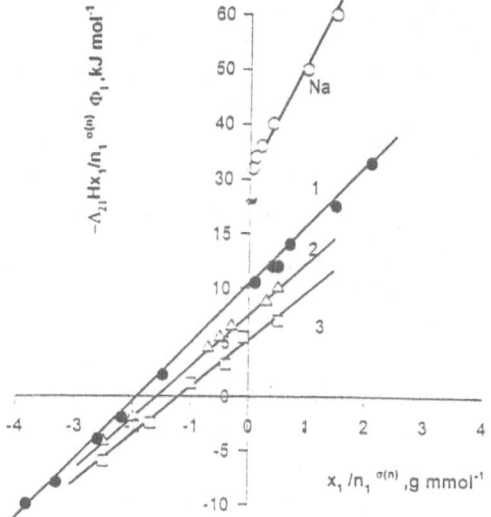

*Figure 16.* Linear enthalpy of wetting functions on Na:sodium montm.,1,2,3:HDP-illites in methanol(1)-benzene(2)mixtures.

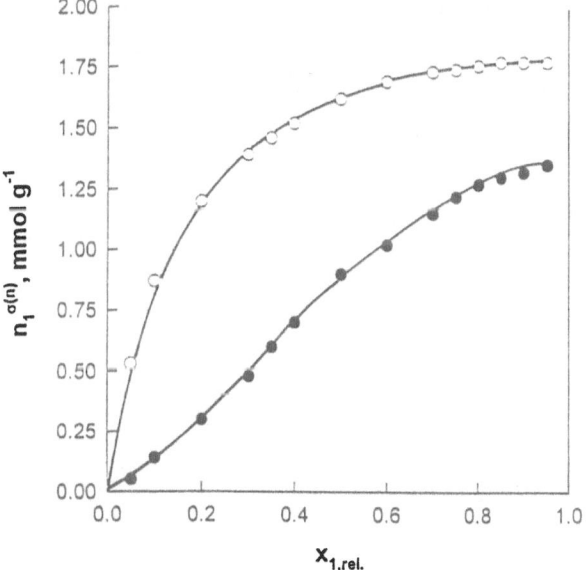

*Figure 17.* Adsorption excess isotherms in n-butanol-water solutions on (o)dodecylammonium-vermiculite und on (●) dodecyldiammonium-vermiculite.

394

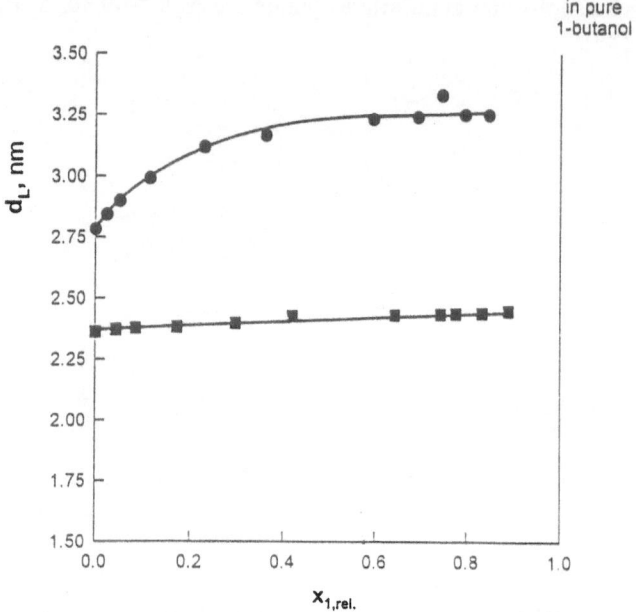

*Figure 18a.* Basal spacings of (●) dodecylammonium-vermiculite und (■)dodecyldiammonium-vermiculite in n-butanol-water solutions

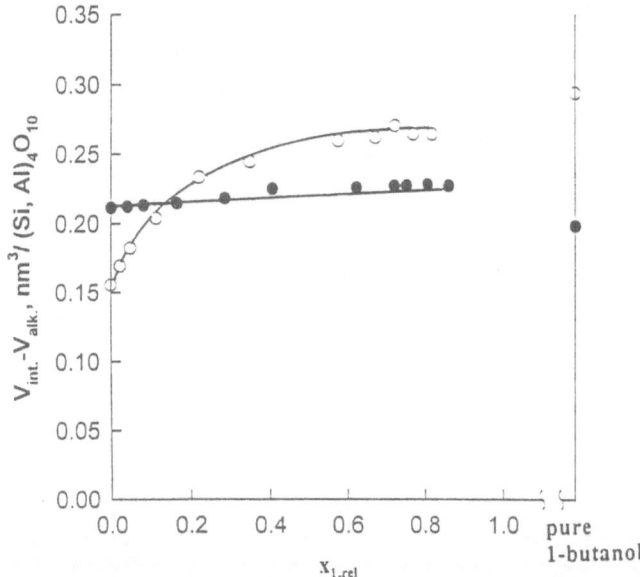

*Figure 18b.* Free interlayer volume of (o) dodecylammonium-vermiculite und (●) dodecyldiammonium-vermiculite in n-butanol-water solutions

Adsorption of n-butanol from water on the surface of vermiculite hydrophobized by dodecylammonium alkyl chains of two different structures is illustrated by Figure 17. It can be established that there is a very large difference between dodecylammonium derivatives with carbon chains of an identical length but having different interlamellar structures. The reason for this is directly shown in Figure 18.a, demonstrating that differences in interlamellar swelling are also quite large and, similarly to adsorption, $C_{12}$-$NH_3^+$-vermiculite swells considerably better than $C_{12}$-$(NH_3^+)_2$-vermiculite. As shown by a detailed discussion in our previous publication [30], due to the "bridges" formed by the alkyl chains the swelling of the diammonium derivative is limited (see Fig. 19) and the distance of the lamellae is nearly constant ($d_L$=2.35-2.45 nm). Thus this sample has a smaller free interlamellar volume ($V_{int}$-$V_{alc}$) than the dodecylammonium derivative (see Fig. 18.b).

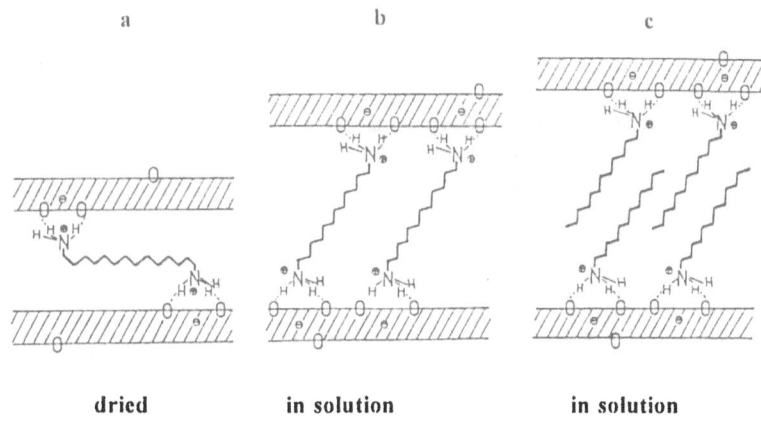

a          b          c

**dried**      **in solution**      **in solution**

*Figure 19.* Schematic representation of the hydrophobic vermiculite at different basal distances a/ dodecyldiammonium-vermiculite in dry state, b/dodecyldiammonium-vermiculite in solution, c/dodecylammonium-vermiculite in bilayer orientation

Figure 20 shows free energy functions $\Delta_{21}G=f(x_{1,r})$ calculated according to equations (15) from the isotherms $n_1^{\sigma(n)}=f(x_{1,r})$. It is obvious that the change in free energy accompanying the adsorption exchange process changes parallel with adsorption. On the other hand, it has to be stressed that the value of $\Delta_{21}G$ is significantly lower in the restrictedly swelling system with lamellae bound together by alkyl chains ($\Delta_{21}G$=4.1 Jg$^{-1}$) than in the ones with lamellae that move independently relative to each other ($\Delta_{21}G$=8.5-10.3 Jg$^{-1}$).

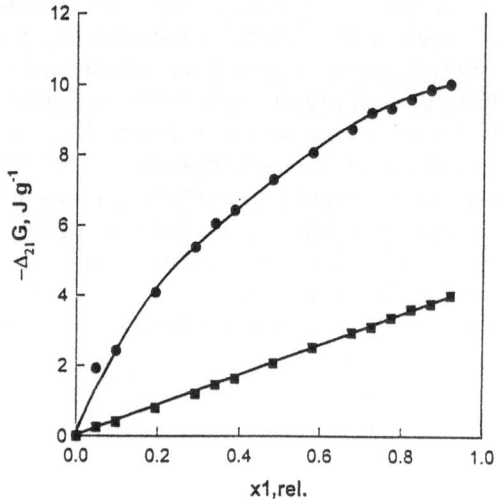

*Figure 20.* Free enthalphy of adsorption in n-butanol-water solutions on (●)dodecylammonium-vermiculite und on (■)dodecyldiammonium-vermiculite.

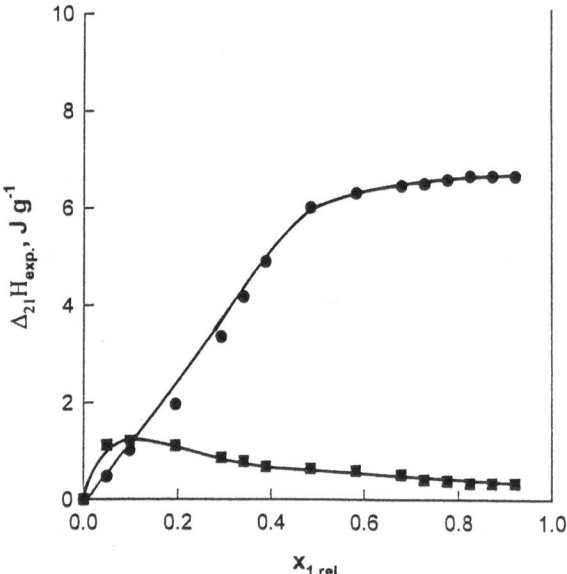

*Figure 21.* Enthalphy of displacement isotherms in n-butanol-water solutions on (●)dodecylammonium-vermiculite und on (■ )dodecyldiammonium-vermiculite.

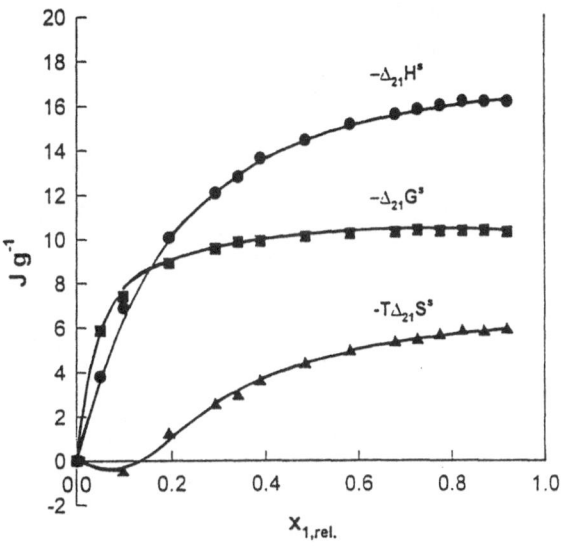

*Figure 22*. Thermodynamic potential functions of the adsorption layer on dodecylammonium-vermiculite in in n-butanol-water solutions.

Enthalpy isotherms $\Delta_{21}H_{exp}=f(x_{1,r})$ determined by the flow technique are presented in Figure 21. The heat effects recorded are found to be endothermic in each cases. Since the adsorption isotherms unambiguously indicate a positive adsorption of n-butanol, the question arises why an exothermic exchange enthaply is not recorded. In our opinion, the reason for this is the endothermic enthalpy of dilution featuring in equations (17) which overcompensates the interlamellar adsorption of butanol. When, knowing the adsorption excesses, the enthalpy isotherm $\Delta_{21}H^s=f(x_{1,r})$ characteristic of the S/L interfacial adsorption layer is calculated on the basis of equation (17), it is indeed the exothermic adsorption enthalpy isotherm specific for the surfacial interaction that is obtained (Fig. 5). It is revealed by a comparison of Figures 21 and 22 that the correction has a tremendous significance for the interpretation of the adsorption process. Figure 22 suggests that the interlamellar adsorption of n-butanol is thermodynamically preferential and is accompanied by the liberation of a very large amount of heat ($\Delta_{21}H^s$=-16.0-16.5 $Jg^{-1}$). The enthalpy isotherm relative to dodecylammonium vermiculite is presented in Figure 23. In agreement with the data on adsorption and swelling, the value of exothermic change in enthalpy is also lower due to the limited adsorption of n-butanol.

$\Delta_{21}G^s$ and $T\Delta_{21}S^s$ functions were also included in Figures 22 and 23, in order to describe the incorporation of the adsorbate also in terms of change in enthalpy. In Figure 22 the entropy term of dodecylammonium vermiculite is decreased due to an increase

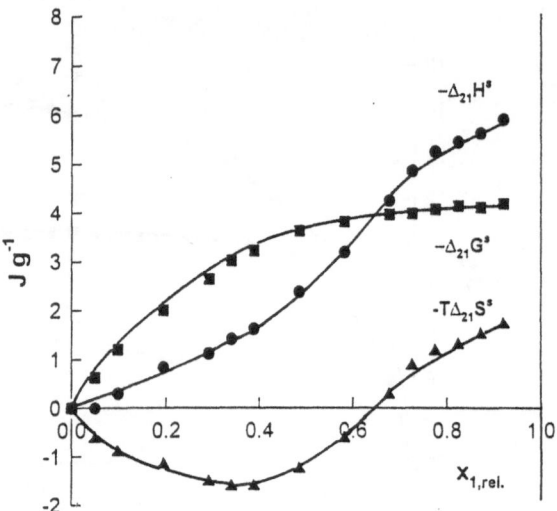

*Figure 23.* Thermodynamic potential functions of the adsorption layer on dodecyldiammonium-vermiculite in in n-butanol-water solutions.

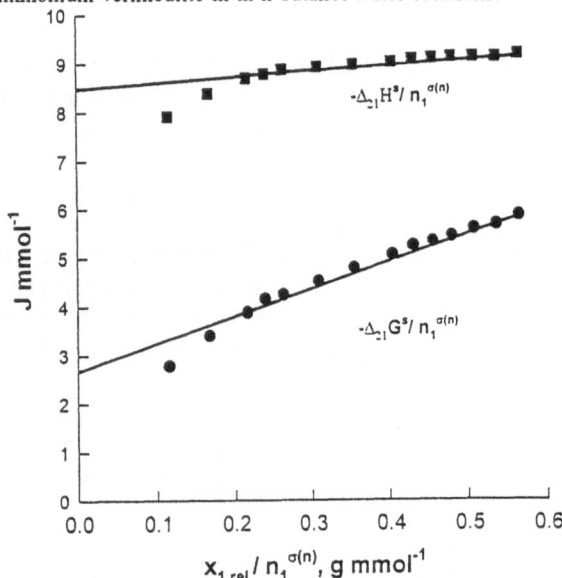

*Figure 24.* Combination of the adsorption excess and the enthalphy Eq.(13), or the free entalphy isotherms Eq.(14) in n-butanol-water solutions on dodecylammonium-vermiculite.

in the adsorption of butanol ($T\Delta_{21}S^s<0$). Conversely, in the case of the restrictedly swelling dodecyldiammonium sample an increase in entropy ($T\Delta_{21}S^s>0$) is observed up to $x_{1,r}\leq0.7$ and it is only in the range of adsorption saturation that a decrease in entropy occurs. This means that in the case of the restrictedly swelling system, water molecules are arranged in the interlamellar space in a more orderly manner (presumably in clusters) than n-butanol molecules.

Figures 24 represent the combined functions calculated according to equations (13). The difference of ($h_1^s$-$h_2^s$/r) calculated from the intersection is quite significant, indicating that the difference between the molar adsorption enthalpies of the components is very large (-2.7 J mmol$^{-1}$). In the case of the diammonium derivative the values of the corresponding data are lower ($h_1^s$-$h_2^s$/r)=-3.8 J mmol$^{-1}$ and ($g_1^s$-$g_2^s$/r)=-2.2 J mmol$^{-1}$, since − due to the linking effect of alkyl chains − the hydrophobicity of the surface is higher.

## 3. Small-angle X-ray scattering of SiO$_2$ particles in binary liquids

Colloidal dispersions containing silica (SiO$_2$) particles highly dispersed in organic liquids are of practical importance. Dispersed particles like these in binary liquid mixtures result in suspensions with widely different stabilities and rheological properties depending on the mixture composition [31]. Adsorption measurements were also made on surface of hydrophobic SiO$_2$ particles (Aerosil R972, Degussa, Germany) dispersed in binary mixtures of benzene(1)--n-heptane(2). The stabilizing role of the adsorption layer will be discussed in suspension and its effect on the interparticle interaction. In this part we attempt to correlate the results of rheological and adsorption measurements on aerosil particles, compare these results with SAXS experiments and establish the role of selective adsorption in the aggregation and structure-forming process.

Figure 25 shows a U-shaped excess isotherm for benzene(1)-n-heptane(2) mixtures. The individual adsorption isotherms ($n_1^s$, $n_2^s$ vs. $x_1$ functions) can be calculated, and these give the adsorbed amount in the interfacial layer. From the series of rheological measurement the concentration of hydrophobic SiO$_2$ suspensions were 2 % (w/v) in benzene(1)-n-heptane(2) liquid mixtures. The Bingham yield stress $\tau_B$ can be extrapolated from the flow curves: $\tau=\tau_B+\eta_{pl}D$, where $\tau$ is the shear stress, (in Pa) D is the shear rate gradient (s$^{-1}$) and $\eta_{pl}$ is the plastic viscosity. The $\tau_B$ as a function of mole fraction can be seen in Figure 26. The optical density for 1 g per 100 cm$^3$ dispersions in the whole liquid composition range were calculated via equation $D_{opt}$=-log$_{10}$(I/I$_o$), where I$_o$ and I are the intensities of incident and transmitted light, respectively. The results of the optical experiment were parallel presented in Figure 26. Comparing Figs. 25 and 26 can be established, that a rather flocculated structure of SiO$_2$ particles was observed, when the mixture was rich in n-heptane, i.e. the composition of the surface layer ($n_1^s$, $n_2^s$) greatly influences the aggregation - characterized by $\tau_B$ and $D_{opt}$ - of the hydrophobic silica particles.

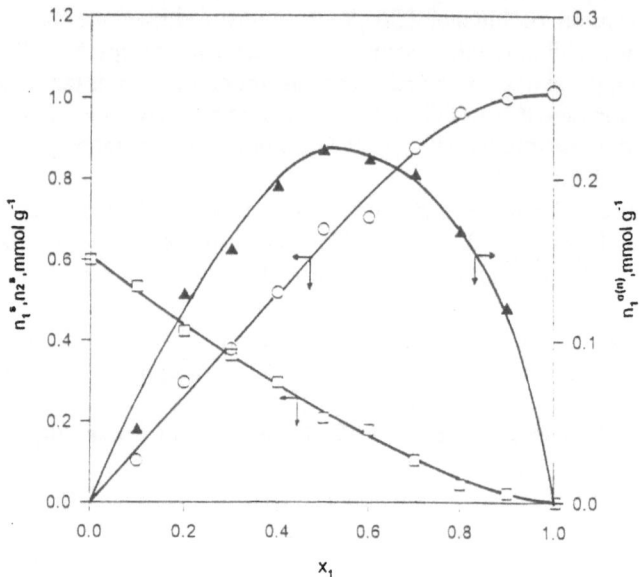

**Figure 25.** Adsorption excess isotherm and the individual isotherms $(n^s_1, n^s_2)$ on hydrophobic silica (R 972)in in benzene(1)-n-heptane(2) mixture.

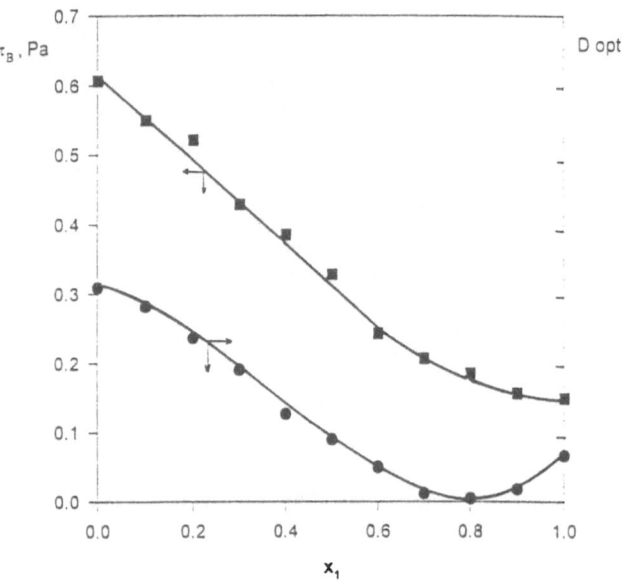

**Figure 26.** Optical density and the Bingham yield stress vs. mole fraction in in benzene(1)-n-heptane(2) mixture.

In benzene(1)-n-heptane(2) mixtures the decrease in Bingham yield stress $D_{opt}$ and the displacement of n-heptane with benzene occur continuously and in parallel. Our results show that the composition of the adsorption layer is crucial in determining the particle/particle interactions. Therefore, it seems fruitful to study the dependence of the structural properties of suspension on the composition of the S/L interfacial layer.

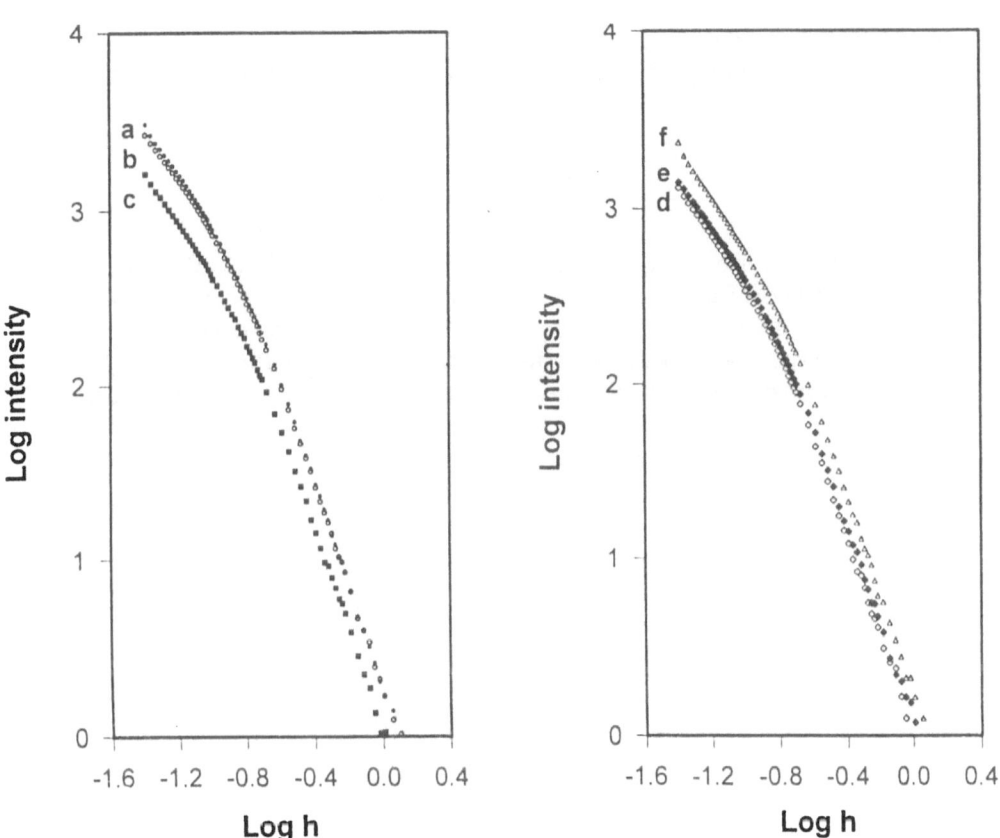

*Figure 27.* SAXS scattering curves of hydrophobic silica (R 972)in benzene(1)-n-heptane(2) mixture in (2%w/v) suspension, a: $x_1=0$ (n-heptane), b: $x_2=0.3$, c: $x_1=0.5$, d: $x_1=0.7$, e: $x_1=0.9$, f: $x_1=1$ (benzene)

When X-rays are scattered by colloidal particles, due to differences in electron density caused by inhomogeneities, the intensity of scattered radiation(I) is a function of scattering vector (h). The invariant Q, calculated by the integration of the scattering function:

$$Q=2\pi^2 w_1 w_2 (\Delta\rho_e)^2 V=\int_0^\infty h^2 I(h)\,dh \qquad (19)$$

where $w_1$ and $w_2$ are the volume fractions of solid and liquid phases, respectively. The $\Delta\rho_e$ is the average electron density difference and V is the volume of the scattering system.

The relative inner surface area of the particles S/V is:

$$S/V=\pi\lim_{h\to\infty} I(h)\,h^4/Q=\pi K_p/Q \qquad (20)$$

where $K_p$ is the tail end constant, calculated from the tail end section of the scattering curve. It the values of the relative specific surface area and the volume fractions are known, the average intersection length characteritic of the individual phases of the colloidal suspension can be calculated: $l_1=4w_1/(S/V)$ and $l_2=4w_2/(S/V)$.

The scattering curves measured in the whole range of the benzene(1)-n-heptane(2) mixtures are given in Figure 27. The $I(h)h^3=f(h^3)$ Porod representation of the scattering curves are given in Figure 28.

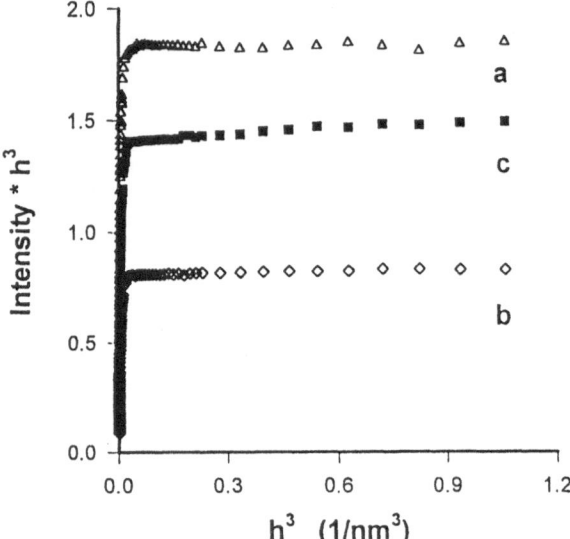

*Figure 28.* The Porod plot of hydrophobic silica R 972 in benzene(1)-n-heptane(2) mixtures, a: $x_1=0.5$, b: $x_1=0.7$, c: $x_1=0.9$.

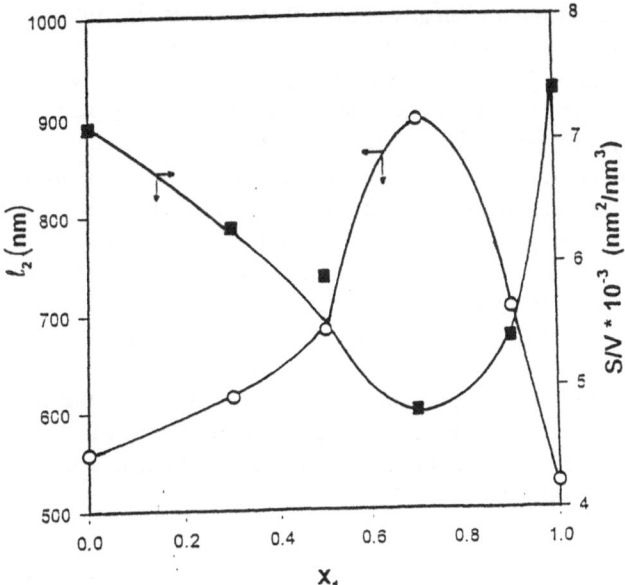

*Figure 29.* The (S/V) inner relative surface and the (l₂) intersection length of the liquid in benzene(1)-n-heptane(2) mixture

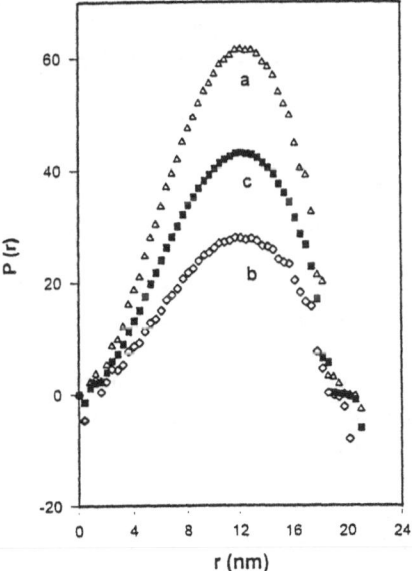

*Figure 30.* Distance distribution function of hydrophobic silica(R 972) particles in benzene(1)-n-heptane(2) mixture in (2%w/v) suspension. a: $x_1$=0.5, b: $x_1$=0.7, c: $x_1$=0.9

404

From the SAXS experiment is clearly shown, that the scattering intensity depend on the bulk mixture composition. A S/V values can be calculated from Eqs. 19,20 and its values are presented on Figure 29. As the mole fraction of benzene increases, however, the value of S/V decreases because of the displacement of n-heptane by benzene from the adsorption layer initiate the dissagregation of the particles. At composition $x_1$=0.7-0.9 the surface of the aggregates has a minimum. The result of our rheological and optical studies are complete agreement with this experiments. Data of $l_2$ characteristic of the distance between aggregates calculated from Eq. (20) are shown in Figure 29. This distance reaches its maxima at $x_1$=0.7 because the interparticle interaction have a minimum value. The P(r) vs.r. distance distribution functions presented in Figure 30 were calculated from I=f(h) function by inverse Fourier transformation. It can be clearly seen, that the numeric values of the functions indicating differences in the density of aggregates in the three liquid mixtures with different polarities and compositions. The interparticle interactions may thus be regulated via the selective liquid sorption process and can be established that the interfacial layer compositon is a crucial factor on the stability of colloidal dispersion.

## 4. References

1.      Schay, G. (1969) Surface and Colloid Science, Vol. 2. 155 p. Ed.: Matijevic, E., Wiley, London.
2.      Schay, G. (1970) Surface Area Determination, Proc. Int. Symp. 1969. 273 p. Ed.: Everett, D.H., Butterworths London.
3.      Redlich, O. and Kister, A.T. (1948) Thermodynamics of nonelectrolyte solutions, *Ind. Eng. Chem.* **40**, 341.
4.      Scatchard, G. and Ticknor, L.B. (1952) Vapour-liquid equilibrium IX. The methanol-carbon tetrachloride-benzene system *J. Am. Chem. Soc.* **74**, 3724.
5.      Dékány, I., Nagy, L.G. and Schay, G. (1978) Effect of surface modification on solid-liquid interfacial adsorption of mixtures, *J. Colloid Interface Sci.* **66**, 197-200.
6.      Dékány, I., Szántó, F., Nagy, L.G. and Schay, G. (1983) Sorption and immersional wetting properties of palygorskite and its hexadecylpyridinium derivatives, *J. Colloid Interface Sci.* **93**, 151.
7.      Schay, G. and Nagy, L.G. (1960) Thermodynamic investigation of the interfacial properties of liquid mixtures, *Periodica Polytechn.* **4**, 45.
8.      Kipling, J.J. (1965) Adsorption from Solutions of Non-Electrolytes. Academic Press, London.
9.      Everett, D.H. (1964) Thermodinamics of adsorption from solution Part 1-Perfect systems, Part 2-Imperfect systems, *Trans. Faraday Soc.* **60**, 1803. (1965) **61**, 2478.

405

10. Everett, D.H. (1979) in Colloid Science. Vol. 3., p. 66. Chem. Soc., London.

11. Oscik, J., Dabrowski, A., Jaroniec, M. and Rudzinski, W., (1976) On formation of multilayers in adsorption from binary liquid mixtures, *J. Colloid Interface Sci.* **56**, 403.

12. Dabrowski, A., Oscik, J., Rudzinski, W. and Jaroniec, M. (1979) Effects of surface heterogeneity in adsorption from binary liquid mixtures II., *J. Colloid Interface Sci.* **69**, 287.

13. Dabrowski, A. and Jaroniec, M. (1979) Application of dubinin-radushkevich and freundlich equations to characterize the adsorption from non-ideal binary liquid mixtures on heterogeneous solid surfaces, *Acta Chim. Acad. Sci. Hung.* **99**, 225.

14. Dabrowski, A. (1983) An isotherm equation for adsorption from binary liquid mixtures on solids involving surface heterogeneity, *Monatshefte f. Chemie* **114**, 875.

15. Rudzinski, W., Zajac, J., Wolfram, E. and Pászli, I. ((1987) Heats of immersion in multilayer adsorption from binary liquid mixtures on strongly heterogeneous solid surfaces, *Colloids and Surfaces* **22**, 317-336.

16. Kern, H.E., Piechocki, A., Brauer, U. and Findenegg, G.H. (1978) Adsorption from solution of chain molecules onto graphite: Evidence for lateral interactions in ordered monolayers, *Progr. Colloid Polym. Sci.* **65**, 118.

17. Liphard, M., Glanz, P., Pilarski, G. and Findenegg, G.H. (1980) Adsorption of carboxylic acids and other chain molecules from n-heptane onto graphite, *Progr. Colloid Polym. Sci.* **67**, 131.

18. Groszek, A.J. (1970) Selective adsorption at graphite/hydrocarbon interfaces, *Proc. Roy. Soc.* **A314**, 473.

19. Denoyel, R., Rouquerol, F. and Rouquerol J. (1983) Adsorption from Solution (Ed.: Ottewill, R.H., Rochester, C.H. Smith A.L.) Academic Press, New York, 225 p.p.

20. Allen, T. and Patel, R.M. (1971) The adsorption of long-chain fatty acids on finely divided solids using a flow microcalorimeter, *J. Colloid Interface Sci.* **35**, 647.

21. Woodbury, G.W., and Noll, Jr.-L.A. (1983) Heat of adsorption of liquid mixtures on solid surfaces: comparison of theory and experiment, *Colloid and Surfaces* **8**, 1.

22. Noll, L.A. , Woodbury, G.W. and Burchfield, T.E. (1984) Dependence of adsorption of cosurfactant on chain length, *Colloids and Surfaces* **9**, 349.

23. Billett, D.F., Everett, D.H. and Wright E.E.H. (1964) Heats of wetting of solids by non-aqueous solutions , *Proc. Chem. Soc.* **216**.

24. Dékány, I., Zsednai Á., Király, Zs., László, K. and Nagy, L.G. (1986) Enthalpy of displacement of binary liquid mixtures on solid surfaces, I. Analysis of U-shaped isotherms, *Colloids and Surfaces* **19**, 47.

25.  Dékány, I., Zsednai Á., László, K. and Nagy L.G. (1987) Enthalpy of displacement of binary liquid mixtures on solid surfaces, II. Analysis of S-shaped excess isotherms, *Colloid and Surfaces* **23**, 41.

26.  Dékány, I., Ábrahám, I., Nagy, L.G. and László, K. (1987) Enthalpy of displacement of binary liquid mixtures on solid surfaces, III. Determination of the adsorption capacity from calorimetric and adsorption data, *Colloids and Surfaces* **23**, 57.

27.  Rusanov, A.I. (1967) Phase Equilibrium and Surface Phenomena. Chimica, Leningrad, Chap. VI.

28.  Dékány, I. and Nagy, L.G. (1991) Immersional wetting and adsorption displacement on hydrophilic/hydrophobic surfaces, *J. Colloid and Interface Sci.* **147**, 119-128.

29.  Király, Z. and Dékány, I. (1990) Enthalpy and entropy effects in adsorption and displacement, *Colloids and Surfaces* **49**, 95-101.

30.  Regdon, I., Király, Z., Dékány, I. and Lagaly, G. (1994) Adsorption of 1-butanol from water on modified silicate surfaces, *Colloid and Polymer Sci.* **272**, 1129-1135.

31.  Machula, G. and Dékány I. (1991) Rheological, adsorption and stability behaviour of hydrophobic aerosil particles in binary liquid mixture, *Colloid and Surfaces* **61**, 331-348.

# LIQUID-SOLID ADSORPTION AS THE BASIS FOR SELECTIVE SEPARATION PROCESSES

K.K. UNGER
*Instiutut für Anorganische Chemie und Analytische Chemie*
*Johannes Gutenberg-Universität*
*Becherweg 24*
*D-55099 Mainz, Germany*

## 1. Introduction

Adsorption at the solid-liquid interphase is the basic phenomenon in processes of biological and technical importance [1,2]. The adsorbents cover porous and non-porous materials as well as colloids. The surface of the solid being in equilibrium with the solution forms an interphase whose properties determine the efficacy of the system. The solutes comprise a wide range of species: non-electrolytes, eletrolytes, low-molecular weight as well as highmolecular weight compounds.

Liquid-solid adsorption is the dominating feature in high-resolution separations on analytical scale in High-Performance Liquid Chromatography (HPLC) [3,4] and in isolation and purification processes on large scale in chemical and pharmaceutical industry [5].

The aim of this article is to review in brief the principles of liquid-solid adsorption and to highlight the most important aspects of separation techniques and processes at analytical, preparative and industrial scale.

## 2. Thermodynamic aspects

### 2.1. THE RIGOROUS THERMODYNAMIC APPROACH

When the adsorbent is in equilibrium with a binary mixture of compound 1 and 2 the adsorption of compound 2 is expressed in terms of the Gibbs surface excess quantities [6], by

$$n_2^{\sigma(n)} = n^0 \left( x_2^0 - x_2^l \right) \tag{1}$$

where $x_2^0$ and $x_2^l$ are the mole fractions of compound 2 in an amount $n^0$ of liquid mixture before and after equilibration with the adsorbent. The reduced surface excess quantity, $n_2^{\sigma(n)}$, can be converted into a volume-reduced form as follows:

$$n_2^{\sigma(v)} = V^0 \left( c_2^0 - c_2 \right) \tag{2}$$

407

*J. Fraissard (ed.), Physical Adsorption: Experiment, Theory and Applications, 407–428.*
© *1997 Kluwer Academic Publishers.*

where $c_2^0$ and $c_2$ are the concentrations of compound 2 before and after equilibration with the adsorbent and $V^0$ the volume of the liquid mixture. Typical surface excess isotherms $n_2^{\sigma(n)} = f(x_2^l)$ of three binary systems are shown in Fig. 1 [7]. It can be seen that the solutes are preferentially adsorbed in the sequence cyclohexane < benzene < 1.2-dichloroethane.

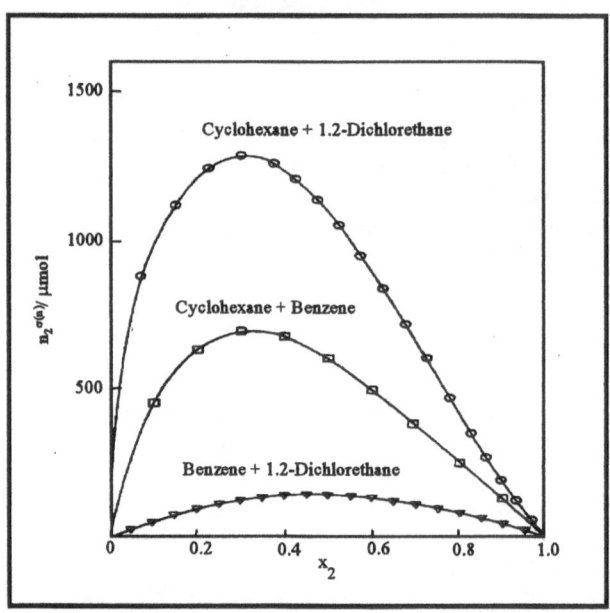

*Figure 1.* Adsorption isotherms of three binary systems on silica gel at 25 °C, $n_2^{\sigma(n)}$ represents the reduced surface excess adsorbed by the total mass of adsorbent, $m_s$ = 1.28 g, $a_s$ = 350 m²/g

(from ref. [7] with permission of the publisher)

The adsorption equilibrium constant $K_{12}$ can be calculated from the surface excess isotherm as

$$\ln K_{12} = a^0 \, {}_0\!\int^1 \Gamma_2^{(n)} / x_2^l \gamma_2^l (1 - x_2^l) \, d(x_2^l \gamma_2^l) \tag{3}$$

where $a^0$ is the standard area per mol which is chosen as the unit area per mol, $\gamma_2^l$ the activity coefficient of compound 2 and $\Gamma_2^{(n)}$ the surface excess concentration being equal to $n_2^{\sigma(n)} / m_s a_s$, where $m_s$ denote the mass and $a_s$ the specific surface area of adsorbent.

## 2.2. INDIVIDUAL AND COMPOSITE ISOTHERMS

Commonly, solute isotherms are plotted as the amount adsorbed per mass of adsorbent as a function of the amount of solute in solution at equilibrium. Single or individual isotherms, ie. isotherms of a one-component solution are measured either by static methods, eg. by means of batch experiments or by dynamic methods, eg. by means of breakthrough curves. A comparison of the different methods with respect to their reliability and applicability is given in reference [8]. A number of models have been developed to simulate the course of the isotherms. The most common is the Langmuir isotherm, the Freundlich isotherm and the approach based on the vacancy solution (VS) theory [9]. As the goal of adsorption is to separate at least two compounds the composite isotherm of the two-component solution on the adsorbent must be known. Composite isotherms can be either experimentally measured or calculated using the individual isotherms. One of the most straightforward approaches is the ideal adsorbed solution (IAS) theory developed by Radtke and Prausnitz [10]. While the IAS theory is applicable to ideally diluted solutions, the vacancy solution theory is preferably applied to simulate the high concentration regime of the isotherm.

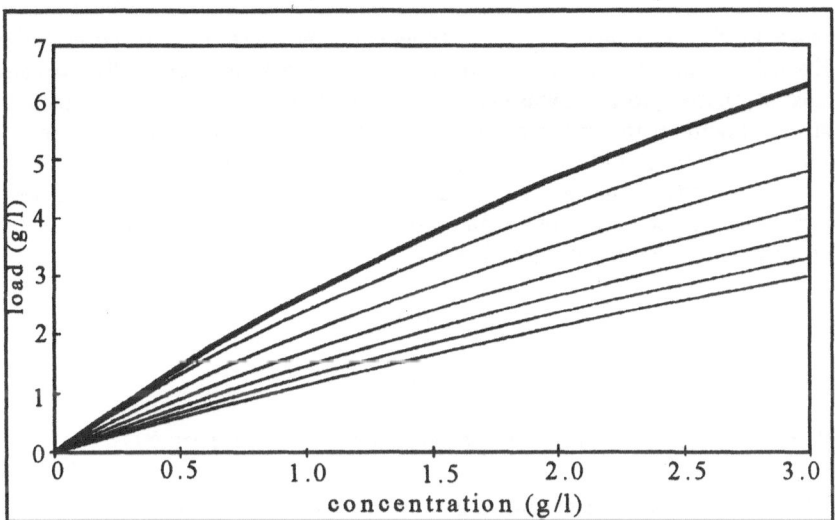

*Figure 2.* Composite isotherm of the (-)- enantiomer of the Troeger base at different concentrations of the (+)+ enantiomer calculated by the IAS theory (from ref. [8] with permission of the publisher)

## 2.3. CHANGES OF THE FREE ENERGY AND THE SELECTIVITY IN LIQUID-SOLID ADSORPTION PROCESSES [11].

The slope of the isotherm corresponds to the distribution coefficient K of the solute. At infinite dilute concentration of solute, ie. at the linear range of the isotherm, the distribution coefficient is constant. At the non-linear range of the isotherm K is dependent on the actual concentration. In case of a preferential adsorption of the solute K is larger than unity.
K is related to the changes in the free energy, $\Delta G$, of the system by

$$\Delta G = \Delta H - \Delta S = - RT \ln K \qquad (4)$$

where $\Delta H$ is the change of the enthalpy, $\Delta S$ the change in entropy, R the gas constant and T the absolute temperature. In case of a preferential adsorption $\Delta G$ becomes negative. Rearranging eq. (4) one obtains

$$\ln K = - \Delta H / RT + \Delta S / R \qquad (5)$$

In liquid-solid adsorption processes can be either enthalpy driven or entropy driven or both. Size Exclusion Chromatography [12] for instance is a typical example for an entropy-controlled separation process. Samsonov and coworkers have examined the separation of proteins on highly permeable crosslinked polymeric ion-exchangers and found both enthalpic and entropic contributions responsible for the selective separation [13].
In order to resolve two compounds 1 and 2 differences in the distribution coefficients $K_1$ and $K_2$ are required. The selectivity of a liquid-solid adsorption system is usually characterized by the selectivity coefficient $\alpha$ being defined by

$$\alpha = K_2 / K_1 \; ; \; (K_2 > K_1) \qquad (6)$$

To resolve two compounds $\alpha$ has to be larger than unity.

## 2.4. HOW IS THE SEPARATION ACCOMPLISHED IN PRACTICE ?

The first step in developing a separation is to choose an adsorbent and a solvent or solvent mixture at which a preferential adsorption of the solutes occurs. In the simplest case, when the difference between the distribution coefficients of the two solutes is very large, a batch experiment can be performed where one compound is adsorbed, the other remains in the solution. This is a typical situation met with immunoadsorbents which exert a high specificity towards individual solutes, eg. enzymes. In most cases, however, the solutes to be separated from a complex mixture closely resemble in chemical structure and the selectivity coefficients range between 1.01 and 1.5. A complete separation cannot be achieved in a single equilibration step.
The repetitive equilibration in liquid-solid adsorption processes can be accomplished by holding the adsorbent stationary and moving the solvent. The concept serves as the basis of Liquid-Solid Chromatography and is illustrated in Fig.3.

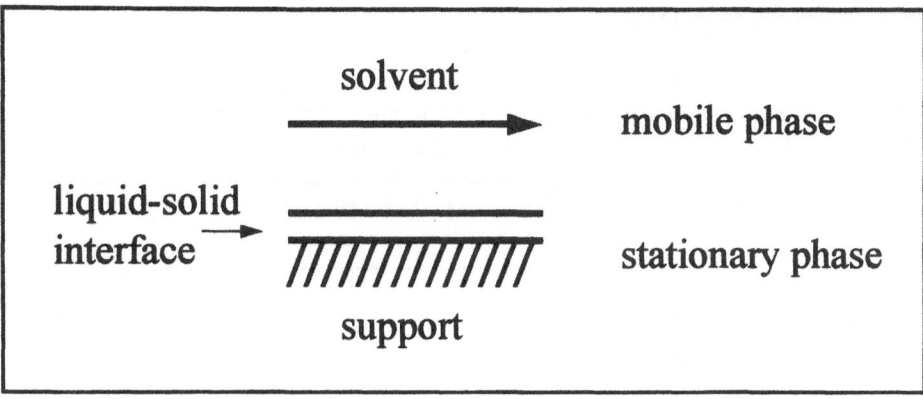

*Figure 3*. Schematic representation of a phase system in Column Liquid Chromatography

Generally, there are two modes in which the adsorbent is kept stationary. Firstly, the adsorbent can be filled in a cylindrical tube (column) with frits on both ends holding the packing (Column Liquid Chromatography [14]). Secondly, the adsorbent is spread as a solid layer on a plate (Thin Layer Chromatography [15]). In Column Liquid Chromatography [3,4,14] the solvent is continuously pumped through the column with a constant flow-rate. The sample is injected as a pulse at the top of the column and migrates through the column with the mobile phase.During the travel of the solute along the column repetitive equilibration takes place between the mobile and the stationary phase. When the solutes are preferentially adsorbed and have slight differences in their distribution coefficients they leave the column outlet with different migration velocities. The detector at the column outlet monitors the concentration of solutes as a function of time. The elution profiles have a Gaussian shape. In case of a preferential adsorption the solutes are eluted later than the solvent. They are retained as compared to the solvent. The retention is expressed by the capacity factor k, as

$$k = t_R - t_m / t_m = V_R - V_m / V_m \qquad (7)$$

where $t_R$ ($V_R$) is the retention time (volume) and $t_m$ ($V_m$) the dead time (dead volume) of the column. The solutes should be eluted with capacity factors in the range between $1 < k < 10$ at constant solvent composition.

The capacity factor k is proportional to the distribution coefficient of solute. The proportionality factor is the volume ratio of the stationary phase and the mobile phase, $\Phi$. The relation between k and K is given by

$$k = v_s / v_m \, K \qquad (8)$$

where $v_s$ ($v_m$) is the volume of the stationary (mobile) phase. The term phase is used in Chromatography in a somewhat less defined sense as it is defined in physical chemistry.

## 3. Kinetic aspects [16]

During the migration of the solute through the column a selective dilution occurs. Furthermore, due to dispersion effects the elution profile is broadened with increasing length of the travel distance (column length). The sample mixture is injected on the column top as a narrow plug. At the column outlet the elution profiles are monitored as peaks with a Gaussian shape. As it can be seen in Fig.4 the peaks become broader the longer the solute is retained.

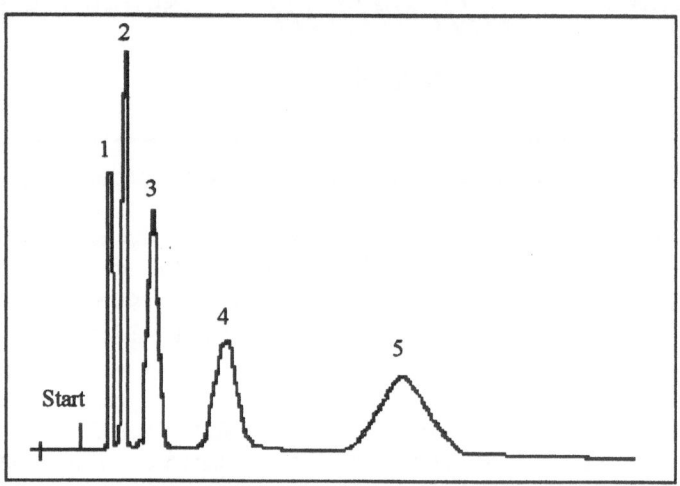

*Figure 4.* Chromatogram of a five component mixture. All components are completely resolved
(from ref. 4 with permission of the publisher)

The broadening of the solute band during migration along the column is caused by several phenomena:

- eddy diffusion (convective mixing of the solute in the interstitial voids of the column)
- longitudinal diffusion (diffusion of solute along the column axis)
- limited mass transfer kinetics of solutes between the stationary and mobile phase

Peak dispersion or peak broadening is characterized by the plate height H of the column expressed in length units. The plate height is defined as the increase in the variance, the square of the standard deviation of the peak in length units, $\sigma^2_L$, divided by the column length, L .

$$H = \sigma^2_L / L \qquad (9)$$

Peak dispersion has a negative impact on separation. The goal in chromatographic separation is to make the plate height as small as possible. Small plate heights mean narrow peak profiles.

The peak dispersion expressed by H is dependent on the flow-rate of the mobile phase. At low flow-rates the plate height is relatively high, because diffusion is the dominating dispersion phenomenon. With increasing flow-rate the impact of diffusion diminishes and the curve goes through a minimum where the plate height is lowest being determined by the eddy diffusion only. Beyond the minimum the plate height increases again due to slow mass transfer kinetics between the mobile and the stationary phase. The parabolic course of the H vs. u function can be expressed by the equation

$$H = A + B/u + Cu \tag{10}$$

where A is the term representing the contribution by the eddy diffusion, B the term reflecting the contribution due to the longidutinal diffusion and C the term comprising the contribution due to the mass transfer kinetics [16 - 19].

In order to compare columns with different particle sizes and different column dimensions dimensionless parameters have been introduced:

$$reduced\ plate\ height \qquad h = H/d_p \tag{11}$$

$$reduced\ linear\ velocity \qquad \nu = u\,d_p/D_{im} \tag{12}$$

where $d_p$ is the average particle diameter of the packing, u the linear velocity and $D_{im}$ the diffusion coefficient of the solute in the mobile phase. It was found that the reduced plate height of a retained solute equals two times of the average particle diameter of the packing at the minimum of the h vs. $\nu$ curve at $\nu \approx 5$. For a well-packed and efficient column the constants in eq.(10) amount to A = 1, B = 2 and C = 0.01.

## 4. Separations based on liquid-solid adsorption at analytical, preparative and process scale

### 4.1. SEPARATIONS AT ANALYTICAL SCALE

The objective of an analytical separation is to achieve an optimum resolution between two solutes or analytes at a minimum time [14, 20]. Resolution of the two peaks is needed for their identification on the basis of their rentention times or retention volumes. Base-line resolution is mandatory for quantitative analysis based on the peak height or peak area measurements.

The chromatographic resolution, $R_s$, between two analytes 1 and 2 is defined as:

$$R_s = 2\,(t_{R2} - t_{R1})/(w_{t1} + w_{t2})\ ;\ (t_{R2} > t_{R1}) \tag{13}$$

where $t_{R1}$ ($t_{R2}$) is the retention time of compound 1 (2) and $w_{t1}$ ($w_{t2}$) the peak width of the elution profile of compound 1 (2) at 10 % at the peak height. Assuming that both peak profiles are similar a complete separation (base-line separation) is achieved at $R_s = 1.5$ (optimum resolution ). Higher values of chromatographic resolution are a waste of time. In this context it is important to know the influence of the chromatographic parameters on the chromatographic resolution. It is given by the following equation

$$R_s = 1/4 \ (\alpha - 1) \ (k \, / \, 1 + k) \ N^{1/2} \tag{14}$$

Chromatographic resolution is affected by three terms: $(\alpha - 1)$ being the selectivity term, $(k \, / \, 1 + k)$ being the retardation term and $N^{1/2}$ being the dispersion term. The most dominating term is the first one: minor changes in the selctivity coefficient $\alpha$ exert a drastic efffect on $R_s$. In order to achieve a separation, $\alpha$ must be larger than unity. The second term is less important than the first one: it has only a notable impact at low k value and approaches to unity at higher k values of the analyte. The chromatographic resolution increase only with the square root of the plate number. The plate number N is defined as

$$N = L \, / \, H \tag{15}$$

N is a dimensionless quantity and related to one metre of column length. Doubling the column length for instance at otherwise constant conditions increases the plate number by a factor of two and the chromatographic resolution by a factor of 1.4. In order to achieve a separation the selectivity coefficient has to be maximized rather than the plate number.

In HPLC, stainless steel columns are used with about 100 to 200 mm in length and 4 and 4.6 mm in inner diameter. They are packed with adsorbents of about 5 μm average particle diameter. A well-packed column generates a plate number N being equal to

$$N = L \, / \, H = L \, / \, 2 \, d_p \tag{16}$$

at optimum flow rate, where $d_p$ is the average particle diameter of the adsorbent. Thus the plate number of the above mentioned column is about 10,000 (100 mm length) and about 20, 000 (200 mm length). With such a column performance the majority of analytical separation problems can be solved in practice. However, one should emphasise that the performance of a column can be drastically diminished by so-called extra-column effects due to large volumes of the connections between the injector and the top of the column and between the column outlet and the detector cell as well as other volume contributions. In other words the equipment must be designed in such a way that a low dead-volume results and a mixing outside of the column is avoided.

The number of analytical HPLC columns consumed per year world-wide amounts to about one million. This indicates the wide-spread application of HPLC in pharmaceutical, clinical, environmental and bioanalysis and in the analysis of products and chemicals. HPLC offers the unique advandage of resolving nearly every kind of soluble analyte from complex mixtures. Depending on the chemical and physical properties of the analytes several modes of HPLC can be applied to achieve a resolution [21]. The most common variants in HPLC are Reversed Phase (RPC), Ion-Pair Reversed Phase (Ion-Pair RPC), Ion Exchange (IEC) and Size Exclusion Chromatography (SEC). The latter is based on an entropic exclusion effect of the solutes from the pores of the adsorbent, whereby enthalpic interactions are suppressed. Table 1 lists the main characteristics of the HPLC variants.

TABLE 1. Characteristics of the most common HPLC variants

| mode | type of adsorbent | solvent /solvent mixture | operation |
|------|-------------------|--------------------------|-----------|
| RPC | silanized porous silicas, cross-linked hydrophobic polymers, porous carbon | aqueous/organic eluent of pH between 1 and 7 | isocratic elution gradient elution increasing content of organic solvent |
| Ion-Pair RPC | silanized porous silica | aqueous buffered eluent organic solvent containing 10 mM ion-pairing salt | isocratic elution |
| IEC | cation- or anion exchanger | aqueous buffered eluent containing an electrolyte of 10 mM to 1 M | gradient elution increasing salt concentration |
| SEC | porous silica, cross-linked polysaccharides, cross-linked polymer gels | aqueous buffered eluent containing an electrolyte, organic solvent | isocratic elution |

The separation can be performed in two operational ways : isocratic or gradient elution [3,4]. At isocratic elution the composition and thus the solvent strength of the eluent is maintained constant through the analysis. The peaks are eluted in a sequence of increasing retention, whereby the the peak height of the analyte becomes lower and the peak broader the longer the analyte is retained. At extended retention times ($k > 30$) the peak disappears at the base-line. For optimum resolution with respect to column efficiency and analysis time the capacity factor should range between $1 < k < 10$.

Isocratic elution is the preferred technique for sample mixtures typically containing maximum up to ten analytes and for quantitative analysis.

In gradient elution the solvent strenght of the eluent is either step-wise or continuously increased. In common the eluent is composed of a mixture ot two solvents, eg. water and acetonitrile in RPC.

The solvent strength is a parameter characterizing the polarity of the eluent relative to the adsorbent employed [22]. Solvents and solvent mixtures are ordered according to their solvent strenght in a series with increasing solvent strength. In RPC for silanized silicas water is the weakest solvent, followed by methanol, acetonitrile, tetrahydrofuran and n-hexane [20]. Changing the solvent or solvent composition from a one with lower to higher solvent strength decreases the retention of analytes due to the increasing interaction between the solvent and the stationary phase relative to the analytes. In other words changing the solvent strength of the eluent is an operational tool in HPLC to shift the analytes into a desired retention window.

In gradient elution, applying a continuous linear increase of the content of the stronger solvent leads to an expansion of the chromatogram for the weakly retained analytes and to a compression of the chromatogram for the strongly retained analytes. The situation is illustrated in Fig.5, comparing the chromatograms of a synthetic mixture resolved at isocratic and gradient elution. As mentioned above the peaks become broader with inceasing retention at isocratic elution. At gradient elution the peaks possess nearly the same peak heigth and width.

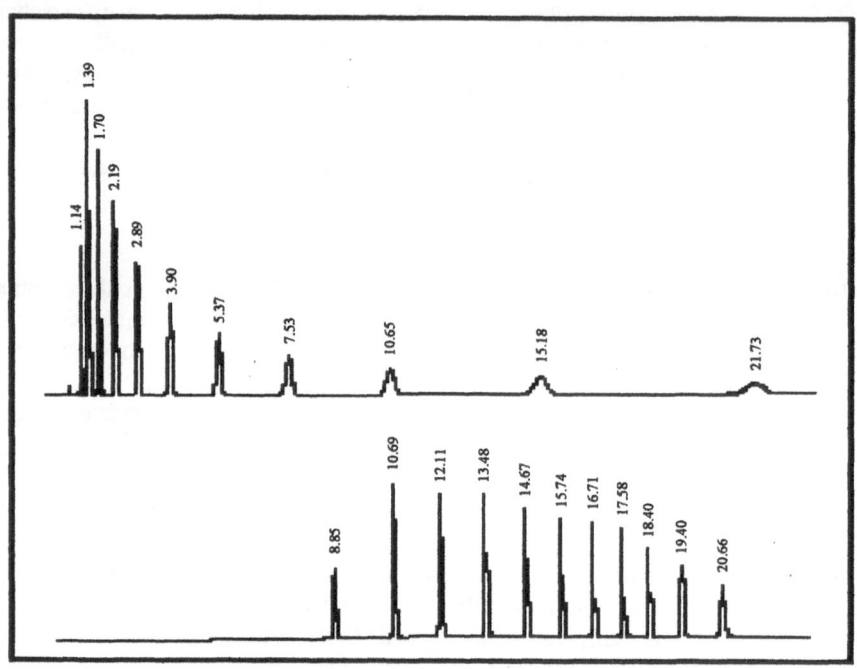

*Figure 5.* Isocratic elution vs. gradient elution of a synthetic multicomponent mixture
Isocratic conditions : column 125 x 4 mm packed with Kromasil 100 C 18, flow-rate:
1.5 ml/min ; eluent : water/acetonitrile 25/75 (V/V)
Gradient elution conditions: column and flow-rate the same;
water/acetonitrile 90/10 (V/V), 5 %/min to 10/90 (V/V) (5 min)
(from ref. [4] with permission of the publisher).

In conclusion, gradient elution is the prefered technique for the analysis of a multicomponent mixture, eg. amino acid analysis. It also serves as a means to survey the composition of an unknown mixture, when developing a separation strategy.

Running a separation isocratically or by gradient elution requires different instruments. In gradient elution one needs a corresponding number ot solvent reservoirs, mixing valves and pumps for mixing depending a low-pressure or high-pressure gradient is performed. The precision of quantitative analysis in HPLC is less in gradient elution than at isocratic operation.

Classical Column Liquid Chromatography was performed on native silica and alumina adsorbents with mixtures of organic solvents of adjusted solvent strength [22]. The elution order at HPLC on a native silica is in the sequence of increasing polarity. There are two main drawbacks when working with silica columns. Firstly, when analytes are resolved with weakly polar organic solvents, eg. n-hexane, the retention is strongly affected by the water content of the silica and the solvent. Reliable measurements can only be performed at a strict adjustment and control of the water content. Secondly, strongly polar analytes such as acids are often irreversibly adsorbed at the silica and thus cannot be eluted and separated. With the introduction of Reversed Phase packings [21] at mid of 1970's these disadvantages were eliminated. In RPC hydrophobic adsorbents are applied with aqueous/organic solvent mixtures. Retention is controlled by the content of the organic solvent eg. acetonitrile. There is quasi-linear realationship between the logarithm of the capacity factor of the analyte and the content of the organic solvent in the aqueous/organic solvent mixture in RPC: log k decreases linearly with the content of organic solvent. The diminution spans several orders of k from zero to 100 % of organic solvent in water. As compared to HPLC on native silica the elutuion sequence in RPC is reversed. The analytes are retained in the sequence of decreasing polarity. This means that ionic and strongly polar analytes elute within the dead volume of the colume, whereas strongly hydrophobic analytes are elutes last.

It should be emphasized that the adsorbent (silanised silica) does not behave totally hydrophobic. The silanization of the silica does not result in a complete removal of the acidic hydroxyl groups at the native silica surface. About 50 % remain after the chemical modification and thus provide a hydrophilic character of the Reversed Phase adsorbent. As a consequence a n-octadecylbonded silica is a bifunctional adsorbent with polar residual hydroxy groups and lipophilic n-alkyl groups. This bifunctional character in the surface composition is reflected by the retention behavior of RPC columns: lipophilic as well as hydrophilic analytes can be resolved on such adsorbents. Even ionized analytes acan be separated provided a pH-value of the eluent is chosen where the solute ionization is depressed. Now-a-days the majority of HPLC analyses are performed on Reversed Phase columns mainly for the reason of their widely spread selectivity to all kinds of analytes. The major disadvatages of silanized silicas as adsorbents is their limited pH-stability at pH above 8.

Ion-pair RPC [24] is based on the formation of neutral ion pairs formed between the negatively (positively) charged analyte and a positively (negatively) charged pairing ion added to the eluent. For the separation of acids tetraalkylammonium salts as ion-pair reagents are employed. Bases require n-alkylsulphonic acids as additive. The pH of the eluent must be set to a range where the analytes are ionized. Reversed Phase adsorbents are employed as packings at which the ion-pair reagents are adsorbed through their hydrophobic entities. Ion-Pair RPC is a powerful tool to resolve basic and acidic analytes and much more less extensive with respect to method development as compared to IEC.

IEC is performed on ion-exchanger columns with aqueous buffered eluents [25,26]. The analytes span low-molecular weight species such as small cations and anions, acids and bases as well as high-molecular weight such as peptides, polypeptide, polynucleotides and nucleic acids. The support material of ion-exchangers are porous silica and various cross-linked polymer gels. Depending on the type of exchanger (cation, anion) they contain functional groups which are acidic or basic. Cation (anion) exchanger differ in the type and the pk of the functional group and can be classified into strong and weak ion exchangers depending on their pk. In order to fully utilize the capacity of an ion exchanger one has to choose pH-conditions where the acidic or basic functional groups are ionized. For the resolution of polypeptides, polynucleotides and nucleic acids large pore diameter adsorbents have to be employed to avoid a steric exclusion and a hindered diffusion of the charged analytes [27].

Typical application areas of ion exchangers are to analyse cations and ions in water and the analysis ot proteins and enzymes. At the latter case gradient elution is the preferred technique. In both cases the ion exchangers employed are tailor-made with respect to their texture, pore structure and surface chemistry [27].

SEC is a HPLC variant which allows one to separate analytes according to their size and shape [12]. The analytes are eluted in the sequence of decreasing size or molecular weight between two limits: large analytes which are too large to penetrate the pores of the packing are exclude and elute with the elution volume being equal to the interstitial column volume, $V_i$. Analytes which are to small to penetrate the whole pore volume of the packing are eluted with a volume corresponding to the interstitial and the total pore volume of the packing (dead vovlume of the column), $V_m$. Analytes which selectively permeate the packing according to their size and shape elute with decreasing molecular weight between these two limits. The resolution of a SEC column is usually characterized by its calibration curve, ie. the plot of the logarithm of the hydrodynamic volume or molecular weight against the elution volume. The calibration curve is experimentelly measured using molecular weight standards. The shape and the slope of the curve is a function of the type and pore size of the packing, the type of polymer used and the eluent.

SEC is based on an entropic exclusion of analytes. Enthalpic interactions which might lead to an additional retention are suppressed by choosing an eluent with a high solvent strength. In practice, SEC is performed in two ways: synthetic polymers are resolved on polymer gels in organic solvents, eg. tetrahydrofuran. Proteins and enzymes are resolved on porous hydrophilic packings and aqueous buffered eluents with an electrolyte added to suppress electrostatic interactions. The decisive parameter of a packing and column is the average pore diameter of the packing or the molecular weight fractionation range of the column being assessed with standards. Commonly, two columns with a mean pore size of 10 and 50 nm, respectively, are sufficient to span a fractionation range between 1.000 and 1.000.000 [12].

In conclusion, each variant of HPLC provides a specific selectivity towards specific types of analytes and different tools to achieve the resolution. Often several alternatives can be used to separate a certain class of compounds. For instance ionic analytes can be resolved by IEC, Ion-Pair RPC and also by means of RPC provided the ionisation is suppressed.

## 4.2. SEPARATION AT A PREPARATIVE SCALE

The knowledge in analytical HPLC serves in part as a basis to develop a strategy for the isolation and purification of compounds. However, there is a fundamental difference between the analytical and the preparative mode. In analytical HPLC one is working in the linear range of the isotherm. This implies that the chromatographic parameters one is using to judge a separation such as the plate height, the capacity factor etc. are constant and independent of the concentration of the sample. In preparative HPLC one is generally working in the non-linear range of the isotherm, where all the parameters are concentration dependent [28]. To separate larger amounts of solutes (isolates) one can choose between two alternatives [29]:

- enlargement of the column diameter while maintaining the analytical conditions
- overloading the column.

The overloading of a column is performed either by increasing the concentration at constant injection volume (concentration overload) or by increasing the injection volume at constant concentration (volume overload). Concentration and volume overload cause a different change of the peak profile of the sample as it is compared to the analytical separation [30]. This is schematically indicated in Fig. 6.

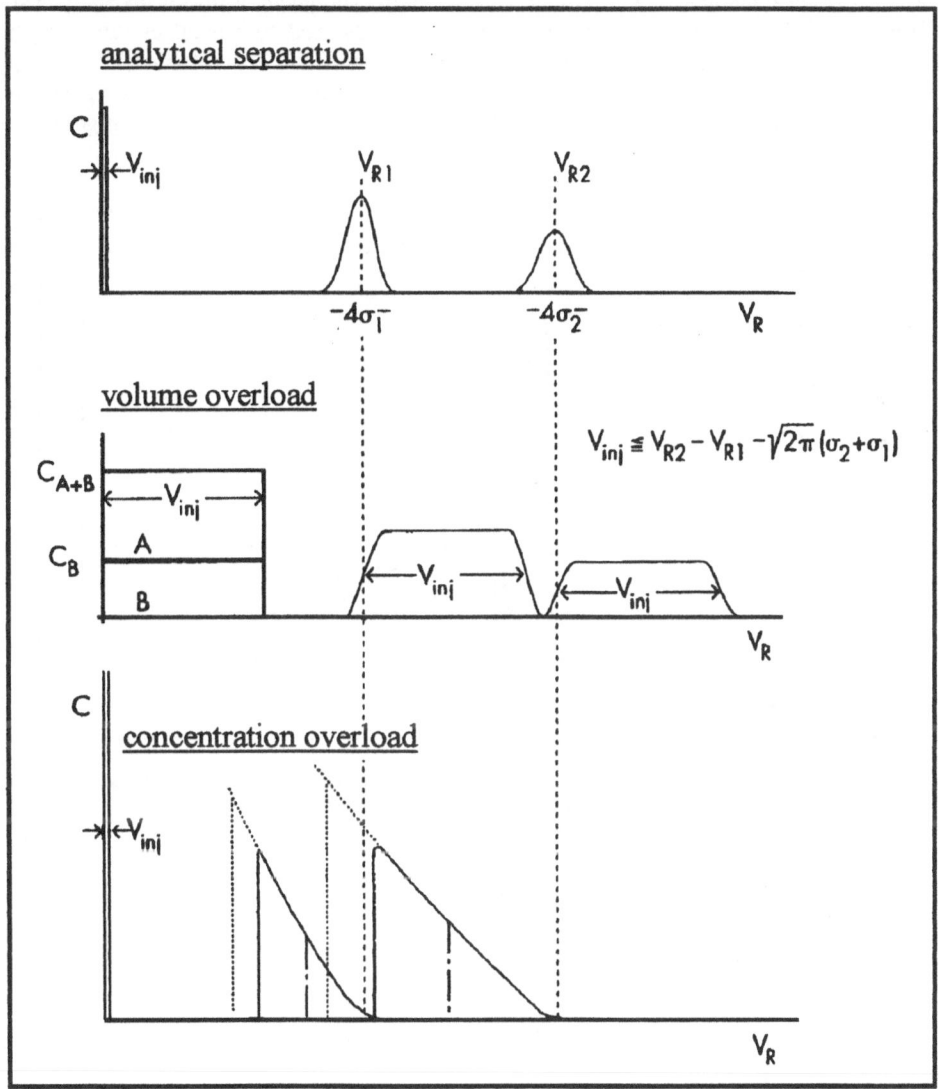

*Figure 6.* Schematic representation of the effect of volume overload and concnetration overload in HPLC (from ref. [30] with permission of the publisher)

Fig. 6 shows the ideal case, when no interaction between the two solutes occurs. In case of competitive adsorption the elution profiles of two solutes change significantly, as it is

420

illustrated in Fig. 7. The first eluting peak is sharpened whereas the second indicates a pronounced tailing. The effect results in a enhanced concentration of the first eluted compound and a dilution of the second compound.. The following two figures exemplify a situation where both compounds of a binary solution are injected in different concentration ratios. In Fig. 8 the ratio is 1:9, at Fig. 9: 9:1. At a higher concentration of the second, more strongly retained compound, the peak of the first eluting compound is drastically sharpened. This effect is due to the displacement of the weakly adsorbed compound by the stronger adsorbed compound (displacement effect). In Fig. 9 where the concentration of the first compound is higher than the second, a pronounced smearing of the second compound is observed (tag-along effect). The peak profiles were calculated on the basis of Langmuirian isotherms.

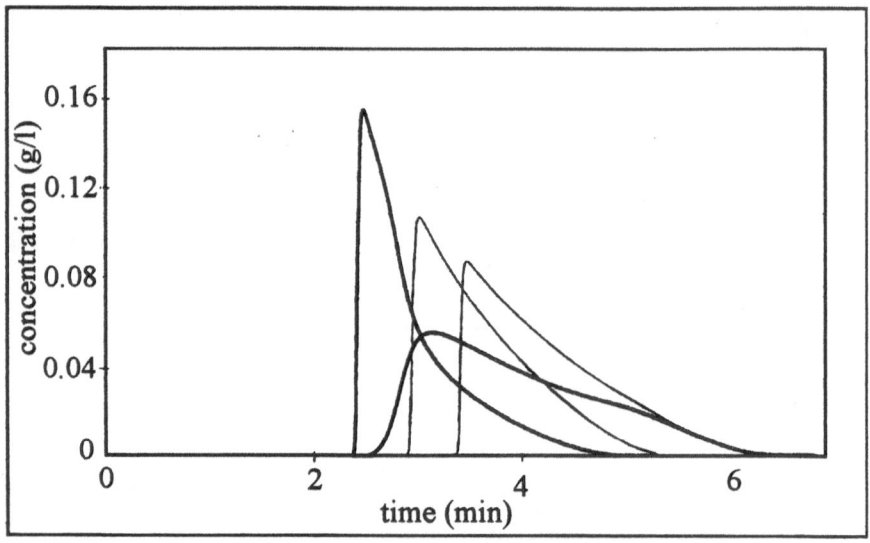

*Figure 7.* Comparison of elution profiles of a binary system with competitive adsorption (thick lines)
and without competitive adsorption (thin lines)
(from ref. [8] with permission of the publisher)

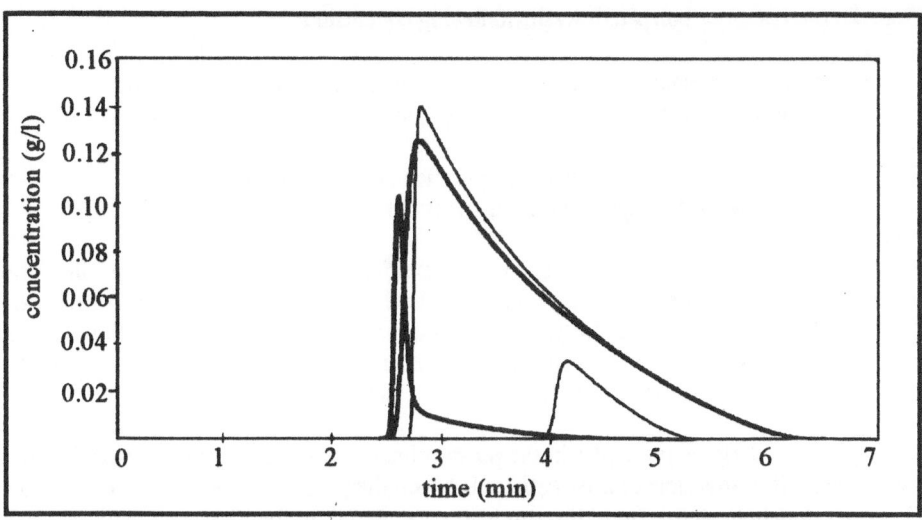

*Figure 8.* Comparison of elution profiles of a binary system with competitve adsorption (thick lines) and without competitve adsorption (thin lines) at a concentration ratio of component 1 to component 2 of 1:9 (from ref. [8] with permission of the publisher)

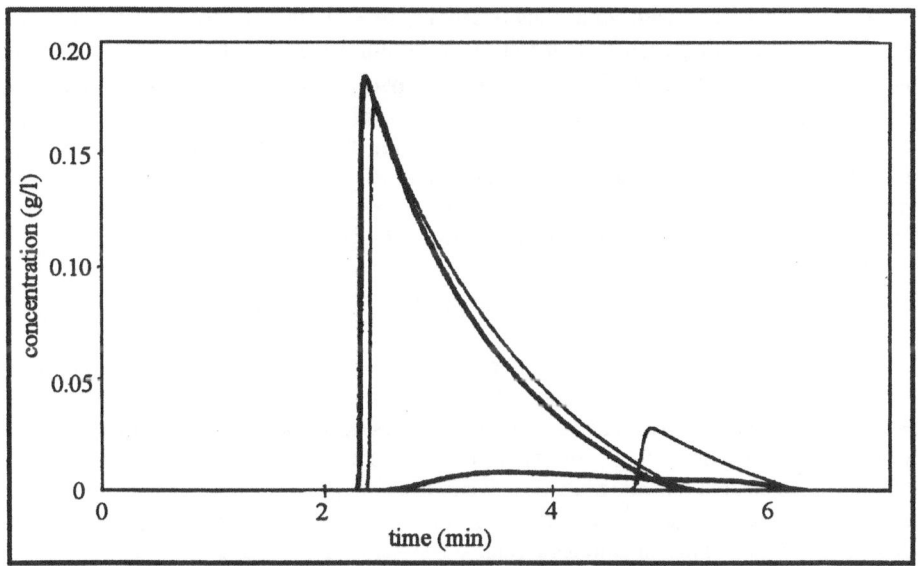

*Figure 9.* Comparison of elution profiles of a binary system with competitive adsorption (thick lines) and without competitive adsorption (thin lines) at a concentration ratio of component 1 to component 2 of 9:1. (from ref. [8] with permission of the publisher)

From the presentations one can draw the following conclusion:

- modelling of isotherms and computing of the peak profiles helps to a large extent to predict the elution behavior of solutes from binary solutions

- it also serves as a tool to set up a separation and to optimize the pararmeters with respect to certain target quantities.

Whereas in analytical separation the target quantities are resolution, analysis time and sensitivity, preparative separation obeys totally different rules. The target quantities here are the purity of product required, the yield of product of desired purity and the through-put, ie. the mass of pure product obtained per unit time. In process scale operation the dominating parameter is the cost of purification and isolation.

Although the underlying principles in the purification of solutes by chromatographic means are the same, it is advisable to differentiates between the preparative and the process mode. The discrimination stems from the amount of material to be separated and purified. Typically, by preparative means, mg to g amounts are purified. Such sample amounts are required for biological and biochemical tests and for the structure elucidation of compounds. The compounds represent by-products from synthetic mixtures, metabolites in biological matrices and natural substances. Preparative HPLC is performed on wide-bore columns of 8 mm to one inch inner diameter at flow rates up to several 100 ml/min. Such columns can be either packed with coarse particles of 30 to 40 μm average particle diameter or with high-performance packings of 5 to 10 μm average particle diameter. The injection volume usually ranges between several hundred μl to a few ml. The instrumentation is either an anlytical equipment, where the column is overloaded or the sample is collected by means of repetitive injections. For running an one-inch inner diameter column one needs a special pump delivering higher flow-rates than analytical pumps and corresponding detectors designed for preparartive purposes. The cost factor is less important.

## 4.3. SEPARATIONS AT PROCESS SCALE [31,32]

As it is named, the separation is performed at process scale where kg amounts are purified and isolated. Due to legislative regulations severe demands are put towards the purity of compounds such as pharmaceuticals. To-day impurities have to be identified and quantitated in such products down to 0.1 %. In case of enantiomers the purity of the two enantiomers and their toxicolocigal efficacy must be evidenced before the drug is registred. It should be noted in this context that more than 50 % of the pharmaceuticals on the market are enantiomers. Often pure enantiomers cannot be obtained by means of a stereoselective synthesis. Then the only solution of the problem is to apply chiral adsorbents and to perform a separation.

Process scale separation of fine chemicals and value-added products is more an engineering task rather than a laboratory work. The columns in such processes are made of stainless steel and have inner diameters between 100 and 800 mm and a length up to one meter. A prefractionation is conveniently performed before the essential separation, depending on the purity of the isolate. The column is a minor part of the whole plant. Devices for storage and evaporation of solvents are mandatory. When using organic solvents the whole unit has to be placed in an explosive-protected environment with the control devices outside. Most importantly, the elution profiles must be monitored during the operation, using on-line analytical equipment in order to cut the correct fractions.

The whole process is no longer a chromatographic technique as applied in analytical HPLC or the preparative mode. For economic reasons the costs for solvents have to reduced. In this respect the elution mode is not an economic means for separation. Recycling techniques with heart cutting of peaks and simulated moving bed technology [33] are the techniques of choice. Furthermore, displacement techniques are also applied. Microparticulate packings which are used in analytical HPLC are not required due to the large overload of the column. Often columns which generate a few hundred plate numbers do the job. The packing required should offer a high selectivity at high loading and a high utilization of the capacity at high column through-put.

## 5. Case studies

### 5.1. ULTRA-FAST SEPARATIONS OF PROTEINS

Analytical HPLC separations are performed within a few minutes under routine work. For the control of chemical and biotechnological processes, however, a fast analytical response is required.

The chromatographic theory predicts the conditions to perform ultrafast analysis. Firstly, one can increase the flow-rate at columns packed with 5 µm particles from 1ml/min to about 5 ml/min. In most cases the increase of flow-rate must be paid with a loss in column performance depending on the quality of the column and column packing. One solution is to further reduce the particle diameter of the adsorbent to 3 to 4 µm. On the basis of optimization calculations taking the column plate number, the analysis time and the pressure drop into account, this ultimate value was already predicted by Knox in 1977 [19]. Reducing the average particle diameter by a factor of two, increases the plate number by a factor of two, enhances the analysis time by a factor of four, but increases the column pressure drop four times. To run the column at reasonable pressure, the column length has to be reduced as the column length is proportional to the pressure drop. A severe problems in the manufacture of 3 to 4 µm particles is to size such materials into narrow size fractions eg. by means of air elutriation techniques and to pack efficient columns with the expected plate number at optimum eluent flow-rate.

Another alternative to fast separations of large molecular weight compounds such as proteins is to use non-porous particles of about 2 µm average particle diameter. Such non-porous spherical silica particles are manufactured by means of hydrolysis and condensation of tetraethoxysilane under defined conditions [34-36]. The particular advantage of the manucfacturing processes of these particles is that particles are obtained with an adjusted and controlled particle size. In other words, no subsequent sizing is required. The particles can be subjected to surface modifications to obtain Reversed Phase adsorbents, ion-exchangers etc [35]. Such particles are packed in columns of 4 mm inner diameter and 30 mm length. The dead volume of such columns typically amounts to about 200 µl. To

424

maintain the performance of the column the extra-column volume contribution must be at minimum. The injection volume usually is about 0.5 μl and the detector cell volume has about the same value. Special detectors eg. UV-detectors must be used with a response time smaller than 30 ms. At such conditions ultrafast separations can be performed within a few seconds a illustrated in Fig. 10.

The concept of non-porous particles being originally developed for the separation of proteins can also be applied for the fast resolution of low-molecular weight compounds such as enantiomers.

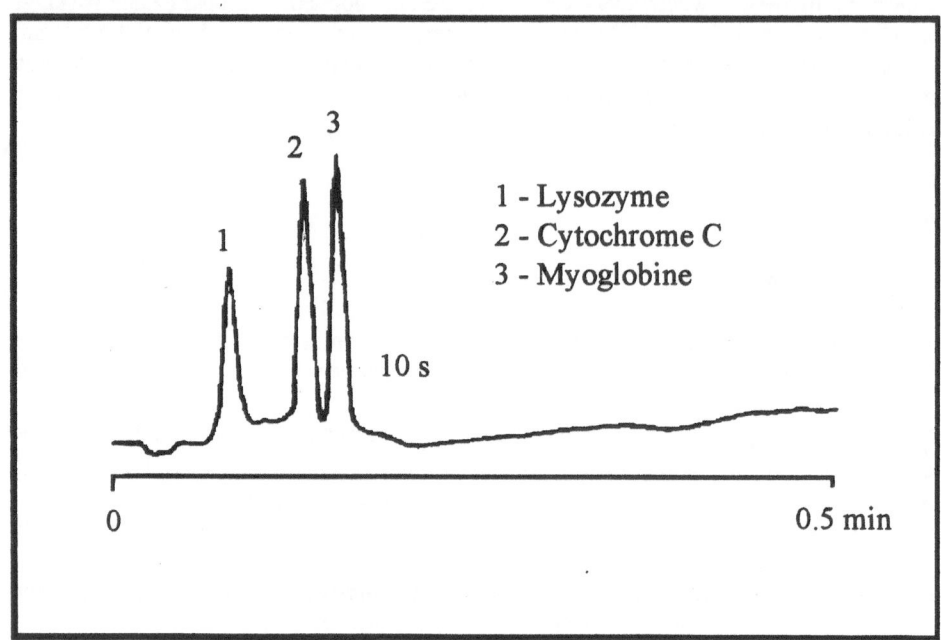

*Figure 10.* Ultrafast separation of three proteins on a column of 20 x 4.6 mm packed with n-octadecyl bonded 1.7 μm non-porous silica particles
Conditions : gradient elution from water/acetonitrile of 72/28 (V/V) to pure acetonitrile, flow-rate: 5 ml/min; detection: UV 210 nm; column temperature: 45 °C; volume of gradient mixer: 240 μl
(from ref. [37] with permission of the publisher)

## 5.2. PURIFICATION AND ISOLATION OF ENANTIOMERS BY RECYCLING AND HEART CUTTING TECHNIQUES

In case of a two-component isolation, eg. enantiomers, recycling has be found to be an effective means as compared to the elution mode. Recycling the sample and running it again through the column simulates a situation where the column is virtually extended. After a certain number of cycles which is monitored by on-line analytical HPLC, two elution profiles move apart in such a way that pure fractions can be cut of the two compounds. The scheme of the equipment needed for recycling is shown in Fig. 11. Fig. 12 displays an industrial scale of two enantiomers an a large-scale column packed with a chiral adsorbent.

In conclusion, as compared to the elution mode, recycling enhances the column through-put and the yield. It drastically reduces the consumption of solvent and packing. Furthermore, the technique can be easily automated.

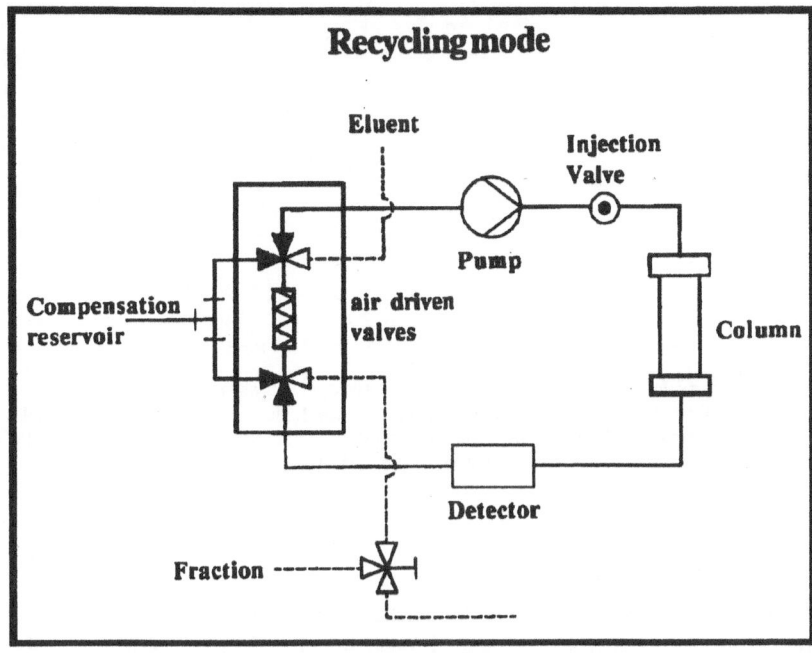

*Figure 11.* Scheme of equipment to be operated in the recycling mode [38]

426

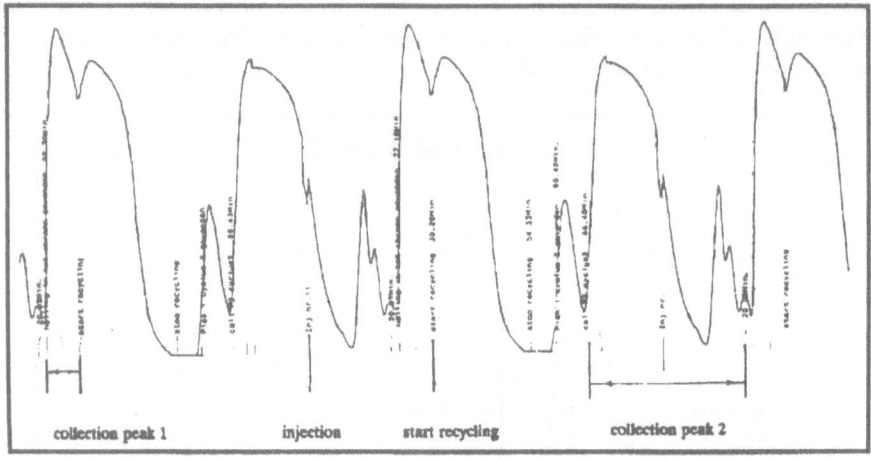

| collection peak 1 | injection | start recycling | collection peak 2 |

*Figure 12.* Industrial scale recycling of enantiomers on a column with a chiral packing [39]
column: Prepbar[R] 400-100 mm, E. Merck, Darmstadt, filled with 1.2 kg of Chiracel OD, 20 μm, Daicel;
sample: 5 g dissolved in 175 ml of eluent; flow-rate of eluent 150 ml/min; eluent: n-hexane / ethanol (80/20 (V/V);
column temperature: ambient

## 6. Conclusion and perspectives

Separation processes based on liquid-solid adsorption have gained a dominanting place in separation technology under environmental, economical and energy saving aspects. They are applied at process scale to separate tons of base chemicals and kg-amounts of value-added products such as pharmaceuticals. At analytical scale HPLC has found its unrevailed place as routine method in nearly every laboratory.

Further developments are directed towards miniaturization using packed capillaries or even chips for the quantitation of trace amounts of analytes. Hybrid methods are under development which combine features of two methods such as Electro-Chromatography being a hybrid between capillary electrophoresis and micro-HPLC.
The most promising area, however, is in the field of cost-effective purification procedures for fine chemicals, intermediate products and value-added organics by means of liquid-solid adsorption processes.

# 7. References

[1]     Kippling, J.J. (1965) *Adsorption from solutions of non-electrolytes* , Academic Press, London.

[2]     Parfitt, G.D., Rochester, C.H. (1983) *Adsorption from solution at the solid/liquid interface*, Academic Press, London.

[3]     Meyer, V.R. (1988) *Practical High-Performance Liquid Chromatography* , Wiley & Sons, Chichester.

[4]     Unger, K.K., Weber , E. (1995) *Handbuch der HPLC, Tei 1 : Leitfaden für Praktiker und Anfänger* , GIT Verlag, Darmstadt (English edition in preparation).

[5]     Johnson, J.A., Oroskov, A.R. (1989) Sorbex technology for industrial scale separation in H.G. Karge and J. Weitkamp (eds), *Zeolites as catalysts, sorbents and detergent builders,* 451-467, Elsevier, Amsterdam,

[6]     Everett, D.H. (1972), *Definitions, termonology and symbols in colloid and surface chemistry,* Pure Appl. Chem. 31, 579

[7]     Köster, F., Findenegg, G.H. (1982) *Adsorption from binary solvent mixtures onto silica by HPLC frontal analysis* , Chromatographia 15, 743-747

[8]     Seidel-Morgenstern, A. (1995) *Mathematische Modellierung der präparativen Flüssigchromatographie,* Deutscher Universitätsverlag, Wiesbaden .

[9]     Fukuchi, K., Kobuchi, S. Aray, Y. (1982) Chem. Engng. Japan 15, 316

[10]    Radke, C.J., Prausnitz, J.M. (1972) , AlChE 18, 761

[11]    Karger, B.L., Snyder, L.R. , Horvath, Cs. (1973) *An introduction to separation science,* John Wiley & Sons, New York.

[12]    Yau, W.W., Kirkland, J.J. , Bly, D.D. (1979) *Modern Size-Exclusion Liquid Chromatography,* John Wiley & Sons, New York.

[13]    Samsonov, G.V. , Kuznetsova, N.P. (1992) *Crosslinked polyelectrolytes in biology,* Adv. Polym. Sci. 104, 1-50.

[14]    Knox, J.H. (1978) *High-Performance Liquid Chromatography,* Edinburgh University Press, Edinburgh.

[15]    Zlatkis, A. Kaiser, R.E. (1977) *HPTLC High Performance Thin-Layer Chromatography* , J. Chromatogr. Libr. 9, Elsevier, Amsterdam.

[16]    Giddings, J.C. (1965) *Dynamics of Chromatography,* Part I, Marcel Dekker, New York.

[17]    Kennedy, G.J., Knox, J.H. (1972) J. Chromatogr. Sci. 10, 549

[18]    Knox, J.H. , Pryde, A. (1975), J. Chromatogr. 112, 171

[19]    Knox, J.H. (1977) J. Chromatogr. Sci. 15, 352

[20]    Snyder, L.R. , Glajch, J.L. , Kirkland, J.J. (1988) *Practical HPLC method development* , John Wiley & Sons, New York.

[21]    Unger, K.K. (1990) *Packings and stationary phases in chromatographic techniques,* Marcel Dekker, New York.

[22]    Snyder, L.R. (1968) *Principles of adsorption chromatography* , Marcel Dekker, New York.

[23]    Unger, K.K. (1979) *Porous silica - its properties and use in Column Liquid Chromatography,* Elsevier, Amsterdam.

[24]    Persson. B.-A., Lagerström. P.-O. (1990) Packings and stationary phase for ion pair chromatography in K.K. Unger (ed.) *Packings and stationary phases in chromatographic techniques,* Marcel Dekker, New York.

[25]    Pietrzyk, D.J. (1990) Ion exchangers in K.K. Unger (ed.) *Packings and stationary phases in chromatographic techniques* , Marcel Dekker, New York, 585-720.

[26]    Dorfner, K. (1991) *Ion exchangers* , Walter de Gruyter, Berlin.

[27]    Gooding, K.M., Regnier, F.E. (1990) *HPLC of biological macromolecules* , methods and applications, Marcel Dekker ,New York.

[28]    Guiochon, G. , Shirazi, S.G., Katti, A.M. (1994) *Fundamentals of preparative and non-linear chromatography* , Academic Press, Boston

[29]    Unger, K.K. (1993) *Handbuch der HPLC, Teil 2: Präparative Säulenflüssig-Chromatographie,* GIT Verlag, Darmstadt.

[30]    Knox, J.J., Pyper, H.M. (1986) , J. Chromatogr. 363 , 1

[31]    Subramanian, G. (1995) *Process scale chromatography* , VCH, Weinheim.

428

[32]     Sofer, G.K., Nyström, L.E. (1989) *Process chromatography - a practical guide,* Academic Press, London.

[33]     Nicoud, R.M. , Baily, M. , Kinkel, J.N. , Devant, R.M., Hampe, Th. R.E., Küsters, E. (1995)
         *Simulated mowing bed (S.M.B.): Applications for enantiomer separations on chiral stationary phases ,*
         SEPAREX, Champigneulles, France.

[34]     Unger, K.K., Giesche, H. , US patent 4,775,520

[35]     Kovatzs , sz.E., Jelinek, L., European patent EPPO.574.642.A1

[36]     Barder, T.J., Bubois, US patent 4.983.369

[37]     Kaiser, Ch., Hanson, M., Giesche, H., Kinkel, J., Unger, K.K. (1996) Nonporous silica microspheres in the
         micron and submicron size range : manufacture,  characterization and application in Pelizetti, E. (ed.)
         *Fine particle science and technology: from micro to nanoparticles,*
         Kluwer academic publishers, Dordrecht, The Netherlands.

[38]     By courtesy of Kinkel, J.N. (1994) E. Merck KGaA, Darmstadt, FRG.

[39]     By courtesy of Dingenen, J, (1994) Janssen Pharmaceutica, Beerse, Belgium .

# MOLECULAR MODELING AS A COMPLEMENT TO EXPERIMENT; APPLICATION TO THE SEPARATION OF NITROGEN AND OXYGEN.

C. Mellot and J. Lignières

*AIR LIQUIDE, Centre de Recherche Claude Delorme, Les Loges-en-Josas, B.P. 126, 78350, Jouy-en-Josas, France.*

## 1. Introduction

Classical cryogenic processes for the separation of oxygen and nitrogen are gradually giving way to a new generation of technologies involving either Pressure Swing Adsorption (PSA) and Vacuum Swing Adsorption (VSA) processes or polymeric membranes permeation processes. The basic principle of $O_2$ production from air by PSA/VSA processes relies on the selective adsorption of nitrogen over oxygen in a zeolitic adsorbent bed. The VSA process consists of adsorption-desorption steps operating between a high pressure, for the production of the oxygen enriched stream, and a lower pressure, for the adsorbent regeneration step. In addition to the operating parameters of the process itself, one of the important factors influencing the performance of the production unit is related to the adsorption properties of the zeolitic material at thermodynamic equilibrium. The physico-chemical properties of the zeolitic material strongly determine the shape of both the $N_2$ and $O_2$ isotherms, especially in terms of $N_2/O_2$ selectivities at the production pressure (~1 bar). A thorough understanding of the dependance of adsorption isotherms upon the microscopic features of the zeolite is therefore of prime importance.

Molecular modeling has recently emerged as an efficient tool for improving the fundamental understanding of basic microscopic phenomena and for helping to solve industrial relevant problems involving chemical or physico-chemical processes. Starting from the atomic scale, molecular modeling is the appropriate tool to develop a qualitative and quantitative understanding of structure-properties relationships in a wide range of systems like molecules, solids, or surfaces [1,2]. One of the great challenge in theoretical and simulation approaches consists of understanding and retrieving useful or even predictive information about these systems. When dealing with complex molecular systems like zeolites, experimental studies have preceeded by

429

*J. Fraissard (ed.), Physical Adsorption: Experiment, Theory and Applications, 429–459.*
© 1997 *Kluwer Academic Publishers.*

far the theoretical ones. Yet, as programming technologies improve and computational power increases, simulation methods are being increasingly used as a complement to experiment. A molecular modeling approach enhances the fundamental understanding of the basic microscopic phenomena and gives a better insight into interactions at a molecular level, especially into the behaviour of adsorbed phase at the solid-gas interface.

In the following sections, we focus on the simulation of $N_2$ and $O_2$ adsorption in zeolitic materials. As the separation mechanism is determined by molecular-level interactions, Grand Canonical Monte Carlo (GCMC) simulations are appropriate for correlating the microscopic features of the zeolite/sorbate system to macroscopic properties of interest that are experimentally determined such as adsorption isotherms and isosteric heats. In this chapter, we intend to show how different molecular modeling tools, like statistical mechanics, molecular mechanics or quantum mechanics, may be used complementarily to work out a relevant understanding of $N_2$ and $O_2$ adsorption phenomenon in zeolites and to support the experimental approach to $N_2/O_2$ separation. Calculations were performed using Biosym softwares package (Biosym Technologies Inc., San Diego, 1994, release 237).

## 2. General considerations on zeolites and their adsorption properties

### 2.1. ZEOLITES STRUCTURES [3,4]

Zeolites are tridimensional aluminosilicates with the following chemical formula in the as-synthesized form :

$$x M_{2/n}O, \ x Al_2O_3, \ y SiO_2, \ n H_2O$$

where M is a cation of valence $n$, which may be of organic or inorganic nature. The main characteristic of a zeolitic material is that the tetraedral $[SiO_4]$ and $[AlO_4^-]$ primary building blocks are linked through oxygens to build a tridimensional microporous network of cages or channels of molecular dimensions. Depending on the number of oxygen atoms that determine the pore aperture of their largest channels, which vary between 5 and 20 Å, zeolites are commonly referred to as ultralarge (>12), large (12), medium (10) or small (8-membered ring) pore structures. The isomorphous substitution of $Si^{4+}$ by the trivalent metal $Al^{3+}$ in the framework induces negative charges that are compensated by cations or protons, the number of which depends on the Si/Al ratio of the framework. When considering both structural and chemical features, the interconnection of cavities and channels confers upon zeolites very high surface areas and pore volumes, which, when combined with the possibility of having active cationic or protonic sites, generate a large panel of adsorption [5] and catalytic properties [6]. A wide range of crystalline structures of the zeolite type but containing tetrahedrally coordinated Si, Al, P atoms, as well as transition metals and many groups elements with the valence ranging from I to V (B, Ga, Fe, Cr, Ti, V, Mn, Co, Zn, Cu...) have been also synthesized and have the generic name of zeotypes.

Figure 1 shows the main characteristics of three different types of structures, widely used in separation and catalytic areas. Zeolites A and X are especially used in air components separation for oxygen production.

## 2.2. SEPARATION PROPERTIES

Separation properties of zeolites rely either on shape selectivity or energetic selectivity of the sorbent/sorbate system. The first type of selectivity, occuring when the zeolite acts as a true molecular sieve, is caused either by the size and/or the shape differences between the sorbate molecules and the pore apertures, or by significant differences in diffusivities leading to a mass transport discrimination. The second type of selectivity is related to differences in sorbate-sorbent interation energies occuring between competing species. When adsorbed within the micropores, sorbate molecules are subjected to a series of interactions, such as van der Waals and electrostatic interactions including induction and various multipole-multipole interaction terms (field-dipole, field gradient-quadrupole, dipole-dipole, dipole-quadrupole, quadrupole-quadrupole interactions). To the extent that adsorption may be regarded as the diffusion of molecules within the micropore by thermally activated jumps from one site to another, it can be easily understood that the local structure and chemical features of adsorption sites are essential in determining the sorbent-sorbate interaction energy, and therefore, adsorption and separation effects.

The van der Waals energy term is of a non-specific nature and is always present whatever the sorbate and sorbent properties are, in contrast to the other types of forces that may or may not be present depending on the properties of the interacting species. Other terms, of an electrostatic and therefore specific nature, will be more determined by the physical properties of the interacting species and particularly of the sorbate molecule itself (charge, dipole and quadrupole moments). In other words, specific sorbent-sorbate interactions may be used to induce energetic discrimination between competing sorbate species, and consequently to perform their separation. In the area of separations, the use of zeolitic sorbents having low framework Si/Al ratios and having a large number of extraframework cations, which generate very high electrostatic fields and field gradients within the microporous volume, leads to selective sorption of molecules having different polarities.

## 2.3. SEPARATION OF $N_2$ AND $O_2$ IN ZEOLITES

The separation of $N_2$ and $O_2$ is performed in cation-containing zeolites. When considering $N_2$ and $O_2$ molecular properties (table 1), the significantly higher permanent quadrupole moment of $N_2$ compared to that of $O_2$ is the main cause of their thermodynamic separation, as $N_2$ interacts more strongly than $O_2$ with the field gradient generated by the extra-framework cations. In fact, the combination of a large variety of framework structures and of their different cationic-exchanged forms leads to a large diversity of adsorption and separation properties. Figure 2 is an attempt to illustrate this latter point, showing the main features of pure $N_2$ and $O_2$ adsorption isotherms in three different structures. In this chapter, $N_2/O_2$ selectivity is estimated through the ratio of adsorption capacities of $N_2$ over $O_2$ at 1 bar. Purely siliceous silicalite is a representative example of a zeotype material without any $N_2/O_2$ selectivity, which is attributed to the lack of extra-framework cations in this material. When considering cation-containing structures such as X-type (or A-type) zeolites, separation effects are observed. The as-synthetized NaX zeolite is characterized by a $N_2/O_2$ selectivity of 3 at 1 bar. The exchanged CaX zeolite is characterized by a clear increase in $N_2/O_2$ selectivity, up to 4 at 1 bar, which can be explained by the stronger interaction of $N_2$ molecules with divalent cations.

The simple difference in physical properties between $N_2$ and $O_2$ molecules, while explaining the separation effect in its basic principle, does not permit to establish a systematic correlation between the structural and chemical features of a specific zeolite and its separation performances in terms of selectivity. The use of statistical mechanics through Monte Carlo simulations is therefore of special interest in order to investigated this correlation, furnishing the link between microscopic phenomena and macroscopic properties such as adsorption isotherms and isosteric heat.

### 3 . Simulation of $N_2$ and $O_2$ adsorption in zeolites

### 3.1. GENERAL CONSIDERATION ON MONTE CARLO SIMULATIONS

The Monte Carlo method is a means of estimating the configurational contributions to the thermodynamic properties of a classical statistical mechanical system, consisting in N molecules in a confined volume V at a given temperature T. This method, developed by Metropolis [7], allows us to generate different types of statistic ensemble, calculating isosteric heats of adsorption of guest molecules in host structures in the $(N,V,T)$ canonical ensemble, adsorption isotherms in the $(\mu,V,T)$ grand canonical ensemble, or gas phase equilibria in the Gibbs ensemble. As for the simulation of adsorption isotherm, the GCMC (grand canonical Monte Carlo) simulation can thus capture the equilibrium between a solid sorbent phase and a liquid or gas phase, and provides as a simulation result the particule population density adsorbed in the sorbent at the defined temperature, pressure and volume. In its basic principle, the Metropolis procedure generates randomly microscopic states, $k$, with a frequency that is proportional to the Boltzmann factor $\exp(-U_k/kT)$. The main steps in a Monte Carlo calculation may be summarized as follows: (i) random generation of a starting configuration. (ii) generation of a new state by the translation and/or the rotation of a randomly chosen molecule. (iii) selection of the new state according to an energetic criteria of acceptance. (iv) creation and deletion tests, which are added in case of a GCMC simulation. The thermodynamic quantities are then calculated using direct averaging over all accepted configurations.

The input of the simulation consists, on the one hand, of valid models for both the host structure and the guest molecule and, on the other hand, of a pair-wise model describing host-guest interactions. In order to simulate an infinite system comparable to a macroscopic system and to take into account the long range character of the interactions, periodic boundary conditions are used, consisting of replicating the fundamental unit-cell in the three directions of space. As all cells contains N molecules in the same relative positions, only the interactions emanating from the fundamental N molecules contained in the fundamental unit-cell have to be followed during the MC calculation. Consequently, in the case of the calculation of adsorption properties, an adsorbed molecule will be interacting with an infinite number of atoms, so that a cut-off within which the interactions are calculated is generally used, especially for van der Waals interactions. A cut-off of half of the unit-cell parameter is generally used. Electrostatic interactions are usually calculated using an Ewald summation.

Aside from the numerical result itself (average number of adsorbed molecules per unit-cell, isosteric adsorption heat, Henry's constant), a significant amount of information may be obtained from graphical and visual exploitation of the results. This first gives information on the suitability of sites accessed during the calculation. To the extent that the simulation is considered as valid and reliable, 3D-displays of energy distribution density grids from Monte Carlo calculations are very helpful for identifying adsorption sites in the microporous network and making a spatial assignment in terms of adsorption energies.

## 3.2. INTERACTION MODEL FOR $N_2$ AND $O_2$ ADSORPTION IN ZEOLITES

When considering the simulation of adsorption isotherms in zeolites, the two critical steps in the GCMC calculations are the choice of the structural model for the zeolite and the choice of the interaction model describing sorbate-sorbent and sorbate-sorbate interactions. Once a full structural representation of the zeolite is fixed, the success of an attempt at modeling sorption properties relies essentially on the potential function describing the interactions involved in the sorbent-sorbate system. The validity of a GCMC calculation has then to be established in comparison with experiment (isosteric heats, isotherms, Henry's constant), as it is certainly the best test of the assumptions regarding the intermolecular interactions.

Our purpose here is to obtain from the GCMC simulations a relevant approach to $N_2$ and $O_2$ adsorption, especially when compared to experimental data, through the use of a simple model that may be easily transferred to any zeolitic structure. A simple model is therefore used to describe the intermolecular interactions, including both sorbate-sorbate and sorbate-sorbent interactions. Considering the sorbate/sorbent system, the total intermolecular interaction energy may be expressed as follows :

$$E_{total} = E_{dispersion-repulsion} + E_{electrostatic} + E_{induction} \qquad (1)$$

As a first approximation, the inductive part of the interaction is deliberately neglected. The induction energy term, often distinguished from other electrostatic contributions while being of an electrostatic nature, results from the distorsion of the molecule under the influence of the external electric field. The strength of this interaction is directly related to the atomic polarizabilities of the interacting species and to the strength of the external electric field gradient generated by extraframework cations. When considering $N_2/O_2$ separation in a cationic zeolite, the neglect of induction energy may at first be justifiable as the $N_2$ and $O_2$ atomic polarizabilities are very similar. The inductive interaction between sorbed molecules and cations is consequently of a non-specific nature, since it gives rise to no separation effect.

The total energy is therefore taken as the sum of a repulsion-dispersion term and an electrostatic term. The dispersive interaction results from the interaction between two instantaneous dipoles and is generally described as a function of the polarizabilities of the interacting atoms. For the repulsion-dispersion term, we use the Lennard-Jones potential expression:

$$E_{ij} = \frac{A_{ij}}{r_{ij}^{12}} - \frac{B_{ij}}{r_{ij}^{6}} = \varepsilon_{ij} \left[ \left( \frac{r^{*}_{ij}}{r_{ij}} \right)^{12} - 2 \left( \frac{r^{*}_{ij}}{r_{ij}} \right)^{6} \right] \qquad (2)$$

where $A_{ij}$ is the repulsion constant and $B_{ij}$, the dispersion constant, with $\varepsilon_{ij}=B_{ij}^{2}/4A_{ij}$, $\sigma_{ij}=(A_{ij}/B_{ij})^{1/6}$ and $r^{*}_{ij}=2^{1/6}\sigma_{ij}$. This potential considers interactions between pairs of atoms and is often referred to as an atom-atom potential. A knowledge of the parameters $(A_{ij}, B_{ij})$ or $(r^{*}_{ij}, \varepsilon_{ij})$ for each interacting pair is therefore required. The equilibrium separation distance $r^{*}_{ij}$ of the interacting pair is expressed as the sum of the van der Waals or ionic radii of the interacting atoms ($r_i$ and $r_j$) :

$$r^*_{ij} = (r^*_{ii} + r^*_{jj})/2 \quad \text{with} \quad r^*_{ii} = 2r_i \tag{3}$$

Concerning the dispersive interaction, there are three commonly used expressions for the estimation of the attractive constant, $B_{ij}$, which may all be summarized by the following formula:

$$B_{ij} = K \, \alpha_i \alpha_j / (\beta_i + \beta_j) \tag{4}$$

with three different ways [8] of defining the $\beta$ term according to London ($\beta = 1/E_i$), Slater-Kirkwood ($\beta = (\alpha/n)^{1/2}$), or Kirkwood-Müller ($\beta = \alpha/\chi$), where $E_i$ is the ionization potential, $\alpha$ the polarizability, n the number of electrons in outer shells, and $\chi$ the magnetic susceptibility. The variety of formulations of the dispersion constant, $B_{ij}$, is a barrier to the development of a quantitative theoretical approach to the *a priori* calculation of adsorption energies. Nevertheless, the problem may be avoided if the development of the forcefield is based either on an evaluation of ($A_{ij}$, $B_{ij}$) from first-principle DFT calculations, or on the adjustment of the $\varepsilon_{ij}$ parameters, using a mixing rule from both formula (3) and (4):

$$\varepsilon_{ij} = \sqrt{\varepsilon_{ii}} \cdot \sqrt{\varepsilon_{jj}} \, \frac{(r^*_{ii})^3 (r^*_{jj})^3}{(r_{ij*})^6} \cdot \frac{2(\beta_i \beta_j)^{1/2}}{(\beta_i + \beta_j)} \tag{5}$$

It is interesting to notice that when both conditions $r^*_{ii} \sim r^*_{jj}$ and $\beta_i \sim \beta_j$ are fulfilled, the commonly used mixing rule $\varepsilon_{ij} = (\varepsilon_i \varepsilon_j)^{1/2}$ is obtained.

A further classical reduction of the number of adjustable parameters consists of neglecting the repulsive and dispersive contributions of silicon and aluminum atoms, as they are screened by the oxygens of the framework. The accessible part of the framework towards sorbed molecules consists mainly of oxygen atoms. Furthermore, the low polarizabilies of Si and Al atoms in comparison with that of oxygen suggest that negligible Lennard-Jones interactions are involved between adsorbed $N_2$ or $O_2$ molecules and T atoms. Consequently, the repulsive-dispersive contribution of the zeolite may be assigned exclusively to oxygens of the framework and cations. Nevertherless, this assumption implies that the influence of Si and Al atoms on the dispersion-repulsion interactions will be included in the fitting of $\varepsilon_{ij}$ parameters of oxygens of the framework.

The electrostatic interaction is estimated using the following classical expression:

$$E = \sum_{i,j} \frac{q_i q_j}{r_{ij}} \tag{6}$$

The zeolite is described as a semi-ionic structure and the partial charges are taken as follows : Si (+2.4), O (-1.2), Al (+1.4), Na(+1) and Ca(+2). The quadrupole moments of $N_2$ and $O_2$ are described by a three point charge model. The two outer sites are separated by a distance $l$ (1.2 Å for $O_2$, 1.1 Å for $N_2$) and have a charge -q (-0.112 for

$O_2$) and (-0.486 for $N_2$). The third midpoint site has a charge $+2q$. Charges are chosen in order to fit the experimental quadruple moments [9] of nitrogen, $4.7.10^{-40}C.m^2$, and oxygen, $1.3.10^{-40}C.m^2$. Electrostatic interactions have been calculated using an Ewald summation.

## 3.3. APPROACH FOR SIMULATING $N_2$ AND $O_2$ ADSORPTION IN ZEOLITES

(i) In a first step, the interaction model is worked out in order to have a complete force field for describing interactions involved in the $N_2,O_2$/zeolite system. First, the pair potential interactions of $N_2$ and $O_2$ molecules with the framework oxygens are investigated through the study of the adsorption in the purely silicous, therefore cation-free, silicalite zeotype material. Secondly, the pair potentials of $N_2$ and $O_2$ molecules with cations such as $Na^+$ and $Li^+$ are extracted from quantum mechanical calculations.

(ii) In a second step, GCMC simulations are investigated on cationic zeolites having known crystalline structures and starting from the forcefield parameters developed in the first section. Simulations are discussed in comparison with experimental data.

(iii) In a third part, concerning the structural model of the zeolite, we present an attempt to predict cationic positioning when experimental diffraction information are lacking, using a simple molecular mechanics approach.

## 4. Working out the interaction model

### 4.1. INTERACTION OF $N_2$ AND $O_2$ WITH THE FRAMEWORK [10]

The silicalite structure was considered in its orthorombic form [11] with a=20.02 Å, b=19.89 Å and c=13.38 Å. The microporous network is made of two different types of channels, both delineated by 10-member-rings. A set of straight channels runs in the direction of the (010) axis and a second set of sinusoidal channels intersects the straight channels and runs in the direction of the (100) axis. GCMC calculations were performed on a 2x2x2 unit-cell with periodic boundary conditions.

The calculations require a knowledge of the dispersive interactions of oxygens of the framework with nitrogen atoms of $N_2$ molecule ($\varepsilon_{OZ-N}$), and oxygen atoms of $O_2$ molecule ($\varepsilon_{OZ-O}$). For that purpose, we first deduced the $\varepsilon_{OZ}$ parameter from the $\varepsilon_{OZ-CH4}$ parameter reported in [12] and $r*_{CH4}$ and $r*_{OZ}$, using the mixing rule (5). A knowledge of $\varepsilon_{OZ}$ therefore allows the extention of the forcefield to the $\varepsilon_{OZ-i}$ parameters for i=N($N_2$) and i=O($O_2$), using the mixing rule (5) in a similar way and the $\varepsilon_{N(N2)}$ and

$\varepsilon_{O(O2)}$ parameters from liquids [13]. Table 2 summarizes the different potential parameters obtained. Table 3 reports the results of calculations performed at 0.2 bar, for $N_2$ and $O_2$ adsorption in the silicalite structure.

Despite the underestimation of adsorption enthalpies, good simulations of $N_2$ and $O_2$ adsorption are obtained, especially as no $N_2/O_2$ selectivity of adsorption between is found. Figure 3 shows the energy distributions for $N_2$ and $O_2$ in silicalite, represented as the number of grid points having accepted configurations of energy $E_i$ as a function of $E_i$. It is noticeable that we obtain very similar distributions in both cases, centered around $E_1$=-10.2 kJ.mol$^{-1}$ and $E_2$=-13.2 kJ.mol$^{-1}$ for $N_2$ adsorption, and around $E_1$=-10 kJ.mol$^{-1}$ and $E_2$=-12.5 kJ.mol$^{-1}$ for $O_2$ adsorption. Figure 4 shows the spatial distribution of $N_2$ molecules in silicalite for $E_1$ (a) and $E_2$ (b). Adsorbed $N_2$ molecules are uniformly distributed over the microporous structure, over straight, sinusoidal channels and intersections. A selection among most energetic configurations (b) shows that local minima of the potential surface are in the confined regions of the sinusoidal and straight channels and at intersections of both sets of channels. Spatial representations for oxygen adsorption would give results similar to nitrogen.

TABLE 2. Potential parameters used for $CH_4$, $O_2$ and $N_2$ and O(zeolite) in GCMC calculations [10].

|  | $CH_4$ | $O(O_2)$ | $N(N_2)$ | O(zeolite) |
|---|---|---|---|---|
| $r^*_{ii}$ (Å) | 4.187 | 3.468 | 3.724 | 3.040 |
| $\varepsilon_{ii}/k_B$ (K) | 147.95 | 44.5 | 36.4 | 139.96 |
| $r^*_{oz\text{-}i}$ (Å) | 3.613 | 3.254 | 3.382 |  |
| $\varepsilon_{oz\text{-}i}/k_B$ (K) | 133.30 | 77.90 | 69.21 |  |

TABLE 3. Adsorption of $N_2$ and $O_2$ in silicalite at 0.2 bar and T=298 K [10].

|  | N(cal) molec/uc | $K_H$(calc) mol/uc/atm | $K_H$(exp) mol/uc/atm | $\Delta$H(calc) kJ.mol$^{-1}$ | $\Delta$H(exp) kJ.mol$^{-1}$ |
|---|---|---|---|---|---|
| $N_2$ | 0.23 | 1.13 | 0.98 | -14.40 | 15.1 ±1.4 |
| $O_2$ | 0.22 | 1.11 | 1.14 | -14.14 | - |
| $CH_4$ | 0.85 | 4.3 | 5.3 - 3.4 | -18.31 | 20 |

This comparison clearly shows that the simulation is successfull in simulating no energetic discrimination between $O_2$ and $N_2$ adsorption as the interactions are essentially of a dispersive nature. In conclusion, a unique set of (r*,ε) parameters is able to provide a good simulation of silicalite adsorption properties, within a restricted pressure end temperature range, toward $N_2$ and $O_2$ sorbate molecules.

## 4.2. INTERACTION OF $N_2$ AND $O_2$ WITH EXTRA-FRAMEWORK CATIONS

Forcefields have been mainly developed in order to reproduce intramolecular structures (bond lengths, bond and torsion angles, hydrogen bonds). Very little information is available in the field of weak intermolecular interactions involving polarisable interacting entities. Consequently, no specific forcefield has been developed for describing interactions between cations and small molecules like $N_2$ and $O_2$. Furthermore, the variety of formulations for the dispersive constant $B_{ij}$ of the Lennard-Jones pair potential is an important barrier to the *a priori* determination of the forcefield parameters. In order to circumvent that problem, quantum mechanics using DFT methods were used to calculate the interaction energy curves between $N_2$ and $O_2$ molecules and isolated alkaline cations. The forcefield parameters - dipersive and repulsive interactions terms - needed for the GCMC calculations were extracted from the quantically calculated energy curves, through a least-square adjustment of the forcefield parameters of the classical Lennard-Jones expression. Besides yielding forcefield parameters, this approach allows us to compare the intrinsic properties of a cation toward $N_2$ and $O_2$ adsorption when extracted from its zeolitic environment, through direct comparison of their respective interaction energies.

### 4.2.1. *Basic principles of Density Functional Theory calculations* [14]

Like any first-principle quantum mechanical calculations, DFT methods rely on the resolution of secular equations emanating from the time-independant Schrödinger equation : $H\phi = E\phi$, where $\phi$ is the total wavefunction of the system. The specificity of DFT methods consists of stipulating that all ground-state properties are functions of the charge density, $\rho$, which is therefore taken as a variable rather than the total wavefunction $\phi$, with $\rho = \sum |\phi_i(r)|^2$. The total energy of the system may be expressed as follows :

$$E_{total}\left[\rho(r)\right] = \underset{\substack{\text{Kinetic energy of non-interacting} \\ \text{particules of density } \rho}}{T\left[\rho(r)\right]} + \underset{\text{Coulombic interaction}}{U\left[\rho(r)\right]} + \underset{\substack{\text{Exchange and correlation} \\ \text{energy}}}{E\left[\rho(r)\right]} \qquad (7)$$

Variational self-consistent solutions to the DFT equations are calculated, expressed in a numerical atomic orbital basis. The solutions provide the molecular wavefunctions and electronic densities, which may be used for evaluating energetic, magnetic or electronic properties of the system. Kinetic and coulombic terms are calculated exactly, while the exchange-correlation term is evaluated through an approximation as its exact expression is unknown. A common one is the local density approximation, which is based on the known exchange-correlation energy of the uniform electron gas. Corrections to systematic errors may be added through the use of more sophisticated non-local functionals.

When considering intramolecular energies, DFT methods generally lead to an overestimation in comparison with the real energy, while other first-principle calculations like Hartree-Fock methods lead to an underestimation. As final energies are very much dependant on the functionals that are chosen for treating either exchange or correlation terms, DFT energy calculations may be improved by the use of more or less (local, non-local) sophisticated functionals. As far as the estimation of intermolecular binding energies is concerned, it is noteworthy that the use of numerical basis sets in DFT methods, rather than analytical ones, facilitates a minimization of basis sets superposition effects, and consequently yields an excellent description of even weak bonds. A further significant advantage of DFT methods is the shorter time required for calculations, which offers the possibility of treating larger systems.

### 4.2.2. Interaction of $N_2$ and $O_2$ with isolated $Li^+$ and $Na^+$ cations

During the calculation, the [cation-molecule] system is in a linear geometry, which is easily shown to lead to the maximum interaction energy between the quadrupolar molecular and the electric field gradient generated around the cation. Geometry optimization calculations are performed for increasing cation-molecule constrained distances. The interaction energy is then calculated by substracting the energies of the single cation and of the isolated molecule from the total energy. We now describe calculations that have been performed using the D_Mol package of Biosym/MSI. In this study, the choice of the functional is only of relative importance as our purpose is having comparative binding energies of similar systems like $[Na^+/N_2]$, $[Na^+/O_2]$, $[Li^+/N_2]$ and $[Li^+/O_2]$, rather than having exact values of energies. All calculations have been perfomed in the local density approximation, using the VWN (Vosko-Wilk-Nusair) functional for either correlation or exchange energy.

Figure 5 shows the energy curves of the $[Na^+/N_2]$, $[Na^+/O_2]$, $[Li^+/N_2]$ and $[Li^+/O_2]$ systems in the linear arrangement, as a function of distance separating the cation from the first atomic center of the $N_2$ or $O_2$ molecule. The stronger interaction energy of the $[Na^+/N_2]$ or $[Li^+/N_2]$ systems in comparison with $[Na^+/O_2]$ or $[Li^+/O_2]$ systems is straightly related to the higher quadrupolar moment of $N_2$ when compared to that of $O_2$. At large distances, both $Li^+$ and $Na^+$ cations give similar interaction energies with either $N_2$ or $O_2$ molecules. It has been shown in a recent extended study of these systems [15] that this part of the curve is well represented by a point charge model, confirming the predominance at large distances of the electrostatic component in the total interaction energy. At short distances, it is clearly seen that $Li^+$ gives more stable systems than $Na^+$. The more stable interaction energies observed in the case of $Li^+$ are systematically associated with shorter equilibrium distances (see table 4). The higher interaction energies observed in the $[Li^+/N_2]$ and $[Li^+/O_2]$ systems when compared to those of $[Na^+/N_2]$ and $[Na^+/O_2]$ are a result of stronger contribution of the electrostatic component of the total interactions in the case of $Li^+$, as repulsion occurs at a shorter distances with $Li^+$ than with $Na^+$. The difference between the [$N_2$/cation] and [$O_2$/cation] binding energies is $\Delta E(N_2-O_2)=19$ kJ.mol$^{-1}$ in the case of $Li^+$, and $\Delta E(N_2-O_2)=16$ kJ.mol$^{-1}$ in the case of $Na^+$, anticipating the increase in the $N_2/O_2$ separation effect when replacing $Na^+$ by $Li^+$.

This simple comparison between $Li^+$ and $Na^+$ cations shows that the intrinsic properties for $N_2$ and $O_2$ adsorption of an alkali cation, i.e. when extracted from its zeolitic environment, may in a first step be correlated to its cationic radius. As a result, the smaller the cationic radius is, the stronger the $N_2/O_2$ separation effect is expected to be. The quantum-calculated curves were then interpolated by the classical expression of energy (Lennard-Jones and coulombic term), in order to be used in our interaction model for the GCMC calcutations. Starting with an estimation of the dispersive constant $B_{ij}$ from the London formula, the repulsive constant $A_{ij}$ was adjusted by means of a least-squares refinement in order to preferentially reproduce the equilibrium geometry and energy.

## 5. Application to cationic zeolites

The simulation of adsorption properties of cationic zeolites is now made possible, as pair potentials are available both for $N_2,O_2$/framework oxygens and $N_2,O_2$/cations interactions. The transferability of these parameters to any zeolitic structure is assumed *a priori*. This assumption is supported by recent DFT clusters calculations reported in [15]. When taking into account the zeolitic environment, the same relative energy ordering for $O_2$ and $N_2$ binding energies was obtained between $Li^+$-type and $Na^+$-type cluster systems as with isolated cations. Starting from this assumption, the simulation of adsorption properties of cationic zeolites like Y or X faujasite-type zeolites is investigated and discussed in comparison with experimental data.

Zeolites X and Y have topologically similar aluminosilicate faujasite-type framework structures, although they are distinct zeolite species with characteristic features. The chemical differences between X and Y zeolites are related to their Si/Al ratios, and therefore to the number and the distribution of cations in the unit-cell. The value of the Si/Al ratio varies from 1 to 1.5 for zeolite X and greater than 1.5 to 3 for zeolite Y. Model structures for X and Y zeolites are used for GCMC calculations. Zeolite Y is taken with the following composition $Si_{128}Al_{64}Na_{64}O_{384}$ (Si/Al=2). Zeolite X is chosen with the following composition $Si_{96}Al_{96}Na_{96}O_{384}$ (Si/Al=1), referred to as LSX (Low Silica X).

## 5.1. INTERPRETATION OF NaY AND LiY ADSORPTION PROPERTIES

### 5.1.1. *Introduction*

In order to perform the GCMC simulation, both NaY and LiY are assumed to have the following composition $Na_{64}Y$ and $Li_{64}Y$, with a Si/Al ratio of 2. Their cationic distributions are taken as follows, in term of composition per unit-cell: 32 cations are located in sites $S_I$ and the 32 others are located in sites $S_{II}$, in each 6-ring window of the supercage. There are four cationic sites $S_{II}$ in each supercage that are accessible to $N_2$ and $O_2$ molecules. The other cations are in confined regions, as the sodalite units are inaccessible to $N_2$ and $O_2$ molecules. The main structural difference between LiY and NaY zeolites concerns the relative positions of the compensating cations in site $S_{II}$ within the 6-ring window [16]. When located in site $S_{II}$, the $Na^+$ cation is more displaced than the $Li^+$ cation by 1.73 Å along the <111> axis towards the center of the supercage, while $Li^+$ nearly lies in the plane of oxygen atoms of the 6-ring window. Furthermore, it has been shown that the smaller cationic radius of $Li^+$ provokes a notable narrowing of the hexagonal window making a further difference between the NaY and LiY zeolites.

On the basis of energetic criteria, LiY is expected to be more selective than NaY, similarily to what is observed when comparing separation performances of LiX and NaX zeolites. Yet, it is shown experimentally that both LiY and NaY zeolites have similar $N_2/O_2$ selectivities at 1 bar (~2) and similar capacites and isosteric heats.

### 5.1.2. *GCMC simulations*

GCMC calculations of $N_2$ and $O_2$ adsorption has been performed for both LiY and NaY zeolites starting with the $P_1$ structures described above, using periodic boundary conditions and the pair potential parameters determined from QM calculations. Both calculations give a good simulation of the experimental data, giving similar selectivities for LiY and NaY zeolites at 1 bar, similar adsorption capacities for $N_2$ and $O_2$, and similar adsorption heats. In order to have a first interpretation of this unexpected result, a comparison of the energy density distribution grids of the [LiY/$N_2$]

and [NaY/N$_2$] sytems was investigated. The location of the adsorption sites of highest energies was compared in NaY and LiY. For NaY, the isosteric adsorption heat of N$_2$ is found to be 14.7 kJ.mol$^{-1}$. The spatial distribution of N$_2$ molecules in NaY selecting adsorption sites of high energy (-23 kJ.mol$^{-1}$) is shown in figure 6. This picture shows that the most energetic adsorption sites are localized around Na$^+$ cations in sites S$_{II}$, building a continuous adsorption area between neighboring S$_{II}$ sites. For LiY, the isosteric adsorption heat obtained for N$_2$ is similar to that in NaY (14.8 kJ.mol$^{-1}$). A similar visualization (E$^y$=-23 kJ.mol$^{-1}$) of the spatial distribution of N$_2$ molecules in LiY shows that the energetic adsorption areas are more closely confined around Li$^+$ cations in sites S$_{II}$ than they were in NaY.

As a matter of fact, this qualitative approach gives an explanation at a molecular level for the similar performances of NaY and LiY for N$_2$/O$_2$ separation. As suggested from quantum calculations, adsorption energies and N$_2$/O$_2$ selectivity should be higher in LiY than NaY. In the special case of LiY, the intrinsic adsorption strength expected from the Li$^+$ cations is cancelled because of the immediate zeolitic environment of the cation. A first interpretation is that the less accessible position of the Li$^+$ cation in site S$_{II}$ when compared to that of Na$^+$ leads to a lower density of energetically favourable adsorption sites for N$_2$ molecules around the Li$^+$ cation, as the repulsive interactions with oxygen atoms of the framework have a higher contribution in the sorbent/sorbate interactions. Moreover, in the dehydrated form of the partially Li-exchanged LiX [16], the Li$^+$ cations are bonded to oxygen atoms of the framework at short distances of 2.07 Å, while Na$^+$-O$_{framework}$ distances are 2.48 Å. Consequently, the more shielded crystallographic position of Li$^+$ when compared to that of Na$^+$ may lead to a stronger screening of the cation by oxygen atoms of the framework, and therefore to a decrease in the apparent charge and in the electric field gradient generated around the Li$^+$ cation.

Monte Carlo simulations, since they take into account the whole framework environment and long range electrostatic interactions, show that the expected increase in N$_2$/O$_2$ separation performances is strictly compensated by the less accessible position of the Li$^+$ cations in the supercage. Through this example, it is clearly shown that, besides the intrinsic properties of the cation for N$_2$/O$_2$ separation that may be anticipated from the QM calculations, the zeolitic structure plays a crucial role in the final performance of the separation.

This is one successful example of the complementary use of theoretical tools (QM for forcefield parameters, MC for simulation of isotherms) leading to a significant description of the zeolite performance and a realistic interpretation, at a molecular level, of their adsorption properties, starting from a simple interaction model.

## 5.2. ADSORPTION OF NITROGEN IN THE CaLSX ZEOLITE

The CaX zeolite is known for its good performances for $N_2/O_2$ separation, which is at first interpreted by the stronger electrostatic interactions of the $N_2$ molecules with divalents cations. For the structure used in the simulations, the Monte Carlo calculations were carried out starting from a recent diffraction refinement of the dehydrated CaLSX zeolite structure [17]. The chemical composition is taken as follows : $Si_{96}Al_{96}Ca_{48}O_{384}$ per unit-cell. The 48 $Ca^{2+}$ were distributed as follows: 16 $Ca^{2+}$ cations in sites $S_I$, and 32 $Ca^{2+}$ cations in sites $S_{II}$., both sites $S_I$ and $S_{II}$ therefore having full occupancies. Similarly to zeolites NaY and LiY, each supercage contains four accessible $Ca^{2+}$ cation symmetrically distributed over sites $S_{II}$, while the other $Ca^{2+}$ cations in sites $S_I$ are in confined regions. In a first step, Monte Carlo simulations were run with a semi-ionic framework and formal charges of +2 on each calcium cation, using Lennard-Jones potential parameters for $N_2/Ca^{2+}$ interactions from interpolations of DFT potential curves. Isosteric heat for nitrogen adsorption were calculated at low temperatures (260K, 280K, 300K) and for increasing loadings, up to 7 molecules of $N_2$ per supercage. The most striking feature is that the simulated adsorption heats are nearly constant over the whole loading range (figure 7a), failing to reproduce the decrease in adsorption heats that is experimentally observed.

As a matter of fact, the decrease in experimental adsorption heats is observed for loadings less than one molecule per cage. The loading of one molecule of nitrogen per supercage is far bellow the saturation limit of adsorption sites in the CaLSX zeolite (i.e. 4 $Ca^{2+}$ per supercage). Consequently, the contribution of repulsive sorbate-sorbate interactions is not involved as this decrease starts from zero adsorbate loading. Such a behaviour in isosteric adsorption heats with loading is generally interpreted in terms of important energetic heterogeneity of the zeolite surface relative to sorbate molecules, leading to a competition between adsorption sites involving different sorbate-sorbant interactions energies. The above calculated adsorption heats are consistent with the uniform distribution of calcium cations that are experimentally reported and used in the model, simulating uniformity of the interior surface of the zeolite relative to adsorbed nitrogen molecules, as all cationic adsorption sites behaves similarly. Figure 8 gives an illustration of an energy distribution density grid for $N_2$ molecules ($E^y$=-30.5 kJ.mol$^{-1}$) and shows that all $Ca^{2+}$ cations are equally surrounded with packed $N_2$ molecules.

In order to simulate energetic heterogeneity, Monte Carlo simulations were performed working with a non-uniform assignment of point charges on calcium cations in sites $S_{II}$. Cations charges were chosen as follows : one Ca(2+) and three Ca(+1.2) per supercage. The appropriate adjustement of the charges of the oxygen atoms of the framework was made in order to ensure the electrical neutrality of the whole unit-cell. Isosteric heats for nitrogen were calculated at low temperatures (260K, 280K, 300K) and for increasing loadings, up to 4 molecules of $N_2$ per supercage. Figure 7b shows that the experimental decrease in isosteric adsorption is now correctly simulated, as it starts from zero loadings. As shown in figure 9, a vizualization of energy distribution density grid for $N_2$ molecules ($E^y$=-30.5 kJ.mol$^{-1}$) shows, as expected, that the most energetically favourable sites are restricted to the remaining calcium cations modelled with high charges (+2).

The meaning of this simulation in terms of chemistry has to be related to the effective accessibility of $Ca^{2+}$ cations in the real material. These simulations show that a decrease in the apparent charge of the $Ca^{2+}$ cation, arising from partial screening owing to its local environment, is a realistic explanation of the experimental adsorption heat behaviour. Experimentaly, the screening of the calcium cation is to be expected either from a partial hydration of the zeolite, or cation hydrolysis. There is general agreement that the formation of hydroxyl groups in multivalent zeolites is due to hydrolysis of the cation and dissociation of the water molecule by the electrostatic field created by the cation. Through this study, simulation is efficient in orienting the experimental approach and further study toward the influence of dehydration and activation on the adsorption performances of the Ca-exchanged X-type zeolite is suggested.

# 6. New perspectives for an *a priori* estimation of cation positions in zeolites

## 6.1. DISORDER IN CATIONIC DISTRIBUTIONS OF ZEOLITES

A knowledge of cation positions in zeolitic materials has always been a key feature either in adsorption or in catalysis areas, since they have a strong influence on adsorption or catalytic properties.

The separation of $N_2$ and $O_2$ from air relies on the difference in the interaction energies of these molecules with the electric field gradient generated around the extraframework cations. The cation location in terms of accessibility towards $N_2$ and $O_2$ molecules is therefore crucial. Consequently, one of the critical step of the simulation deals with the choice of the cation distribution, especially as a $P_1$ triclinic structure is required for GCMC calculations. When considering structure refinements, cations are referred to in terms of a definite number of cationic sites defined with their crystallographic positions, coordinates and population parameters. Provided the crystallographic cationic sites are known through X-ray or neutron diffraction analysis and their respective occupancies are 100%, the cation distribution may be fixed with no ambiguity in the $P_1$ symmetry required for GCMC simulations. In the case of population parameters differing from full occupancies, a single cationic site may be looked at as a statistical distribution of cations among a number of possible sites. This statistical ground leads to an ambiguity about the choice of the cation distribution in $P_1$ symmetry, and is a real barrier to the precise knowledge of the cation distribution. This is a problem that has to be faced with most of the existing cationic zeolites.

In order to circumvent this problem, a molecular mechanics approach was investigated with J. Newsam (Biosym/MSI) in San Diego. A simple methodology for obtaining relevant cation distributions in zeolites is presented here, starting with no prior structural or chemical hypothesis about the cation positions. In order to test the methodology, only zeolites having well-known cationic site occupancies with only minor statistic distribution ambiguities, such as $Na_{12}A$ and $Na_4Ca_4A$ zeolites, have been studied. These zeolites may be considered as model structures, as reliable experimental diffraction data are available. In the following section, results for both $Na_{12}A$ and $Na_4Ca_4A$ zeolites are discussed in comparison with the experimental crystal structures. The ability of a simple molecular mechanics approach to predict a valid cationic distribution is outlined.

## 6.2. CRYSTALLINE STRUCTURES OF NaA AND NaCaA ZEOLITES

Zeolite A has the following general composition $[(SiO_2)_{12}(AlO_2)_{12}]^{12-}[M^{n+}_{12/n}]^{12+}$, where $M^{n+}$ is a cation of valence n. As the Si/Al is precisely equal to 1, the framework

446

is made of a strict alternation of [SiO$_4$] and [AlO$_4^-$] tetrahedra. The pseudo unit-cell is cubic and consists in the arrangement of 8 β-cages (figure 1) connected through their 4-membered rings. Each β-cage is centered at each corner of a cube of 12.3 Å on edge, enclosing a larger cavity having a free aperture diameter of 6.6 Å. The center of the unit-cell is therefore a large cavity, referred to as the α-cage, which has a free internal diameter of 11.4 Å. There are three different crystallographic sites for cations (figure 10). Sites S$_I$ are centered in the 6-membered rings separating β and α-cages. Sites S$_{II}$ are located in the 8-membered rings separating two α-cages, slightly off the center of the ring. Sites S$_{III}$ are located near 4-membered rings in the α-cage.

Diffraction data for dehydrated Na$_{12}$A [18] show that the 12 Na$^+$ cations are distributed as follows (figure 10). Eight cations are in site S$_I$, with an occupancy factor of 1. Three cations are in site S$_{II}$, with an occupancy factor of 1/4: each 8-membered ring is occupied only with one compensating cation, as strong electrostatic exclusion would occur if two cations were located in the same window. As a consequence, each 8-membered ring is occupied with only one Na$^+$ cation shared with the two neighbouring α-cages. The remaining Na$^+$ cation is in site S$_{III}$, with an occupancy factor of 1/12. It is noteworthy that, despite the fact that all the occupancies are not 100%, the cation distribution of Na$_{12}$A has little statistic disorder.

In the case of Na$_4$Ca$_4$A [19], both cations occupy 6-ring positions, with Na$^+$ inside the β-cage and Ca$^+$ on the opposite side in the α-cage, over the eight sites S$_I$ of the α-cage. This arrangement results in a completely open structure capable of admitting molecules with diameters of around 4.3 Å. As shown in figure 10, the most energetically favourable distribution should consist of an alternating distribution of Na$^+$ and Ca$^{2+}$ cations over the 8 edges of the cube.

## 6.3. METHODOLOGY OF THE SIMULATION

In a first step, a valid structure for the aluminosilicate framework is constructed. We start from the published X-ray structure of the NaA zeolite of Pluth and Smith. As periodic boundary conditions are to be employed for energy calculations, a single unit-cell of the structure is constructed. A reduction of the a=24.61 Å supercell of Pluth and al. is made to a P$_1$ triclinic a=12.305 Å sub-cell. In a second step, the procedure consists of Monte Carlo driven insertions of cation positions into the framework host structure. This docking procedure is based on the random selection of positions, followed by a calculation of the trial interaction energy for the cation/framework system. If this energy is below a definite threshold, the docked configuration is accepted and saved for the following step. This energetic criterion, chosen to avoid clash positions, avoids the selection of non-relevant starting distributions. It is noteworthy that the Monte Carlo procedure used here differs from the well-known sampling of Metropolis, which selects each structure depending on the previously accepted conformations. The final step consists of minimizing each Monte Carlo docked structure, using the cvff_aug zeolite/cations forcefield. This step allows cations to optimize their interaction with the host structure. The framework atom coordinates are fixed in the optimization calculation and only the coordinates of the $Na^+$ and $Ca^{2+}$ cations are allowed to vary. The methodology described here is illustrated in figure 11.

The minimization itself uses non-bonded Lennard-Jones terms between cations and framework atoms, and an electrostatic term, with a semi ionic distribution of charges on the framework ($Si^{+2.4}$, $Al^{+1.4}$, $O^{-1.2}$, $Na^{+1}$, $Ca^{+2}$). Periodic boundary conditions and Ewald summation are used.

When performing the random insertion of cations into the host structure, the starting structures are often in highly energetic configurations, and are therefore prone to becoming trapped in local minima during the energy minimization procedure. Dynamical simulated annealing allows cations to overcome potential energy barriers and relax to more stable configurations. For that purpose, the temperature is raised over 10000 K and progressively lowered to ambient temperatures. High temperature dynamics are performed at each temperature step, computing non-bond interactions with quartic expressions rather than with the classical Lennard-Jones ones. This allows atoms to pass very close to one another without producing infinite repulsive energies. When lowering the temperature, interaction terms are progressively re-introduced, and finally, energy minimization is employed to produce the final low energy configurations. This procedure has been used in the case of $Na_4Ca_4A$ zeolite.

## 6.4. RESULTS

The result of the whole procedure is the generation of 30 different minimized structures for zeolites $Na_{12}A$ and $Na_4Ca_4A$. Tables 5 and 6 report a classification of the minimized structures, for $Na_{12}A$ and $Na_4Ca_4$ respectively, according to decreasing energy. The final minimized structures are compared to the experimental ones in terms of the number of $Na^+$ and $Ca^{2+}$ cations in sites $S_I$, $S_{II}$ and $S_{III}$ and eventually others sites.

As expected, the distribution of the cations into the different possible sites will be largely determined by long range electrostatic interactions, as the cations will adopt the arrangement leading to the lowest coulombic interaction energy.

As far as the simulation of $Na^+$ positions in $Na_{12}A$ zeolite is concerned, the most remarkable feature is that the first 11 trials having lower energies, gathered within a range of 12.54 kJ.mol$^{-1}$, are the only ones having the correct cation distribution. Their final cation distributions are consistent with the experimental one, with correct occupancies of 8 $Na^+$ in sites $S_I$, 6 $Na^+$ in $S_{II}$ and the remaining $Na^+$ in site $S_{III}$. When compared with other trials, a gap of +41.2 kJ.mol$^{-1}$ is observed between trials 3 and 15 and is concomitant with the occupancy of non-valid positions, leading to a series of non-valid structures having even higher relative energies.

Similarly, simulations for the $Na_4Ca_4A$ zeolite show that the 6 most energetic structures, gathered within a range of 12.54 kJ.mol$^{-1}$, correspond to the only valid cationic distributions, made of 4 $Na^+$ and 4 $Ca^{2+}$ cations, all in sites $S_{II}$. Two gaps in the final energies (between trials 14 and 5, then trials 16 and 4) are observed and are correlated to the occupancies of non-valid positions. The first gap of +20.12 kJ.mol$^{-1}$ is concomitant with the location of $Na^+$ cations in sites $S_{II}$. The second gap of +23.55 kJ.mol$^{-1}$ is associated to the further occupation of sites $S_{II}$ and $S_{III}$ by $Na^+$ and $Ca^{2+}$ cations.

These results may be looked at as an interesting perspective for developing an *a priori* study of cations positions in zeolites. The further ability of this mechanics approach to predict cation positions in other zeolitic structures is expected to be dependant on the refinement of both potential parameters and the methodology itself. Nethertheless, these first results show that this methodology offers the possibility of overcoming the statistical limitation of the structural information retrieved from diffraction analysis, allowing an exploration of disorder effects, which may be of major interest when dealing with sorption or catalytic properties. Furthermore, it is noteworthy that the procedure presented here is easy to use and requires only a short time to be carried out, as routine molecular modeling tools are used, such as energy minimization or molecular dynamics.

## 7. Conclusions

In the particular field of the separation of nitrogen and oxygen in zeolites, a molecular modeling approach allowed us to improve the fundamental understanding of the basic microscopic phenomena and to establish correlations between microscopic features of the sorbent/sorbate systems and their macroscopic properties such as isotherms and adsorption heats.

Various molecular modeling tools were used. Quantum mechanics were carried out on $N_2,O_2$/cation systems, yielding the forcefield parameters required for the atomistic GCMC simulations of adsorption isotherms and allowed us to investigate the intrinsic properties of alkali cations towards $N_2$ and $O_2$ molecules. As far as GCMC simulations are concerned, due to the large size of the systems and the softness of the interactions involved in $N_2$ and $O_2$ adsorption, obtaining quantitative results is a difficult task. As the results of the GCMC calculations are dependant of the interaction model and of the forcefield parameters, an accurate quantitative description of adsorption properties of zeolitic materials probably requires more sophisticated

## Zeolite A

## Zeolite X

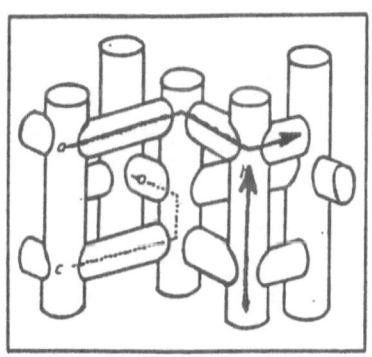

CHEMICAL COMPOSITION: $Na_{12}$ $[(AlO_2)_{12}(SiO_2)_{12}]$
SYMMETRY: cubic
SPACE GROUP: Pm3m (pseudo-cell), Fm3e (true cell)
UNIT-CELL CONSTANT: a= 24.64 Å (12.32Å for pseudo-cell)
DENSITY: 1.99 cc/g
FRAMEWORK DENSITY: 1.27 g/cc
PORE STRUCTURE: three dimensional 8-ring parallel
        to <100>, free aperture of 4.1 Å.

CHEMICAL COMPOSITION: $Na_{86}[(AlO_2)_{86}(SiO_2)_{106}]$
SYMMETRY: cubic
SPACE GROUP: Fd3m
UNIT-CELL CONSTANT: a= 25.02 Å to 24.86 Å
DENSITY: 1.93 cc/g
FRAMEWORK DENSITY: 1.31 g/cc
PORE STRUCTURE: three dimensional
        free apertures of 7.4 Å

## Zeolite ZSM-5

CHEMICAL COMPOSITION: $(TPA)_2O*48SiO_2$
SYMMETRY: orthorombic
SPACE GROUP: Pnma or $Pn2_1a$
UNIT-CELL CONSTANT: a= 20.06 Å, b= 19.80 Å, c= 13.36 Å
FRAMEWORK DENSITY: 1.76 g/cc
PORE STRUCTURE: intersecting 10- rings 5.7*5.1 Å, and 5.4 Å.

*Figure 1.* Views of three zeotype structures LTA (A), FAU (X) and MFI (ZSM-5), with their main structural characteristics [4].

450

TABLE 1. Molecular properties of nitrogen and oxygen.

| | Lennard-Jones | | Polarizability | Quadrupole Moment |
|---|---|---|---|---|
| | $\sigma(\text{Å})$ | $\varepsilon/k$ (K) | $\alpha(10^{-24}$ cm$^3)$ | $Q(10^{-40}$ C.m$^2)$ |
| $O_2$ | 3.467 | 106.7 | 1.58 | -1.3 |
| $N_2$ | 3.798 | 71.4 | 1.74 | -4.7 |

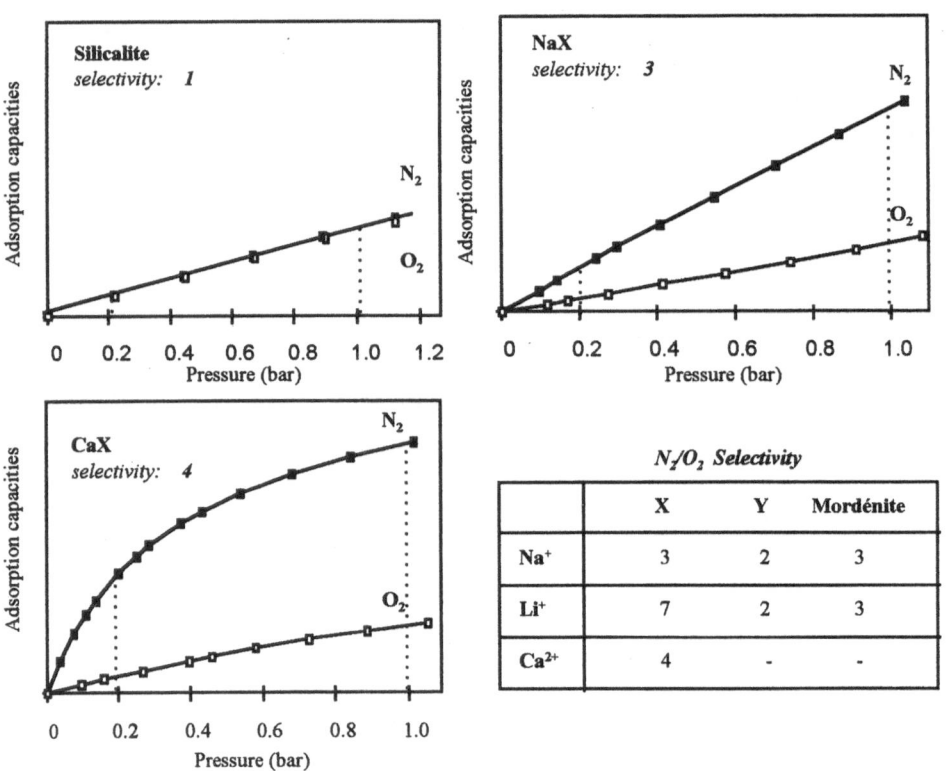

| $N_2/O_2$ Selectivity | | | |
|---|---|---|---|
| | X | Y | Mordénite |
| Na$^+$ | 3 | 2 | 3 |
| Li$^+$ | 7 | 2 | 3 |
| Ca$^{2+}$ | 4 | - | - |

*Figure 2*. Nitrogen and oxygen adsorption isotherms for different zeotype materials.

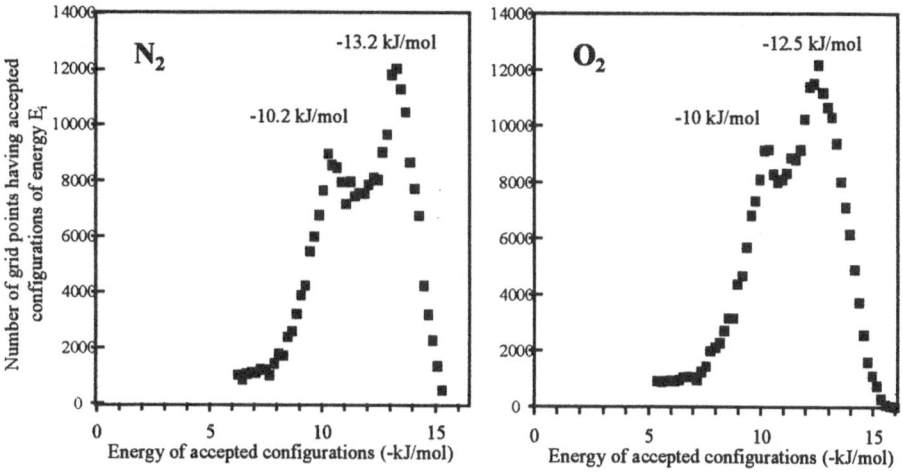

*Figure 3.* Energy distribution of accepted configurations for $N_2$ and $O_2$ in silicalite (number of grids points having accepted configurations of energy $E_i$ as a function of $E_i$).

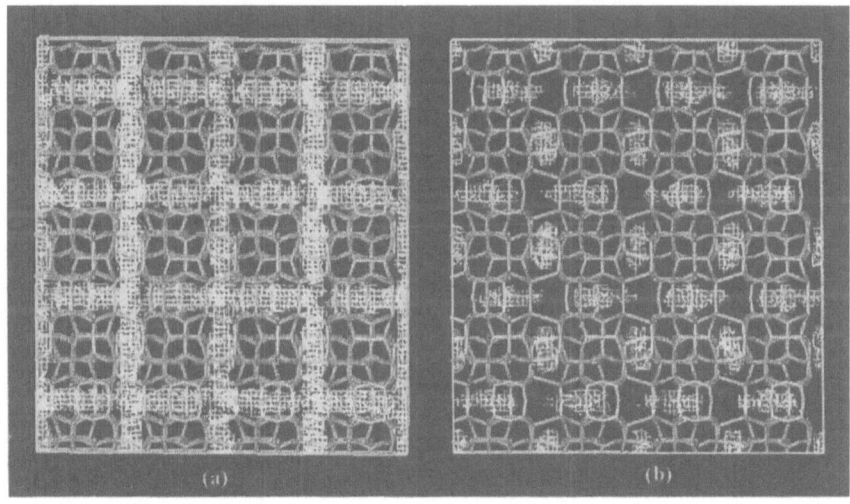

*Figure 4.* Energy distribution density grids for nitrogen in the silicalite channels. **(a)** $E_1 = -10.2$ kJ.mol⁻¹ **(b)** $E_2 = -13.2$ kJ.mol⁻¹.

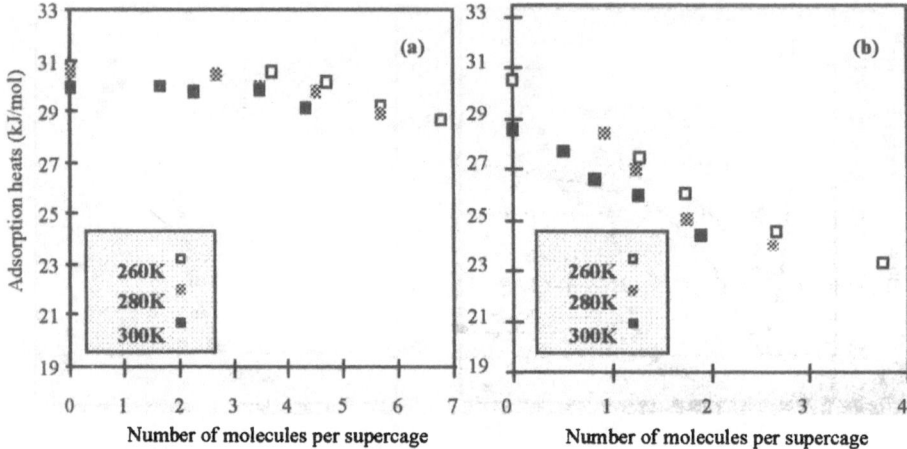

*Figure 7.* Simulated isosteric adsorption heats for $N_2$ adsorption in CaLSX as a function of increasing loadings. **(a)** with uniform distribution of four $Ca^{+2}$ cations over sites $S_{II}$. **(b)** with an heterogeneous distribution of one $Ca^{+2}$ and three $Ca^{+1.2}$ over sites $S_{II}$.

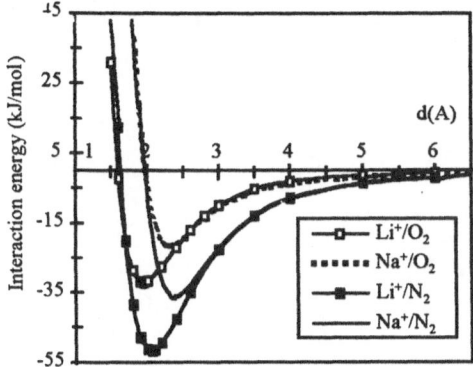

| | Na⁺ | | Li⁺ | |
|---|---|---|---|---|
| | $d_{eq}$ (A) | $E_{eq}$ (kJ/mol) | $d_{eq}$ (A) | $E_{eq}$ (kJ/mol) |
| $N_2$ | 2.4 | -36 | 2.05 | -51 |
| $O_2$ | 2.3 | -20 | 1.94 | -32 |

*Figure 5.* Potential energy curves for cation/$O_2$ and cation/$N_2$ in a linear arrangement, for increasing distances between the cation and the first atomic center of the $O_2$ or $N_2$ molecule.

TABLE 4. Equilibrium distances and energies for cation/$O_2$ and cation/$N_2$ in a linear arrangement.

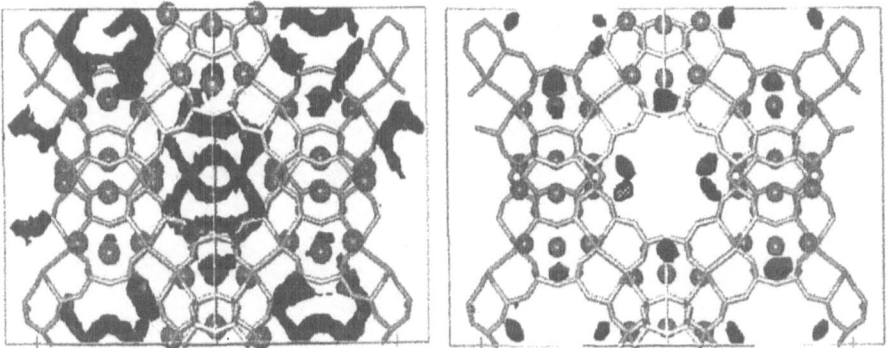

*Figure 6.* Energy distribution density grids for $N_2$ molecules (in black) adsorbed in NaY (left) and LiY (right) zeolites. Only configurations corresponding to adsorption energies of -23 kJ.mol⁻¹ are represented in either case.

454

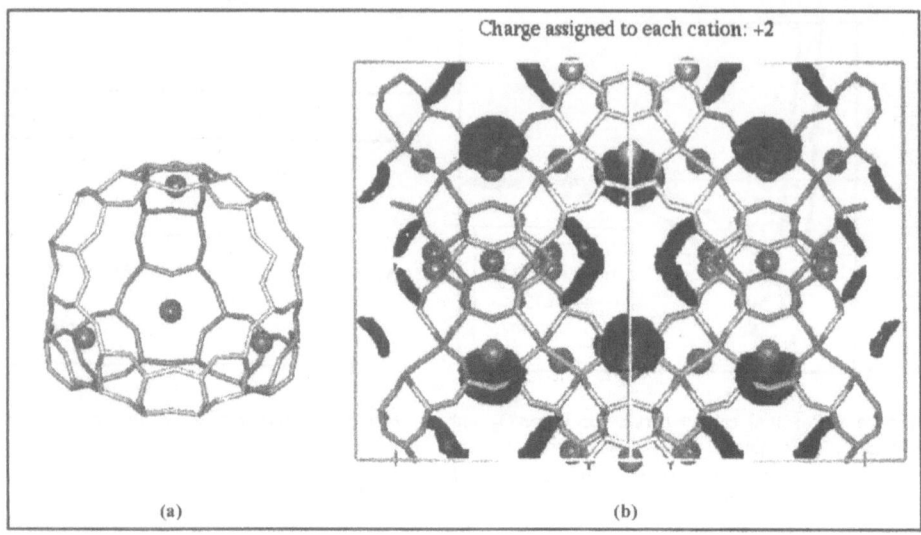

*Figure 8.* **(a)** : Representation of an isolated supercage in the CaLSX zeolite, with a uniform distribution of $Ca^{2+}$ cations over sites $S_{II}$. **(b)** : Energy density grid of nitrogen molecules ($E^y$= -30.5 kJ.mol$^{-1}$) for a uniform cationic distribution.

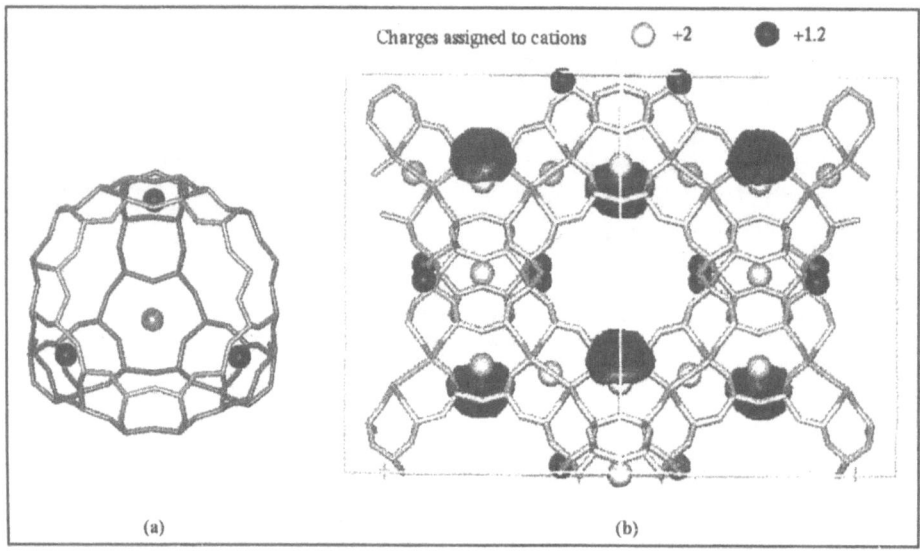

*Figure 9.* **(a)** : Representation of an isolated supercage in the CaLSX zeolite, with an hetegeneous cation distribution modelled with 1 $Ca^{2+}$ and 3 $Ca^{1.2+}$ per supercage. **(b)** : Energy density grid of nitrogen molecules ($E^y$= -30.5 kJ.mol$^{-1}$) for an heterogeneous cationic distribution..

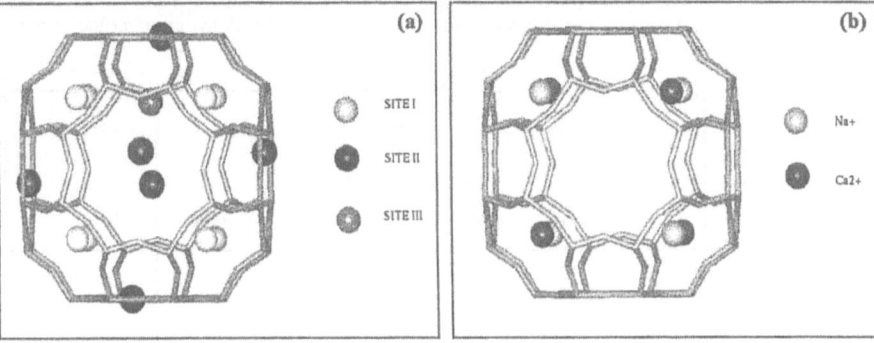

*Figure 10.* Representations of the α-cage in zeolite A. **(a)**: cation positions in the $Na_{12}A$ zeolite **(b)**: cation positions in the $Na_6Ca_6A$ zeolite.

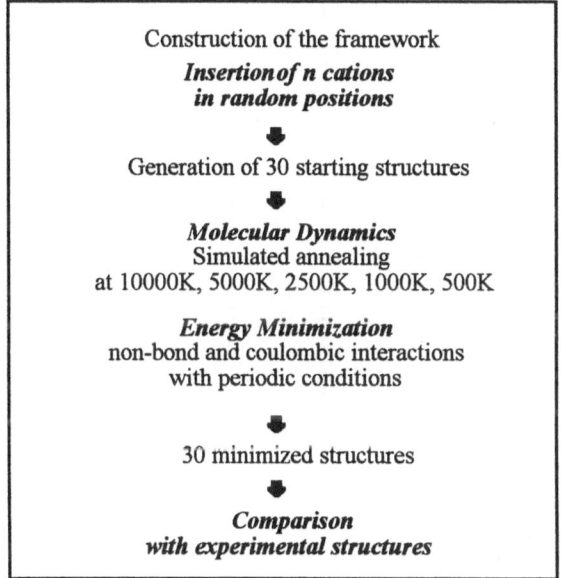

*Figure 11.* Molecular Mechanics methodology for simulating cation positions in zeolite A.

TABLE 5. Result for automatic placement of 12 Na$^+$ cations in a subcell of the A zeolite. Trials are classified according to decreasing energies and compared with the experimental cationic distribution.

| Trials | Na$^+$ | | | | | Energies |
| --- | --- | --- | --- | --- | --- | --- |
| | Windows occupancies | | | | | |
| | Site I | Site II | Site III | D4R | others | kJ/mol |
| Exp | 8 | 3 | 1 | | | |
| 2 | 8 | 3 | 1 | | | -10737.92 |
| 29 | 8 | 3 | 1 | | | -10737.53 |
| 30 | 8 | 3 | 1 | | | -10737.37 |
| 10 | 8 | 3 | 1 | | | -10737.16 |
| 28 | 8 | 3 | 1 | | | -10737.11 |
| 4 | 8 | 3 | 1 | | | -10733.01 |
| 20 | 8 | 3 | 1 | | | -10733.01 |
| 9 | 8 | 3 | 1 | | | -10731.94 |
| 5 | 8 | 3 | 1 | | | -10731.25 |
| 27 | 8 | 3 | 1 | | | -10731.24 |
| 3 | 8 | 3 | 1 | | | -10729.37 |
| 15 | 8 | 3 | | 1 | | -10688.17 |
| 26 | 8 | 3 | | 1 | | -10685.40 |
| 14 | 7 | 1 | | | 4 | -10658.03 |
| 25 | 7 | 3 | 1 | 1 | | -10648.22 |
| 11 | 7 | 3 | 1 | 1 | | -10638.41 |
| 23 | 8 | 2 | | 1 | 1 | -10633.05 |
| 1 | 6 | 2 | 4 | | | -10625.05 |
| 21 | 7 | 2 | 2 | 1 | | -10623.37 |
| 19 | 6 | 3 | 2 | 1 | | -10613.70 |
| 17 | 8 | 2 | | 2 | | -10610.35 |
| 8 | 6 | 3 | 2 | 1 | | -10607.16 |
| 12 | 7 | 3 | | 2 | | -10598.51 |
| 7 | 7 | 3 | | 2 | | -10597.92 |
| 16 | 7 | 3 | 1 | 1 | | -10956.71 |
| 18 | 7 | 3 | | 2 | | -10589.08 |
| 22 | 5 | 2 | 3 | 1 | 1 | -10553.46 |
| 24 | 5 | 2 | 2 | 1 | 2 | -10550.70 |
| 13 | 5 | 2 | 2 | 2 | 1 | -10331.13 |
| 6 | 6 | 1 | 3 | 2 | | -10325.74 |

TABLE 6. Results for automatic placement of 4 Na$^+$ and 4 Ca$^{2+}$ cations in a sub-cell of zeolite A. Trials are classified according to decreasing energies and their cationic distribution compared to the experimental one.

| Trials | Na$^+$ | | | Ca$^{2+}$ | | | Energies |
|---|---|---|---|---|---|---|---|
| | *Windows occupancies* | | | *Windows occupancies* | | | *kJ/mol* |
| | *Site I* 6-ring | *Site II* 8-ring | *Site III* 4-ring | *Site I* 6-ring | *Site II* 8-ring | *Site III* 4-ring | |
| *Exp* | 4 | | | 4 | | | |
| 10 | 4 | | | 4 | | | **-11240.30** |
| 17 | 4 | | | 4 | | | **-11239.73** |
| 18 | 4 | | | 4 | | | **-11239.73** |
| 7 | 4 | | | 4 | | | **-11226.61** |
| 28 | 4 | | | 4 | | | **-11226.04** |
| 14 | 4 | | | 4 | | | **-11225.33** |
| 5 | 3 | 1 | | 4 | | | -11205.21 |
| 29 | 3 | 1 | | 4 | | | -11204.20 |
| 13 | 3 | 1 | | 4 | | | -11220.23 |
| 21 | 1 | 3 | | 4 | | | -11195.33 |
| 16 | 3 | 1 | | 4 | | | -11190.55 |
| 4 | 2 | 2 | | 3 | 1 | | -11167.25 |
| 6 | 2 | 2 | | 3 | 1 | | -11158.40 |
| 27 | 3 | 1 | | 3 | 1 | | -11150.50 |
| 30 | 3 | 1 | | 3 | | 1 | -11146.72 |
| 3 | 3 | 1 | | 3 | 1 | | -11142.30 |
| 15 | 2 | 2 | | 3 | 1 | | -11139.39 |
| 8 | 2 | 2 | | 3 | 1 | | -11139.19 |
| 1 | 3 | 1 | | 2 | 2 | | -11133.51 |
| 2 | 3 | 1 | | 3 | 1 | | -11133.16 |
| 24 | 3 | 1 | | 3 | 1 | | -11131.87 |
| 20 | 3 | 1 | | 2 | 2 | | -11125.80 |
| 9 | 3 | 1 | | 2 | 2 | | -11123.95 |
| 12 | 2 | 2 | | 3 | 1 | | -11122.23 |
| 23 | 3 | 1 | | 2 | 2 | | -11117.64 |
| 11 | 3 | 1 | | 2 | 2 | | -11116.63 |
| 26 | 1 | 2 | 1 | 3 | 1 | | -11112.87 |
| 22 | 3 | 1 | | 1 | 2 | 1 | -11104.08 |
| 19 | 4 | | | 1 | 3 | | -11102.08 |
| 25 | 3 | 1 | | 2 | 2 | | -11099.56 |

forcefields and extended interaction models. Nevertheless, using a simple model has proved to give relevant interpretations of the experimental behaviour of cationic zeolites towards $N_2$ and $O_2$ adsorption. As it may be easily transferred to any zeolitic structure, our approach offers, especially in the perspective of industrial research, the possibility to be easily used by experimentalists and theoretical chemistry non specialists. This section is an attempt to show that molecular modeling tools can be used as a complementary tool to experimental studies, as far as they succeed in reproducing experimental tendancies and in yielding relevant molecular-level explanation of the macroscopic adsorption properties. In some cases, theoretical methods can even predict experimental observations and guide experimental work in developing molecular systems with improved properties.

## References

[1] Allen, M.P. and Tildesley, D.J. (1987) *Computer Simulation of Liquids*, Clarendon Press, Oxford.

[2] Ciccotti, G., Frenkel, D. and McDonald, J.R. (1987) *Simulation of Liquids and Solids*, North Holland.

[3] Breck, D.W. (1984) *Zeolite Molecular Sieves*, R.E. Krieger Publishing, Malabar.

[4] (a) Meier, W.M. and Olson, D.H. (1988) *Atlas of Zeolite Structure Types*, Butterworth & Co., Cambridge. (b) Mortier, W.J. (1982) *Compilation of Extra Framework Sites in Zeolites*, Butterworth & Co., Leuven.

[5] Ruthven, D.M. (1984) *Principles of Adsorption and Adsorption Processes*, Wiley, New York.

[6] (a) A. Corma (1995) *Chem. Rev.* 95, 3, 559. (b) H. Hattori (1995) *Chem. Rev.* 95, 3, 537.

[7] Metropolis, N., Rosenbluth, A.W., Rosenbluth, M.N., Teller, A.H., and Teller, E. (1953) *J. Chem. Phys.* 21, 1087 (1953).

[8] (a) London, F. (1940) *Zeit. Phys. Chem.* B1, 222. (b) Slater, J.C. and Kirkwood, J.G. (1931), *Phys. Rev.* 37, 682. (c) Müller, A. (1936) *Proc. Roy. Soc.* A154, 624.

[9] Steele, W. (1993) *Chem. Rev.* 93, 2355.

[10] Mellot, C., Lignieres, J., Pullumbi, P., Guilard, R. (1996) *Revue de l'Institut Français du Pétrole* 51, 1, 81.

[11] Van Köningsveld, H., Van Bekkum, H. and Jansen J.C. (1987) *Acta Crystallogr.* B43, 127.

[12] Goodbody, S.J., Watanabe, K., MacGowan, D., Walten, J.P.R.B. and Quirke, N. (1991) *J.Chem.Soc. Faraday Trans.* 87, 1951.

[13] (a) Razmus, D.M. and Hall, C.K. (1991) *AiChe Journal* **37**, 771. (b) Murphy C.S., Singer, K., Klein, M.L. and McDonald, I.R. (1980) *Molec. Physics* **41**, 1387.

[14] Politzer, P. and Maksic, Z.B. (1995) *Modern Density Functional Theory*, Elsevier, Amsterdam.

[15] Papai, I., Goursot, A., Fajula, F., Plee, D. and Weber, J. (1995) *J.Phys. Chem.* **99**, 34, 12925.

[16] Shepelev, Y.F., Anderson, A.A. and Smolin, Y.I. (1990) *Zeolites* **10**, 61.

[17] Vitale, G., Bull, L.M., Morris, R.E., Cheetham, A.K., Toby, B.H., Coe, C.G., MacDougall, J.E. (1995) *J. Phys. Chem.* **99**, 16087.

[18] Pluth, J.J. and Smith,J.V. (1980) *J. Am. Chem. Soc.* **102**, 4704.

[19] Siegel, H., Schöllner, R., Van Dun, J.J. and Mortier, W.J. (1987) *Zeolites* **7**, 148.

# INORGANIC MEMBRANES : PORE STRUCTURE CHARACTERIZATION

A. K. STUBOS[+], TH. A. STERIOTIS[*],
A. CH. MITROPOULOS, G. E. ROMANOS[*]
AND N. K. KANELLOPOULOS[*]

*Institute of Physical Chemistry, NCSR Demokritos, 15310 Ag. Paraskevi Attikis, Greece

+Institute of Nuclear Technology and Radiation Protection, NCSR Demokritos, 15310 Ag. Paraskevi Attikis, Greece

Keywords: Porous membranes, Structural characterisation, Pore connectivity, Pore size distribution, Supercritical liquefaction, Capillary condensation.

## 1. Inorganic Membranes

The membrane separation technology has been developed over the last 35 years. The first commercial application of membranes with asymmetric morphology is the cellulose acetate membrane systems used extensively for water desalination.

Inorganic membranes present several advantages over the organic ones; they can withstand high temperatures and high pressures and they are resistant to corrosive solutions. For this reason they have been extensively used for liquid phase separations via microfiltration (with pores greater than 0.1 μm), ultrafiltration (with pores ranging from 3 nm up to 0.1 μm) and nanofiltration ( with pores less than 3 nm). The first application of the inorganic membranes is the Uranium isotopic enrichment by gaseous diffusion, developed by US and French companies.

Professor R. M. Barrer and his group, were the first to investigate the possibility of gas separation by supercritical liquefaction [1,2]. In this

J. Fraissard (ed.), Physical Adsorption: Experiment, Theory and Applications, 461–484.

separation scheme, the micropores of the active layer of the asymmetric ceramic membrane are blocked by condensation of the heavy component (sulphur dioxide) of the gas mixture and the membrane becomes impermeable to lighter component of the mixture .

Asymmetric inorganic oxide disk-shaped membranes have been developed at the University of Twente, whereas other groups attempted the growth of zeolitic crystals within the surface pores or at the surface of a macroporous support [3].

Several other novel techniques are promising for the production of asymmetric membranes. Plasma treatment of Langmuir-Blodgett films made from metal salts of fatty acids or from substances from polysilixane backbone, followed by thermal desorption, can lead to a formation of a continuous inorganic layer [4].

Finally asymmetric carbon membranes, suitable for supercritical liquefaction separations, have been prepared jointly by BP, UK and SCT, France [5]. Phenolic resin is precured, grinded and sorted according to size. The membrane substrate was formed by extrusion into tubes or monoliths from the large size powder and the thin surface layer of the order of 10 nm from the smaller size powder; the carbon membranes are produced by carbonisation in nitrogen atmosphere and activation with $CO_2$.


## 2. Characterisation of the active layer of the asymmetric inorganic membranes.

The evaluation of the commercial potential of the ceramic membrane separation is heavily dependent on the development of a reliable model predicting the flow of gases and condensed vapours through the microporous active layer of the composite membrane as a function of the pore structure. To this end an accurate evaluation of the pore structure of this layer is of vital importance and the application of several standard techniques is necessary, such as scanning electron microscopy, bubble point measurement, sorption of gases and vapours, NMR, permporometry etc. Several other techniques are under development, based on studies of the microporous active layer, partially blocked by a condensed vapour.

## 2.1 EQUILIBRIUM METHODS

*2.1.1 Evaluation of micropore structure by sorption measurements.*

This method is based on the comparison of the pore structure analysis of surface chips of the ceramic membrane (comprised of the microporous surface layer and meso-macroporous substrate) and the meso-macroporous substrate.

While for mesopores and micropores there exist a lot of more or less established characterisation methods, the assessment of microporosity is much less advanced despite the recent interest on microporous systems like zeolites, activated carbons and clay minerals [6]. The pore structure analysis of microporous materials is hampered by experimental difficulties, arising from the long equilibration times required at the liquid nitrogen temperatures. In order to overcome this limitation the micropore structure evaluation can be based on isotherms of carbon dioxide or of other vapours obtained at higher temperatures, provided that a suitable equilibrium model for the sorption of non spherical molecules is available. The currently employed molecular (nitrogen) adsorption method is based on the thermodynamic approach of Dubinin, who assumed that the micropore filling process is governed by a so called adsorption potential that characterises the adsorbed molecules, and that the micropore size distribution is Gaussian. The Dubinin-Radushkevich (DR) equation relates the adsorbed amount per unit of micropore volume to the temperature, relative pressure the characteristic energy and affinity coefficients (which are in turn related to the isosteric heat of adsorption). The DR method has been subject to criticism mainly because the very mechanism of molecular adsorption in micropores is still under active debate. In fact, several studies employing both simulation and density functional theory have added to the accumulating evidence that none of the conventional adsorption methods of pore characterisation (such as Kelvin or DR analysis) is entirely satisfactory [7]. Improved approaches to the problem, based on molecular level theories, should be developed.

The Monte Carlo technique has been found to be a promising tool in the study of adsorption of pure or multi-component gases in zeolites and other microporous solids [8, 10, 11]. In this work, the method is used in its grand ensemble variant in combination with experimental isotherm data to characterise microporous structures and obtain the corresponding pore size distribution (PSD). Specifically, the mean $CO_2$ density inside a single slit shaped graphitic pore of given width is found on the basis of Grand Canonical Monte Carlo simulations for a pre-defined temperature and different relative pressures. Starting from an initial PSD guess, it is then possible to produce a computed $CO_2$ sorption isotherm and compare it to the measured one. After a few iterations, the procedure results in a PSD which, if desired, can be further refined at the cost of additional computational effort. Pore size distributions of activated carbon membranes obtained by a Nitrogen porosimeter (with Krypton upgrade) via the conventional (DR) approach are employed for the sake of comparison with the present method.

The Grand Canonical Monte Carlo (GCMC) method is ideally suited to adsorption problems because the chemical potential of each adsorbed species is specified in advance [9]. At equilibrium, this chemical potential can then be related to the external pressure making use of an equation of state. Consequently, the independent variables in the GCMC simulations are the temperature, the pressure and the micropore volume, i.e. a convenient set since temperature and pressure are the adsorption isotherm independent variables. Therefore, the adsorption isotherm for a given pore can be obtained directly from the simulation by evaluating the ensemble average of the number of adsorbate molecules whose chemical potential equals that of a bulk gas at a given temperature and pressure.

Three types of trial have been used, i.e. attempts to move (translate or reorient) particles, attempts to delete particles and attempts to create particles in the simulation box. A decision is made on whether to accept each trial or to return to the old configuration based on a probability which in the case of an attempted move takes the form:

$$P_{move} = min\left[ exp\left( -\frac{\Delta U}{kT} \right) \right] \qquad (1)$$

where $\Delta U = U_{new} - U_{old}$ is the difference in the potential energies of the new and old configurations. A detailed presentation of the method is given in [10, 11]. Periodic boundary conditions have been applied in the directions other than the width of the slit. For a given simulation, the size of the box (i.e. the two dimensions other than H) have been selected such that ca. 200-400 molecules are present in the simulation. Statistics were not collected over the first $2 \times 10^6$ configurations to assure adequate convergence of the simulation. The uncertainty on the final results (ensemble averages of the number of adsorbate molecules in the box and the total potential energy) is estimated to be less than 3%. Typical calculations for a single point require between 0.5 and 2 hours of CPU time on a Convex 3820.

A microporous carbon membrane has been tested using the Quantachrome AUTOSORB-1 Nitrogen porosimeter equipped with Krypton upgrade. The total pore volume has thus been found. It compares favourably to the value estimated from the measured $CO_2$ isotherm (at 195.5 K). To start with, a skewed triangular pore size distribution has been postulated making sure that the total pore volume equals to the experimentally determined value. The micropore range (from 0.5 to 2.0 nm) has been subdivided in equidistant spaces with 0.1 nm width. The most probable pore size (slit width) and the standard deviation of the assumed distribution have been varied systematically. For each case, the pore volume associated with each class of pores (the aforementioned subdivisions of the overall pore range) has been calculated (such that the total pore volume remains equal to the measured one) and the amount of gas ($CO_2$) adsorbed in every class at a certain pressure has been computed using the GCMC code. In this way, a computed isotherm has been reconstructed up to relative pressures of 0.4 (at higher values the mesopore effect comes into play). By comparing this segment of the $CO_2$ isotherm to the corresponding experimental counterpart, the most suitable micropore size distribution was selected. The most probable pore size (0.75 nm) found is in good agreement with the independent estimation of 0.7-0.9 nm. Currently, the above described trial and error procedure is attempted to become free of the need to specify a certain (readily manipulated) distribution function. The solution (by least squares) of a minimisation problem under certain constraints will provide the optimal distribution that fits best the selected segment of the measured isotherm data.

### 2.1.2 Small Angle Neutron Scattering of partially blocked membranes

According to Small Angle Scattering (SAS) theory, the intensity I(h) (h is the scattering vector) scattered by a two phase system is related to the electron densities $\rho_1$ and $\rho_2$ of the phases in terms of the expression

$$I(h) \approx \left(\rho_1 - \rho_2\right)^2 \qquad (2)$$

According to the scattering theory, the Contrast Variation (incomplete contrast matching) between the solid matrix of a porous medium and the pore fluid can be changed by the use of different concentrations of H-isotopes. Since the Scattering Length Densities (s.l.d.) of $H_2O$ and $D_2O$ are known, and the s.l.d. of pores equals to zero, adsorption in conjunction with SANS can be achieved by using an appropriate mixture of $H_2O/D_2O$ as an adsorbate. Apparently, when $\rho_1 = \rho_2$ Contrast Matching is reached and the scattering intensity will become zero. The condensed cluster of pores will cease to act as a scatterer, and only the remaining empty pores will produce a measurable scattering. In terms of neutron scattering Contrast Matching reduces the solid/film/pore system to a binary one.

In this way, an interface of the two methods can be achieved by selecting an adsorbate with electron density (SAXS) or scattering length density (SANS) similar to that of the solid matrix. Then, by determining a number of scattering curves at various P/Po, corresponding to both the adsorption and desorption branches, a correlation of the two methods could be possible. If the predictions of the Kelvin equation are in accordance with the SAS analysis, a reconstruction of the adsorption isotherm can be obtained from the SAS data, as it has been illustrated (Fig. 1) for the case of Alumina membranes [12].

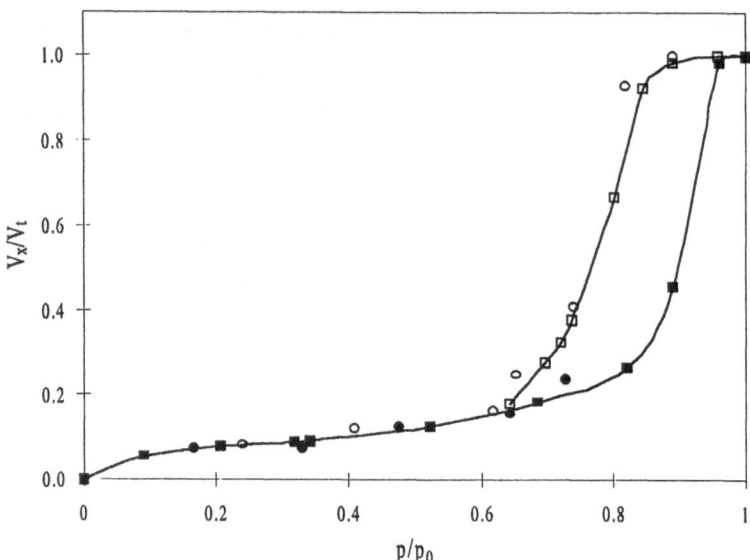

Fig. 1    Adsorption-Desorption Isotherm of $H_2O/D_2O$ mixture at 295 K for an
Al$_2$O$_3$ membrane.

(•) Adsorption and (o) desorption from SANS data,

(▪)Adsorption and(□) desorption volumetrically.

## 2.2 DYNAMIC METHODS OF CHARACTERISATION

The dynamic methods rely on the study of fluid flow properties of porous media, which are extremely sensitive functions of the pore size distribution (PSD) and of additional pore structural characteristics, such as the degree of the pore connectivity. For this reason, the elucidation of the pore structural characteristics from dynamic data alone is very difficult and this accounts for the scant use of the flow pore structural characterisation methods. On the contrary, if the dynamic data are analysed in combination with the static measurements, they can provide important structural information, relevant to the evaluation of performance of membranes. Hence, the dynamic methods comprise a very powerful supplement of the aforementioned static characterisation techniques, provided that the appropriate pore structure model is used for the interpretation of the relevant data. In the following section an outline of the basic models will be presented along with the description of the most promising dynamic method, the gas relative permeability.

### 2.2.1 Pore structural models

*Homoporous capillary tube model.* The most simplified model of porous media used for the formulation of the gas flow, is the homoporous model of capillaries with equivalent radius $r_e$, defined as: $r_e = 2 \ \varepsilon/\Lambda$, where $\varepsilon$ is the porosity and $\Lambda$ the internal specific area of the porous medium.

The steady-state flux along the x axis of a long cylindrical capillary (radius r) with diffuse reflection, for the case of non adsorbed gas, which is of interest in the methods outlined hereunder, may be written as:

$$J = -\pi r^2 D_K \frac{\partial C}{\partial x} = \frac{4\pi}{3} r^3 \left( \frac{2RT}{\pi M} \right)^{1/2} \frac{C_o - C_\ell}{\ell} \qquad (3)$$

where $C_0$ and $C_\ell$ are the concentrations at x=0 and x=$\ell$ respectively and $D_K$ is the Knudsen diffusion coefficient.

The Knudsen -flow permeability, $P_K$ is defined as:

$$P_K = \frac{J \cdot \ell}{\pi r^2 \cdot (C_o - C_\ell)} \qquad (4)$$

468

and is valid for the limiting case of $\dfrac{r}{\lambda} \to 0$ (Knudsen regime), where $\lambda$ is the mean free path of the gas molecules.

For higher pressures the expression for pressure dependence of the flow through a single cylinder is given by Weber [13] as the sum of three components, namely:

a) Poiseuille flow component (erg/sec)

$$J_g^P = -\left(\frac{3\pi}{64} \cdot \frac{r}{\lambda} \cdot \frac{dp}{dx}\right)\pi r^2 P_K \tag{5}$$

b) Slip flow component which contribute in erg/sec

$$J_g^S = -\frac{\pi}{4}\left(\frac{2\,{}^r\!/_{\!\lambda}}{1+2\,{}^r\!/_{\!\lambda}}\right)\frac{dp}{dx}\pi r^2 P_K \tag{6}$$

c)and the Self-Diffusion component

$$J_g^{SD} = -\left(\frac{1}{1+2\,{}^r\!/_{\!\lambda}}\right)\frac{dp}{dx}\pi r^2 P_K \tag{7}$$

The total flux is given by $J_g = J_g^P + J_g^S + J_g^{SD}$ so the permeability, P, is given by:

$$P = P_K\left\{\frac{3\pi}{64}\frac{r}{\lambda} + \frac{\pi}{4}\left[\frac{2r/\lambda}{1+2r/\lambda}\right] + \left[\frac{1}{1+2r/\lambda}\right]\right\} \tag{8}$$

so that:

a) for $\dfrac{r}{\lambda} \to \infty$ $\qquad P \to P_K\left(\dfrac{3\pi r}{64\lambda}\right)$ viscous flow (Poiseuille)[14]

b) for $\dfrac{r}{\lambda} \to 0$ $\qquad\qquad P \to P_K$ molecular flow (Knudsen)

As it is shown in Fig. 2, the total permeability starts from the Knudsen limiting value for low pressures and decreases to a minimum. The decrease is attributed to the gradual elimination of long free paths $\lambda$, due to the increase of the molecule-molecule collisions, which are negligible in the Knudsen flow regime. As the pressure increases, the molecule-molecule collisions become predominant and cause the co-operative linear increase of the flow with pressure (viscous flow regime).

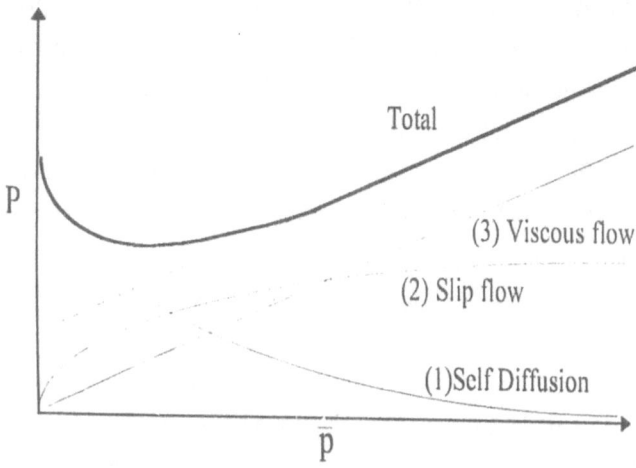

**Fig. 2** Presentation of the 3 Weber flow components [15] and the overall

permeability P vs. mean pressure $\bar{p} = \dfrac{p_0 + p_\ell}{2}$ .

In this simple formulation, the porous structure is pictured as a bundle of uniform size capillaries of radius $r_e$ and length $\tau\ell$, where the tortuosity $\tau$ takes into account the increase of the flow path of the molecules due to the constrictions of the pore structure.

*Heteroporous model of tortuous capillaries.* This is a more realistic advance of the previous model to include the heteroporosity, by incorporating in the model the pore size distribution (PSD) function of the capillary radii, f(r). The distribution is defined in the range of $r_a < r < r_b$ and is subject to the normalising condition:

$$\int_{r_a}^{r_b} f(r)dr = 1 \tag{9}$$

where $r_a$ and $r_b$ are the lower and upper bounds of r respectively. f(r) is readily determined from the sorption branch of the equilibrium isotherm.

*Network model.* The capillary network models constitute a significant improvement over the aforementioned mentioned tortuosity model, which is considered to be inadequate in describing satisfactorily phenomena in porous media partially blocked by condensed vapours,  such as the gas relative permeability measurement, $P_R$. $P_R$ is defined  as the permeability of a fluid through a porous medium partially blocked by a second fluid, normalised by the permeability through the same porous solid, when the pore space is free of this second fluid. In most cases, the gas permeability diminishes at the "percolation threshold", at which a significant portion of the pores are still conducting; however in the heteroporous tortuous capillary model the percolation threshold arises only when all the pores are blocked by sorption and condensation. On the contrary, the network models can  provide a satisfactory analysis of percolation threshold problem, without increasing the number of the model parameters. The parameter of tortuosity is replaced by the network connectivity, $n_T$, which is more  useful for the evaluation of the membrane performance and amenable to  evaluation from equilibrium and dynamic methods.

The network model  consists of a regular cubic array of nodes, joined together by cylindrical tubes [16]. The radius of each tube is assigned from the PSD, obtained from the treatment of equilibrium data. In the simplest version the nodes are assigned volume zero. In a more advanced model spherical cavities are distributed over the nodal points, connected by narrow cylindrical pores.[17]. Similar models are used for the simulation of oil recovery processes [18]. Finally, more realistic models are under development, based on random sphere modelling approach,  suitable for the simulation of adsorbents made up of spherical or  spheroidal particles [19]. In these models the dense random sphere packing is considered as an assembly of tetrahedral subunits, whose vertices lie at the centres of the spheres (Fig. 33). Each cavity is connected with the four neighbouring cavities triangular windows (Fig. 3). The amount sorbed  is estimated as the sum of (I) the volume the sorbed layer of the spherical surface (ii) the volume of condensate in the form of  pendular rings between neighbouring spheres   (Fig. 3). At higher pressures the tetrahedral cavities are filled by condensate. The filling is assumed to occur when the average of the six pendular radii exceeds the characteristic radius of the cavity within the tetrahedron as reduced in size by the adsorbed layer and the pendular rings. Similarly the triangular windows close by condensation , when the average of the three pendular ring radii becomes greater than the characteristic radius calculated for the window, as reduced in size by the presence of the sorbate.  The aforementioned characteristic radii are estimated on the methods of (i) the inscribed sphere of circle and (ii) equivalent sphere of circle. A two-dimensional network model is employed for the simulation of the sorption-desorption isotherm and relative permeability in dense equal-sphere random packs (of porosity of 0.37) [20-21]. A more realistic simulation is being advanced lately, wherein the random packing of spheres at different porosities are divided into tetrahedral subunits; and a similar simulation of the sorption-desorption isotherm and the relative permeability is developed.

*2.2.2 Characterisation of inorganic membranes by gas relative permeability.*
*Formulation of the gas relative permeability model.* The theoretical studies for gas permeability through mesoporous media are rather extensive. For porous plugs formed by compaction of spherical particles, a two-dimensional network, composed of divergent-convergent flow channels has been developed to study the relative permeability as function of the amount of sorbed vapour Fig. 3[21]. Although the complex geometry of the flow channels connecting adjacent tetrahedral cavities between the spheres cannot be fully represented, the network structure proposed in [21] retains its important characteristic, namely the alternate convergent-divergent shape of each channel as the fluid passes from one cavity to the next through the intervening window. In addition, simpler regular three-dimensional networks, consisting of capillaries of randomly varying radius have been employed [22]. By application of the Effective Medium Theory (EMT) to the stochastic network, an analytical expression relating the Knudsen gas relative permeability of a partially blocked mesoporous solid to the moments of the PSD and the connectivity, $n_T$, of the pores [23] is developed. In the effective Medium Approximation, the actual network is replaced by an
"effective" of uniform radius. The "effective" radius is estimated by requiring equal flux to result from both the effective and the actual networks at the same pressure difference. The effective radius is obtained by solving the following integral equation:

$$J = \int_{x_a}^{x_b} \frac{(P_m - x^3) \cdot f_c(x)dx}{x^3 + v \cdot P_m} = 0 \qquad (10)$$

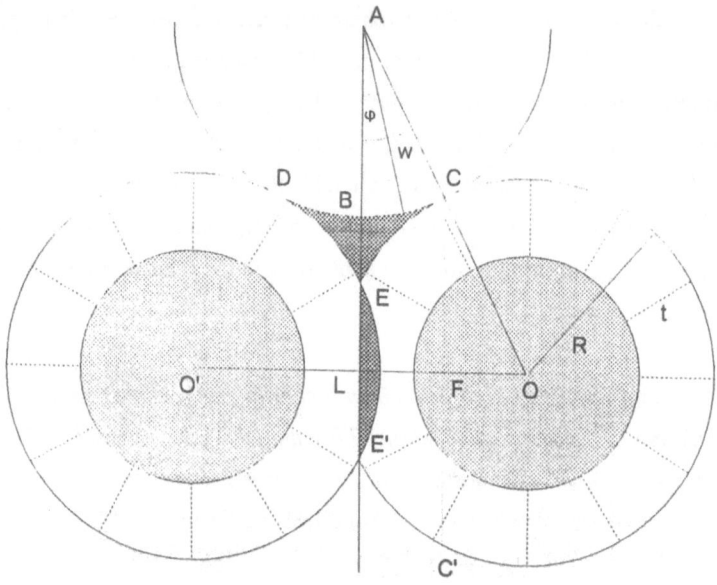

**Fig. 3**   An excluded cap from the adsorption

During adsorption, the pores follow a probability distribution $f(x)$ with $x=r-t$ (r is the pore radius and t is the thickness of the adsorbed layer). Upon condensation, all the pores with radii smaller than a critical value $x_k(x_k=r_k-t)$ are filled with condensed vapour so that $f_c(x) = f_a \cdot \delta(x) + f(x_k \leq x \leq x_b)$

where $\delta(x)$ is the Dirac function, $f_c(x)$ is normalised ($\int_0^\infty f_c(x)dx = 1$),

$f_a = \int_0^{x_k} f_c(x)dx$ is the fraction of the blocked pores and $f_b = \int_{x_k}^\infty f_c(x)dx$ is the fraction of the open pores ($f_b+f_a=1$).

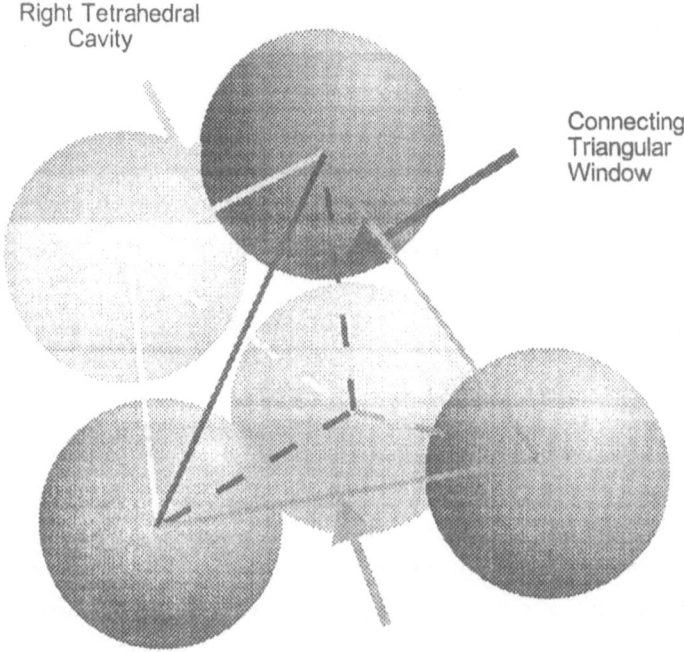

Right Tetrahedral
Cavity

Connecting
Triangular
Window

Left Tetrahedral
Cavity

**Fig. 4**  Tetrahedral Sub-units

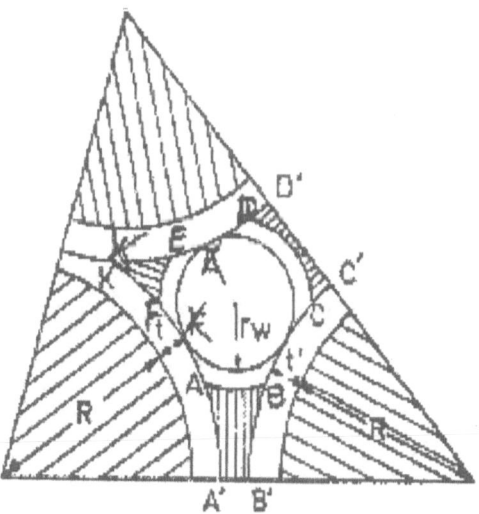

**Fig. 5**  Irregular triangular window with three spheres at its vertices of radius
R+1. $r_w$ is (a) the radius of the inscribed circle and (b) the radius of a
circle

474

From EMT,

$$\int_0^\infty \frac{(P_m - c \cdot x^3) \cdot f_c(x) dx}{x^3 + v \, P_m} = 0 \tag{11}$$

with $v = \dfrac{n_T}{2} - 1$, $n_T$:connectivity $(2 \le n_T \le \infty)$, $P_m$:permeability,

$c \cdot x^3$:conductance (Knudsen regime). $\tilde{f}(x)dx = \dfrac{f(x)}{f_b}dx$ is the normalised

distribution of the open pores and $\bar{x}_c = \int_{x_k}^\infty x \cdot \tilde{f}(x)dx$ is the average radius of

the open pores.

By suitable moment expansion of the aforementioned EMA equation the relative permeability can be related explicitly to the relevant microstructural network parameters, namely the suitable moments of the pore size distribution and the network connectivity. The relative permeability, $P_R$, is then given by:

$$P_R = \frac{\bar{x}_c^3 \cdot (1 + \gamma_1) \cdot f_b \cdot (\gamma_{M1} + \gamma_{M2})}{\bar{r}_c^3 \cdot (1 + a_1) \cdot \gamma_M} \cdot \left( \frac{f_b - \lambda}{1 - \lambda} \right) \tag{12}$$

with

$$\gamma_{M1} = \frac{1}{f_b} - \lambda \cdot \alpha_2' + \lambda^2 \cdot \alpha_3' - \lambda^3 \cdot \alpha_4' + \lambda \cdot \alpha_1'^2$$

$$\gamma_{M2} = (2\lambda^3 - \lambda^2 f_b) \cdot \alpha_2'^2 + (2\lambda f_b - 3\lambda^2) \cdot \alpha_1' \cdot \alpha_2'$$

$$\gamma_M = 1 - \lambda \cdot a_2 + \lambda^2 \cdot a_3 - \lambda^3 \cdot a_4 + \lambda \cdot a_1^2 + (2\lambda^3 - \lambda^2) \cdot a_2^2 + (2\lambda - 3\lambda^2) \cdot a_1 .$$

$$\alpha_n' = \frac{\alpha_n}{f_b^n} \ , \quad \alpha_n = \int_{x_k}^\infty \alpha^n \cdot \tilde{f}(x)dx \ , \quad \alpha = 3w + 3w^2 + w^3, \quad w = \frac{x - \bar{x}_c}{\bar{x}_c}, \quad \lambda = \frac{2}{n_T}$$

$$a_n = \int_0^\infty a^n f(r)dr, \quad a = 3z + 3z^2 + z^3, \quad z = \frac{r - \bar{r}_c}{\bar{r}_c}, \quad \bar{r}_c = \int_0^\infty r \cdot f(r)dr$$

This expression gives a satisfactory approximation of the numerical network relative permeability estimation and, at the same time, provides an insight to the relative permeability dependence on the PSD characteristics and the network connectivity [23].

*Experimental gas relative permeability measurements.* Contrary to the theoretical modelling studies, the experimental investigations of relative permeability are very scant [24]. Although relative permeability is a sensitive function of the microscopic structural characteristics and an important supplement of the vapour sorption isotherm measurements in elucidating the pore structure, its widespread use is hindered mainly due to the experimental complications.

In order to reduce the difficulties in obtaining relative permeability measurements, one can resort to the use of the simplified single-component relative permeability method [25] Alternatively a more sophisticated novel design multiple- membrane rig can be employed [26].

*Single component relative permeability.* A new simplified relative permeability technique has been demonstrated recently, in which certain experimental complications encountered by the conventional procedure are avoided. The method is an extension of the pre-adsorption method [27]. The experimental rig is shown schematically in Fig. 6. After equilibrating the porous membrane with a strongly sorbed vapour, both the upstream and the downstream sections of the cell are evacuated. Upon evacuation, helium is admitted in the upstream section and allowed to permeate to the downstream section. If the desorption rate is reasonably low, the linearity of the permeability curve remains unaffected by the desorption process, for the period of a few minutes that is normally required to complete the permeation experiment. In this way, the presence of a gas mixture (consisting of the permeating gas and the blocking vapour) circulating along both sides of the porous sample is not needed, thus rendering the present procedure much simpler to execute.

The simplified gas relative permeability method provides satisfactory data for microporous membranes . Indeed, comparison of the present and the conventional relative permeability techniques show good agreement for the case of microporous carbon membranes with pore radii of the order of 1 nm blocked by benzene. The difference of the two methods is of the order of 10% , even for relative pressures as high as P/Po=0.6.

For the case of mesoporous membranes, the application of the simplified relative permeability method is possible, if the temperature of the membrane bath is appropriately low to reduce the desorption rate to an acceptable level; namely to a desorption rate that will not affect the linearity of the pressure uptake as a function of time in the low pressure section of the permeability rig. The relative permeability data obtained by the present method and shown in Fig. 8, have been analysed by employing the analytical expression[25]; the corresponding moments, $\alpha_2$, $\alpha_3$ and $\alpha_4$, were estimated from the pore size distributions obtained by liquid nitrogen porosimetry (Fig. 7). By fitting the data , a reasonable value for the connectivity $n_T=8$ is obtained.

*Novel Design of the multiple gas permeability rig.* The three-membrane apparatus is shown in Fig. 9. The apparatus is made of stainless steel and can operate from high vacuum up to 70 bars. Each of the three membranes, accommodated by the facility, is connected to the high and low pressure sections through four valves MiH, MiL, MiHF, MiLF (i=1, 2, 3). Both sections are connected to the high vacuum line (HV, LV), the gas line (LG, HG) and the Gas Chromatographer loop (GCHi, GCLi, i=1, 2). The 5 liters and 0.5 liters high pressure sample cylinders are used to minimise the high pressure section pressure drop and the low pressure section pressure uptake during permeation measurements[28].

The gas composition uniformity is maintained by circulating the gas mixtures past the surfaces of the membranes in the high and the low pressure sections. The circulation is effected by means of two gas circulators (CH, CL). In order to analyse low pressure gas samples (pressure ⟨ 1 bar), the high pressure side of a differential manometer (4...20 mA, 0...1250 mbar) is connected with the GC outlet while the low pressure side is constantly kept under vacuum, and thus a pressure readout is available for each sample.

An differential manometer (0...1250 mbar) is used to monitor the pressure difference between high and low pressure sections, while the absolute high and low sections pressure values, are continuously monitored by means of pressure transducers.

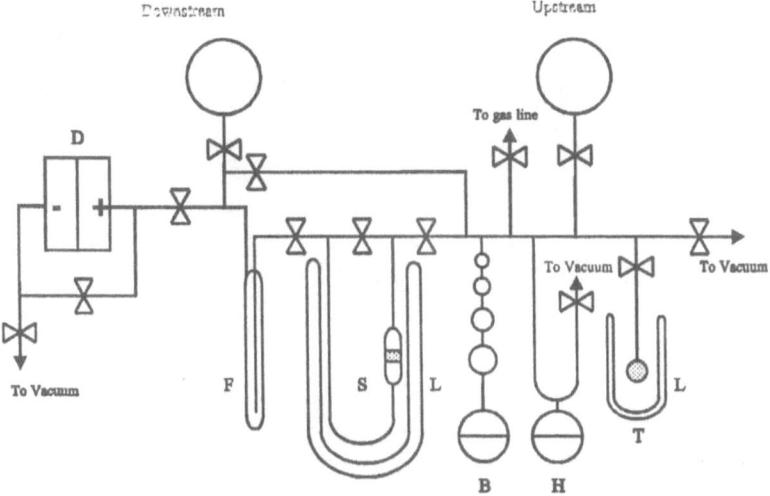

**Fig. 6** Single phase gas relative permeability apparatus.
D: Differential Manometer, F:Liquid Nitrogen finger, S: Sample, L: Deawar Vessel, B: Gas Burrette, H: Hg-Manometer, T: t-curve apparatus.

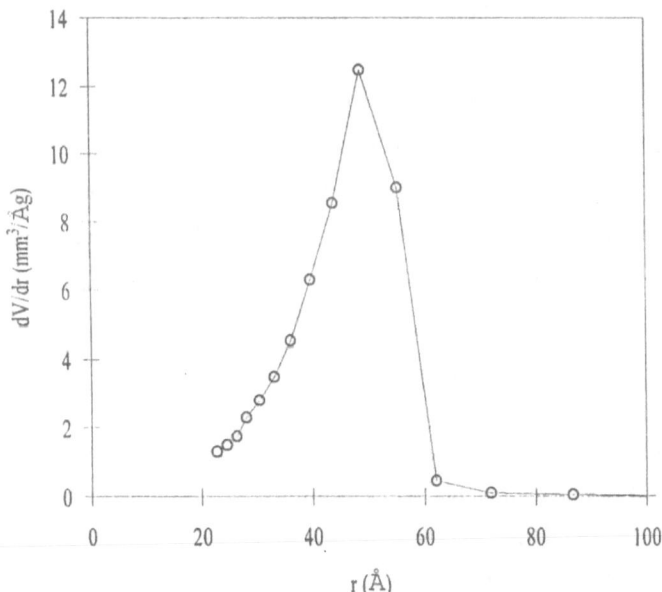

**Fig. 7** PSD for $Al_2O_3$ pellet obtained from the desorption branch of N2 (77 K) adsorption isotherm.

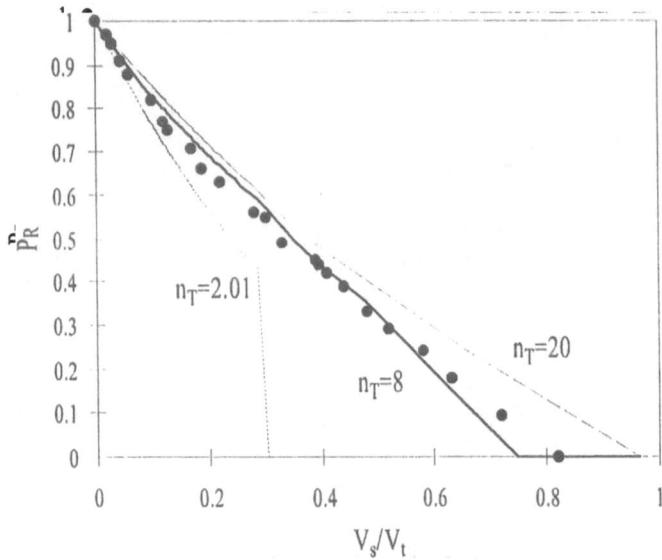

**Fig. 8**  $P_R$ curves calculated for different $n_T$ values. The curves are compared with experimental data for $Al_2O_3$ membrane (●).

In order to minimise the temperature variation effects on the differential manometer pressure measurements, the whole apparatus is immersed in an oil bath maintained in constant temperature, by means of a PID temperature controller; the bath temperature variations are minimised by insulating the space over the bath and keeping its temperature constant to $\pm$ 0.1 °C by means of a PD temperature controller and an air circulation fan. In addition, the room temperature over the insulating cover is kept constant to $\pm$ 1 °C.

The specific experimental procedures that can be performed with the described facility are:

<u>a. Sorption isotherms:</u> The sorption isotherms of gases and vapours can be obtained by volumetrically estimating the amount sorbed in the porous structure. The high and low pressure sections are connected through the HL valve and the equilibrium pressure can be monitored by the pressure transducer (high vapour pressure-permanent gases) or by the differential manometer (low vapour pressure-vapours) maintaining vacuum on the low side of the differential manometer. Sorption kinetics can also be deduced by

connecting the membranes with the whole apparatus volume and monitoring the small pressure changes by means of the differential manometer.

b. Integral permeability: The high pressure section of the apparatus can be pressurised to 1-70 bar (limited of course by the membrane's strength) with the appropriate gas, while the low pressure section and the membranes are kept under vacuum. By opening MiH, MiHF the high pressure section of the membrane is exposed to the gas which permeates to the low pressure section. The pressure uptake is measured by means of the differential manometer (the high pressure section of the manometer is isolated and pressurised to 1 bar), while the pressure head is measured by means of the absolute pressure transducer. When a sorbed gas is used, time lags along with steady state permeation rates can easily be calculated.

c. Single component differential permeability: All the membranes are equilibrated at the same pressure (0-70 bars) on both sides, by pressurising the whole apparatus to the appropriate pressure, through HG and LG. The adsorption process is monitored roughly by means of the absolute pressure transducer and in detail by reading the small pressure changes on the differential manometer, after isolating its high pressure section. The membranes are isolated by closing the valves MiH, MiL, MiHf, MiLF and the high pressure section is pressurised to approximately 1 bar higher than the low pressure section. Subsequently the valves adjacent to each membrane are opened and gas is allowed to permeate through the porous structure. The pressure increase is monitored by the differential pressure manometer.

d. $P_R$ measurements: The high and low pressure sections are connected by opening the connecting valve (HL) and both sides of the membranes are equilibrated to a pre-specified vapour pressure. After attainment of equilibrium, all the 12 membrane valves (MiH, MiL, MiHF, MiLF) and the connecting valve HL are kept closed and helium is introduced in the high pressure section. The mixture is homogenised by the operation of the circulator (CH). The degree of homogenisation is monitored by sampling the gas mixture at appropriate time intervals and feeding the samples to the GC. After achieving satisfactory levels of homogenisation, the four valves adjacent to each membrane MiH, MiHF, MiL, MiLF are opened consecutively and the gas mixture is fed past the high pressure face of each membrane. Helium flows in the low pressure section, where the gas mixture is homogenised by means of the second circulator (CL). The helium concentration in the low pressure section is monitored by G.C. sampling at appropriate time intervals. Alternatively, the constant temperature conditions obtained by the elaborate temperature control allow for the determination of the helium permeability, by measuring the pressure increase in the low pressure section with the differential pressure manometer. Pore connectivity can be derived by processing such data according to the analytical computational procedure of the present authors.[26]

e. Integral Selectivity Measurements: The 12 membrane valves (MiH, MiL, MiHF, MiLF) are closed, after evacuation of the membranes. The gases are successively introduced in the high pressure section and the concentration of the mixture is calculated after GC sampling or alternatively, using their partial pressures and the appropriate compressibility factors. The homogenised mixture is fed past the high pressure face of the membrane and selectivity is determined by measuring the composition of the permeate section by GC sampling.

To summarise, this experimental apparatus is characterised by the unique capability of performing a broad range of porous membrane characterisation and evaluation measurements, namely: equilibrium isotherms, absolute (integral and differential) and relative permeabilities and selectivity. Proper processing of the obtained data using conventional [29] or recently proposed [25] procedures, results in valuable information on the pore structure, pore volume, specific area, pore size distribution, pore connectivity and gas separation efficiency. The above are accompanied by considerable saving of experimental time, because (a) the membranes are degassed simultaneously (microporous membranes may require more than two weeks of degassing) (b) the equilibration of the membranes to a certain gas pressure , which for microporous membranes requires several hours, can be realised for all the membranes simultaneously, prior to the permeability experiments and (c) the homogenisation of the gas mixtures is accomplished simultaneously for all the membranes.

481

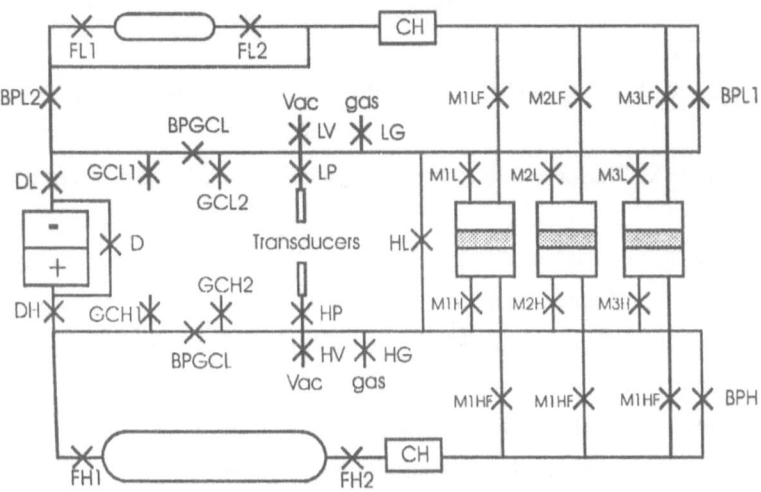

Fig. 9    3 membrane apparatus (Abbreviations: H-High Pressure Section, L-Low
Pressure Section, D-Differential manometer, C-Circulator, M-Membrane,
BP-By Pass, F-Flask, GC-Gas Chromatographer, V-Vacuum Line, G-Gas
Line, P-Pressure Transducer)

References

---

1 . R. Ash, R.M. Barrer and C.G.Pope, Proc. Roy. Soc., London, A271, 19 (1963).

2 . R. Ash, R.M. Barrer and R.T. Lowson, J.Chem. Soc., Faraday Trans., 1, 2166 (1969).

3 . D. Uzio, J. Peureux, A. Giroir-Fendler, J.A.Dalmon and J.D.F. Ramsay, Characterization of Porous Solids III,87, 411.

4 .A.A. Kalachev, K. Mathauer, U. Hohne, H. Mohwald and G. Wegner, submmited to Thin Solid Films.

5 . F. Katsaros et al., High Pressure Gas Permeability of Microporous Carbon Membranes, submitted to Microporous Materials.

6 . K. Kaneko, J. Membrane Sci., 96, 59-89 (1994).

7 . D. Nicholson, J. Chem. Soc. Faraday Trans., 90(1), 181-185 (1994).

8 . R.F. Cracknell et al., Mol. Phys., 80(4), 885 (1993).

9 . D. Nicholson and N.G. Parsonage, Computer Simulation and the Statistical Mechanics of Adsorption, Academic Press (1982).

10 . D.M. Razmus and C.K. Hall, AIChE J., 37, 5, (1991).

11 . R.F. Cracknell et al., Mol. Sim., 13, 161 (1994).

12 . A. Mitropoulos, J.M.Haynes, R.M.Richarson and N.K.Kanellopoulos, Phys.Rev. B, 10035 (1995).

13 . Barrer, R.M. Diffusion in porous media. Applied Mater. Res. 1963,2,129.

14 . Kozeny J., *Wasserkr. Wasserwirt*, 22, 67 (1927).

15 . Barrer R. M., *Appl. Mater. Res.*, 2, 129 (1963).

16 . D. Nikolson and J.H. Petropoulos, J.Phys. D, 1971,4,181.

17 . D.K. Efremov and V.B. Fenelov, React. Kinet. Lett.,40, 1,177.

18 . I.Chatzis, F.A.L. Dulien, International Chemical Engineering, 25,1,47 (1985).

19 . S. Sasloglou et al. in preparation.

20 . N.K. Kanellopoulos, J.K. Petrou and J.H. Petropoulos, J.Colloid and Interf. Sci. 96, 1, 90,(1983).

21 . N.K. Kanellopoulos, J.K. Petrou and J.H. Petropoulos, J.Colloid and Interf. Sci. 96, 1, 101, (1983).

22 . N.K. Kanellopoulos and J.K. Petrou, J. Membrane Sci. 37, 1 (1988).

23. N.K. Kanellopoulos and J.H. Petropoulos, J.Chem. Soc., Faraday Trans. 1, 79, 517 (1983).

24 . 7. J.H. Petropoulos, J.K. Petrou and N.K. Kanellopoulos, Chem. Eng. Sci. 44, 2967 (1989).

25. Steriotis T. A., F. K. Katsaros, A. Mitropoulos, A. K. Stubos and N. K. Kanellopoulos, Characterisation of Porous Solids by Simplified Gas Relative Permeability Measurements, J. of Porous Materials, 2, 281-285 (1995).

26. T.Steriotis, F.K.Katsaros, A.K.Stubos, A.Ch.Miropoulos, N.Zouridakis, P.Galliastatou, N.K.Kanellopoulos, Rev.Sci.Instruments, 2,67(1996).

27. P.J.M. Carrott and K.S.W. Sing in *Characterisation of porous solids,* edited by K.K. Unger et al.(Elsevier Science Publishers B.V., Amsterdam, 1988) p. 77.

28 . Ash R., R. M. Barrer and R. Sharma, Sorption and Flow of Carbon Dioxide and some Hydrocarbons in a Microporous Carbon Membrane, J. Membr. Sci., 17,1 (1976).

29 . S.J. Gregg, K.S.W. Sing, "Adsorption, Surface and porosity" 2nd ed., Academic Press 1982.

## List Of Symbols

| | |
|---|---|
| $\Delta U$ | Difference in the Potencial Energy |
| K | Boltzman constant |
| p | probability |
| I(h) | Scattering intensity |
| $\rho$ | electron density |
| Po | saturation pressure |
| C | concetration |
| M | molecular weight |
| J | flux |
| D | diffusion coefficient |
| r | radius |
| Pk | Knudsen permeability |
| $\epsilon$ | porosity |

| | |
|---|---|
| A | area |
| $\ell$ | length |
| $\lambda$ | mean free path |
| $\tau$ | statistical thickness |
| $n_\tau$ | connectivity |
| Pm | Permeability |
| a | moment |
| M | membrane |
| L | Low pressure section |
| H | High pressure section |
| F | Flask |
| V | Vacuum line |
| G | Gas line |
| GC | Gas Chromatographer |
| C | Circulator |

# CERAMIC MEMBRANES: INDUSTRIAL APPLICATIONS

A.K. STUBOS[+], A. MITROPOULOS[*], T. STERIOTIS[*],
F.K.KATSAROS[*], G. E. ROMANOS[*] AND N.K.
KANELLOPOULOS[*]

*Institute of Physical Chemistry, NCSR Demokritos, 15310 Ag.
Paraskevi Attikis, Greece*

+*Institute of Nuclear Technology and Radiation Protection,
NCSR Demokritos, 15310 Ag. Paraskevi Attikis, Greece*

Keywords: Porous membranes, Gas adsorption, Gas permeability, Gas
separation, Capillary condensation, Supercritical Liquefaction.

## Abstract

Pores, and especially mesopores (with sizes between 2 and 50 nm) and
micropores (with sizes less than 2 nm), play an essential role in physical and
chemical properties of industrially important materials like adsorbents,
membranes, catalysts, soils, biomaterials, etc. Apart from their structural
characterisation (see the first lecture), the description of transport phenomena
in porous materials has received generally much attention due to its
importance in many applications, such as drying, moisture transport in
building materials, filtration, etc. Although widely different, these topics share
many common aspects, such as heat and mass transfer, capillarity and
multiphase (vapour-liquid) flow, all in a porous media environment. In
particular, transport in mesoporous structures and the associated
phenomenology of multilayer adsorption and capillary condensation have
been investigated, mainly as a separation mechanism for gas mixtures [1-6].
Barrer and co-workers [1,2] were the first to observe the capillary

*J. Fraissard (ed.), Physical Adsorption: Experiment, Theory and Applications,* 485–510.
© *1997 Kluwer Academic Publishers.*

condensation effect studying various gas mixtures in microporous carbon membranes. Since the late 1980's, the publications and marketing by manufacturers in USA, Japan and Europe of membrane systems based on substrates coated by thin mesoporous layers have resulted in the growth of interest in such phenomena. The reason is that the new products proved substantially more effective in selectivity than the commonly used techniques. More recently, novel developments in manufacturing and controlling thin film structures have boosted industrial interest in the study of the underlying mechanisms for the case of micropores too [7-15].

In this lecture, reference is made to discrete approaches for the modelling of gas/condensate flow through mesoporous and microporous structures. The former case is dealt with in more detail and a network model is presented along with typical predictions and sensitivity studies. The situation for micropores is far less advanced and a general outline of an approach that is currently investigated by our group is given.

## 1. Mesoporous Membranes

Although the static behaviour of the condensate in the capillaries is relatively well understood [16-20], the dynamic behaviour of both the adsorbed layers and the condensed phase that determines the overall separation and transport efficiency of the process, has not been adequately described yet. The measured permeability curves were found to exhibit a maximum at some relative pressure, a fact that has been attributed to the occurrence of condensation (and blocking) in the main body of the pores [3,8,12]. The transition between the surface flow and condensate flow regimes depends on the pore sizes and the nature of the fluid. Hysteresis aspects similar to those in adsorption and desorption cycles were also observed. A limited number of quantitative models have been reported in the literature [8,12,13,21] with some success in predicting the membrane permeability as a function of relative pressure and average pore size. The macroscopic descriptions in single tube geometries [3,7,8] can not address the influence of the pore structure and the micromechanics of surface flow and condensation in porous media. In the present work, it is recognised that essential porous media phenomena involve an interaction over an ensemble of pores, and that a relevant theory is required for the interpretation of data on the basis of a model that captures many of the unique aspects of vapour-liquid flow in mesoporous structures.

A discrete (network) approach to modelling transport of condensable vapours in meso- and macro-porous structures is adopted. Such models possess the potential for improving the understanding of the mechanisms responsible for the observed transport behaviour. Tube networks are considered primarily. A simple network simulation of the porous structure can be constructed in the form of a square or cubic lattice of nodes. Each node is connected to $n_T$ neighbouring nodes by $n_T$ tubes (bonds). The sizes of the bonds can be distributed (randomly or in a correlated fashion) as specified from the experiments, or by following a desired pore size distribution that introduces randomness and non-uniformity. Realistic pore sizes can be obtained by one of the standard static methods, whereas $n_T$ for porous solids of uniform pore structure can be estimated by dynamic flow data, such as relative permeability measurements [22]. Connectivity varies between two limiting cases. Thus, for $n_T=2$ the network reduces to a serial arrangement of tubes while for $n_T$ equal to infinity the parallel arrangement of tubes is obtained [23]. The network structure is defined by its connectivity, $n_T$, and the pore size distribution of the bonds.

The gas or liquid flow though a porous solid can be simulated considering the conductance, g, of each bond connecting two neighbouring nodes:

$$g = B \, r^a \qquad (1)$$

where short tube or tube end effects are neglected, r is the radius of the cylindrical bond (or the half width in the cases of slit-shaped pores) and the constants B and a depend on the flow regime and the pore shape. The steady state pressure distribution in the network follows from the usual Kirchoff's law considerations. Mass is not permitted to accumulate at any node and flow through a bond is equal to the product of the bond conductivity and the pressure drop $(P_i - P_j)$ between the nodes i and j connected by the bond:

$$\sum g_{ij} \cdot \left(P_i - P_j\right) = 0 \qquad j = 1, n_T \qquad (2)$$

This set of simultaneous algebraic equations for all the nodes is solved by an iterative method and the permeability of the network follows from the total flux through the network for a given pressure drop.

Depending on the specified pressure differential across the network, different flow regimes may develop inside the individual pores. At low enough mean pressures, the observed mass flow rate, J, is considered to be made up of non-adsorbed molecules moving in the free pore space (gas-phase component $J_g$) and of adsorbed molecules moving along the pore wall surface (surface flow component $J_s$):

$$J = J_g + J_s \qquad (3)$$

The mechanism of gas-phase flow varies from diffusive to viscous depending on the gas concentration (or equivalently pressure $P_g$). In the former case, $J_g$ is formulated in terms of a Knudsen type diffusion coefficient while in the latter Poiseuille's equation with a constant viscosity coefficient $\eta_g$ is used. More specifically, the relative magnitude of pore radius and mean free path $\lambda$ of the gas is employed in the present study of steady state permeation to decide about the prevailing flow mechanism:

(i) In case the mean free path $\lambda$ of the gas molecules in a single pore is such that (r-t)<0.05$\lambda$, Knudsen diffusion applies:

$$J_g = \left( \frac{32 \cdot \pi \cdot M}{9 \cdot R \cdot T} \right)^{1/2} \cdot (r - t)^3 \cdot \frac{\Delta P_g}{l} \qquad (4)$$

(ii) In the intermediate range 0.05$\lambda$<(r-t)<50$\lambda$, slip flow comes into play:

$$J_g = \left( \frac{\pi^3 \cdot M}{8 \cdot R \cdot T} \right)^{1/2} \cdot (r - t)^3 \cdot \frac{\Delta P_g}{l} \qquad (5)$$

(iii) For (r-t)>50$\lambda$, Poiseuille flow is considered predominant:

$$J_g = \frac{\pi \cdot \rho_g}{8 \cdot n_g} \cdot (r - t)^4 \cdot \frac{\Delta P_g}{l} \qquad (6)$$

The mean free path is obtained by the relation:

$$\lambda = \frac{\sqrt{2} \cdot R \cdot T}{\pi \cdot d^2 \cdot N_A \cdot (P_i + P_j)} \qquad (7)$$

In the above, $\rho_g$ is the density of the gas, l is the length of the bond, M is the molecular weight, R is the universal gas constant, T is the absolute temperature, d is the collision diameter of the penetrant gas, $N_A$ is the Avogadro number, and $P_i$, $P_j$ are the upstream and downstream pressures across each single capillary. The thickness of the adsorbed layer t is calculated from isotherm adsorption data, the specific area of the solid and the specific volume of the fluid. The foregoing expressions provide for a sound theoretical formulation of $J_g$ over the entire pressure range of interest.

The gas flows mentioned in the above paragraphs are considered to be uninfluenced by the forces of interaction between the penetrant gas and the solid material. It is therefore necessary to consider the flow resulting from the mobility of the layer of the adsorbed gas on the pore wall, namely the surface flux. The situation concerning its theoretical evaluation is far less satisfactory. Among several propositions [5,10,25,26], the formulation developed in [27] in terms of the spreading pressure and a constant surface flow resistance coefficient $C_R$ is applied in this work:

$$J_s = \frac{2 \cdot \pi \cdot r \cdot \chi^2 \cdot R \cdot T}{P_m \cdot C_R \cdot S^2} \cdot \frac{\Delta P_g}{1} \qquad (8)$$

where $P_m=(P_i+P_j) / 2$ is the mean pressure in the bond, S is the specific area of the medium and $\chi$ represents the amount of fluid sorbed per unit of solid mass. The limitations of this approach should be kept in mind. In fact, representing $J_s$ by equation (8) with constant resistivity of the medium over the entire relative pressure range may be questioned physically, although some success is reported in correlating empirically a variety of surface flow data [27]. Anyhow, in the present simulations, the contribution of surface flow to the overall permeation rate diminishes after the occurrence of capillary condensation effects and the flow rate given by (8) is significant at low and medium relative pressure values only.

Following multilayer adsorption on the pore wall, capillary condensation occurs at high enough pressures, as indicated by the modified form of the Kelvin equation for the case of adsorption:

$$\frac{P}{P_o} = \exp\left(\frac{\sigma \cdot \cos \vartheta}{\rho \cdot R \cdot T \cdot (r - t)}\right) \qquad (9)$$

where $P_o$ is the saturation pressure, $\sigma$ is the surface tension and $\theta$ is the contact angle. This form implies that the onset of the phenomenon denotes an instability of the film interface due to the pore wall curvature [28], a mechanism which is also responsible for a single pore hysteresis, where the phase change (from liquid to vapour) during desorption occurs at an effective radius smaller by a factor of 2 than that found by (9). The steady state viscous condensate flow rate, $J_c$, is assumed to obey Poiseuille's formulation:

$$J_C = \frac{\pi \cdot r^4 \cdot \rho_1}{8 \cdot n_1} \cdot \frac{\Delta P_1}{1} \qquad (10)$$

Equation (10) may be recast in terms of the gas phase pressure by making use of the thermodynamic relation (neglecting gas non-ideality):

$$\frac{dP_l}{\rho_l} = \frac{dP_g}{\rho_g} = \frac{R \cdot T}{M} \cdot d \ln P_g \qquad (11)$$

Thus, (10) becomes:

$$J_c = \frac{\pi \cdot r^4 \cdot \rho_l}{8 \cdot n_l} \cdot \frac{\rho_l \cdot R \cdot T}{M \cdot P_m} \cdot \frac{\Delta P_g}{1} \qquad (12)$$

The discrepancy between liquid and gas pressure drops implied by (11) can only be supported by a significant difference in the curvatures of the menisci at the high and low pressure sides of the bond, as shown schematically in fig.1. Based on this configuration, equation (10) can be rewritten as:

$$J_C = \frac{\pi \cdot r^4 \cdot \rho_l}{8 \cdot n_l} \cdot \frac{\left(\Delta P_g - \Delta P_c\right)}{1} \qquad (13)$$

where $\Delta P_c$ represents the capillary pressure gradient along the single bond and can be assumed equal to the following relation, under equilibrium conditions:

$$\Delta P_c = -\frac{\rho_l \cdot R \cdot T}{M} \cdot \left[ \ln \frac{P_1}{P_0} - \ln \frac{P_2}{P_0} \right] = -\frac{\rho_l \cdot R \cdot T}{M} \cdot \ln \frac{P_1}{P_2} \qquad (14)$$

Substituting (14) into (13) gives:

$$J_c = \frac{\pi \cdot r^4 \rho_l}{8 \cdot n_l} \cdot \left[ \frac{\Delta P_g}{1} + \frac{\rho_l \cdot R \cdot T}{1 \cdot M} \cdot \ln \frac{P_1}{P_2} \right] \qquad (15)$$

which is an expression that renders the system of equations to be solved non-linear. For $\Delta P_g < P_m$ a simplification can be performed leading to:

$$J_c = \frac{\pi \cdot r^4 \rho_l}{8 \cdot n_l} \cdot \left[ 1 + \frac{\rho_l \cdot R \cdot T}{M \cdot P_m} \right] \cdot \frac{\Delta P_g}{1} \qquad (16)$$

This expression is quite similar to (12) and has been used in the present simulations. When compared to Poiseuille flow (10), equation (16) is characterized by an enhancement factor ($\rho_l$ R T/ M $P_m$) that is physically attributed to capillary pressure gradients [8]. Indeed, an additional driving force occurs due to the presence of menisci between nodes and bonds filled with condensate. This capillary action is gradually diminishing as the mean pressure increases for a given bond. The reason is that the menisci begin to flatten as the pressure is raised above Kelvin equilibrium conditions and up to the saturation vapour pressure of the liquid, at which point the liquid surfaces at the ends of the bond are planar [4]. This gradual flattening of the menisci is taken properly in account by the present approach. Capillary effects should vanish completely when both nodes connected with the bond are liquid occupied. For the latter to happen, it is necessary that all bonds emanating from a specific node are filled with condensate. Elimination of the driving force due to capillarity has been assumed to take place by other investigators [8,12,21] as a result of thickening adsorbed layers, i.e. when t (as calculated from BET equation) becomes larger than the bond radius. Such a view is contradictory to the postulation of adsorbed layer instability on which Kelvin equation is based. Following the loss of menisci, liquid Poiseuille flow is considered to be the transport mechanism in the bond for pressures up to saturation.

## 2. Results and Discussion

In an attempt to determine the effect of different parameters on the overall vapour transport behaviour, the permeability of freon-113 through model mesoporous membranes under varying conditions is simulated using a two-dimensional (50x50) network of cylindrical pores shown in fig.2. Their radii follow given distribution functions that can be selected among several possibilities (uniform, skew triangular, gaussian). The pore connectivity is also variable (currently 4 or 8, i.e. planar networks consisting of square elements with or without diagonals are considered). The lateral boundaries of the network are treated as periodic. The adsorbed layer thickness is calculated from the BET equation with coefficients estimated using isotherm data from [8].

Figures 3-7 present a parametric study of permeability over the entire range of relative pressure, $P_r$, which is defined using the mean between the feed and permeate side (i.e. differential tests are simulated). In all simulations, unless explicitly stated, a value of 4 for connectivity has been used. It is observed that in all cases the experimentally found maximum in the permeability vs. relative pressure curve [8,12] is reproduced at $P_r$ values that vary depending on the prevailing temperature and the geometric characteristics of the network. Similar results have been obtained for the case of moisture transport in building materials [21] in a study where surface flow has been neglected and loss of capillary action is based on different considerations than presently (cf. previous section).

The effect of pore size distribution is shown in fig.3, where different functions (uniform, gaussian, skew triangular), with the same mean radius and standard deviation are employed. The very small radii of the gaussian and skew triangular distributions (below 1 nm) have not been considered for the simulations. Due to the relatively narrow mesoporous distributions examined, the respective permeability curves behave quite similarly. The uniform distribution allows for larger relative percentage of small pore radii and the corresponding permeability curve starts rising sharply at slightly lower $P_r$. Varying the mean pore radius causes dramatic changes in both the magnitude and the location of the peak (fig.4). Evidently, the flow enhancement due to capillarity occurs at lower relative pressures as the pore size decreases. On the other hand, smaller pore radii correspond to smaller area available for flow leading to reduced flow rates in the individual pores and permeability maxima. Significant increase in the permeation rates is attained by using networks with higher pore connectivities (fig.5), i.e. by employing media characterized by highly interconnected porous structure. Capillary enhancement effects start operating at lower $P_r$ as connectivity increases and conducting clusters with condensate filled pores are more probable.

As indicated by fig.6, reducing the medium resistivity to surface flow (expressed by $C_R$) by a factor of 3, affects significantly the permeability curve in the range of relative pressures before the occurrence and during the initial stages of capillary condensation. On the contrary, the part of the curve around the maximum remains practically uninfluenced due to the domination of the capillary mechanism there. It should be stressed again that the postulation for a constant coefficient of resistance to surface flow may not sound physically correct (a dependence on the degree of pore filling among others should be more realistic [5]) but nevertheless it leads to safe qualitative predictions as far as the relative contributions of Knudsen, surface and condensate flow to the overall permeation rate are concerned. The basic $C_R$ value used in the present simulations is an experimental finding from freon-113 transport through (mesoporous) Vycor glass [8].

The effects of temperature are presented in fig.7. The permeability decreases along the entire $P_r$ range with increasing temperature, in agreement with the results of [21]. Rates of transport in all flow regimes considered (Knudsen, surface, capillary enhancement, Poiseuille) depend on temperature either straightforwardly or through the variation of physical properties of the fluid/solid system (mainly saturation pressure, viscosity, adsorbed amount per unit solid mass and vapour density). The reduction in permeability is more pronounced in the capillary condensation regime despite the decrease of liquid viscosity with temperature. The influence of temperature on both the Kelvin radius and the capillary enhancement factor (through $P_m$) proves stronger and leads to the observed behaviour.

The difference between integral and differential permeation tests is depicted in fig.8. In both cases the mean pressure is the same but the actual total pressure drop across the membrane is small (differential test) or large (integral test). The significant reduction of permeability during integral tests is attributed to the partial elimination of capillary condensation effects in the low pressure side of the network. The result can be used to interpret experimental permeability data exhibiting wide discrepancies at the same mean pressure values [8].

An extension of the code to three dimensions is currently underway permitting more realistic simulations. The model will be used to study further the effect of pore shape (slits versus cylinders), pore constrictions, macroscopic inhomogeneities in pore size (to explain observed differences in flow rates when the direction of flow is reversed) and dual porosity systems (fractured networks). Comparisons with experiments are also performed for a rigorous testing of the specific approach followed and a validation of the results obtained. Finally, permeability hysteresis during desorption and relative permeability curves are investigated.

## 3. Microporous Membranes

Contrary to the above, the situation in the case of microporous media is far less satisfactory. The static and dynamic behaviour of the fluid in the micropores, which determine the overall separation and transport efficiency of the system, are the subject of several theoretical studies [18,24]. However, there is a noteworthy lack of experimental data in the literature to support the theoretical developments, especially as far as high pressures of operation are concerned.

In our group, a series of new microporous carbon membranes have been characterised using nitrogen sorption and gas flow methods. The pore size distribution and other features of the structure (pore constrictions, activated diffusion energy) are thus estimated. In a subsequent step, the $CO_2$ permeability of the membranes is measured in a high pressure gas permeation rig (fig.9) for temperatures around the critical. To perform the single component differential permeability experiments reported in [29], all the membranes are equilibrated at the same pressure (0-70 bars) on both sides, by pressurising the whole rig to the appropriate pressure, through HG and LG (fig.9). The adsorption process can be roughly monitored by means of the absolute pressure transducer and in detail by reading the small pressure changes on the differential manometer, after isolating its high pressure section. The membranes are isolated by closing the valves MiH, MiL, MiHf, MiLF and the high pressure section is pressurised to approximately 1 bar higher than the low pressure section. Subsequently, the valves adjacent to each membrane are opened and gas is allowed to permeate through the porous structure while the pressure increase is monitored by the differential pressure manometer.

$CO_2$ permeation data have been obtained at a temperature very slightly below the critical (310 K). Interestingly, the available experimental pressure capability covers the full relative pressure range (from 0 to 1) as the critical pressure of $CO_2$ is just above 70 bars. Permeance variation with pressure is shown typically in fig.10. The measurements depict the existence of a maximum around 30-35 bars (relative pressure of approximately 0.5) in close analogy with data from mesoporous media. The finding is quite repeatable and is intended to offer useful information for the validation of theoretical approaches. Remarkably, the measured permeability curves smooth out considerably at a temperature higher than the critical.

Based on the above results, a theoretical study is underway for the prediction of gas permeability in microporous membranes. Specifically, the mean (equilibrium) fluid density inside a single (cylindrical or slit shaped) pore of given size is found on the basis of Grand Canonical Monte Carlo simulations for a predefined temperature and different relative pressures. The Grand Canonical Monte Carlo (GCMC) method is ideally suited to adsorption problems because the chemical potential of each adsorbed species is specified in advance [30]. At equilibrium, this chemical potential can then be related to the external pressure making use of an equation of state. Consequently, the independent variables in the GCMC simulations are the temperature, the pressure and the micropore volume, i.e. a convenient set since temperature and pressure are the adsorption isotherm independent variables. Therefore, the adsorption isotherm for a given pore can be obtained directly from the simulation by evaluating the ensemble average of the number of adsorbate molecules whose chemical potential equals that of a bulk gas at a given temperature and pressure.

Three types of trial have been used, i.e. attempts to move (translate or reorient) particles, attempts to delete particles and attempts to create particles in the simulation box. A decision is made on whether to accept each trial or to return to the old configuration based on a probability which in the case of an attempted move takes the form:

$$p_{move} = min\ [exp(-\Delta U/kT);\ 1]$$

where $\Delta U = U_{new} - U_{old}$ is the difference in the potential energies of the new and old configurations. A detailed presentation of the method is given in [31,32]. Periodic boundary conditions have been applied in the directions other than the width of the slit. For a given simulation, the size of the box (i.e. the two dimensions other than H) have been selected such that ca. 200-400 molecules are present in the simulation. Statistics were not collected over the first $2 \times 10^6$ configurations to assure adequate convergence of the simulation. The uncertainty on the final results (ensemble averages of the number of adsorbate molecules in the box and the total potential energy) is estimated to be less than 3%. Typical calculations for a single point require between 0.5 and 2 hours of CPU time on a Convex 3820.

A microporous carbon membrane has been tested using the Quantachrome Autosorb-1 Nitrogen porosimeter equipped with Krypton upgrade. The total pore volume has thus been found. It compares favorably to the value estimated from the measured $CO_2$ isotherm. To start with, a skewed triangular pore size distribution has been postulated making sure that the total pore volume equals to the experimentally determined value. The micropore range (from 0.5 to 2.0 nm) has been subdivided in equidistant spaces with 0.1 nm width. The most probable pore size (slit width) and the standard deviation of the assumed distribution have been varied systematically. For each case, the pore volume associated with each class of pores (the aforementioned subdivisions of the overall pore range) has been calculated (such that the total pore volume remains equal to the measured one) and the amount of gas ($CO_2$) adsorbed in every class at a certain pressure has been computed using the GCMC code. In this way, a computed isotherm has been reconstructed up to relative pressures of 0.4 (at higher values the mesopore effect comes into play). By comparing this segment of the $CO_2$ isotherm to the corresponding experimental counterpart, the most suitable micropore size distribution was selected. The most probable pore size (0.75 nm) found is in good agreement with the independent estimation of 0.7-0.9 nm. Currently, the above described trial and error procedure is attempted to become free of the need to specify a certain (readily manipulated) distribution function. The solution (by least squares) of a minimization problem under certain constraints will provide the optimal distribution that fits best the selected segment of the measured isotherm data.

The above outlined procedure can then used to compute transport (flow) of the fluid through a microporous structure. Following investigations of the viscous behaviour of fluids in confined spaces [24], an effective viscosity coefficient (depending on the assumed type of flow and the pore size) can be calculated from the density profile. This information is then introduced to a network simulator which provides the permeability of the porous structure for the whole pressure range of interest. Comparisons with the experimental findings should validate the approach the development of which is currently underway.

## Acknowledgements

This work has been partly supported by the BRITE-EURAM programme of the European Commission, Contract No BRE2-CT92-0568.

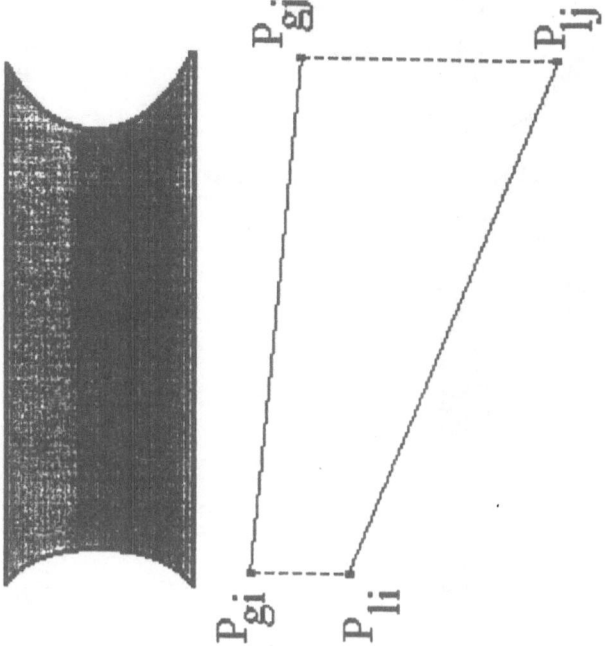

Fig. 1:     Menisci curvatures across a pore and associated pressure drops

498

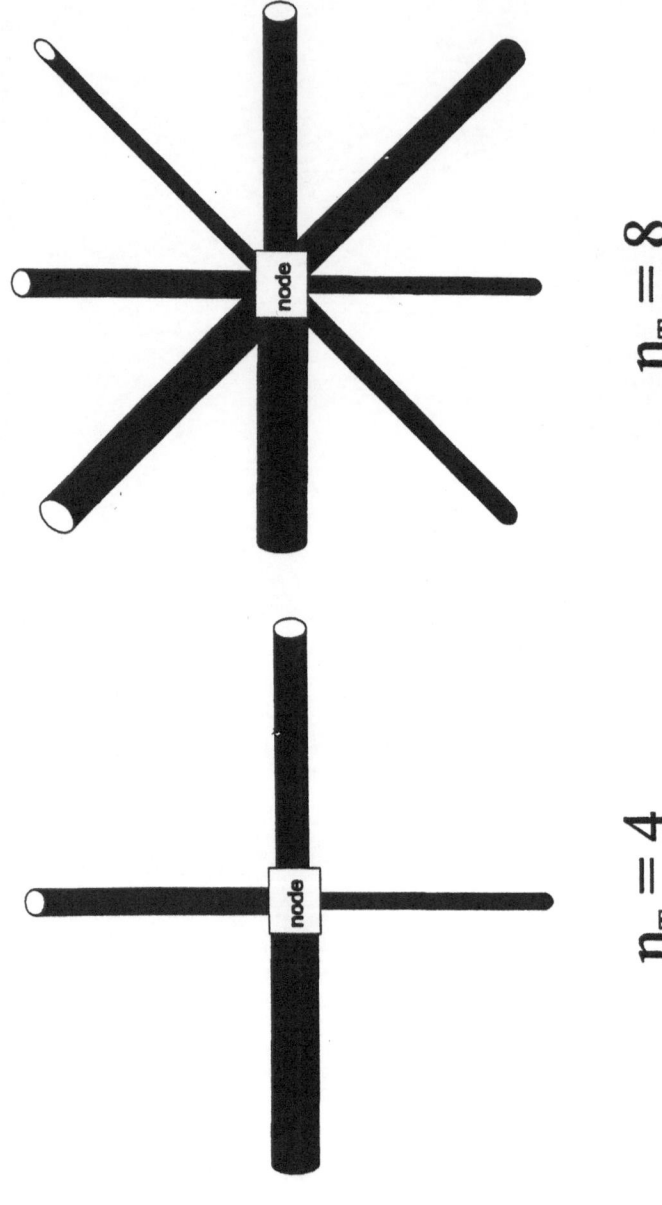

$n_T = 4$

$n_T = 8$

Fig. 2: Typical elements of network structure with connectivities 4 and 8

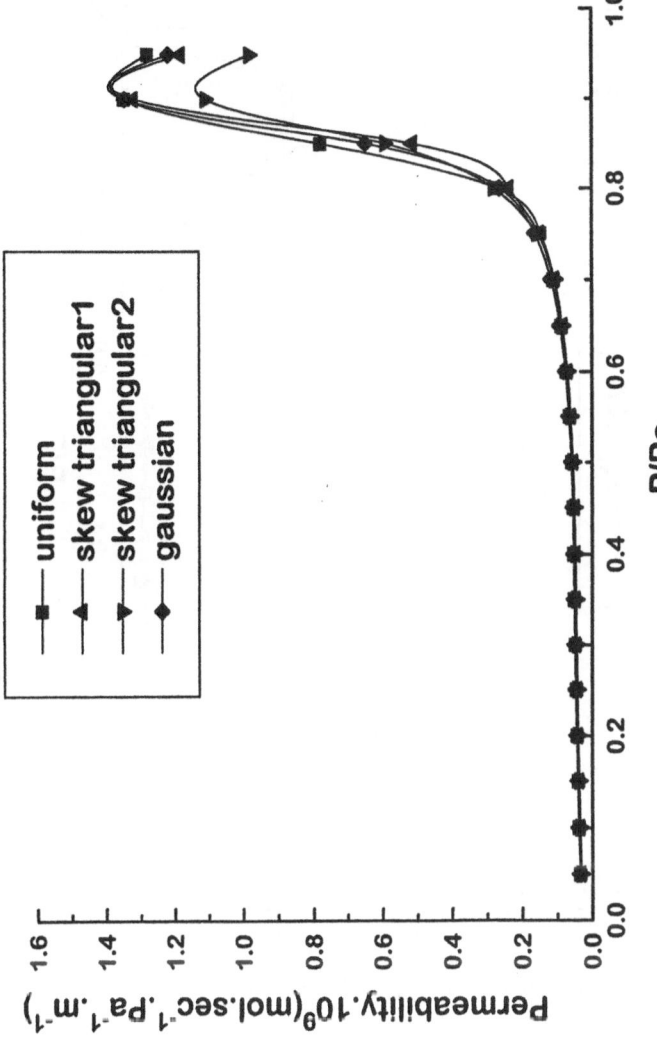

Fig. 3:   Effect of pore size distribution functions on permeability curve

500

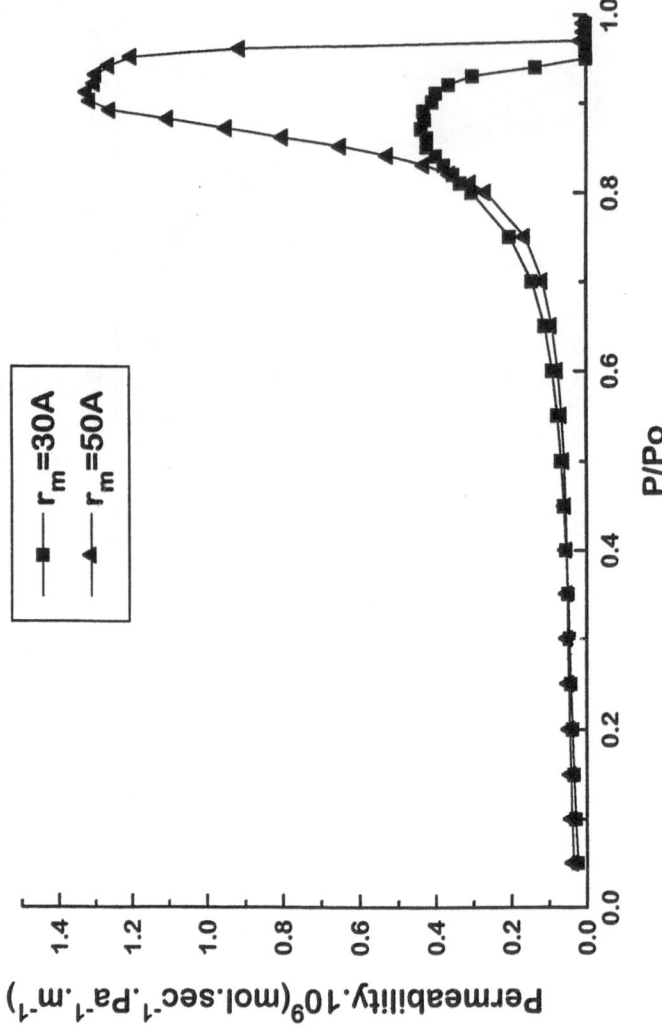

Fig. 4:    Effect of mesn pore radius on permeability curve

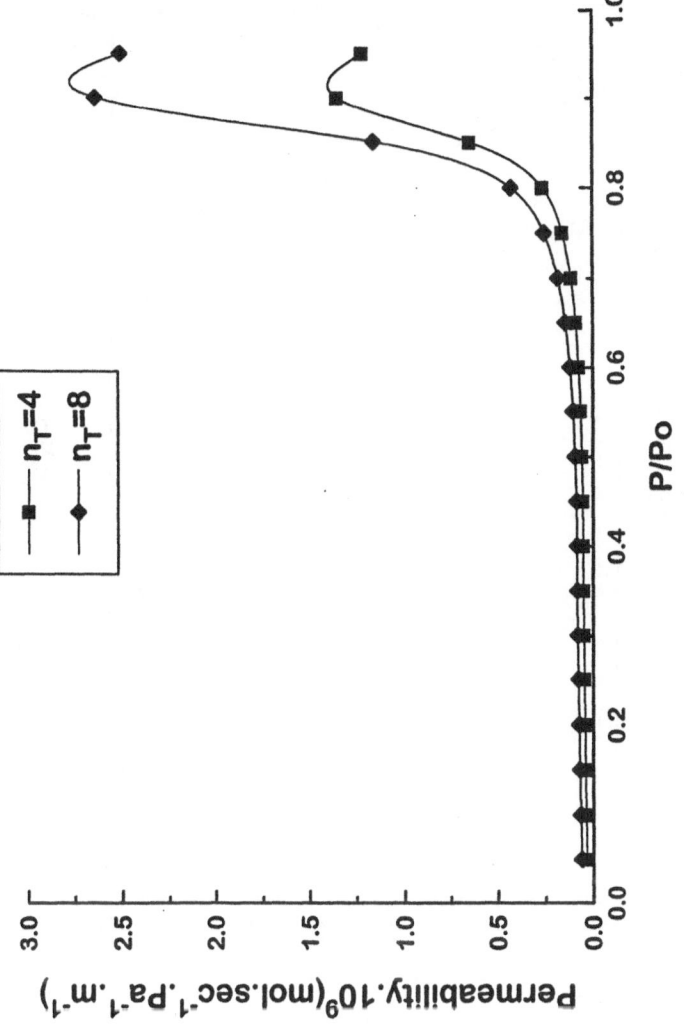

Fig. 5:    Effect of pore connectivity on permeability curve

502

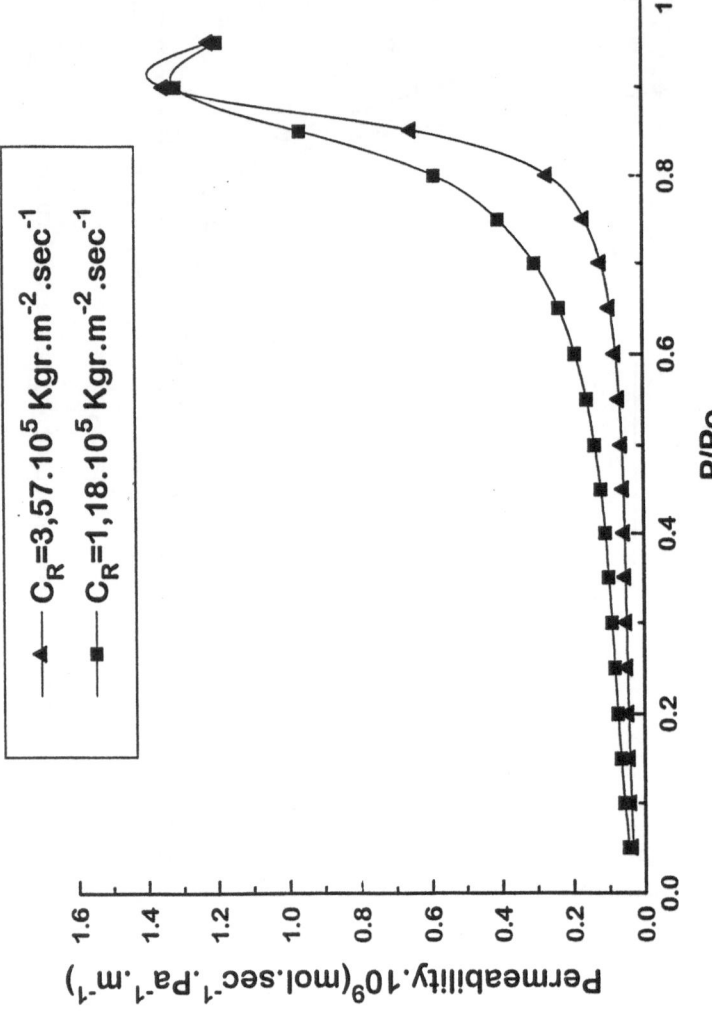

Fig. 6:    Effect of surface flow resistivity on permeability curve

503

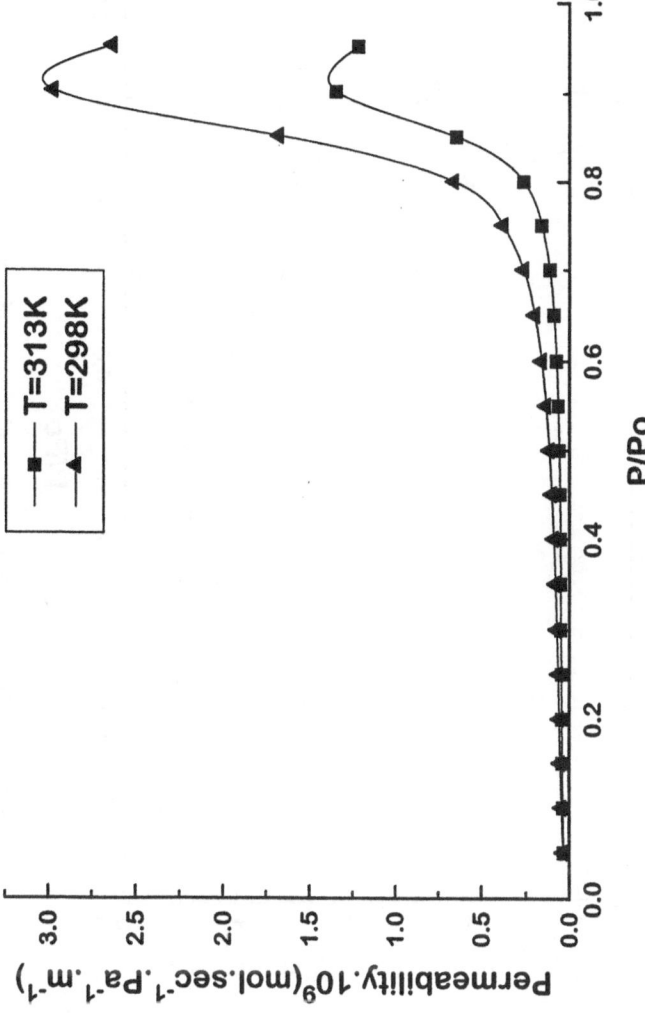

Fig. 7: Effect of temperature on permeability curve

504

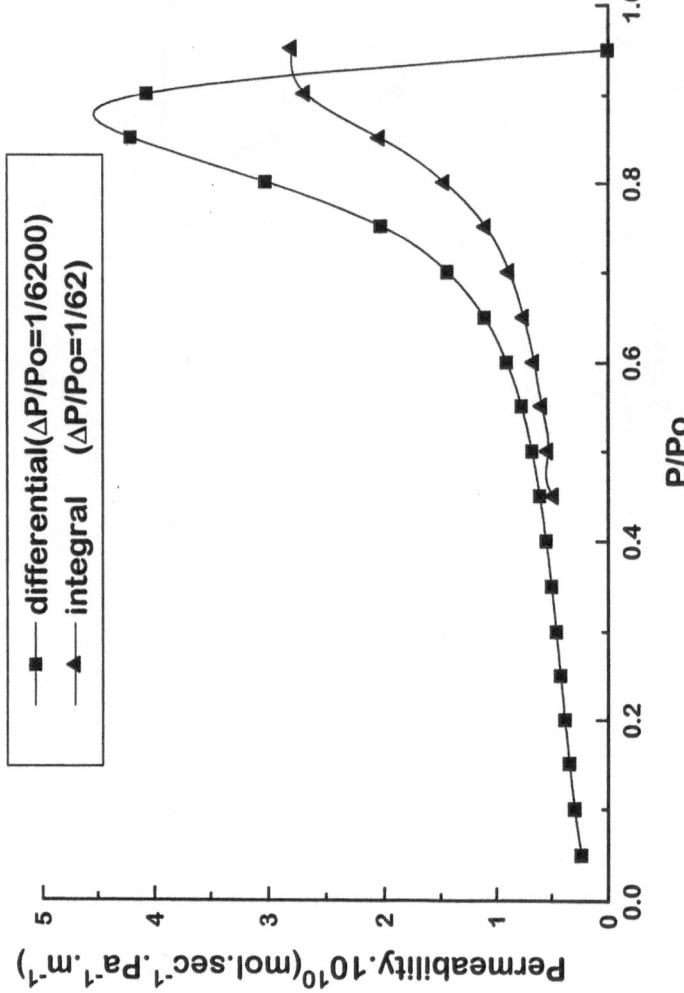

Fig. 8: Intergral versus differential permeability curves

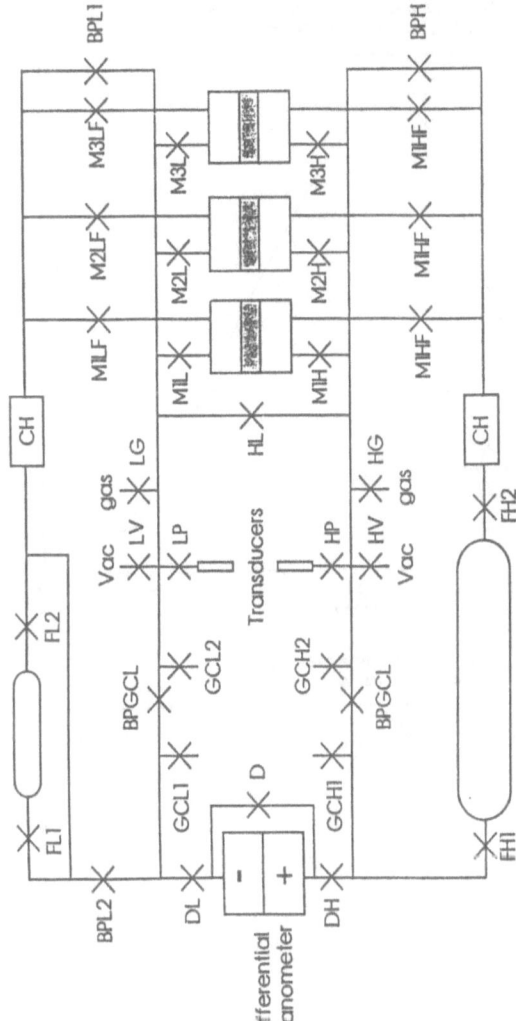

**Fig. 9:** High pressure permeability rig (Abbreviations : H–High Pressure Section, L–Low Pressure Section, D–Deltabar, C–Circulator, M–Membrane, BP–By Pass, F–Flask, GC–Gas Chromatographer, V–Vacuum Line, G–Gas Line, P–Pressure Tranducer)

506

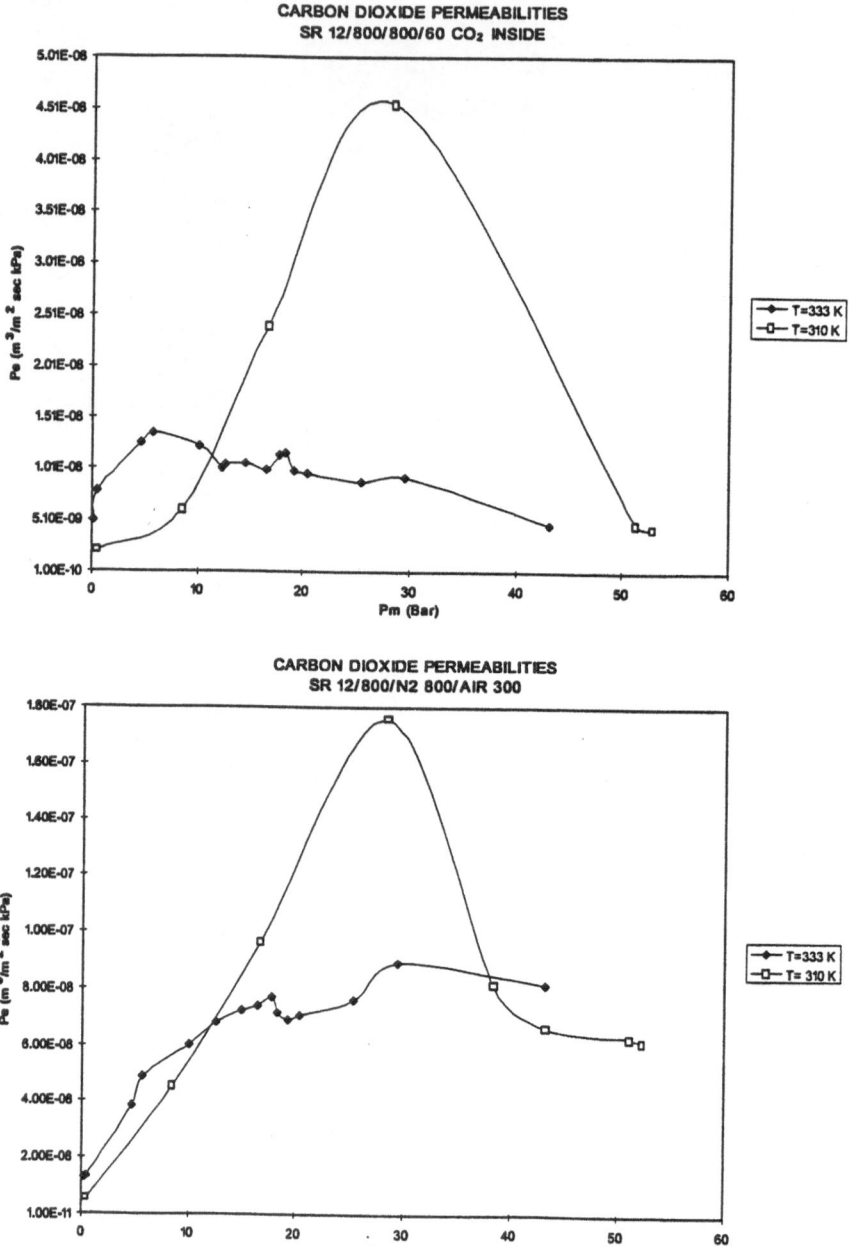

Fig. 10: CO$_2$ Permeabilities of microporous carbon membranes

References

1. R. Ash, R.M. Barrer and C.G. Pope, Proc. Roy. Soc. A, 271 (1963) 1.

2. R. Ash, R.M. Barrer and C.G. Pope, Proc. Roy. Soc. A, 271 (1963) 19.

3. H. Rhim and S-T. Hwang, Transport of capillary condensate, J. Colloid Interface Sci., 52(1) (1975) 174.

4. D.H. Everett and J.M. Haynes, Model studies of capillary condensation: I. Cylindrical pore model with zero contact angle, J. Colloid Interface Sci., 38(1) (1972) 125.

5. J.M. Haynes and R.J.L. Miller, Surface diffusion and viscous flow during capillary condensation, in: J.Rouquerol and K.S.W.Sing (Eds.), Adsorption at the gas-solid and liquid-solid interface, Elsevier, Amsterdam, 1982, pp. 439-447.

6. D. Nicholson, Molecular theory of adsorption in pore spaces, J. Chem. Soc. Faraday Trans. 1, 71 (1975) 238.

7. H. Tamon, M. Okazaki and R. Toei, Flow mechanism of adsorbate through porous media in presence of capillary condensation, AIChE J., 27(2) (1981) 271.

8. K-H. Lee and S-T. Hwang, The transport of condensable vapours through a microporous Vycor glass membrane, J. Colloid Interface Sci., 110(2) (1986) 544.

9. K. Keizer, R.J.R. Uhlhorn, R.J. Van Vuren and A.J. Burggraaf, Gas separation mechanisms in microporous modified $\gamma$-$Al_2O_3$ membranes, J. Membrane Sci., 39 (1988) 285.

10. H.J. Sloot, C.A. Smolders, W.P.M. Van Swaaij and G.F. Versteeg, Surface diffusion of hydrogen sulfide and sulfur dioxide in alumina membranes in the continuum regime, J. Membrane Sci., 74 (1992) 263.

11. D.P. Sperry, J.L. Falconer and R.D. Noble, Methanol-hydrogen separation by capillary condensation in inorganic membranes, J. Membrane Sci., 60 (1991) 185.

12. R.J.R. Uhlhorn, K. Keizer and A.J. Burggraaf, Gas transport and separation with membranes. Part I: Multilayer diffusion and capillary condensation, J. Membrane Sci., 66 (1992) 259.

13. A. Elkamel and R.D. Noble, A statistical mechanics approach to the separation of methane and nitrogen using capillary condensation in a microporous membrane, J. Membrane Sci., 65 (1992) 163.

508

14. M.B. Rao and S. Sircar, Nanoporous carbon membranes for separation of gas mixtures by selective surface flow, J. Membrane Sci., 85 (1993) 253.

15. M-D. Jia, B. Chen, R.D. Noble and J.L. Falconer, Ceramic-zeolite composite membranes and their application for separation of vapor/gas mixtures, J. Membrane Sci., 90 (1994) 1.

16. G. Mason, The effect of pore space connectivity on the hysteresis of capillary condensation in adsorption-desorption isotherms, J. Colloid Interface Sci., 88(1) (1982) 36.

17. M. Parlar and Y.C. Yortsos, Nucleation and pore geometry effects in capillary desorption processes in porous media, J. Colloid Interface Sci., 132(2) (1989) 425.

18. R. Evans, U.M.B. Marconi and P. Tarazona, Capillary condensation and adsorption in cylindrical and slit-like pores, J. Chem. Soc. Faraday Trans. 2, 82 (1986) 1763.

19. H. Liu, L. Zhang and N.A. Seaton, Determination of the connectivity of porous solids from nitrogen sorption measurements - II: Generalisation, Chem. Eng. Sci., 47(17/18) (1992) 4393.

20. C. Satik and Y.C. Yortsos, A pore network model for adsorption in porous media, presented at the 20th Stanford Geothermal Workshop, Stanford CA, Jan. 24-26, 1995.

21. D. Quenard and H. Sallee, Water vapour adsorption and transfer in cement-based materials: a network simulation, Materials and Structures, 25 (1992) 515.

22. N.K. Kanellopoulos and J.H. Petropoulos, Study of gas relative permeability in a mesoporous alumina pellet, J. Chem. Soc. Faraday Trans. I, 79 (1983) 517.

23. N.K. Kanellopoulos and J.K. Petrou, Relative permeability of parallel and serial capillary models with various radius distributions, J. Membrane Sci., 35 (1987) 21.

24. I. Bitsanis, T.K. Vanderlick, M. Tirrell and H.T. Davis, J. Chem. Phys., 89(5) (1988) 3152.

25. L. Riekert, The relative contribution of pore volume diffusion and surface diffusion to mass transfer in capillaries and porous media, AIChE J., 31(5) (1985) 863.

26. S. Sircar and M.B. Rao, Estimation of surface diffusion through porous media, AIChE J., 36(8) (1990) 1249.

27. E.R. Gilliland, R.F. Baddour and J.L. Russell, Rates of flow through microporous solids, AIChE J., 4(1) (1958) 90.

28. S.L. Gregg and K.S.W. Sing, Adsorption, Surface Area and Porosity, Academic Press, New York, 1982.

29. F.K. Katsaros, T.A.Steriotis,A.K.Stubos,N.K.Kanellopoulos, S. Tennison, "High presure gas permeability of microporous materials", submitted to Microporouw Materials.

30. D. Nicholson and N.G. Parsonage, Computer Simulation and the Statistical Mechanics of Adsorption, Academic Press (1982).

31. D.M. Razmus and C.K. Hall, AIChE J., 37, 5, (1991).

32. R.F. Cracknell et al., Mol. Sim., 13, 161 (1994).

## List of Symbols

| | |
|---|---|
| $C_R$ | surface flow resistance coefficient |
| d | collision molecular diameter |
| f(r) | pore size distribution function |
| g | bond (pore) conductance |
| J | mass flow rate |
| l | bond (pore) length |
| M | molecular weight |
| $N_A$ | Avogadro number |
| $n_T$ | pore connectivity |
| P | pressure |
| $P_0$ | saturation pressure |
| R | gas constant |
| r | pore radius |
| S | specific area of porous medium |
| T | fluid temperature |
| t | thickness of the adsorbed layer |
| $\eta$ | viscosity |
| $\theta$ | contact angle |

510

| λ | mean free path |
|---|---|
| ρ | fluid density |
| σ | surface tension |
| χ | amount of fluid sorbed per unit of solid mass |

*Subscripts*

| c | capillary |
|---|---|
| g | gas phase |
| l | liquid phase |
| m | mean |
| r | relative |
| s | surface flow |

# ADSORPTION ENGINEERING
*Hydrodynamics of PSA Columns*

## G. HORVATH[1], K. KUTICS[1], M. SUZUKI[2]
[1] *Department of Chemical Engineering, University of Veszprem*
*8201 Veszprem P.O.B. 158, Hungary*
[2] *Institute of Industrial Science, University of Tokyo*
*7-22-1 Roppongi, Minato-ku Tokyo 106 Japan*

## 1. Introduction

Considering some PSA (Pressure swing adsorption) or RPSA (Rapid pressure swing adsorption) technologies, there are usually fast pressurization steps and later depressurizations. In this technological solution the energy introduction is carried out in the form of mechanical work. The hydrodinamics of flow through a packed bed has been investigated and detailed accounts may be found in many engineering handbooks. Except the so called closed end column, because the experimental apparatuses were not able to measure adequate data. To measure a fast pressurization process a measuring apparatus was planned and built. From the results of former studies and references [1], [2] it can be concluded, that an Ergun-type equation can be used for simulation.

## 2. Theory

For a packed column the packing material may be spheres, cylinders or various kind of packings. It is supposed that the packing is uniform and the diameter of packings <0,1 diameter of the column. The basis of this model is the imagine that the packed bed is just a tube with very complicated cross sections.

Let us consider the variables found in the Ergun-equation. The definition of a friction factor for a tube with D diameter and L lenght:

$$F = f \cdot A^* \cdot K = f \cdot D\pi \cdot l \cdot \frac{1}{2}\rho \, v^2 = \Delta \, p \frac{D^2 \pi}{4} \tag{1}$$

$F$     friction force
$f$     friction factor
$A^*$    friction surface
$K$     characteristic kinetic energy
$\Delta p$    pressure drop

*J. Fraissard (ed.), Physical Adsorption: Experiment, Theory and Applications,* 511–524.
© 1997 *Kluwer Academic Publishers.*

512

$\dfrac{D^2\pi}{4}$  cross section

$$f = \frac{1}{4}\frac{D}{L}\frac{\Delta p}{\frac{1}{2}\rho v^2} \tag{2}$$

where $v$ is the "superficial velocity". The meaning of this velocity, average linear velocity, fluid would have in the column, if no packing were present. For laminar flow in a circular tube the superficial velocity (for an empty column the superficial velocity is same as the intersticial velocity) can be calculated:

$$0 \leq r \leq R_o \qquad 2R_o = D \tag{3}$$

$$-\mu\frac{dv}{dr}\cdot 2r\pi L = \Delta p r^2 \pi$$

after the integration, the velocity profile

$$v_r = \frac{\Delta p R^2}{4\mu L}\left[1-\left(\frac{r}{R_o}\right)^2\right] \tag{4}$$

the intersticial (superficial) velocity

$$u = \frac{\displaystyle\int_0^{2\pi}\!\!\int_0^{R_o} v_r \cdot r\, dr\, d\varphi}{\displaystyle\int_0^{2\pi}\!\!\int_0^{r} r\, dr\, d\varphi} = \frac{\Delta p R_o^2}{4\mu L}\frac{\displaystyle\int_0^{2\pi}\!\!\int_0^{R_o} r\left(1-\frac{r^2}{R_0^2}\right) dr\, d\varphi}{\displaystyle\int_0^{2\pi}\!\!\int_0^{R} r\, dr\, d\varphi} = \frac{1}{8}\frac{\Delta p R_0^2}{\mu L} \tag{5}$$

This equation multiplied by the cross section gives the Hagen-Poiseuille equation. When we imagine a very complicated packed bed, a hydraulic radius $R_h$ has to be introduced. The definition:

$$R_h = \frac{cross\ section}{wetted\ perimeter} \tag{6}$$

For example in a tube

$$R_h = \frac{R^2\pi}{2R\pi} = \frac{R}{2} \tag{7}$$

Thus the intersticial velocity

$$u = \frac{1}{2}\frac{\Delta p\, R_h^2}{\mu L} \qquad (8)$$

The next step is

$$R_h = \frac{volume\ of\ fluid}{wetted\ surface} = \frac{volume\ of\ fluid\ /\ volume\ of\ bed}{wetted\ surface\ /\ volume\ of\ bed} = \frac{\varepsilon}{a} \qquad (9)$$

Having introduced the specific surface $a_s$ namely the total particle surface/volume of particles

$$a = a_s(1-\varepsilon) \qquad (10)$$

and the value of superficial velocity

$$v = u \cdot \varepsilon \qquad (11)$$

Thus subsituting equation (9), (10) and (11) to (8) the next expression can be obtained

$$v = \frac{\Delta p \cdot \varepsilon^3}{2\mu L \cdot a_s^2 (1-\varepsilon)^2} \qquad (12)$$

the $a_s$ for a sphere particle

$$a_s = \frac{4 R_0^2 \pi}{\frac{4}{3} R_0^3 \pi} = \frac{6}{d_p} \qquad (13)$$

so

$$v = \frac{\Delta p \cdot \varepsilon^3 \cdot d_p^2}{72 \mu L (1-\varepsilon)^2} \qquad (14)$$

Having based on experimental values the real superficial velocity is

$$v = \frac{\Delta p\, d_p^2\, \varepsilon^3}{150 \mu\, L (1-\varepsilon)^2} \qquad (15)$$

So the pressure drop for unit lenght in laminar flow is

$$\frac{\Delta p}{L} = \frac{150 \mu (1-\varepsilon)^2}{d_p^2 \varepsilon^3} v \qquad (16)$$

In case of turbulent flow the friction factor depends only on the roughless of tube wall. Using the same model for turbulent flow (equation (2))

$$\frac{\Delta p}{L} = 4 f_t \frac{1}{D} \frac{1}{2} \rho u^2 \qquad (17)$$

and substituting the following equations

$$D = 2R = 4R_h = 4\frac{\varepsilon}{a}$$

the (11), (10) into (17), we obtain:

$$\frac{\Delta p}{L} = \frac{1}{2} f_t \frac{6}{d_p} \frac{1-\varepsilon}{\varepsilon^3} \rho v^2 \qquad (18)$$

On the basis of experimental results the corrected equation is:

$$\frac{\Delta p}{L} = 1,75 \cdot \frac{1}{d_p} \frac{1-\varepsilon}{\varepsilon^3} \rho v^2 \qquad (19)$$

The pressure drop for the whole range

$$\frac{\Delta p}{L} = \frac{150 \mu (1-\varepsilon)^2}{d_p^2 \varepsilon^3} v + 1,75 \frac{1}{d_p} \frac{1-\varepsilon}{\varepsilon^3} \rho v^2 \qquad (20)$$

In infinitesimal form the equation is

$$-\frac{dp}{dx} = \frac{(1-\varepsilon)}{\varepsilon^3} \frac{\rho}{d_p} \left[ \frac{150}{Re} + 1,75 \right] v^2 \qquad (21)$$

and

$$Re = \frac{v d_p \rho}{\mu (1-\varepsilon)} \qquad (22)$$

To evaluate the pressurization time the model of isothermal gas flow in porous materials has been used [3], [4], [5], [6]. The model is based on the following assumptions:

— the gas temperature is constant,
— radial gradients in the pressure and in the macroscopic velocity are negligible (the plug flow condition),
— the inertia forces are negliglible,
— the friction forces obey the Ergun equation,

- an ideal gas is assumed,
- the gravitational forces are negligible,
- the dynamic viscosity does not depend on pressure.

The mathematical model consists of a mass balance

$$\varepsilon \frac{\partial c}{\partial t} + \frac{\partial}{\partial x}(\varepsilon u c) = 0 \tag{23}$$

and a momentum balance which contains only two terms expressing pressure and friction forces.

$$-\frac{\partial p}{\partial x} = \lambda \rho \frac{(1-\varepsilon)}{\varepsilon d_p} u^2 \tag{24}$$

For the friction coefficient $\lambda$, the Ergun equation has been used:

$$\lambda = \frac{A}{Re} + B \tag{25}$$

The state equation for an ideal gas connects the molar concentration $c$ and pressure $p$, $c = p / RT$. The density of a gas is the product of the molar concentration, $c$, and the molecular weight, $\rho = Mc$. Instead of the velocity, $u$, we introduced the mass flux density in the form $\dot{M} = uc$. With the constant temperature assumption, the (23) and (24) can be rearranged to the form

$$\frac{1}{RT} \frac{\partial p}{\partial t} + \frac{\partial \dot{M}}{\partial x} = 0 \tag{26}$$

$$\frac{\partial p}{\partial x} = -A\mu \, RT \left( \frac{1-\varepsilon}{\varepsilon d_p} \right)^2 \frac{\dot{M}}{p} - B \frac{1-\varepsilon}{\varepsilon d_p} MRT \frac{\dot{M}^2}{p} \tag{27}$$

There are two dependent variables, $p$ and $\dot{M}$, and two independent variables $x$ and $t$. Under the assumptions mentioned above, all other quantities are constant. In addition to Eqs. (26) and (27), the two initial conditions

$$p(x,0) = p_I \tag{28}$$

$$\dot{M}(x,0) = 0 \tag{29}$$

and two boundary conditions are required. The model describes the pressurization of a fixed bed closed at one end. The pressure at the other end is suddenly fixed at a different pressure. At the closed end of the column, the boundary condition

$$\dot{M}(x = H, t) = 0 \qquad t > 0 \tag{30}$$

516

applies, and at the open end

$$p(x=0,t)=p_F \qquad t>0 \qquad (31)$$

Equations (26) and (27) were transformed into dimensionless form by means of the following dimensionless variables

$$p^* = \frac{p-p_I}{p_F-p_I} \qquad (32)$$

$$M^* = M \frac{RT}{p_I^2} \left(\frac{1-\varepsilon}{\varepsilon\, d_p}\right)^2 HA\mu \qquad (33)$$

$$\eta = \frac{x}{H} \qquad (34)$$

$$\tau = t \left(\frac{\varepsilon\, d_p}{1-\varepsilon}\right)^2 \frac{p_I}{H^2 A\mu} \qquad (35)$$

and parameters

$$\Delta p^* = \frac{p_F-p_I}{p_I} \qquad (36)$$

$$\delta = \frac{B}{A^2} \left(\frac{\varepsilon\, d_p}{1-\varepsilon}\right)^3 \frac{p_I^2 M}{H\mu^2 RT} \qquad (37)$$

The transformed Eqs. (26), (27) become:

$$\Delta p^* \frac{\partial p^*}{\partial \tau} + \frac{\partial \dot{M}^*}{\partial \eta} = 0 \qquad (38)$$

$$\Delta p^* \frac{\partial p^*}{\partial \eta} = -\frac{\dot{M}^*}{1+\Delta p^* p^*}(1+\delta M^*) \qquad (39)$$

with the initial conditions

$$p^*(\eta,\tau=0)=0 \qquad (40)$$

and the boundary conditions

$$\dot{M}^*(\eta,\tau=0)=0 \qquad (41)$$

$$M^*(\eta = 1, \tau) = 0 \qquad (42)$$

$$p^*(\eta = 0, \tau) = 1 \qquad (43)$$

Equations (41) to (42) has been solved numerically. Dependence of the pressurisation time on the complex parameter $\delta$ can be seen in FIGURE 1. For $\delta$ approaching 0, the flow approaches Darcy's region, while for $\delta \gg 1$ it is in the high velocity range.

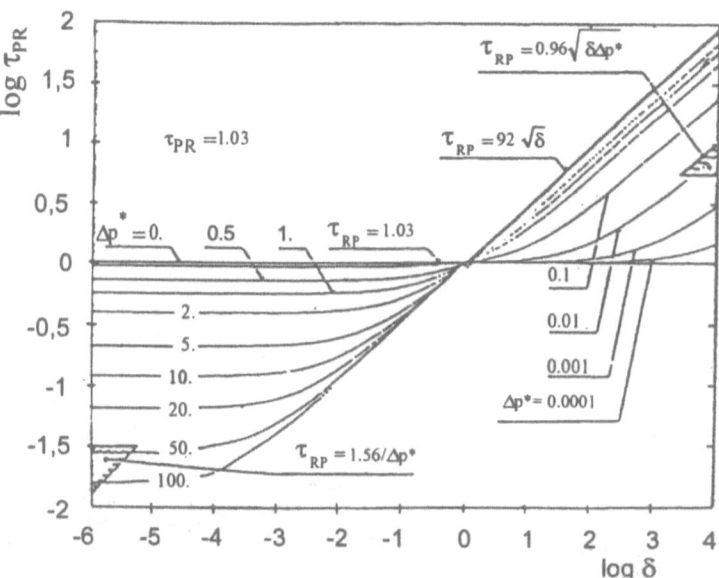

*Figure 1*. The pressurization time vs. $\delta$ for different value of the inlet pressure difference.
(Reproduced with permission by Rousar, I., Ditl, P., Kotsis, L., Kutics K.:
Transient Flow of a Compressible Fluid Through Beds of Solid Particles or Porous
Materials, Chemical Engineering Communications, 112, 67-83, 1992.)

## 3. Experimental

To measure fast pressurization transients, an experimental apparatus shown schematically in FIGURE 2. has been constructed. The applied column has a diameter of 41,5 mm, with a length of 1000 mm. The piesoresistive pressure sensors produced by INTERBIP have a pressure range 0-11 bars nonlinearity ±0,5 %, temperature sensitivity ±0,1 %/°C frequency transmission 0,5 kHz, the outlet signal -1 +10 V (1 V/bar).

Since there is no displacement other than the compression of the crystal, these devices are a borderline case between a force-balance and a motion balance devices.

This measuring system is new in this field in the sense of high frequency as well as data aquisition.

For data processing the real time version of LT/CONTROL from LABTECH was used. The choosed sampling frequency was 0,3 kHz/chanell.

The frequency is supposed to be high enough for getting of adequate digital signs. For compressible fluids the parameter $X$ is used to predict chooking. The ratio of pressure drop to absolute upstream pressure is called $X$

$$X = \frac{p_u - p_F}{p_u} \tag{44}$$

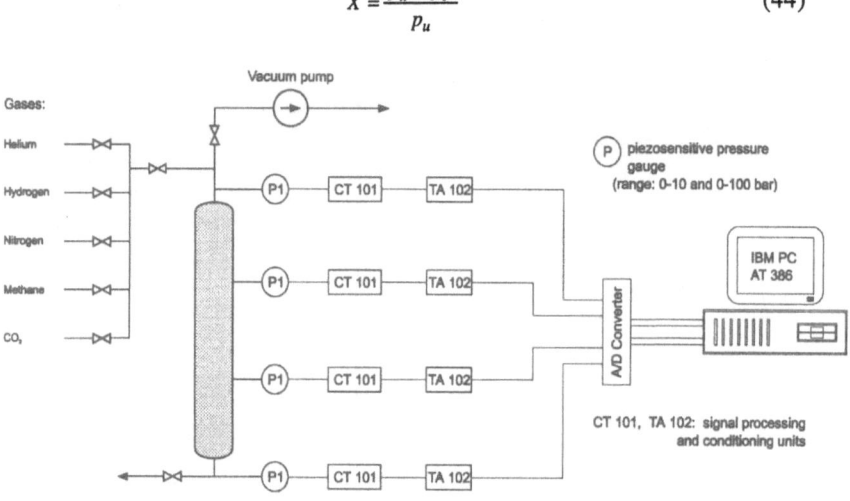

*Figure 2.* Experimental apparatus for the investigation of fast adsorption processes

where $p_u$ and $p_F$ are the upstream and downstream pressure, respectively. If $p_u$ is held constant and $p_F$ reduced, flow will increase with the pressure drop until sonic velocity is developed at some point in the valve. This value is the so called critical pressure ratio. When $X<0,5$ the mass flow has a constant value which is independent from $X$.

| Used materials: | Gases | Packings |
|---|---|---|
| | $H_2$ | Empty |
| | He | Glass $\varnothing$4-4,5 |
| | $CH_4$ | Zirk $\varnothing$2-2,5 |
| | $N_2$ | Zirk $\varnothing$1-1,5 |
| | Ar | Glass $\varnothing$0,5-0,9 |
| | $O_2$ | Glass $\varnothing$0,08-0,23 |
| | $CO_2$ | Glass $\varnothing$0,01-0,055 |
| | | Zeo 5A $\varnothing$2-2,5 |
| | | Zeo 4A $\varnothing$2-2,5 |

With the help of this method the constant inlet massfluxes could be ensured.

In the our case the upstream pressure values were choosed in the range of 5-6 bars. For the analysing of pressurization and adsorption, the downstream processes up to 2 bars were taken into consideration.

## 4. Results

The constants of the Ergun-type equation were determined from steady state as well as dynamic experiments. FIGURE 3. shows an example for the A value determination. The constant A was found to be 212-218, Ergun proposes 150. From the dynamic experiments this value somewhat higher 220. With the same method the B values can be determined, comparison to Ergun this value is also higher, 2,2 insted of 1,75. Comparison of experiments and numerical simulation is shown in FIGURE 4. One can conclude that the larger the particles are, the greater are the deviations. When the average particle diameter is higher than 0,2 mm, the error is more than 20 %. Over 1 mm the model does not follow the experimental data.

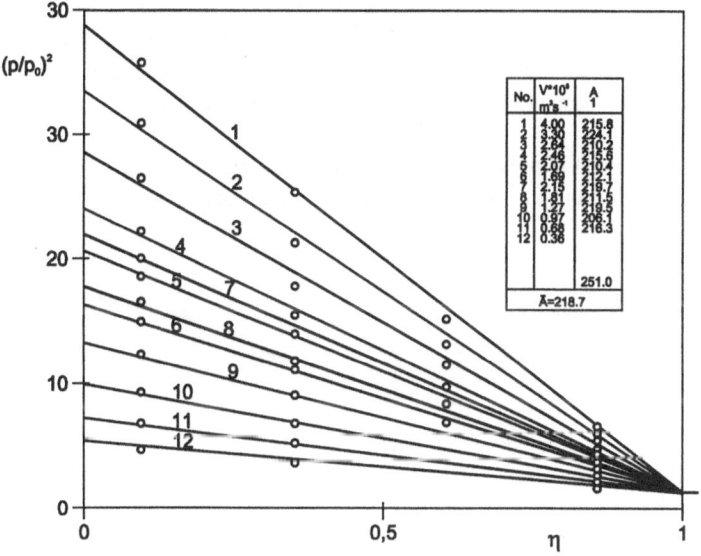

*Figure 3*. Pressure profiles along the fixed bed with small particles ($d_p$=0,027 mm) at the stationary flow.

(Reproduced with permission by Rousar, I., Ditl, P., Kotsis, L., Kutics K.: Transient Flow of a Compressible Fluid Through Beds of Solid Particles or Porous Materials, Chemical Engineering Communications, 112, 67-83, 1992.)

520

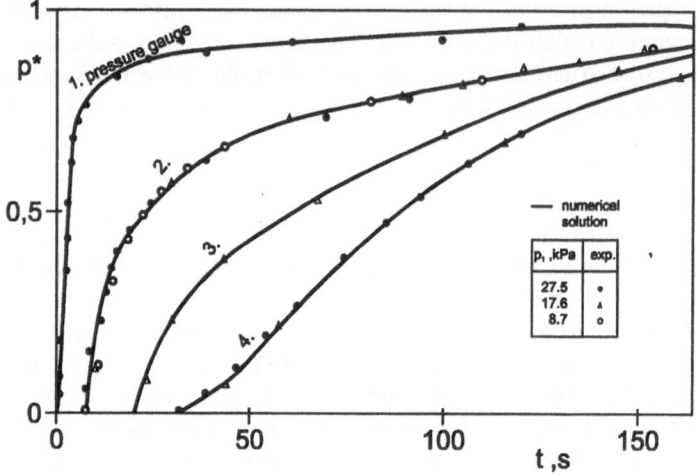

*Figure 4.* The course of pressurization at four different heights of column, filled with small particles
(*d*ₚ=0,027 mm)
(Reproduced with permission by Rousar, I., Ditl, P., Kotsis, L., Kutics K.:
Transient Flow of a Compressible Fluid Through Beds of Solid Particles or Porous
Materials, Chemical Engineering Communications, 112, 67-83, 1992.)

FIGURE 5 shows the pressurization of a column filled with large particles. FIGURE 6. composes the pressure drops in a column filled with active and inactive particles.

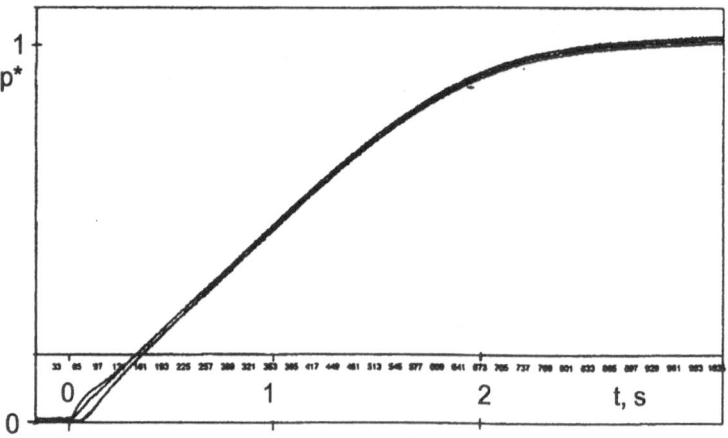

*Figure 5.* The course of pressurization at four different heights of column, filled with large particles
(*d*ₚ=2,2 mm)

For the estimation of necessary engineering data, we used a simple method. Having get the time, necessary for the pressurization up to 2 bars and the volume of the column; the inlet massflow can be calculated. Concerning the inert packings, the void fraction calculated from the pressurization time – divided by the pressurization time for the same gas in empty column – practically is independent from the gases. Having based on these empirical results the massflow for the pressurization and the massflow of adsorption have been decided. Supposing that He does not adsorb, the void fractions were calculated from He experiments. The numerical results for the pressurization up to 2 bars and the calculated void fractions are shown in TABLES 1-3.

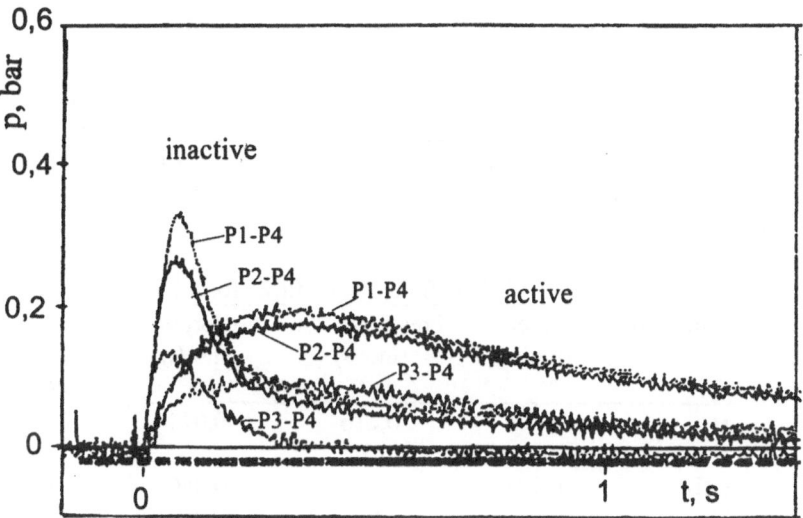

*Figure 6.* The courses of pressurization -pressure drops- at four different heights of column filled with large active and inactive particles ($d_p$=0,027 mm)

TABLE 1. Calculated void fractions

| Gas | Column packing | Empty | Glass Ø4 | Zirk Ø2-2,5 | Zirk Ø1-1,5 | Zeo 5A | Zeo 4A |
|---|---|---|---|---|---|---|---|
| $H_2$ | Time(s) | 0,38 | 0,18 | 0,22 | 0,22 | 0,35 | 0,27 |
| | Void fraction | 1 | 0,47 | 0,58 | 0,58 | | |
| He | Time(s) | 0,46 | 0,25 | 0,27 | 0,28 | 0,32 | 0,32 |
| | Void fraction | 1 | 0,54 | 0,59 | 0,6 | 0,69 | 0,69 |
| $CH_4$ | Time(s) | 1,03 | 0,44 | 0,61 | 0,58 | 3,68 | 0,69 |
| | Void fraction | 1 | 0,42 | 0,59 | 0,57 | | |
| $N_2$ | Time(s) | 1,24 | 0,6 | 0,75 | 0,74 | 3,11 | 0,88 |
| | Void fraction | 1 | 0,48 | 0,6 | 0,59 | | |
| Ar | Time(s) | 1,4 | 0,63 | 0,795 | 0,84 | 2,12 | 0,96 |
| | Void fraction | 1 | 0,45 | 0,57 | 0,60 | | |
| $O_2$ | Time(s) | 1,4 | 0,75 | 0,8 | 0,8 | 2,12 | 0,93 |
| | Void fraction | 1 | 0,43 | 0,57 | 0,57 | | |
| $CO_2$ | Time(s) | 1,8 | 0,8 | 1,06 | 1,01 | 4,8[*] | 1,96 |
| | Void fraction | 1 | 0,44 | 0,59 | 0,61 | | |

[*] 0,1 bar

TABLE 2. Different mass flows for zeolite 5A

| Gas | Mass flow Inlet (mol/s) | Pressurization (mol/s) | Adsorption (mol/s) | Specific Adsorption (mol/s/dm³) |
|---|---|---|---|---|
| $H_2$ | 0,29 | 0,216 | 0,074 | 0,056 |
| He | 0,24 | 0,24 | | |
| $CH_4$ | 0,107 | 0,021 | 0,086 | 0,065 |
| $N_2$ | 0,088 | 0,024 | 0,064 | 0,059 |
| Ar | 0,078 | 0,036 | 0,042 | 0,032 |
| $O_2$ | 0,078 | 0,036 | 0,042 | 0,032 |
| $CO_2$ | 0,061 | 0,0008 | 0,053 | 0,04 |

TABLE 3. The different mass flows for zeolite 4A

| Gas | Mass flow Inlet (mol/s) | Pressurization (mol/s) | Adsorption (mol/s) | Specific Adsorption (mol/s/dm³) |
|---|---|---|---|---|
| $H_2$ | 0,29 | 0,28 | 0,01 | 0,007 |
| He | 0,24 | 0,24 | | |
| $CH_4$ | 0,107 | 0,107 | | |
| $N_2$ | 0,088 | 0,087 | 0,001 | 0,0007 |
| Ar | 0,078 | 0,077 | 0,001 | 0,0007 |
| $O_2$ | 0,078 | 0,077 | 0,001 | 0,0007 |
| $CO_2$ | 0,061 | 0,038 | 0,023 | 0,017 |

## 5. Conclusion

One can conclude from the results of the above study that Ergun-type equation can be used for the simulation of slower pressurization processes (i.e. in case of small particles $d_p<0,2$ mm) provided, that the constants of the equation are determined preliminarily. For high velocity transients an experimental apparatus and a simple method have been worked out. With the help of these equipments PSA and RPSA processes can be examined. This system is new in this field in the sense of high frequency application as well as the data acquistion.

## 6. Notation

| | | |
|---|---|---|
| $a$ | wetted surface/volume of bed | $[m^2/m^3]$ |
| $a_s$ | particle surface/volume of particles | $[m^2/m^3]$ |
| $A, B$ | constants of an Ergun-type equation | |
| $A^*$ | friction surface | $[m^2]$ |
| $c$ | gas concentration | $[kmol/ m^3]$ |
| $d_p$ | particle diameter | $[m]$ |
| $D$ | diameter | $[m]$ |
| $F$ | friction force | $[N]$ |
| $f$ | friction factor | |
| $f_t$ | turbulent friction factor | |
| $H$ | bed height | $[m]$ |
| $K$ | characteristic kinetic energy | $[J]$ |
| $L$ | length | $[m]$ |
| $M$ | molar weight | $[kg/kmol]$ |
| $M(x,t)$ | mass flux | $[kmol/ m^2/s]$ |
| $M^*(\eta,t)$ | dimensionless mass flux | |
| $p(x,t)$ | pressure | $[Pa]$ |
| $p_I$ | initial pressure | $[Pa]$ |
| $p_F$ | feed pressure | $[Pa]$ |
| $p_0$ | atmospheric pressure | $[Pa]$ |
| $p^*(\eta,t)$ | dimensionless pressure | |
| $\Delta p^*$ | dimensionless pressure difference | |
| $p_i^*(t)$ | dimensionless pressure at the ith gauge | |
| $p_u$ | upstream pressure drop | $[Pa]$ |
| $R$ | gas constant | $[kJ/(kmol/K)]$ |
| $Re$ | Reynolds number | |
| $R_0$ | maximal radius | $[m]$ |
| $R_h$ | hydraulic radius | $[m]$ |
| $r$ | radius | $[m]$ |

524

| | | |
|---|---|---|
| $S$ | column cross section | [m$^2$] |
| $t$ | time | [s] |
| $T$ | tempreature | [K] |
| $u$ | gas velocity | [m/s] |
| $v$ | superficial gas velocity | [m/s] |
| $v_r$ | velocity profile | [m/s] |
| $V$ | volume of column | [m$^3$] |
| $V$ | volumetric gas flow-rate | |
| $x$ | axial co-ordinate | [m] |
| $X$ | ratio of pressure drop | |
| $\delta$ | ratio of lin. and non-lin. friction forces | |
| $\varepsilon$ | void fraction | |
| $\varphi$ | polar angle radian | |
| $\eta$ | dimensionless co-ordinate | [m] |
| $\kappa$ | adiabatic compression factor | |
| $\lambda$ | drag coefficient | |
| $\mu$ | gas viscosity | [kg/m/s] |
| $\rho$ | gas density | [kg/ m$^3$] |
| $\tau$ | dimensionless time | |
| $\tau_{PR}$ | dimensionless time of pressurisation | |

## 7. Acknowledgement

We thank the New Energy and Industrial Technology Development Organization for supporting this collaborative work.

## 8. References

1. Ergun, S. (1952) Fluid Flow Through Packed Columns, *Chem. Eng. Prog.*, **48**, 89-94.
2. Bird, R.B., Stewart, W.E., Lightfoot, E.N.: (1960) *Transport Phenomena*, New York.
3. Horváth, G. et al.: (1990) *Method for the separation of helium-neon mixtures*, Hungarian patent, HU 195 690.
4. Arányi, L. et al.: (1980) Mathematical Simulation of Rapid Cyclic Processes, *Magy. Kém Lapja*, **36**, 39.
5. Kutics, K.: (1981) *Hydrodinamic Study of Packed Beds*, M. Eng. Thesis, University of Veszprém.
6. Rousar, L., Ditl, P., Kotsis L., Kutics, K.: (1992) Transient Flow of a Compressible Fluid Through Beds of Solid Particles or Porous Materials, *Chemical Engineering Communications*, **112**, 67-83.

# THE AIR LIQUIDE COMPACT VSA ™
## *A view on an innovating adsorber*

C. MONEREAU
*AIR LIQUIDE, Engineering Division*
*94503 Champigny sur Marne, France*

## 1. Introduction

Pressure Swing Adsorption units are used for oxygen generation both on a small scale for the production of medical oxygen, and increasingly on an industrial scale, as an alternative to bulk supply and cryogenic plants.

AIR LIQUIDE is actively involved in both of these fields, and its objective for industrial application is to cover the whole range between 1 and 150 t/day, by means of a new generation of plants: the COMPACT VSA™.

*Figure 1.* Large Size  compact VSA ™

*J. Fraissard (ed.), Physical Adsorption: Experiment, Theory and Applications,* 525–541.
© *1997 Kluwer Academic Publishers.*

One of the most innovating aspects of this concept, at least as far as technology is concerned, lies in the geometry of the adsorbers which had been further developed at this occasion.

## 2. From basic principles to the first-generation VSA

Regardless of their size, these units bring into play the same principles: a succession of adsorption and regeneration phases.
During the adsorption phase, the nitrogen of the air is blocked in priority to oxygen. During the regeneration phase, this nitrogen is extracted from the adsorber, which becomes as a result free for a new production phase.
Extraction of nitrogen is not effected by an increase in temperature, but by pressure reduction in the adsorber, which is the characteristic feature of P.S.A. cycles.

Figure 2a illustrates this concept by showing the useful nitrogen adsorption capacity for a zeolite unit operating between two pressure values.

In practice, the introduction of air into the adsorber is stopped before the nitrogen front passes out of the bottle. Figure 2b shows a schematic representation of the development of the nitrogen concentration profiles in the adsorber during the production and purge steps.

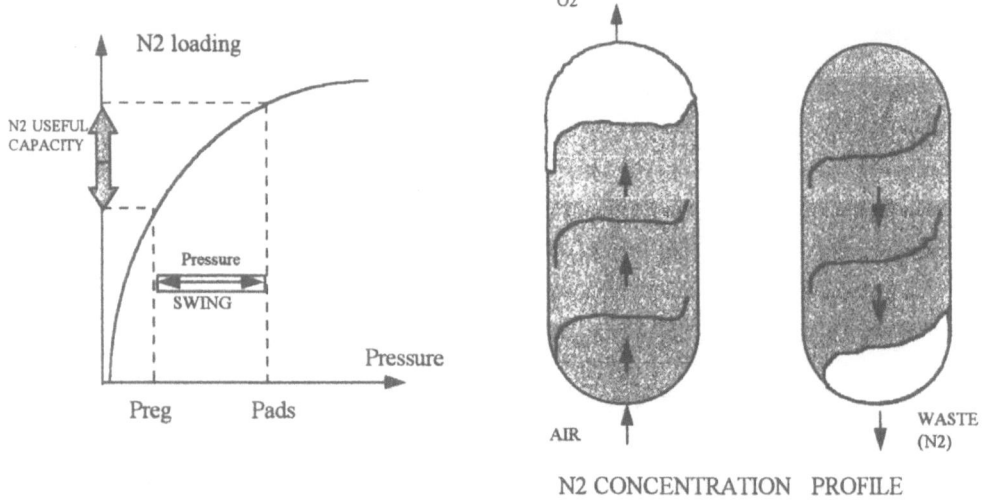

2a           2b

*Figure 2.* PSA concept

## 2.1. PURITY OF OXYGEN

It is mentioned that with the zeolites used, argon has the same behavior as oxygen and is therefore not separated from oxygen in these type of units.
The maximum purity of the oxygen is in the order of 95,5 %. However, the price to be paid in order to eliminate the last traces of nitrogen is significant, so that industrial units rarely produce oxygen of a percentage exceeding 93 %.
On the other hand, production of oxygen of a purity of less than 88 % has only a limited effect on performance. This phenomenon is due to the stiffness of the nitrogen front at the adsorber outlet.
For this reason, most industrial plants operate within a range of 88 to 93 % purity.

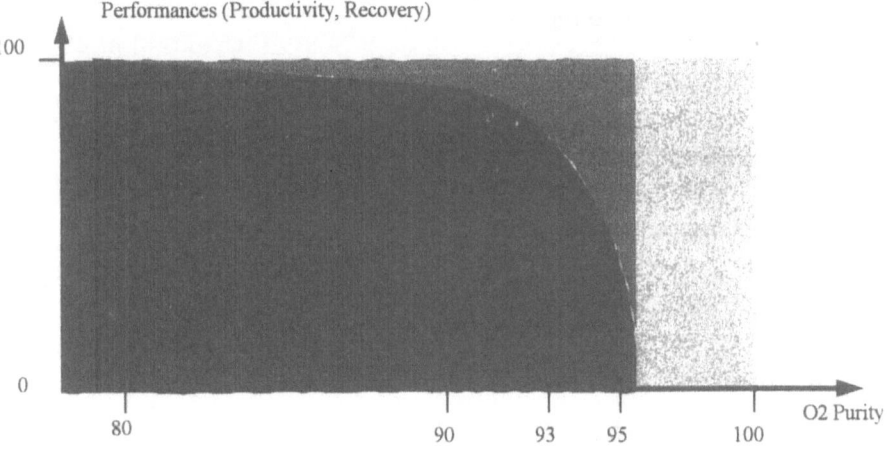

*Figure 3.* Performance vs purity

## 2.2. THE ADSORBENTS

N2/O2 separation is carried out with zeolite of the 13 X or 5 A type. The final elaboration of this zeolite may or may not comprise cation manipulation stages. As these adsorbents are poisoned by traces of water or carbon dioxide, a pre-bed is often included to remove these components. In this case, the two beds are run in conjunction during the same pressure cycle.

528

## 2.3. THE CYCLES

Several types of cycles are used, which are characterized by the production pressure and regeneration pressure.

We distinguish thus

- the P.S.A.s, all of which operate at a pressure above atmospheric pressure;

- the MPSAs, which produce oxygen at a pressure above atmospheric pressure and which are regenerated under vacuum;

- the VSAs, which are run between atmospheric pressure and a lower vacuum level than the above mentioned units.

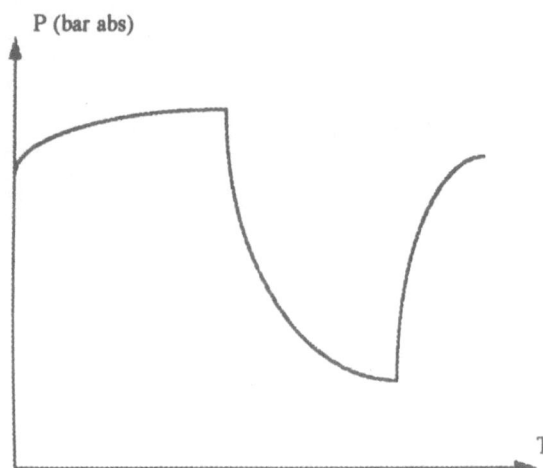

| PSA | MPSA | VSA |
|---|---|---|
| 3 | 1.5 | 1 |
| | | |
| | | |
| Atm.P | Vacuum | vacuum |
| | ~0.45 | ~0.30 |

Figure 4. Pressure cycles

The pressure levels, taken into account in figure 4, to illustrate each one of these cases are indicated as examples only. The final values are determined with regard to economical considerations, taking into account local conditions, especially energy cost, and may vary according to the type of sieve used.

## 2.4. THE FIRST-GENERATION V.S.A

For its largest ON-SITE units for oxygen production, AIR LIQUIDE has from the very beginning opted for the VSA process with 3 adsorbers. The reason for this is the lower energy consumption of this cycle as compared to units with 2 adsorbers or still a single adsorber.
A further characteristic of the VSA cycle is that it assures continuous oxygen production without requiring buffer capacity. One of the three adsorbers is always in the production phase, while the second one is purged and the third one in the pressure buildup phase.

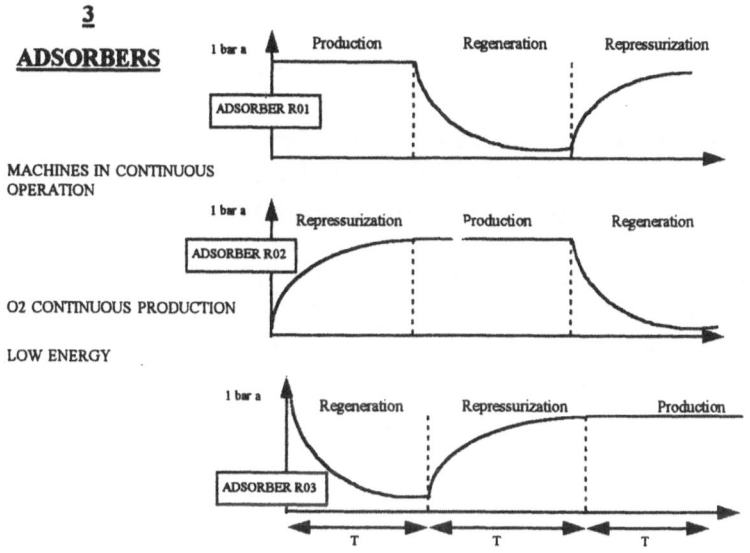

*Figure 5.* 3 adsorbers VSA cycle

As schematically represented in figure 6, each adsorber includes 3 or 4 valves connecting it to the air supply, vacuum and oxygen circuits.
The air circuit is fed by a simple fan which allows to compensate the pressure losses induced by the passage through piping and the adsorber, as oxygen is produced at atmospheric pressure.
The vacuum circuit is connected to a group of vacuum pumps, which is generally made of one, two or even three stages of the Roots type mounted in series.

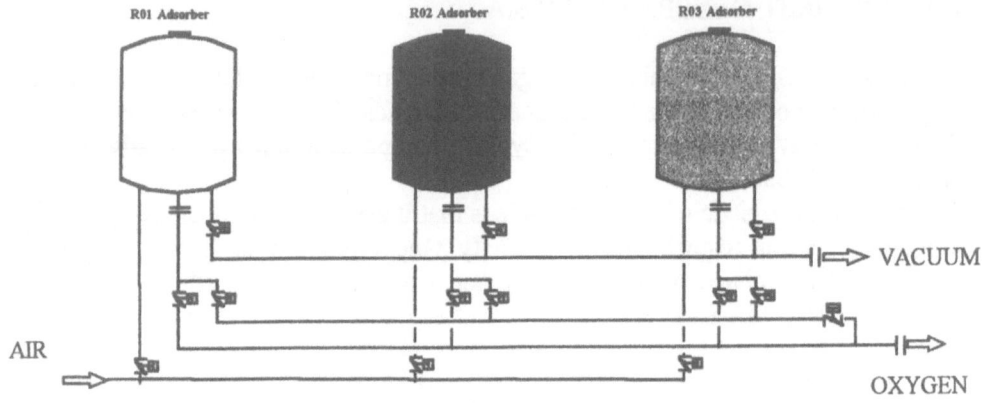

*Figure 6.* Typical VSA unit

The connections on the oxygen side are more or less complex, depending on the supplementary steps which may be added to the basic cycles as described above, and on the level of "intelligence" required for the valves.

Initially, the adsorbents were installed inside cylindrical adsorbers with a horizontal axis. L'AIR LIQUIDE has thus put up units composed of adsorbers reaching diameters of up to 4 meters and a length of 10 meters. This way, a site including 3 identical lines is composed of 9 adsorbers of this type.

*Figure 7.* 3 x 50 t/d first generation VSA

This unit with a total oxygen production of 150t/d for the iron and steel industry, was at that time the largest unit in Europe operating according to this process. Two lines of these first-generation VSAs of larger oxygen production capacity have since been installed in the United States.

## 3. A view on the oxygen on-site market

### 3.1. A LONG WAY TO SUCCESS

The introduction of the PSA process is commonly attributed to Skarstrom of ESSO, Guérin de Montgareuil and Domine of AIR LIQUIDE [1]. This goes back to the end of the fifties.
But real marketing of the oxygen PSA had only started in the seventies, mainly in connection with waste water treatment The following step in the eighties corresponds to the first utilization of these units in the iron and steel industry in Japan. The nineties have witnessed a real explosion of the market, due to new applications, such as the treatment of wood pulp in the Pulp and Paper industry, non ferrous metal, water purification, glass melting and oxy- combustion.

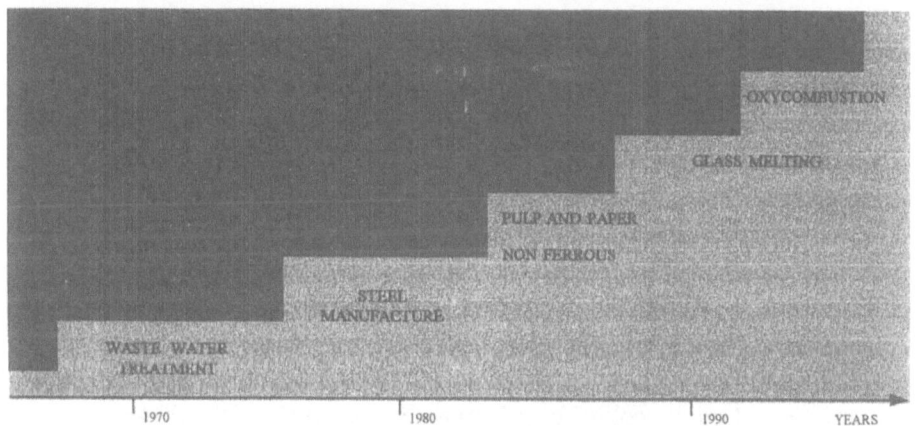

*Figure 8.* O2 PSA worldwide market
Main customers

## 3.2. THE REASONS FOR SUCCESS

The reasons for this success dwell unquestionably in the constant reduction of the cost of PSA-produced oxygen over the last 25 years. Two dominating features explain this development:

- the progress made in the field of adsorbents
- the optimal application of the processes

The best adsorbent would evidently be the one which retains the maximum of nitrogen in the working pressure range of the PSA while holding back the minimum quantity of oxygen.
These properties may be characterized on one hand by the effective nitrogen adsorption capacity and, on the other hand by the N2/O2 selectivity which can easily be defined according to the adsorption pressure, by the ratio:

$$\frac{qN2}{qO2} \times \frac{P\,O2}{P\,N2}$$

where P O2 and PN2 correspond respectively to the partial pressures of the two components.

With regard to useful capacity and selectivity, the zeolites of the Li X family currently synthesized are two times more efficient than the Na X used in the first PSA units.

These improvements entail, at identical process conditions, a reduction of the adsorbent volume and an increased efficiency. In other words, they signify lower investment and production energy cost.
It is possible to take even greater advantage of the superior characteristics offered by the new generation of zeolites by adapting the very cycle of the PSAs specifically to each one of these new materials. The under vacuum desorption has thus been generalized for all medium and large capacity units, leading to a further substantial reduction of the adsorbent volume and to saving of energy necessary for separation.

Numerous alternatives exist for the basic cycles of the MPSA or VSA types allowing for a more precise adaptation according to the capacity required or to the local cost of utilities.

Specific energy for industrial units currently lies in the range between 0.30 and 0.45 Kwh/Nm3O2, which corresponds to an average decrease of a factor of 3 to 4, as compared to the units built during the seventies. Global oxygen production cost over the same period fell more or less in the same proportion.

## 3.3. THE NEW EXPECTATIONS OF THE MARKET

It is obvious that a cutback in the cost of oxygen in the future will not be the only issue expected by the market. Three other requirements, clearly expressed by our clients, are the reason why the units of AIR LIQUIDE have been significantly improved:

- the delivery time of oxygen must be reduced
- impact on the client's site has to be minimized
- the production must be adjustable to the demand

In practice, this means that the unit required by the market has to be pre-assembled and must no longer be built on site, that it should occupy very little space, necessitate a minimum of utilities, respect the environment and be designed to be movable.

Our former conception of VSAs did no longer really satisfy these criteria.

## 4. Air Liquide answer to market expectations = the compact VSA ™

On the basis of these newly defined requirements, the results of the work of a multidisciplinary group featuring representatives of the Marketing, Research and Engineering Departments of AIR LIQUIDE has led to a complete re-evaluation of the first-generation VSA.

The changes issued from these efforts affected the process, technology and engineering of the units. The final results of the development was the concept of the Compact VSA™ :

Standard-size units composed of skidded modules, associated with adsorbers of small diameter, all of it perfectly answering to the basic requirements concerning time-limits, the on-site impacts and the evolution of oxygen production.

534

STANDARD UNITS

SKIDDED MODULES

COMPACT ADSORBERS

EASY TO

- TRANSPORT
- ASSEMBLE
- PUT INTO SERVICE
- DISMOUNT
- CHANGE

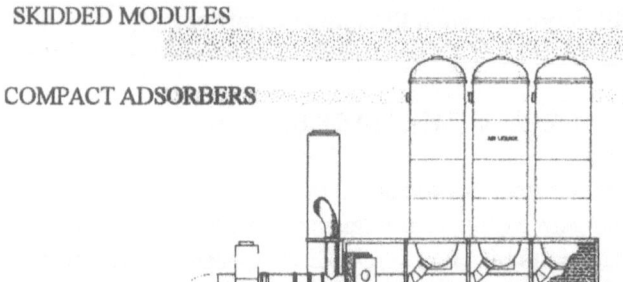

*Figure 9.* Compact VSA ™ concept

*Figure 10.* Medium size compact VSA ™

We shall limit the remainder of this presentation to describing the new adsorbers used, which are probably being the most innovative features of the new concept, and to illustrate by two examples the typical problems which had to be solved by the specialists of AIR LIQUIDE before the final geometry of the equipment could be defined.

## 5. The adsorber of compact VSA™

The basic principle adopted was horizontal circulation through the adsorbents, which allowed to settle the problem of attrition or lifting the bed. This latter point is all the more critical, as the reduced cycle time applied to the new units is combined with a higher gas velocity, as well as with adsorbents of a smaller diameter.

Two adsorbers of this type, corresponding to different sizes, have been developed and patented. The diagrams are represented in figure 11.

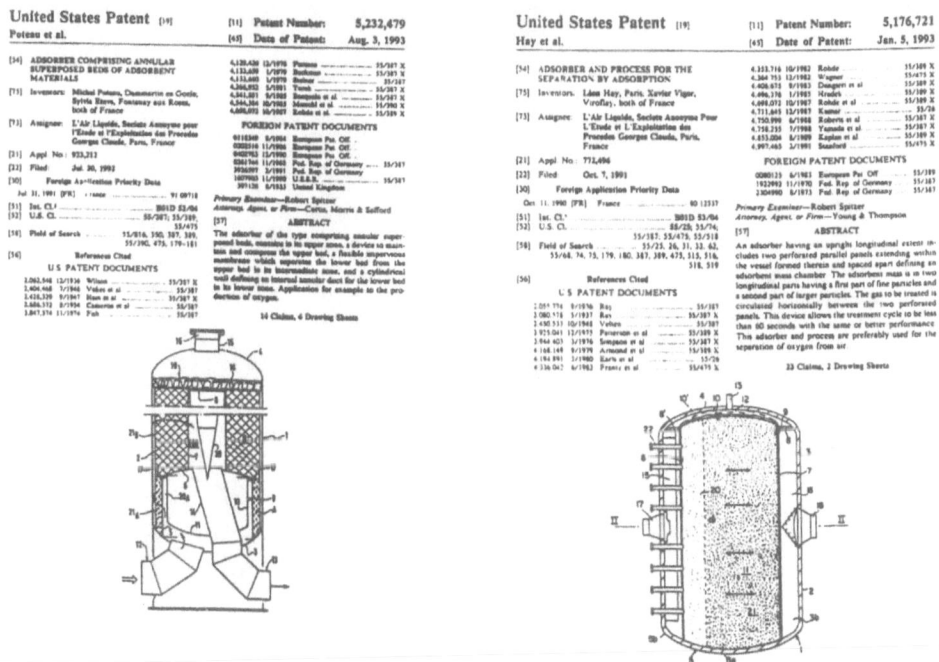

*Figure 11.* Horizontal circulation adsorbers

The radial adsorber is used for units where oxygen production exceeds approximately 10 tons a day.

536

## 5.1. THE RADIAL ADSORBER

As can be seen in figure 12, the radial adsorber comprises two superposed annular beds. The first one consists of alumina and serves to retain water and CO2 of the air, the second one, consisting of zeolite, is used for O2/N2 separation.

The adsorbents are held in position between a common perforated external grid, which supports the two beds, and individual perforated internal grids.

At the lower part, a special part fulfills the threefold task to separate the two adsorbents, to reduce the dead volume of air and to contribute to a satisfactory distribution of the entering gaseous flow.

This hermetic volume is crossed by a pipe allowing the connection of the central cylindrical collector with the oxygen network.

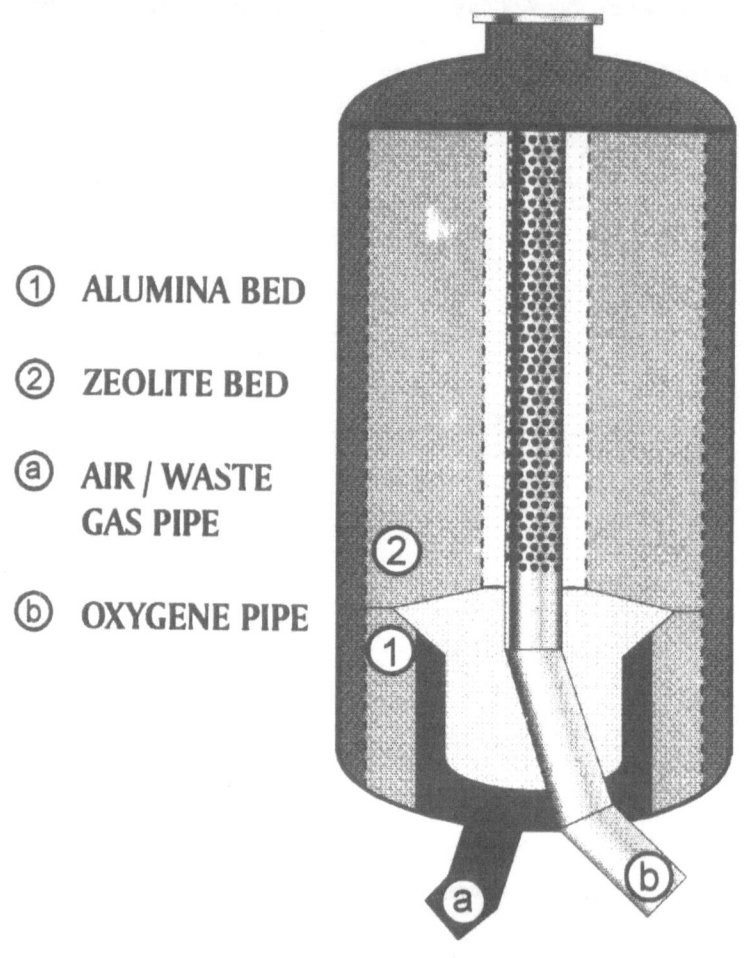

① ALUMINA BED

② ZEOLITE BED

ⓐ AIR / WASTE GAS PIPE

ⓑ OXYGENE PIPE

*Figure 12.* The radial adsorber

Sealing at the upper part of the zeolite bed is obtained by means of an elastomer diaphragm. This diaphragm is permanently applied to the adsorbent by exerting continuous pressure on its upper side.

In the production phase, air entering at the lower part crosses radially the alumina bed, rises along the intermediate volume contained between the external grid and the adsorber shell and passes then radially through the zeolite bed. Nitrogen is adsorbed, oxygen and argon, which are not retained, pass into the central collector and subsequently into the oxygen pipe.

During the regeneration phase, the nitrogen, released from the zeolite by vacuum pumping, crosses under low pressure the alumina bed permitting thus the desorption of water and $CO_2$ retained during the production phase.


## 5.2. GASEOUS DISTRIBUTION IN THE RADIAL ADSORBER

Specific studies have been dedicated to the distribution of the gaseous flow across the two adsorbent beds. The goal of these studies was to ensure a homogeneous circulation of these flows during the various steps of the cycle, with a minimum of pressure drop and dead air volume, in order not to increase energy consumption.

The simulation of the gaseous distribution within the adsorbents has been carried out in two stages. At first, a simple "in-house" model was developed affording to determine the geometric conditions a priori required in order to remain within the authorized limits of maldistribution, which have been established previously by empirical study.

This first "non-scientific" program shows in fact the different pressure levels corresponding to various pathes, assuming as an hypothesis a perfect distribution of the fluid in circulation.
From the resulting pressure differences, it was then possible to deduce the approximate local flow rate differences to be expected in real operating conditions.

The geometry of the adsorber defined through this method was analyzed with the help of three-dimensional computer programs on fluid mechanics, available on the market. With these programs it was possible to simulate the effects on distribution due to the actual position of piping, inlet and outlet, and geometrical singularities.

It is mentioned that the studies based on these three-dimensional programs were carried out for crossing streams without taking into account adsorption phenomena.

As far as the simplified model is concerned, it allows an estimate of the adsorption effects upon distribution; the flow rate differences between inlet and outlet may be taken into account as an option within the scope of pressure loss calculations.

By the combined utilization of a simple tool and more complex three-dimensional programs it was possible to establish the ratios of velocity which must be respected between free sections and adsorbent beds in order to ensure that the residual maldistributions rest within the limits acceptable for the process.

The determination of these velocity ratios allows the engineering divisions of AIR LIQUIDE to select the most favorable geometry of adsorbers of various sizes, without having to recalculate each time the corresponding gaseous distribution.

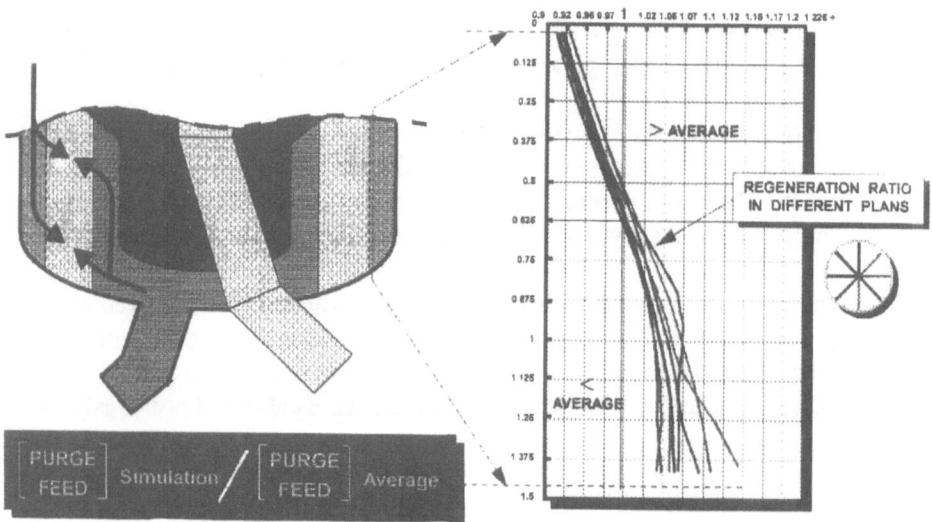

*Figure 13.* Alumina regeneration ratio from 3D simulation

Figure 13 illustrates the differences between the theoretical regeneration ratio of the alumina - to which a value 1 has been attributed, and the actual local ratio.

For the cycle concerned, the regeneration power of the alumina is such that the calculated distribution defects are still acceptable.

## 5.3. THERMAL PROFILE IN THE RADIAL ADSORBER

The second issue which has been studied in detail is the impact of geometry on the thermal profile which develops in the adsorbent beds as the cycles progress.

It is a well-known fact that adsorption of a component is accompanied by the liberation of a certain quantity of heat increasing the temperature of the adsorbent and that, on the other hand, desorption has a cooling effect. The adsorbent beds are therefore generating temperature fluctuations during a cycle.

When nitrogen of the air is adsorbed in a VSA or a MPSA, this fluctuation may reach 10 °centigrade.

As the cycles progress, a second effect is added to this phenomenon.

Heat transfer by convection due to gas circulating first in one and then in the other direction, create a thermal disequilibrium within the adsorbent mass.
This causes cooling in the zone around the inlet and heating in the area around the outlet.

If the adsorber is perfectly adiabatic, this effect develops until an equilibrium is reached

In the case of oxygen PSAs, this phenomenon is very noticeable, and the temperature differences between various points of the zeolite bed may reach an order of magnitude of 50 ° centigrade.

For example, we observed, at certain runs of our very first unit of the former generation, - a unit which was particularly well instrumented - a cold point inferior to - 60 degrees centigrade, while the feed temperature of the adsorber was at + 25° centigrade.

On the other hand, if thermal exchange with the external environment is significant, this thermal profile does not further develop. This is the case for laboratory columns with a diameter on the order of several centimeters.

The existence of this profile is, of course, exerting an influence on the performance, as local modification of the operating temperature of the zeolite changes very sensibly its adsorption properties.

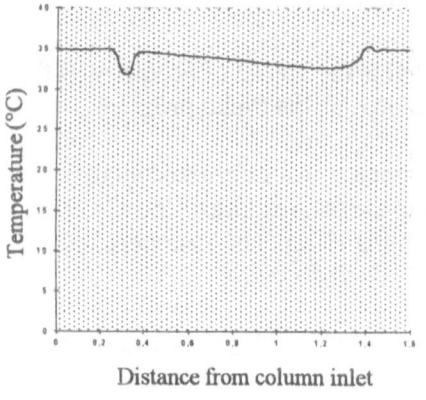

Distance from column inlet

PREVAILING WALL HEAT TRANSFER

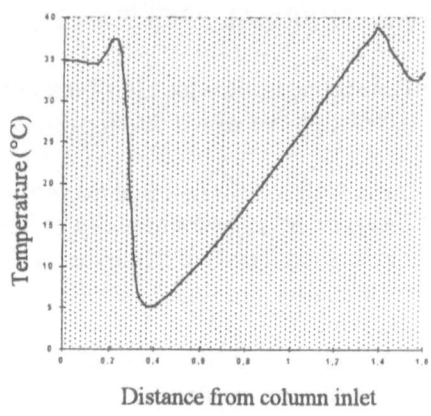

Distance from column inlet

ADIABATIC ADSORBER

*Figure 14.* Temperature profile

More advanced adsorption simulation software, such as SIMBAD 95, developed by AIR LIQUIDE, allows to identify these thermal phenomena and to evaluate their impact on the performances of the cycle.

It should be noted that without using such software, only the creation of pilot units which are sufficiently large to permit the approach of the degree of adiabaticity of industrial units, would provide an accurate assessment of the performances to be expected for a specific cycle or a new adsorbent.

The configuration of the Compact VSA™ allows to control the entry of heat between the two adsorbent beds due to the natural exchange surface constituted by the external shell of the adsorber. This possibility, which is the subject of a complementary patent, allows to control, at least partially, the thermal profile and to utilize the various adsorbents, in the most favorable temperature zone.

## 6. Conclusion

Since the performances of the first new-generation unit had been in conformity with previsions, as from its very start-up in 1993, the range of COMPACT VSA™ has developed rapidly and counts today more than 40 units.

This range comprises units of the MPSA and of the VSA types; the selection of the type depending on the production capacity required and on economical local conditions.

Especially well adapted, due to the specificity of its adsorbers with horizontal circulation, to the utilization of the highest performance molecular sieves and to the new processes which are currently being developed in our Research Center, the COMPACT VSA™ will surely experience an even more significant level of development in the future.

*References :*
1. Ruthven, D., Farooq, S., Knaebel, K. (1994) Pressure Swing Adsorption

# POSTER COMMUNICATIONS

# The Molecular Potential Approach to Helium Adsorption by Carbonaceous Subnanospace

Norihiko Setoyama and Katsumi Kaneko
Department of Chemistry, Faculty of Science, Chiba University,
1-33 Yayoi, Inage, Chiba 263, Japan

## INTRODUCTION

A helium atom is the smallest spherical monoatomic molecule and nonspecifically interacts with solid surfaces. It is expected as a suitable molecular probe for analysis of micropore structures, especially of ultramicropore structures (micropore width $\leq 0.7$nm). In recent years, we have devoted to establish analytical method for helium adsorption isotherm on microporous solids such as activated carbons at 4.2K [1-3]. An empirical thermodynamic analysis according to Dubinin [4] can be extended to helium adsorption isotherms on activated carbons at 4.2K. However, its applicability should be limited to only for carbonaceous materials. A new analysis of more general applicability is needed.

In this work, we propose a new analysis for the helium adsorption isotherm on microporous solids at 4.2K by use of participation of helium molecules according to the molecule–surface potential.

## THEORY

The helium adsorption on the flat surface at 4.2K can be described by Steele's multilayer adsorption equation [5], based on the Dole equation. Steele's equation is expressed as follow,

$$\theta = \frac{W}{W_m} = \frac{\left|1 + zx\dfrac{2-x}{(1-x)^2}\right|}{1 + cx + \dfrac{zcx^2}{1-x}} cx \qquad (1)$$

where c and z are the adsorption partition functions of first and second layers, respectively. $x = P/P_0$. $\theta$ is the number of layers of helium adsorbed. W and $W_m$ are the amount adsorbed at x and the monolayer capacity, respectively. The lateral interaction is neglected. The c value has the same meaning as BET theory, and z determines the second layer adsorption process of helium. It is noted by,

$$z = \frac{j_2}{j_{liq}} \exp\left(\frac{U_2 - U_{liq}}{RT}\right) \qquad (2)$$

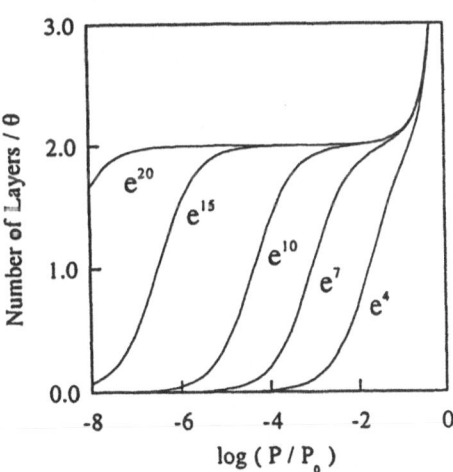

Figure 1. Calculated adsorption isotherms of helium for different z values

$U_2$ is the interaction energy between surface and second adsorbed layer. $U_{liq}$ is expressed as the intermolecular interaction energy in the bulk liquid phase. $j_2$ and $j_{liq}$ are the molecular partition

545

J. Fraissard (ed.), Physical Adsorption: Experiment, Theory and Applications, 545–548.

546

functions for second adsorbed layer and liquid; $j_2 = j_{liq}$ is presumed. In Figure 1, some of generated isotherms with changing z value are shown. c was held to be a large constant value. The adsorption process at low pressure region (below $P/P_0 \leq 0.1$) varied with z, suggesting that the adsorption of helium at 4.2K is largely dominated by the z value. Acceleration of helium adsorption on the microporous system at 4.2K should also be described as a function of micropore sizes, through the z value.

The z in micropore systems, especially for carbonaceous slit-shaped micropores, was calculated from the surface-molecule interaction potential ($\Phi(r)$). The method for calculation of z is schematically shown in Figure 2. The first

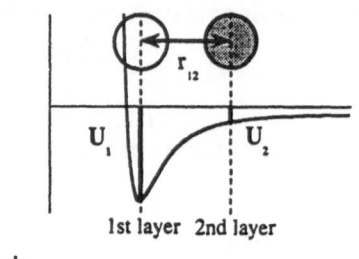

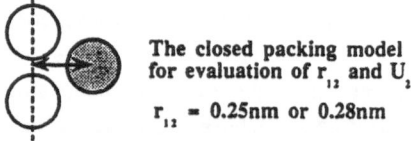

The closed packing model for evaluation of $r_{12}$ and $U_2$

$r_{12} = 0.25nm$ or $0.28nm$

Figure 2. Evaluation of $r_{12}$ and $U_2$

adsorbed layer should be formed at the minimum of $\Phi(r)$, and thereby the interaction potential of second layer ($U_2$) was calculated using the model shown in Figure 2. As distance between the first and second layers ($r_{12}$) dominates the value of $U_2$, $r_{12}$ is quite important. We used $r_{12} = 0.25$ and 0.28nm, which were estimated from the adsorbed density of helium at 4.2K [6], for further calculation of z. The interaction potential $\Phi(r)$ of helium with the graphitic slit pores was calculated using the 10-4-3 potential by Steele [7], as given by,

$$\phi(r) = 2\pi\rho_s\varepsilon\sigma^2\Delta\left\{0.4\left(\frac{\sigma}{r}\right)^{10} - \left(\frac{\sigma}{r}\right)^4 - \frac{\sigma^4}{\left[3\Delta\left(0.61\Delta + r\right)^3\right]}\right\} \quad (3)$$

$$\Phi(r) = \phi(r) + \phi(H - r) \quad (4)$$

We used $\sigma_{gs} = 0.298nm$, $\varepsilon_{gs} = 16.9K$, $\rho_s = 114$ atom/nm³, and $\Delta = 0.335nm$.

The relationship for z and pore width (H) is shown in Figure 3. H is defined as an inter–surface distance. The $\Phi(r)$ becomes greater with the decrease in H, giving a greater z value, as shown in Figure 3. Because we can determine the z value from helium adsorption isotherm, the pore width H can be obtained from the relation of Figure 3.

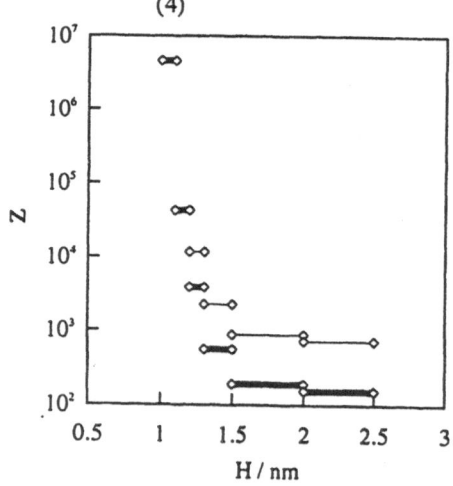

Figure 3. Relationship of H and z

## EXPERIMENTAL

Kynol-based activated carbon fiber (ACF) was used. Adsorption of helium was determined by a McBain type gravimetric apparatus at 4.2K [1-3]. The sample was evacuated at 383K in a high vacuum for 2hrs prior to the adsorption measurement.

## RESULTS AND DISCUSSION

We presume the bimodal pore size distribution; $W_{mS}$ and $W_{mL}$ are the monolayer capacity for smaller and larger micropores, respectively. Then, the Steele's equation is expressed by two term formula for the bimodal pore system, as given by eq. (5).

$$W = W_{mS} \frac{\left| 1 + z_S x \dfrac{2-x}{(1-x)^2} \right|}{1 + cx + \dfrac{z_S cx^2}{1-x}} cx + W_{mL} \frac{\left| 1 + z_L x \dfrac{2-x}{(1-x)^2} \right|}{1 + cx + \dfrac{z_L cx^2}{1-x}} cx \qquad (5)$$

Here $z_S$ and $z_L$ are the z values for smaller and larger micropores, respectively. The eq. (5) was fitted to the experimental isotherm: In particular, the adsorption at the low pressure region ($P/P_0 \leq 10^{-1}$) was carefully used for the best fitting. The fitting result is shown in Figure 4. The fitting is quite good except for the multilayer adsorption region above $P/P_0 = 10^{-1}$. As the Steele's equation describes the multilayer adsorption process, we must neglect the multilayer adsorption region for this micropore filling. The best fitting gave $z_S=5000000$, $V_{mS}=51$mg/g, $z_L=1000$, and $V_{mL}=13$mg/g. The values of $z_S$ and $z_L$ correspond to $H_S=1.0-1.1$nm and $H_L=1.5-2.0$nm or $H_L=1.3-1.5$nm, respectively. The difference of $H_L$ stems from the slight change of $r_{12}$. Although $H_S$ is almost constant regardless of the $r_{12}$ change from 0.28nm to 0.25nm as shown in Figure 3, $H_L$ changes sensitively with such change even in the micropores of $H \geq 1.3$nm.

Recently, the Dubinin-Stoeckli (DS) equation was applied to determination of the pore size distribution (PSD) from helium adsorption isotherms on activated carbons [2,3]. However, the DS approach always provided too simplified distribution of a single peak, which should come from the strong averaging. We can also obtain the PSD from the molecular potential approach described above. The micropore width (H) in this study is not strictly close to that with DS analysis, hence we must reduce the contribution of the hard sphere diameter of carbon walls consisting the slit pores. H is related to effective micropore

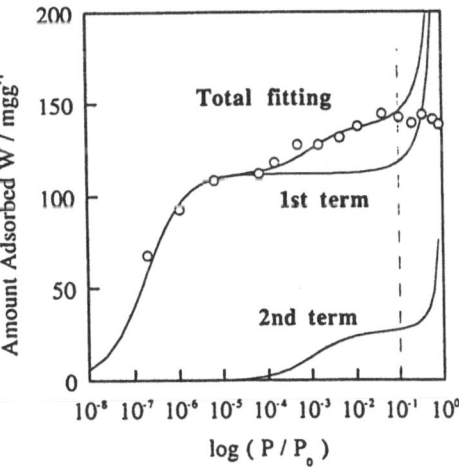

Figure 4. The fitting to experimental adsorption isotherm

548

width ($w$) defined as follow [8],

$$w = H - 2\sigma_{gs} - \sigma_{gg} \qquad (6)$$

where $\sigma_{gg}$ =0.256nm. The PSD obtained in this study is shown in Figure 5, including the results by the model of $r_{12}$ =2.5nm and 2.8nm. Strictly, the fraction of micropore volumes between smaller and larger micropores cannot be evaluated from the obtained $W_{mS}$ and $W_{mL}$ to determine the PSD, but the fractions for the micropore volumes should correspond to $W_{mS}$ and $W_{mL}$. The height (dW/dw) for the PSD is estimated using these values. It is presumed that the contribution of larger micropores will be underestimated by this way. We also show the PSD of Gaussian distribution by the DS analysis in Figure 5 for comparison.

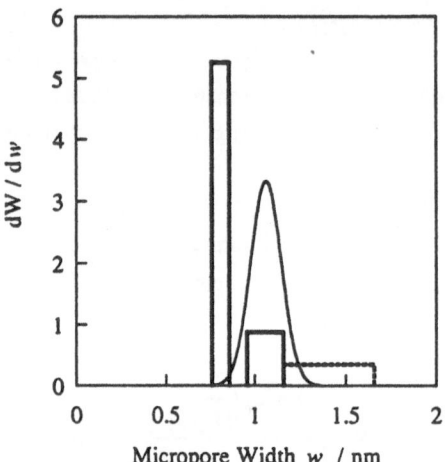

Figure 5. The PSD for ACF by helium adsorption. Normal distribution; DS analysis, Bar chart; this work. Broken bar at larger micropore is obtained from $r_{12}$ =0.25nm

The major peak shifts to the smaller $w$, but minor peak becomes more wider one especially for $r_{12}$ =2.5nm. In the previous studies, it is suggested that the PSDs for several ACFs show the bimodal PSD, containing smaller and larger micropores [9]. The approach proposed in this study also can interpret to this fact. Therefore PSD obtained by DS analysis should be represented as an averaged one for the real bimodal PSDs.

## REFERENCES

1. H. Kuwabara, T. Suzuki, and K. Kaneko, *J. Chem. Soc., Faraday Trans.* **87**, 1915(1991).

2. N. Setoyama, M. Ruike, T. Kasu, T. Suzuki, and K. Kaneko, *Langmuir* **9**, 2612(1993).

3. N. Setoyama, K. Kaneko, and F. Rodriguz-Reinoso, *J. Phys. Chem.*, in press.

4. M.M. Dubinin, *Chem. Rev.* **60**, 235(1960).

5. W.A. Steele, *J. Chem. Phys.* **25**, 819(1956).

6. N. Setoyama and K. Kaneko, *Adsorption* **1**, 165(1995).

7. W.A. Steele, *Surf. Sci.* **36**, 317(1973).

8. K. Kaneko, R. Cracknell, and D. Nicholson, *Langmuir* **10**, 4606(1994).

9. M. Jaroniec, R.K. Gilpin, K. Kaneko, and J. Choma, *Langmuir* **7**, 2719(1991).

# PHYSICAL ADSORPTION OF HYDROCARBONS FROM WATER SOLUTIONS ON CARBONS

N.A.Eltekova
Institute of Physical Chemistry Russian Academy of Sciences,
Leninskii prospekt 31, Moscow 117915, Russia

## INTRODUCTION

Water utiliters are faced with many problems in trying to produce the highest quality water. Contaminated water sources are a few of these problems. Granular activated carbons are the important sorbents for uptake of various organic impurities from contaminated water media with the proper choice of the conditions for the correct operation of adsorbents. Activated carbons are characterized with well-developed system of micro- and macroporespores in which the molecules of organic compounds are adsorbing very strongly.

For adsorption occuring on microporous active carbons the theory of volume filling of micropores (TVFM) has been generally accepted [1-3]. The general equation of this theory is the DR - equation [4]. The DR equation was deduced fifty years ago and since it is successful in descripting vapor adsorption on the activated carbons with homogeneous microporous structure as it has been verified by great number of researchers. There were many attempts to apply the DR equation for the description of liquid-phase adsorption by granular activated carbons [5-11].

The main purpose of this work was to test the TVFM equations applicability for the description of equilibria of aromatic hydrocarbons adsorption from water solutions on two commercial activated carbons and for the modeling of adsorption process of pure water production with using granular activated carbons.

## EXPERIMENTAL

The structural characteristics of commercial activated carbons AG3 and BAC (Khimreaktiv, Russia) used are summarized in Table 1. Before the adsorption measurements the carbon sorbents were washed consecutively with benzene, acetone and distilled water and then evacuated at 150°C for 6 h in vacuum. Benzene and toluene were provided by Khimreaktiv (Russia). Benzene and toluene were used without further purification. As a solvent bidistillated water was used. The adsorption measurements were made at 25°C by conventional batch adsorption method. The equilibrium concentration of solute in water after adsorption experiment was determined with a spectrophotometer SP 8000 (Pye Unicam, U.K.). Adsorption amount n of solute was calculated using the formula

$$n = (Co - C) v / M \tag{1}$$

where Co and C are initial and final concentrations of aromatic compounds in water solution, v is solution volume in an adsorption flask and M is an adsorbent amount in the experiment.

For the description of isotherms of adsorption from water solutions on active carbons we used the modified DR equation in following form [1]

$$n = Wo/V \exp [ - (A / \beta Eo)^2] \tag{2}$$

and in form known as DS (Dubinin - Stoeckli) equation [2]

$$n = (Wo/2BV) \exp (- mb^2A^2/B^2) [1 + erf (b /\sqrt{2} \delta B)] \tag{3}$$

*J. Fraissard (ed.), Physical Adsorption: Experiment, Theory and Applications, 549–552.*
© 1997 *Kluwer Academic Publishers.*

where $A = RT \ln (Cs / C)$, $\beta$ is affinity coefficient, $m = 1/\beta$ b Eo ,volume V is a molar volume of aromatic compound, $B = (1+2m \ \delta^2 A^2)^{1/2}$, $Wo = Wo1 + Wo2$ expresses the micro- and supermicropores volumes, b is half-breadth of slit micropores, $\delta$ is a standard deviation, C/Cs is the relative concentration.

For dynamical measurements of adsorption the set-up was used. This set-up consisted of the next units namely: a pump, a jacketed fixed bed adsorber, a UV-detector. The main unit of the set-up is the adsorber column (15 mm ID, 200 mm height) filled with activated carbon particles. The dynamical runs were carried out with toluene solution in water on two carbon sorbents AG3 and BAC. Toluene concentration in water solution was 30.7 mg/l. Breakthrough curves were measured on both columns at flow rate u = 0.42 mm/c.

## RESULTS AND DISCUSSION

The adsorption isotherms (Tables 2 and 3) demonstrated also good accordance of DR and DS equations with experimental adsorption isotherms for both activated carbons in the range of $0.04 < x < 0.4$ ( $x = Cs / C$ ). The comparison DR and DS equations for $0.01 < x < 0.4$ showed that DR equation described the adsorption of toluene in water solutions on both activated carbons covering wide range of equilibrium concentrations. The comparison DR and DS equations for the adsorption of benzene from water solutions was made early [11] and showed also good accordance.Therefore the activated carbons AG3 and BAC should relate to the adsorbents with heterogeneous microporous structure. The comparison of Wo values calculated by DR and DS equations (Table 4) suggests that the contribution of aromatic hydrocarbons adsorption on mesopore surface in total sorption process plays the marked role. The kinetics coefficients were calculated from experimental data by using the modified Wicke equation. Equilibria constants and kinetics coefficients were used in mathematical model of adsorption process for water purification. The results obtained are important not only for the theory of physical adsorption but also for the industrial sorption processes of the separation, evolution and purification. Accurate isotherms expressions facilitate the use of measured adsorption equilibrium data in dynamical fixed bed model while also allowing for an analysis of thermodynamical quanities important in extrapolating and evaluating data.

In the past years several mathematical models, with different descriptive capabilities, have been developed for the adsorption water purification. The simplest model for multicomponent liquid phase systems is based on the equilibrium theory. This is useful in many cases for constant or proportionate pattern and for constant or variable separation factors, if the kinetic effects can be neglected. Undoubtful this is a good tool to study the general trends and give a qualitative undestanding of the processes. In our model concept we suppose, - based on the theory of volume filling of micropores, - that the micropores are filled by water solution of hydrocarbon, in this way the adsorbent has a constant total capacity analogously to sorption by zeolites of X-type [12]. The effect of ultraporosity (sieve effect) is absent. But this model is not good enough when accurate quantitative fitting is sought.The parameters characterizing the adsorption dynamics were calculated using the Shilov equation [13]

$$\tau = k Z - \theta \qquad (4)$$

where $\tau$ and $\theta$ are the times of protection action and action waste, respectively, Z is the layer bed height and $k = n / uC$. From this equation the expression for $\tau$ as the work time of adsorber was obtained [14] as

$$\tau = k Z - 2 R k \, \varepsilon / (1 - \varepsilon) - R^2 \varepsilon / 6D (1 - \varepsilon) \qquad (5)$$

where 2R is granule size, $\varepsilon$ is the adsorber porosity and D is the diffusion coefficient. The equation (5) was applied for the calculation of $\tau$ in systems carbons - toluene - water. The experimental and calculated $\tau$ values were practically coinsided.

CONCLUSION

The DR and DS equations have been applied for the description of aromatic hydrocarbons from water solutions by granular activated carbons. The results obtained have shown that both carbons posess the ability for removal the organic impurities from contaminated water. the proposed model of adsorption dynamics was used for the description of removal toluene from water by carbons.

REFERENCES

1. M.M. Dubinin, J.Coll.Interface Sci., 23,487 (1967); Carbon, 23,373 (1985).
2. M.M. Dubinin, H.F. Stoeckli, J. Coll. Interface Sci., 75, 34 (1980).
3. H.F.Stoeckli, AST , 10, 3 (1993).
4. M.M. Dubinin, L.V. Radushkevich, Dokl. Akad.Nauk SSSR, 55, 331 (1947).
5. A.M.Koganovski, T.M.Levchenko, Zh.Fiz.Khim., 46, 1789 (1972).
6. A.M.Stadnik, Yu.A.Eltekov, Zh. Pricl. Khim., 48, 146 (1975).
7. A.M.Stadnik, Yu.A.Eltekov, Zh.Fiz.Khim., 49, 771 (1975).
8. A.Davrowski, M.Jaronec, Adv.Colloid.Interface Sci., 27, 211 (1987).
9. A.M.Tolmachev, Langmuir, 7, 1400 (1991).
10. A.Davrowski et.all., AST, 10, 35 (1993).
11. N.A.Eltekova, Yu.A.Eltekov, AST, 10, 203 (1993).
12. Yu.A.Eltekov, A.V.Kiselev, Molecular Sieves (Ed. Symes), Butterworths, L.(1967) p.267.
13. N.A.Shilov et all., Zh.Russ.Khim.Soc., 61, 1017 (1929).
14. A.M.Stadnik and Yu.A.Eltekov, Zh.Fiz.Khim., 51, 2997 (1977).

Table 1 - Structural Characteristics of Carbons

| Adsorbent | Granule size R, mm | Apparent bulk density kg/dm3 | Micropore volume dm3/kg | Mesopore volume dm3/kg | Macropore volume dm3/kg |
|---|---|---|---|---|---|
| AG3 | 2.0±0.5 | 0.90 | 0.26 | 0.09 | 0.41 |
| BAC | 1.5±0.5 | 1.72 | 0.23 | 0.09 | 1.19 |

Table 4 - Parameters Characterizing the Porous Structure of Carbons

| Solute | Carbon | DR | | | | DS | | | | |
|---|---|---|---|---|---|---|---|---|---|---|
| | | Wo cm3/g | b nm | Eo kJ/mol | RSD % | Wo cm3/g | b nm | Eo kJ/mol | $\delta$ nm | RSD % |
| Benzene | AG3 | 0.27 | 1.39 | 8.6 | 2.6 | 0.33 | 1.60 | 7.5 | 0.28 | 1.6 |
| | BAC | 0.26 | 1.24 | 9.6 | 2.0 | 0.29 | 1.35 | 8.9 | 0.19 | 7.7 |
| Toluene | AG3 | 0.33 | 1.25 | 9.6 | 0.4 | 0.33 | 1.26 | 9.5 | 1.10 | 1.6 |
| | BAC | 0.26 | 1.07 | 11.2 | 0.8 | 0.26 | 1.08 | 11.1 | 0.09 | 3.8 |

Table 2 - Comparison of experimental and calculated values for toluene adsorption on AG3 at 25°C.

| No. | C/Cs | n,mmol/g | DR | dev.,% | DS | dev.,% |
|-----|------|----------|------|--------|------|--------|
| 1 | 0.00240 | 0.3050 | 0.3083 | -1.1 | 0.3118 | -2.2 |
| 2 | 0.00344 | 0.4180 | 0.4036 | 3.4 | 0.4052 | 3.1 |
| 3 | 0.00388 | 0.4340 | 0.4400 | -1.4 | 0.4409 | -1.6 |
| 4 | 0.00530 | 0.5520 | 0.5457 | 1.1 | 0.5446 | 1.3 |
| 5 | 0.00660 | 0.6330 | 0.6302 | 0.4 | 0.6276 | 0.9 |
| 6 | 0.01300 | 0.9390 | 0.9456 | -0.7 | 0.9390 | 0.0 |
| 7 | 0.01700 | 1.0930 | 1.0925 | 0.0 | 1.0847 | 0.8 |
| 8 | 0.02400 | 1.2400 | 1.2977 | -4.7 | 1.2892 | -4.0 |
| 9 | 0.02500 | 1.3170 | 1.3230 | -0.5 | 1.3145 | 0.2 |
| 10 | 0.03900 | 1.5870 | 1.6114 | -1.5 | 1.6037 | -1.1 |
| 11 | 0.06500 | 1.9890 | 1.9588 | 1.5 | 1.9547 | 1.7 |
| 12 | 0.06900 | 2.0120 | 1.9997 | 0.6 | 1.9962 | 0.8 |
| 13 | 0.10100 | 2.2780 | 2.2569 | 0.9 | 2.2579 | 0.9 |
| 14 | 0.17400 | 2.5860 | 2.5975 | -0.4 | 2.6067 | -0.8 |
| 15 | 0.22700 | 2.7900 | 2.7442 | 1.6 | 2.7576 | 1.2 |
| 16 | 0.31200 | 2.9210 | 2.8960 | 0.9 | 2.9141 | 0.2 |
| 17 | 0.41700 | 2.9890 | 3.0076 | -0.6 | 3.0296 | -1.4 |

Table 3 - Comparison of experimental and calculated values for toluene adsorption on BAC at 25°C.

| No. | C/Cs | n,mmol/g | DR | dev.,% | DS | dev.,% |
|-----|------|----------|------|--------|------|--------|
| 1 | 0.00250 | 0.4610 | 0.4474 | 2.9 | 0.4530 | 1.7 |
| 2 | 0.00390 | 0.5350 | 0.5715 | -6.8 | 0.5727 | -7.0 |
| 3 | 0.00518 | 0.6710 | 0.6616 | 1.4 | 0.6599 | 1.7 |
| 4 | 0.00610 | 0.7510 | 0.7173 | 4.5 | 0.7140 | 4.9 |
| 5 | 0.01140 | 0.9760 | 0.9545 | 2.2 | 0.9459 | 3.1 |
| 6 | 0.01930 | 1.0930 | 1.1794 | -7.9 | 1.1685 | -6.9 |
| 7 | 0.02900 | 1.3660 | 1.3640 | 0.1 | 1.3534 | 0.9 |
| 8 | 0.05290 | 1.6490 | 1.6426 | 0.4 | 1.6357 | 0.8 |
| 9 | 0.05690 | 1.6610 | 1.6761 | -0.9 | 1.6699 | -0.5 |
| 10 | 0.06130 | 1.7230 | 1.7102 | 0.7 | 1.7047 | 1.1 |
| 11 | 0.08300 | 1.8600 | 1.8459 | 0.8 | 1.8442 | 0.8 |
| 12 | 0.12700 | 1.9870 | 2.0246 | -1.9 | 2.0292 | -2.1 |
| 13 | 0.19300 | 2.2410 | 2.1802 | 2.7 | 2.1917 | 2.2 |
| 14 | 0.24000 | 2.3090 | 2.2509 | 2.5 | 2.2660 | 1.9 |
| 15 | 0.38100 | 2.3360 | 2.3729 | -1.6 | 2.3945 | -2.5 |

# ADSORPTION FROM BINARY LIQUID MIXTURES ON ACTIVATED CHARCOAL SURFACES

M.TUNÇAY ,S.GÖKTÜRK ,A.MARDİNLİ ,M. MAHRAMANLIOĞLU

*University of Istanbul Department of Chemistry Avcılar 34850    Istanbul / TURKEY*

## ABSTRACT
Adsorption   equilibrium  for  binary  mixtures  of  Benzene  and  Chlorobenzene ;
Benzene and Ethanol ; Chlorobenzene and Ethanol with   activated   charcoal has  been
studied  at
25° C.The composition change in these binary liquid phase adsorption studies has been
calculated from refractive index change.The  Composite  and  Individual  Isotherms  for
adsorption  of  benzene  from  Benzene - Ethanol and  Benzene - Chlorobenzene solutions
and  Chlorobenzene  from Chlorobenzene - Ethanol and Chlorobenzene - Benzene solution
at 25° C were compared. The composition of  adsorbed layer was studied by Schay & Nagy's
model.

## INTRODUCTION
The mechanism of adsorption from a binary liquid mixture onto a solid adsorbent
has been studied by many investigators.(1-6) Various factors has been found to influence,
upon adsorption ; nature of the adsorbent , chemical nature of the solute molecules ,
interfacial tension of components at the liquid / solid Interface ,(6,7) volatility,(4)  molecular
size and shape .(8) When a binary mixture is involved,the adsorbed layer has in general,a
different composition from that of the bulk phase in equilibrium with it. Adsorption from a
binary liquid mixture is expressed in terms of composite isotherms (that is the specific
surface  excess  concentration).  The  state  of  adsorbate  may  be  judged  directly from
experimentally determined composite isotherms.The composite isotherms is the functional
relation between the surface excess of a component and its concentration in the equilibrium
bulk phase . (9)
According to the course of the function between these two end points , two main
types of isotherms may be discerned; U- shaped , a maximum for one of the component and a
minimum for the other and   S- shaped , it is a maximum- minimum curve for both
components.Nagy & Schay have proposed a further subdivision .(10) The type of composite
isotherm has been widely used for comparision   the adsorption behavior of binary
mixtures.(11-13)The purpose of this paper is the compare adsorption behavior of a
certain component in different binary mixtures on the same adsorbent , in order to see
the influence of the other component. Three pairs of binary mixtures of Chlorobenzene
(ClB) ; Benzene (B) ; Ethanol (E)  were chosen as to  derive  information about the mutual
effect of components which have different molecular size , surface tension, polarity and
volatility, on the adsorption of each other. The Schay & Nagy's Model  had been applied to
the data and the results were discussed.

*J. Fraissard (ed.), Physical Adsorption: Experiment, Theory and Applications, 553–557.*
© 1997 *Kluwer Academic Publishers.*

## EXPERIMENTAL

### *Chemicals.*

Ethanol (99.8 % pure) , Benzene (99.5 %),Chlorobenzene (99 % pure ) were obtained from Fluka A.G. The activated Charcoal was microporous E.Merck (No.2184) sample and was further purified by extracted (Soxhlet) for several hours with benzene.The specific surface area of the sample was 800 $m^2$ \ g.The ratio of liquid to adsorbent kept as low as possible to obtain maximum readings (4 milliliters per gram were chosen for all of the systems).

### *Procedure.*

A binary mixture was prepared for each experimental point ; its concentration was analyzed refractometrically using Carl Zeiss Jena Refractometer.A weighed amount of adsorbent was then added to the binary mixture.The system consisting of the binary mixture and the adsorbent was shaken for 17-19 hr in a screw - capped erlenmeyer flask, using a reciprocating shaker. Measurements were made over the whole range of concentrations lying between pure components.The temperature was thermostatically controlled to 25 ($\pm$ 0.1) $^0$C. The equilibrium composition of the bulk liquid ,$x_1$ , was then analyzed refractometrically ; the calibration curves needed for the refractometric analyses were prepared by us.The refractive index composition curves of each of the binary mixtures were almost straight line over the entire range of concentration.

## RESULTS and DISCUSSION

The following binary mixtures were studied at 25$^0$C : (a) B (1) +ClB (2) ; (b) B (1) + E (2) ; (c) ClB (1) + E (2) on activated charcoal. The composite isotherms for adsorption from binary mixtures (a,b and c) at 25°C were shown in Figure 1 .These isotherms would be expected to through further liquid on the influence of other component.The composite adsorption isotherms of B from E and from ClB were S-shaped and type IV curves as classified by Schay & Nagy (8). The proportion of B in the adsorbed phase is higher than in the bulk phase over part of the concentration range only approximately (0-0.8) and (0-0.9) from ClB and E respectively. The position being reversed for the rest of the concentration range ; that is charcoal preferentially adsorbed B from its binary mixtures with E and ClB up to relatively high concentrations of B ; whereas showing slight preferential adsorption of the other component when it is present at low concentrations.The composite adsorption isotherm of ClB-E were U-shaped and almost type I curve,which has a maximum or minimum in the middle range of concentration. Surface excess of ClB exhibit a minimum beyond which an increase in solution concentrations of ClB leads to an increase in the relative amount of ClB adsorbed.The cause is clearly competing process between E and ClB. It was observed that E is relatively preferentially adsorbed on charcoal at all

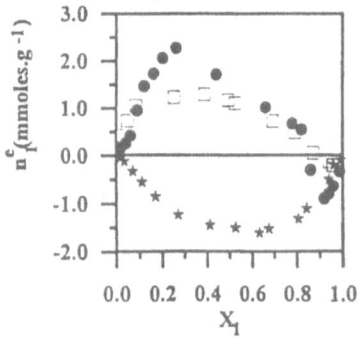

Fig.1: Composite isotherms for adsorption of binary mixtures of B(1)+ClB(2) ●; B(1)+E(2) ☐ and ClB(1)+E(2) ★on activated chacoal at 25$^0$ C.

concentrations.

The individual adsorption isotherms have been obtained from the composite isotherms using the expression (11)

$$n_1^{\sigma} = \frac{Sx_1 + n_1^e A_2}{A_2 + \left(A_1 - A_2\right)x_1} \qquad (1)$$

$$n_2^{\sigma} = \frac{Sx_2 - n_1^e A_1}{A_2 + \left(A_1 - A_2\right)x_1} \qquad (2)$$

Where $n_1^{\sigma}$ , $n_1^e$ are the number of moles and surface excess of component (1) adsorbed from the binary mixture per gram of adsorbent respectevely ; 'S' the specific surface area of the adsorbent ; and $A_1$ and $A_2$ the cross-sectional areas of the adsorbed molecule (1) and (2) at the interface respectively. The cross sectional areas of ClB , B and E were taken to be 33, 30 and 23 $A^{o2}$/molecule, respectively, the values reported by Mc Lellan and Harnsberger[14] for these molecules as obtained from molecular models. The individual adsorption isotherms of binary mixtures of a, b and c were shown in Figure 2.

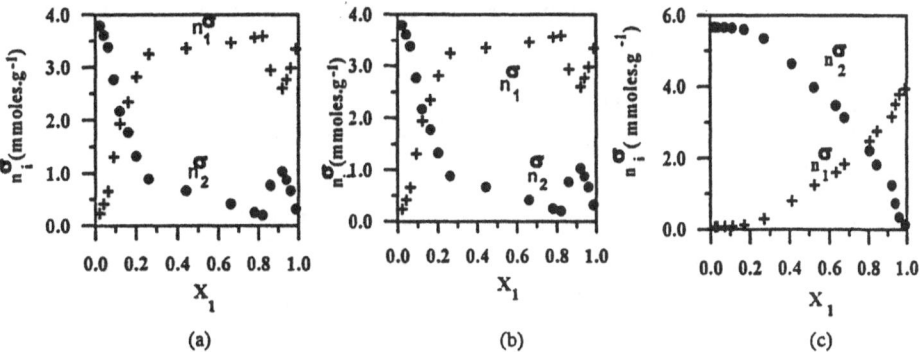

(a)              (b)              (c)

Fig.2: Individual isotherms for adsorption on activated charcoal from B(1)+ClB(2) (a) ; B(1)+E(2) (b) ; ClB (1)+E(2) (c) solutions at 25°C.

It was observed that though the presence of the other component was not totally excluded at the interface , B was relatively preferentially adsorbed in the presence of ClB and E whereas E was preferentially adsorbed in the presence of ClB.

The composition of adsorbed layer was determined by Schay & Nagy's model (8) in which the composition of adsorbed layer was calculated from the expression .

$$x_{1(SN)}^{\sigma} = \frac{Sx_1 + n_1^e A_2}{S + \left(A_2 - A_1\right)n_1^e} \qquad (3)$$

The $x_1^{\sigma}$ values were plotted as a function of $x_1$ in Figure 3 for binary mixtures a , b and c respectively. As seen in figure 3, $x_1^{\sigma}$ values were higher than $x_1$ for the preferentially adsorbed component.. The deviation on the curves occured at the high concentrations of component (1) ,showing the preferential adsorption of the component (2) ;that is $x_1^{\sigma}$ became smaller than $x_1$ when $x_1 > 0.82$ and $x_1 > 0.90$ for system a and b respectevely. On the contrary, $x_1^{\sigma}$ values were smaller than $x_1$ for the binary mixtures of c . The three systems studied were although showing somewhat different adsorption characteristics,their

556

individual behavior may be accounted for reasonably satisfactorily in terms of simple monolayer theories of adsorption.

The adsorption behavior of these binary mixtures were also compared from adsorption equilibrium constants , K , which is related to the interfacial tension difference (12). If the bulk liquid and adsorbed phase are considered to be ideal mixture , K can be calculated from equation

$$\frac{x_1 x_2}{n_1^e} = \left(\frac{1}{m}\right) x_1 + \frac{K}{(1-K)m}$$ (4)

as proposed by Sircar and Myers.[9] For each system, the experimental data plotted in the form of the above equation ,formed a straight line at intermediate concentrations. The K values derived from these plots using the slope and intercept were 0.076 , 0.0106 and 1.59 for systems a , b and c respectively. When component (1) that is , (B) was preferentially adsorbed as in the case of systems a and b , K< 1 and when component (2) that is (E) was preferentially adsorbed as in the case of systems c , K > 1. These results are in agreement with Sircar and Myers.[9] Thus comparision of K values permit prediction of the component which will produce the most decrease of interfacial tension disregarding other effects.

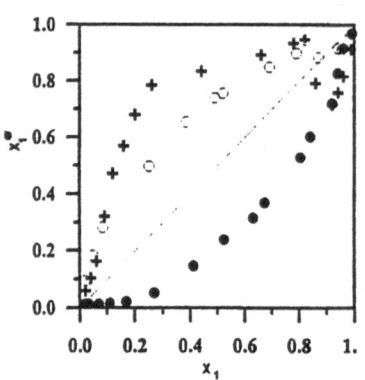

Fig.5: Composition of adsorbed layer $X_1^\sigma$ versus $X_1$ for B (1)+ClB(2) + ;B(1)+E(2) o and ClB (1)+E(2) ● at $25^0$ C.

As a conclusion , B was preferentially adsorbed from its binary mixtures with ClB. In this system , surface tension , polarity ,volatility and molecular area of B are smaller than that of ClB. Thus , all these factors are in favour of preferential adsorption of B. B was also preferentially adsorbed from its binary mixtures with E in which only polarity of B is smaller than that of E . So , polarity is clearly the primary importance to that of the others. From the binary mixtures of ClB-E. E was preferentially adsorbed.Since only polarity of E is higher than that of the others , the result can be interpreted as polarity is secondary importance to that of volatility and free surface energy decrement for this system. The present work provides evidence that intermolecular forces between adsorbate and adsorbent responsible for selectivity were considered to be influenced by the presence of the other component.

**REFERENCES**

1. Lloyd L.C., and Harris B.L., (1954) Binary Liquid Phase Adsorption , *J.Chem.Soc.* **58**,899

2.Chipalkatti H.R., Giles C.H. and (the late) Vallance D.G.M.,J(1954) Adsorption at Organic Surfaces.PartI.Organic Compounds by Polyamide and Protein Fibres from Aqueous and Nonaqueous Solutions. *Chem.Soc* . 4375

3.Kippling J.J. and Peakall D.B.,(1956) Adsorption from Binary Liquid Mixtures on Activated Alumina.*J.Chem.Soc*. 4828

4.Blacburn A., Kippling J.J. and Tester D.A., (1957) Adsorption on Carbons from Binary Liquid Mixtures :Some Surface Activity Coefficients. *J.Chem.Soc*. 2373

5.Berezin G.I., Kiselev A.V., Sagatelyan R.T. and Sinitzyn V.A., (1972) A Thermodynamic Evaluation of the state of the Benzene and Ethanol on a Homogeneous of a Nonspecific AdsJorbent. *Colloid Interface Sci.* **38**, 2

6.Ash G.S., Bown R., and Everett D.H., (1975) Thermodynamics of adsorption from Solution.Adsorption by Graphon from Binary Mi,xtures of Benzene,Cyclohexane and n -Heptane. *J.C.S. Faraday Trans.* 1 ,71,123

7.Elton , (1954) J.Chem.Soc. 3813

8.Schay G., (1969) Adsorption of Solutions of non - electrolytes , in *"Surface and Colloid Science"* (E.Matijevic, Ed.), p.179 , **Vol.2** ,Interscience ,New York

9.Sircar S ., and Myers A.L., (1970) Statistical thermodynamics of Adsorption from Liquid Mixtures on Solids.I.Ideal Adsorbed Phase.*The J.Phy.Chem.* 74 ,No.14 ,2828

10.Nagy L.G., and Schay G., (1961)*J.Chm.Phys*.140

11.Suri S.K., and Patel M ., (1979) Adsorption from Binary Solutions of Benzene and Cyclohexane on Cobalt-Chalcogenide Surfaces. *J.Colloid Interface Sci.* 69 ,347

12.Everett H.D., and Podoll T.R., (1981) Adsorption of Near-Ideal Binary Liquid Mixtures by Graphon.*J.Colloid Interface Sci*. 82,1,14

13.Suri S.K., and Patel M ., (1981) Studies on Chromium Oxide Catalyst Surfaces Obtained from the Hydrous Gels: Adsorption from Binary Mixtures of Benzene and Cyclohexane *Colloid Interface Sci.* 84,1,36

14.Mc.Clellan A.L., and Harnsberger H.F., (1967) *J.Colloid Interface Sci.* 23, 577

# MOLECULAR SIMULATIONS FOR CHARACTERIZATION AND EQUILIBRIUM ADSORPTION PREDICTION ON ACTIVATED CARBON

Vladimir Gusev, James O'Brien, and Nigel Seaton[#]

Department of Chemical Engineering, Yale University
New Haven, CT 06520, USA

[#] Department of Chemical Engineering, University of Cambridge
Cambridge CB2 3RA, United Kingdom

## Introduction

Developments in the theory of strongly inhomogeneous confined fluids have driven considerable progress in adsorption characterization of activated carbons based on Density Functional Theory (DFT) and nitrogen adsorption at 77 K [1,2].

We use the less approximate Monte-Carlo (MC) simulations of adsorption [6] and supercritical adsorption measurements of methane at 308.2 K to extract the pore size distribution of activated carbon.

## Experimental

We used a volumetric custom-built apparatus [3] to measure adsorption at 308 K, 333 K, and 373 K and pressures from 0.001 to 3 MPa. The volumetric measurements were done by two MKS Baratron pressure gauges and Omega model 100W platinum resistance thermometers with digital read-outs. Two thermostats were used. The accuracy of a single adsorption measurement was about 1%.

We used BPL 6x16 microporous carbon (Calgon Carbon Corporation, Pittsburgh) outgassed at 10 microns of mercury pressure and 373 K for 24 hours. The methane and ethane, National Compressed Gases, Inc., were of 99.97% purity.

## GCMC Simulations

In GCMC calculations the chemical potential of the gas phase is specified, as is its temperature. To generate configurations with the correct limiting probability, a Markov chain of consecutive trials is realized: particles are attempted to be moved, created, and deleted using Adams' [4] method. Details are available elsewhere [5,6].

*J. Fraissard (ed.), Physical Adsorption: Experiment, Theory and Applications, 559–564.*

The interactions between fluid molecules were described by truncated Lennard-Jones (LJ) potentials:

$$u_{ff}(r) = \begin{cases} 4\varepsilon_{ff}\left[\left(\dfrac{\sigma_{ff}}{r}\right)^{12} - \left(\dfrac{\sigma_{ff}}{r}\right)^{6}\right] & , r < R_c \\ 0 & , r > R_c \end{cases}$$

The LJ fluid parameters were from the literature second-virial-coefficient data

| Fluid molecule | $\sigma_{ff}$, nm | $\varepsilon_{ff}/k_B$, K | l, nm | Ref. |
|---|---|---|---|---|
| 1LJ center methane | 0.381 | 148.1 | - | [7] |
| 2LJ centers ethane | 0.3512 | 139.81 | 0.2353 | [8] |
| 1LJ center ethane | 0.395 | 243.0 | - | [7] |

Fluid particle - single pore wall interactions were described by Steele's 10-4-3 potential [7]. The Lorentz-Berthelot combination rules were used to estimate the value of solid fluid parameters for all molecular models.

## Results and Analysis

GCMC adsorption isotherms were simulated in 40 pores ranging from 1.65 to 15 $\sigma_{met}$ at 308.15 K at intervals of either 0.01, 0.1, 0.5 or 1 and at the same pressures as the points of the experimental isotherm.

The lower end of this range, 1.65 $\sigma_{met}$, effectively represents the smallest pore in which adsorption of model molecules was still possible. It was used to estimate the excluded pore width for adsorption of methane molecules at all pore sizes as 0.248 nm, and, consequently, the Gibbs excess adsorption.

Absolute density profiles of the adsorbate in the pores (Fig. 2) at our highest pressure (3 MPa) suggest that the maximum pore width of H = 15 $\sigma_{met}$ studied was indeed sufficiently large to be considered as two independent walls rather than a single pore.

## Pore Size Distribution (PSD)

We fitted simulated isotherms to the experimental one using the least-squares solution of the equation

$$\mathbf{R}_{mn} \cdot V_n = \mathbf{A}_m$$

where $\mathbf{R}_{mn}$ is a matrix of adsorbate simulated excess densities, $V_n$ is a vector (the solution) of pore volumes, and $\mathbf{A}_m$ is the experimental data vector.

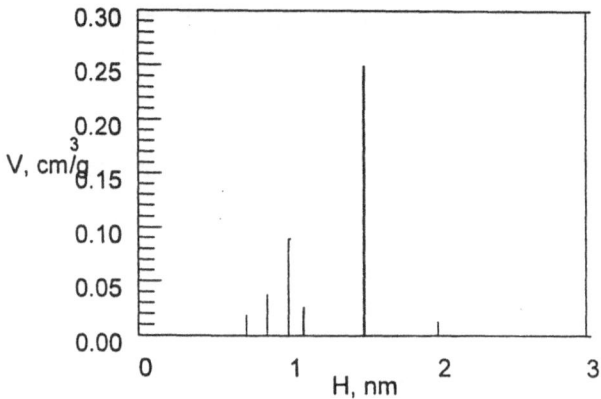

**Fig. 1.** The pore size distribution, obtained from a fit (see Fig. 2) of the experimental adsorption isotherm for methane on BPL-6 carbon at 308 K to a linear combination of GCMC-simulated isotherms. Note that many of the pores emerge with zero volume.

The natural ill-posedness of the problem is treated by applying non-negativity constraints and Singular Value Decomposition [9] within the Non-Negative Least Squares [10] method SVDNNLS.

The PSD (Fig. 1) shows that the BPL-6 activated carbon is a predominantly microporous adsorbent having an overall micropore volume of about 0.42 $cm^3$/g.

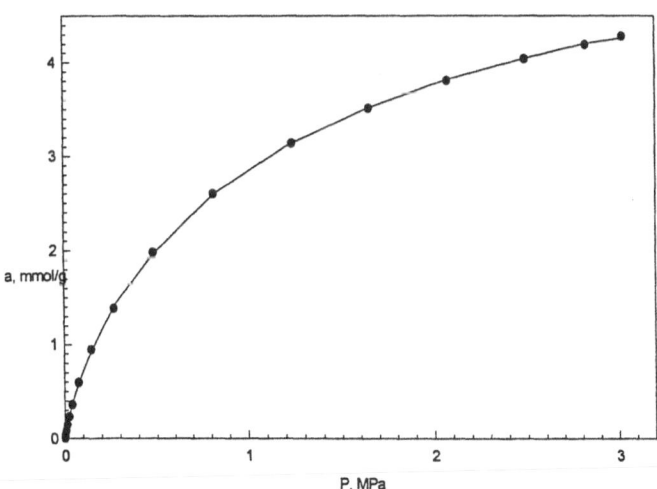

**Fig. 2.** Fit, using the SVDNNLS technique, of methane experimental adsorption isotherm with a set of GCMC isotherms simulated in slit pores at 308 K.

562

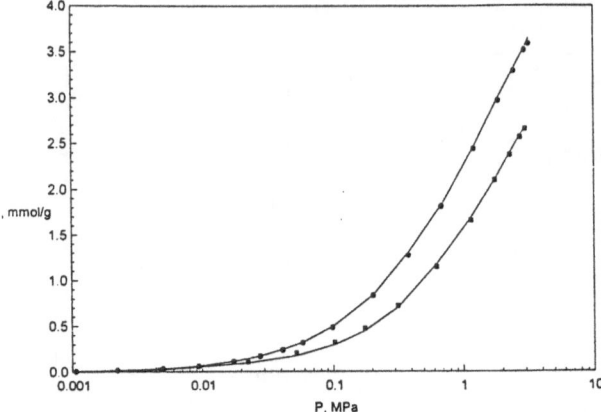

**Fig. 3**. GCMC prediction of methane adsorption on BPL-6 activated carbon at 333 K and 373 K using the pore size distribution derived from the 308 K adsorption isotherm. GCMC integral isotherms: lines; experiment: squares and circles.

## Prediction of Adsorption

The GCMC simulation was used once again to calculate adsorption in the pores comprising the PSD of the BPL-6 activated carbon and to generate the integrated methane adsorption isotherms at two higher temperatures 333 K and 373 K (Fig. 3) and ethane at 308 K, 333 K and 373 K (Fig. 4)

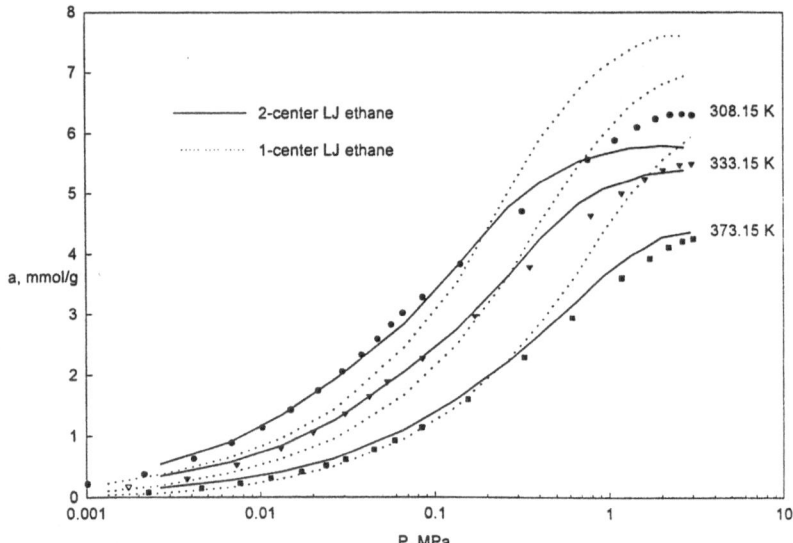

**Fig. 4**. Gibbs excess adsorption of ethane on BPL-6 activated carbon. Filled symbols: experiment; solid lines: GCMC-based prediction using two LJ center molecular model of ethane; dotted lines: GCMC-based prediction using one-LJ-center model of ethane. Data are shown at temperatures 308.2 K, circles, 333.2 K, diamonds, 373.2 K, triangles

In the case of methane quantitative agreement is excellent and serves as a stringent test of self-consistency of the method. The GCMC prediction based on the two-center model of ethane molecule is quantitative for a large part of the pressure range.

High-pressure (above 1 MPa) discrepancies may be due to:

- neglect of the quadrupole of the ethane molecule
- insensitivity of the supercritical adsorption characterization method at P<3 MPa and T=308 K to pores wider than 2.3 nm
- incorrectness of the slit model of the carbon pores

## Conclusions

We demonstrate the potential for quantitatively describing adsorption on industrially-relevant adsorbents using a combination of molecular simulation and a simple description of the adsorbent pore structure. Despite the assumptions used, the method performs remarkably well for this commercial adsorbent. Note that no adjustable solid-fluid interaction parameters were used; only a single methane adsorption isotherm, along with the (independently available in the literature) molecular properties of the bulk solid and fluids, was required.

Future work will proceed on several complementary fronts:

(i) a larger set of pores and a larger pressure range, in order to increase the precision of the initial pore size distribution;

(ii) more sophisticated models of the adsorbate and the adsorbent pores, as well as more complicated adsorbate species...

## Acknowledgments

This work is supported in part by the U.S. National Science Foundation through Grant No. CTS-9215604. V.G. thanks T. Schaefer for useful discussions.

## Literature

1. Seaton, N. A., Walton, J. P. R. B., and Quirke N. A New Analysis Method for the Determination of the Pore Size Distribution of Porous Carbons From Nitrogen Adsorption Measurements. *Carbon,* **27**, 853 (1989)
2. Lastoskie, C.; Gubbins, K. E.; Quirke, N. Pore Size Distribution Analysis of Microporous Carbons: A Density Functional Theory Approach. *J. Phys. Chem.* **97**, 4786 (1993)
3. Gusev V., O'Brien J., Jensen C., and Seaton N. A Theory for Multicomponent Adsorption Equilibrium: the Multi-Space Adsorption Model. *AIChE Journal*, 1996 (in press)
4. Adams, D. J. Chemical Potential of Hard Sphere Fluids by Monte Carlo Methods. *Mol. Phys.*, **28** (5) (1974)
5. Gusev V., O'Brien J., and Seaton N. A Self-Consistent Method for Characterization of Activated Carbons Using Supercritical Adsorption and Grand Canonical Monte-Carlo Simulations. *Langmuir*, submitted.

6. Gusev V., O'Brien J. Can Molecular Simulations Be Used To Predict Adsorption On Activated Carbons? *Langmuir*, submitted.

7. Steele W.A. *The Interactions of Gases with Solid Surfaces*. Pergamon. Oxford. (1974)

8. Fischer J., R. Lustig, H. Breitenfelder-Manske, and W. Lemming. Influence of Intermolecular Potential Parameters on Orthobaric Properties of Fluids Consisting of Spherical and Linear Molecules. *Mol. Phys.* **52**, 485-497 (1984)

9. Press, W. H.; Flannery, B. P.; Teukolsky, S. A.; Vetterling, W. T. *Numerical Recipes in Pascal. The Art of Scientific Computing*. Cambridge University Press. P. 61. (1989)

10. Lawson, C. L. and Hanson, R. J. *Solving Least Squares Problems*. Englewood Cliffs, N. J., Prentice-Hall. (1974)

# INTERACTION ENERGY OF WATER AND METHANOL WITH NaZSM-5 FROM ADSORPTION EXPERIMENTS

M. HEUCHEL[1], B. HUNGER[1] and W.-D. EINICKE[2]
[1] Institute of Physical and Theoretical Chemistry,
[2] Institute of Technical Chemistry,
University of Leipzig, D-04103 Leipzig, Germany

## 1. Introduction

Zeolites of MFI type have a great potential for the separation of alcohols from aqueous solutions (see e.g., [1,2]), because of their pore structure and their hydrophobicity. In this respect it is remarkable that there are only few contributions in literature dedicated to determinations of strength of water and alcoholes interaction with different adsorption sites in ZSM-5 zeolites [3-6]. Therefore, the motivation for this work was applying temperature-programmed desorption (TPD) and vapour adsorption to obtain detailed information about the adsorption behaviour of water and methanol on NaZSM-5.

## 2. Experimental

The NaZSM-5 was a commercial material "HS30" ($Na_6[Al_6Si_{90}O_{192}]$) supplied by Chemiewerk Bad Koestritz GmbH, Germany. The micropore volume was 0.155 $cm^3$/g ($N_2$ adsorption at 77 K). The saturation capacities were determined to 5.0 mmol/g (water) and 3.8 mmol/g (methanol). The total desorbed amount was for water $2.8 \pm 0.4$ mmol/g (16-17 molecules per u.c.) and for methanol $2.5 \pm 0.3$ mmol/g (14-15 molecules per u.c.).

**Temperature-programmed desorption (TPD)** was carried out in a convential flow device (He carrier gas (3 l/h), linear heating programm (10 K/min)). For evolved gas detection both a thermal conductivity detector (TCD) and a quadrupole mass spectrometer (Leybold, Transpector CIS System) with a capillar-coupling system were used.

**Adsorption Isotherms** were determined gravimetrically on a McBain balance fitted to a vacuum system ($\leq$ 1 Pa) at 298 K. Pressure was measured with a mercury pressure gauge ($\pm$10 Pa) at 298 K.

J. Fraissard (ed.), Physical Adsorption: Experiment, Theory and Applications, 565–569.
© 1997 Kluwer Academic Publishers.

## 3. Results and Discussion

The desorption curves of water and methanol from NaZSM-5 are shown in Fig. 1. A kinetic analysis of the TPD data assuming a rate law with constant pre-exponential factor and constant desorption energy was not possible. This simple model failed also for the experimental separated desorption peaks (for details see [5,6]). Therefore, a rate law of first order with a distribution function $f(E)$ of desorption energy $E$ was considered [7]:

$$r_d = -\,d\theta/dt = A \int_{E_{min}}^{E_{max}} \theta_l(E,T)\, \exp(-E/RT)\, f(E)\, dE \qquad (1)$$

where $r_d$ is the overall desorption rate, $\theta$ is the overall degree of coverage, and $A$ the pre-exponential factor. $\theta_l$ is the local coverage of sites characterized by a desorption energy $E$. $E_{min}$ and $E_{max}$ are the limits of the range of desorption energy. From the mathematical point of view, eqn (1) is a linear Fredholm integral equation of the first kind. The calculation of $f(E)$ from eqn (1) is a numerically ill-posed problem [8]; i.e., small changes in $r_d$ caused by experimental errors can distort significantly the calculated distribution function. In the current paper, the program INTEG [8], which involves a regularization method for solving such numerically instable integral equations was used.

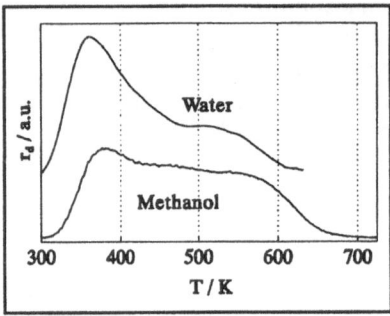

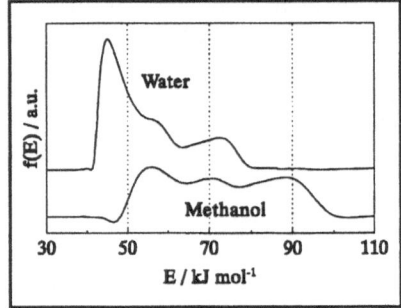

Figure 1. TPD results

Figure 2. Desorption energy distribution f(E)

Fig. 2 represents distributions $f(E)$ of desorption energies for NaZSM-5 calculated from the desorption profiles of Fig. 1. Water and Methanol show three distinct ranges of desorption energy $E$ with maxima at about 45, 54, and 70 kJ/mol for water, and 55, 68, and 89 kJ/mol for methanol. Both distribution functions clearly express energetic heterogeneity of the water/methanol interaction with the zeolite. In [5,6] it was shown, how the peaks can be attributed to unspecific interaction with the zeolitic framework and specific interaction with the cations. The unspecific interaction of methanol with

NaZSM-5 (maximum at 55 kJ/mol) is about 10 kJ/mol higher than for water and reflects the contribution of $CH_3$-group to the unspecific interaction. The energy values for this interaction are in good agreement with heats of adsorption in literature [9]. The observed bimodal distribution for the specific interaction at lower coverage can be explained by two different adsorption states of water and methanol on $Na^+$ ions. This result was in accordance with recent experimental observations [10,11].

In addition, adsorption isotherms were measured at room temperature (see Fig. 3). The isotherms show two branches: first a stronger increase in adsorption and further a smoother one. A comparison with TPD results shows that the amount of the second branch belongs to weakly bounded water or methanol, which desorbs during flushing with He at room temperature before the TPD was started.

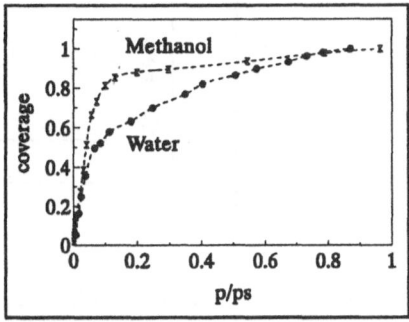

Figure 3. Adsorption isotherms at 298 K

The heterogeneity of adsorption sites in NaZSM-5 is also responsible for the failure of the simple isotherm fits to describe the experimental data. Therefore, an adsorption model was applied for an energetically heterogeneous surface with an adsorption energy distribution $f(U)$:

$$\theta(P,T) = \int_{U_{min}}^{U_{max}} \theta_l(P,T,U)\, f(U)\, dU \qquad (2)$$

where $\theta(P)$ is the measured overall coverage, $\theta_l(P,T,U)$ is the energy-dependent local adsorption isotherm and $U_{min}$ and $U_{max}$ are the limits of the range of adsorption energy. For the numerical calculation of the adsorption energy distribution functions $f(U)$ the same regularization method was used as for the desorption energy distributions $f(E)$. As model for the energy dependent local adsorption isotherm, first, the simple Langmuir form was used. This assumption was successful for water. The resulting adsorption energy distribution (see Fig. 4) shows two peaks. The peak between 20-40 kJ/mol belongs to the weakly

568

bounded water not detected in TPD. The maximum of the second peak is in agreement with the unspecific part of the desorption energy distribution. For methanol it was not possible to describe the experiments in the range of the experimental error with the Langmuir model as local adsorption isotherm. With the Fowler-Guggenheim equation, a local model including lateral interaction, the calculated adsorption energy distributions (see Fig. 5) showed similar coverage dependence like the respective desorption energy distribution for methanol. Despite the nice agreement regarding shape and wideness, in contrast to water, the numerical values of desorption energy $E$ and adsorption energy $U$ differ remarkable. A clarification of this discrepancy requires further theoretical and experimental studies.

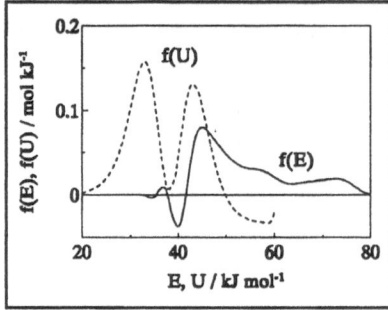

Figure 4. Distribution functions for water

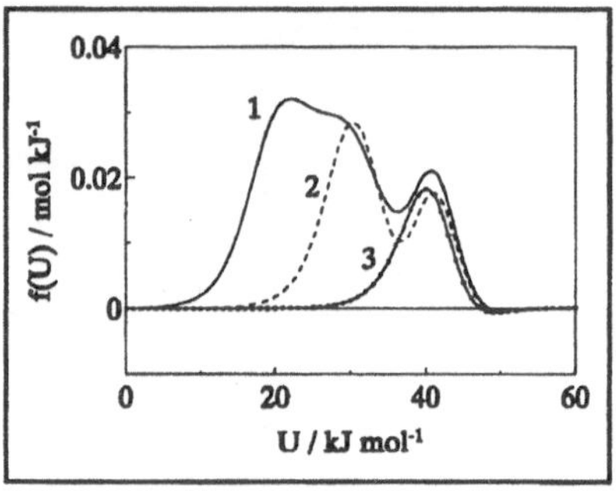

Figure 5. f(U) for methanol: coverage up to 0.62 (1), 0.33 (2), 0.13 (3) mmol/g

## 4. Acknowledgement

The authors gratefully acknowledge the partial support of the Fonds der Chemischen Industrie and the Deutsche Forschungsgemeinschaft, Physical Chemistry of Interfaces - Graduate College.

## 5. References

1. Einicke, W.-D., Heuchel, M., v. Szombathely, M., Bräuer, P., Schöllner, R. and Rademacher, O. (1989) Liquid-Phase Adsorption of Binary Ethanol-Water Mixtures on NaZSM-5 Zeolites with Different Silicon/Aluminium Ratios, *J. Chem. Soc., Faraday Trans. I* **85**, 4277-4285.
2. Einicke, W.-D., Gläser, B., Lippert, R. and Heuchel, M. (1995) Adsorbed Phase Composition in Liquid-Phase Adsorption of Organic Compounds from Aqueous Solution on Hydrophobic Zeolites, *J. Chem. Soc., Faraday Trans.* **91**, 971-974.
3. Pope, C.G. (1993) Adsorption of Methanol and Related Molecules on Zeolite H-ZSM-5 and Silicalite, *J. Chem. Soc. Faraday Trans.* **89**, 1139-1141.
4. Messow, U., Quitzsch, K. and Herden, H. (1984) Heats of Immmersion of ZSM-5 Zeolite in n-Alkanes, 1-Alkenes and 1-Alcohols at 30° C, *Zeolites* **4**, 255-258.

5. Hunger, B., Heuchel, M., Matysik, S., Beck, K. and Einicke, W.-D. (1995) Adsorption of Water on ZSM-5 Zeolites, *Thermochim Acta* **269/270**, 599-611.
6. Hunger, B., Matysik, S., Heuchel, M. and Einicke, W.-D. (1996) Adsorption of Methanol on ZSM-5 Zeolites, *Thermochim. Acta*, submitted.
7. Hunger, B., v. Szombathely, M., Hoffmann, J. and Bräuer, P. (1995) Characterization of the Acid Properties of Zeolites by Means of Temperature-Programmed Desorption (TPD) of Ammonia - Calculation of Distribution Function of the Desorption Energy, *J. Thermal Anal.* **44**, 293-303.
8. v. Szombathely, M., Bräuer, P. and Jaroniec, M. (1992) The Solution of Adsorption Integral Equations by Means of the Regularization Method, *J. Comput. Chem.* **13**, 17-32.
9. Vigné-Maeder, F. and Auroux, A. (1990) Potential Maps of Methane, Water and Methanol in Silicalite, *J. Phys. Chem.*, **94**, 316-322.
10. Ohgushi, T. and Kawanabe Y. (1994) Properties of Na Ions in NaZSM-5 Zeolite, *Zeolites* **14**, 356-359.
11. Esemann, E., Förster, H., Geidel, E. and Krause, K. (1996) Exploring Cation Siting in Zeolite ZSM-5 by Infrared Spectroscopy, EXAFS and Computer Simulation, *Microp. Mater.*, in press.

# AN APPLICATION OF $^{129}$XE NMR OF ADSORBED XENON TO THE STUDY OF SILICA-BASED CATALYSTS

V.V. TERSKIKH, V.M. MASTIKHIN, L.G. OKKEL',
V.B. FENELONOV and K.I. ZAMARAEV
*Boreskov Institute of Catalysis*
*Novosibirsk 630 090, Russia*

H. HU and I.E. WACHS
*Zettlemoyer Center for Surface Studies, Departments of Chemistry and*
*Chemical Engineering, Lehigh University*
*Bethlehem, PA 18015, USA*

## 1. Introduction

$^{129}$Xe NMR of adsorbed xenon was found to be a useful tool for the characterization of microporous solids (zeolites, clathrates, molecular sieves) which have pore sizes close to the xenon free diameter of 4.4 Å [1-3]. Much less attention has been paid to the study of porous solids with the pore sizes much larger than the diameter of the xenon atom.

Recently, $^{129}$Xe NMR spectroscopy of adsorbed xenon was used to characterize mesoporous silicas [4-6]. A correlation between the $^{129}$Xe NMR chemical shift δ (ppm) and the pore diameter D (Å) in the range of 20-400 Å was obtained for silica samples with well-defined structure and narrow pore size distribution [6]:

$$\delta = 115/(1+D/131) \qquad (1)$$

In the present communication, $^{129}$Xe NMR is extended to the study of mesoporous silica-based catalysts.

## 2. Experimental

The silica chosen as the support material was Cabosil EH5. A toluene solution of titanium (IV) isopropoxide, a 1-propanol solution of zirconium (IV) propoxide, a methanol solution of aluminium acetylacetanate and a $CH_2Cl_2$ solution of tetrabutyl tin were used to prepare the corresponding samples by the incipient-wetness impregnation method. The preparations were performed inside a glove box to avoid preoxidation by atmospheric moisture. After impregnation, each catalyst was dried at room temperature with subsequent calcination at 500°C.

*J. Fraissard (ed.), Physical Adsorption: Experiment, Theory and Applications,* 571–574.
© 1997 *Kluwer Academic Publishers.*

The pore structures of the samples were characterized by nitrogen adsorption and capillary condensation (77K) on a Micromeritics ASAP 2400. The mesopore size distributions (Fig.1) were calculated from desorption branches of isotherms in the range of capillary-condensation hysteresis by the Broekhoff and de Boer method for cylindrical pores [7].

Xenon with a natural abundance of $^{129}$Xe isotope (26%) was used for the NMR experiments. The NMR spectra were recorded on a Bruker MSL-400 spectrometer at a frequency of 110.6 MHz. The pulse width was 10 μs with a 2 s delay, the number of scans was from 5000 to 10000. The chemical shifts were measured relative to gaseous xenon at a low pressure with an accuracy of ±1 ppm.

## 3. Results and discussion

The xenon adsorption isotherms for the parent silica and the supported metal oxide catalysts, measured up to 500 Torr of xenon pressure (293 K), were found to obey the Henry's law with nearly identical Henry constants of $(1.9\pm0.3)\cdot10^{17}$ atoms/Torr·g. Henry's type of isotherms is responsible for the independence of the $^{129}$Xe chemical shift upon xenon pressure for all the samples under study that is in agreement with [4, 6].

The $^{129}$Xe NMR spectra for the silica and the supported catalysts (293 K, 110-150 μmol/g of adsorbed xenon) are presented in Fig.2. It follows from these spectra that the supporting of the metal oxides affects the $^{129}$Xe NMR chemical shifts and line widths.

The $^{129}$Xe NMR spectrum of xenon adsorbed on $SiO_2$ shows along with a line at 0 ppm from gaseous xenon between silica grains also a line at 71 ppm due to an adsorbed xenon (Fig.2-a). According to Eq. (1), this value of shift corresponds to the mean pore diameter $D_{NMR}$=81 Å (Table 1). Two other diameters: ($D_{BET}$) - determined from the BET region of the nitrogen isotherm ($D_{BET}$=4$V_P$/$A_{BET}$), and ($D_{des}$) - determined from the desorption branch of the hysteresis loop are also collected in Table 1.

Supporting of 10% $TiO_2$ has a small effect upon both the porosimetry and $^{129}$Xe NMR data (Figs.1, 2-b). It can be concluded that the supported titanium oxide species essentially do not change the interior structure of silica, while the distribution of the high-dispersed titania over the silica surface is relatively uniform.

More pronounced changes are observed for the 10% $Al_2O_3$/$SiO_2$ catalyst. In this case, the $^{129}$Xe chemical shift decreases compared with that of silica (Fig.2-c). Porosimetry data indicate a decrease in the content of both type of pores with a simultaneous increase of both $D_{NMR}$ and $D_{des}$ (Table 1). This indicates the filling of the silica pores by large enough alumina species. $^{27}$Al MAS NMR spectrum of this sample also suggests the formation of large alumina particles with a regular structure that perturb the texture of the silica support.

For the 10% $SnO_2$/$SiO_2$ sample, the $^{129}$Xe NMR line slightly moves to low field (Fig.2-d). The surface area and the pore volume remain practically unchanged as compared with initial silica (Table 1). The slight growth of δ could be attributed to a small decrease of the mean pore diameter due to an increase in the percentage of small pores. This is confirmed by porosimetry data which indeed indicate an increase of the relative amount of pores with D ca. 60 Å (Fig.1-d). The $^{119}$Sn NMR spectrum of this catalyst shows that tin oxide forms disordered and probably two-dimensional surface species. However, supporting 25% $SnO_2$ on $SiO_2$ decreases, rather than increases, the

573

TABLE 1. Morphology of the MeO$_x$/SiO$_2$ catalysts as determined by nitrogen porosimetry and $^{129}$Xe NMR

| Sample | $A_{BET}$ (m$^2$/g) | $V_P$ (cm$^3$/g) | $D_{BET}$ (Å) | $D_{des}$ (Å) | $D_{NMR}$ (Å) |
|---|---|---|---|---|---|
| SiO$_2$ | 355 | 1.39 | 156 | 184 | 81 |
| 10% TiO$_2$/SiO$_2$ | 309 | 1.24 | 161 | 188 | 81 |
| 10% Al$_2$O$_3$/SiO$_2$ | 236 | 1.11 | 189 | 197 | 133 |
| 10% SnO$_2$/SiO$_2$ | 344 | 1.26 | 146 | 186 | 70 |
| 25% SnO$_2$/SiO$_2$ | 202 | 1.01 | 199 | 201 | 112 |
| 10% ZrO$_2$/SiO$_2$ | 332 | 1.15 | 139 | 176 | 57 |
| 30% ZrO$_2$/SiO$_2$ | 338 | 0.91 | 108 | 160 | 28 |

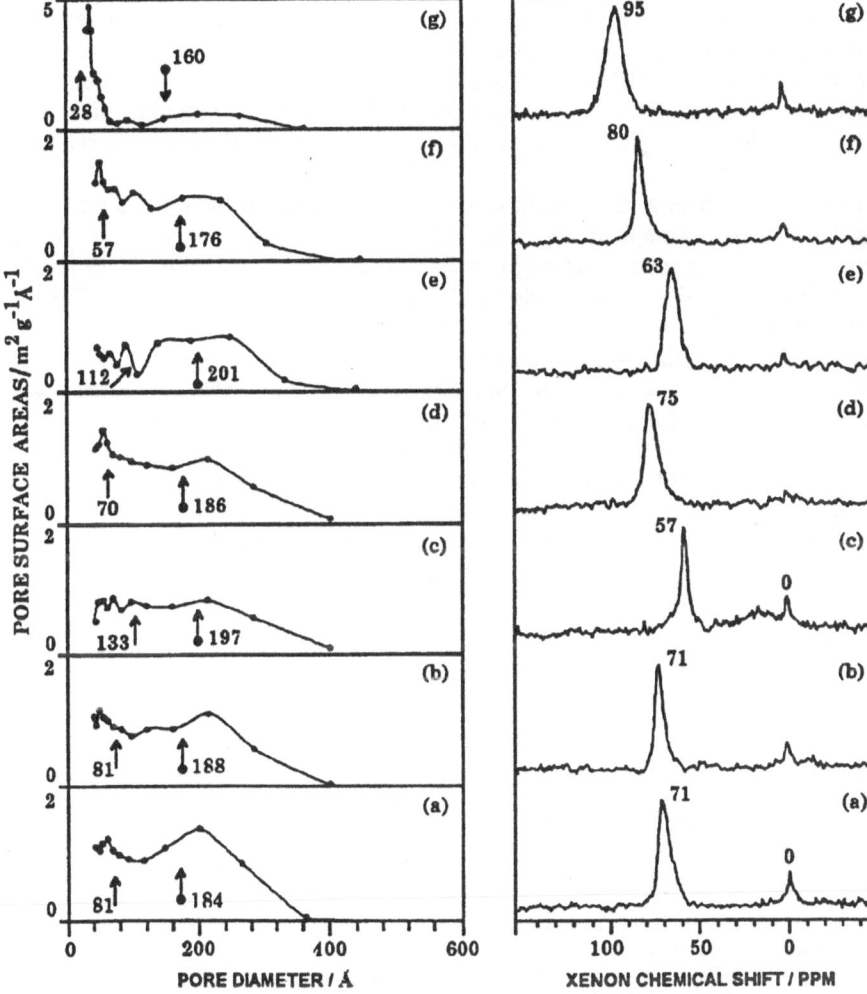

Figure 1. Desorption pore size distributions ($\uparrow$ - D$_{NMR}$, $\uparrow$ - D$_{des}$)

Figure 2. $^{129}$Xe NMR spectra of adsorbed xenon

(a) - SiO$_2$; (b) - 10% TiO$_2$/SiO$_2$, (c) - 10% Al$_2$O$_3$/SiO$_2$, (d) - 10% SnO$_2$/SiO$_2$, (e) - 25% SnO$_2$/SiO$_2$ (f) - 10% ZrO$_2$/SiO$_2$, (g) - 30% ZrO$_2$/SiO$_2$.

$^{129}$Xe NMR chemical shift (Fig.2-e). This indicates a decrease of the amount of small pores and an increase of the mean pore diameter, which is also supported by porosimetry data (Fig.1-e, Table 1). It can be concluded, therefore, that in this sample the oxide species occupy significantly the pores of silica due to the formation of large tin oxide particles. This conclusion also follows from $^{119}$Sn NMR measurements which show that regular oxide particles with a structure close to bulk $SnO_2$ are formed for the 25% $SnO_2/SiO_2$ sample.

For the 10% $ZrO_2/SiO_2$ catalyst, the $^{129}$Xe NMR measurements demonstrates an increased $\delta$-value compared with that of silica (Fig.2-f). At the same time, porosimetry data show a considerable increase in the content of pores in the 40-60 Å range (Fig.1-f). The enhancement of the $ZrO_2$ content up to 30% makes this effect more pronounced (Fig.1-g). The growth of the chemical shift up to 95 ppm is in excellent qualitative agreement with the increase of the amount of narrow pores and with the decrease of the mean pore diameter $D_{des}$=160 Å and $D_{BET}$=108 Å (Fig.1-g, Table 1). Thus, both $^{129}$Xe NMR and nitrogen porosimetry measurements suggest that the zirconia oxide species create a highly developed pore structure within the silica pores that has a large impact on $^{129}$Xe NMR spectra.

The use of $^{129}$Xe as a probe-atom provides information on the porous structure of the catalysts which complements that obtained with the conventional nitrogen adsorption and capillary condensation methods. The discrepancy between $D_{NMR}$ and both $D_{BET}$ and $D_{des}$ seems due to a broader and bimodal distribution of pores (Fig.1), in contrast with the silica samples studied earlier [6]. Despite the fact that Eq. (1) should be valid for samples with unimodal pore size distributions, the tendency in variation of the calculated pore diameter $D_{NMR}$ qualitatively agrees with the tendency in variation of the pore diameters found from nitrogen adsorption/desorption data (Table 1).

Recently, the structure of heteropoly acid $H_3PW_{12}O_{40}$ (HPA) supported on silica was studied by $^{129}$Xe NMR [8]. The NMR spectral parameters (number of lines, their widths and chemical shifts) were strongly dependent on the HPA loading and on the state of the surface HPA species. For these catalysts, the microporosity accessible for xenon penetration was found in accord with nitrogen porosimetry.

Acknowledgment: H. Hu and I.E. Wachs gratefully acknowledge financial support by the National Science Foundation (Grant CTS-9417981).

## 4. References

1. Fraissard, J. and Ito, T. (1988) $^{129}$Xe NMR study of adsorbed xenon: a new method for studying zeolites and metal-zeolites, *Zeolites* 8, 350-361.
2. Barrie, P.J. and Klinowski, J. (1992) $^{129}$Xe NMR as a probe for the study of microporous solids: a critical review, *Progress in NMR Spectroscopy* 24, 91-108.
3. Raftery, D. and Chmelka, B.F. (1994) Xenon NMR spectroscopy, *NMR Basic Principles and Progress* 30, 111-158.
4. Conner, W.C., Weist, E.L., Ito, T. and Fraissard, J. (1989) Characterization of the pore structure of agglomerated microspheres by $^{129}$Xe NMR spectroscopy, *J.Phys.Chem.* 93, 4138-4142.
5. Terskikh, V.V., Mudrakovskii, I.L., Mastikhin, V.M. and Simonova, L.G. (1993) $^{129}$Xe NMR: porous structure of silica gels, *React.Kinet.Catal.Lett.* 49, 13-20.
6. Terskikh, V.V., Mudrakovskii, I.L. and Mastikhin, V.M. (1993) $^{129}$Xe nuclear magnetic resonance studies of the porous structure of silica gels, *J.Chem.Soc. Faraday Trans.* 89, 4239-4243.
7. Broekhoff, I.C.P. and de Boer, I.H. (1967) Studies on pore systems in catalysis, *J.Catal.* 9, 8-27.
8. Terskikh, V.V., Mastikhin, V.M., Timofeeva, M.N., Okkel', L.G. and Fenelonov, V.B. (in press) $^{129}$Xe NMR study of 12-tungstophosphoric heteropoly acid supported on silica.

# STRUCTURAL CHARACTERISTICS AND ADSORPTION PROPERTIES OF MICROPOROUS ALUMINOBORATE OXYFLUORIDES

V.V. BREI, S.Ya. BRICHKA, A.A. CHUIKO
*Institute of Surface Chemistry*
*Prospekt Nauky 31, 252022 Kyiv, Ukraine*

New microporous aluminoborate oxyfluorides have been synthesized and described whose structural elements are octahedral aluminium(III) ions, trigonal and tetrahedral boron(III) ions. They differ substantially in their adsorptional properties. Their specific pore volume is 0.0085 and 0.065 cu cm/g for AlBF1 and AlBF2 respectively. The adsorption properties of the materials obtained are affected by fluorine ions.

## 1. Introduction

Crystalline microporous oxide systems are materials which find use in catalysis, adsorption, and ionic exchange [1]. In recent papers [2,3] new zeolitelike aluminoborate compounds are reported which have anion-exchange properties and good catalytic activities in the selective synthesis of 2-butene from ethanol. It should be noted that in nature there are aluminoborate minerals: sinhalite $MgAlBO_4$, eremeyevite $Al_6[BO_3]_5(OH)_3$. This paper deals with the development and study of a new class of microporous materials - aluminoborate oxyfluorides - and investigation of their adsorption properties.

## 2. Experimental

The hydrothermal synthesis of the above-mentioned substances was carried out using a reaction mixture with molar ratios of $1Al_2O_3$ - $3B_2O_3$ - $3NH_3$ - $24HF$ - $127H_2O$ and $1Al_2O_3$ - $2.6B_2O_3$ - $1.8CaO$ - $6.0HF$ - $280H_2O$ in teflon beakers in an autoclave with a volume of 25 cm$^3$ under autogenous pressure at 160 °C for 3 days. The products were isolated from the mother solution through centrifugation, washed up to the absence of halogen ions, and dried at a temperature of 100 °C. The samples under study were characterized by methods of X-ray diffraction (*DRON-1M*), IR spectroscopy (*Perkin-Elmer 325*), $^{27}Al$, $^{11}B$ MAS NMR spectroscopy

575

*J. Fraissard (ed.), Physical Adsorption: Experiment, Theory and Applications, 575–577.*

(*Bruker CXP-200*). Chemical shifts of $^{27}$Al and $^{11}$B nuclei were measured relative to $B(OCH_3)_3$ and $[Al(H_2O)_6]^{3+}$ Chemical analysis of the substances obtained was made by X-ray fluorescence spectroscopy (*Philips PW-1400*). Sorption isotherms of water vapor and hexane were recorded using a high-vacuum sorption plant at a temperature of 20 °C after activation of samples at 200 °C.

## 3. Results and Discussion

### 3.1. STRUCTURAL CHARACTERISTICS

As a result of the synthesis at various pH values of starting solutions the crystalline substances were obtained whose compositions were $1Al_2O_3$ - $0.5B_2O_3$-$6.0HF$-$0.5H_2O$ (AlBF1) and $1.0Al_2O_3$-$0.1B_2O_3$-$1.5CaO$-$5.8HF$- $-3.5H_2O$ (AlBF2). Our results of X-ray diffraction studies corroborate the conclusion that the substances synthesized are novel crystalline compounds. According to the interpretation of powder patterns all the spectral lines are correctly identified in the tetragonal crystal system with the unit cell parameters being a=6.365 A° and b=10.850 A° for AlBF1 and in the hexagonal crystal system with the unit cell parameters being a=3.337 A° and b=7.987 A° for AlBF2.

  In the $^{27}$Al NMR spectra there are signals at -38 ppm and -2 ppm with the half-width $\Delta v_{1/2}$ = 11 kHz for AlBF1 and $\Delta v_{1/2}$ = 14 kHz for AlBF2 which are assigned to octahedrally coordinated aluminium ions. The broadening of the peaks and their shift towards the strong field can be attributed to the decrease in the symmetry of the electron clouds of aluminium ions due to the substitution of fluorine atoms for a part of oxygen atoms in the $Al^{3+}$ coordination sphere. In the infrared absorption spectra for the materials obtained one can observe stretching vibrations bands of Al–O at 700-650 cm$^{-1}$ that are characteristic of aluminium ions in the octahedral environment [4].

  The strong stretching vibration band at 1490-1432 cm$^{-1}$ and the weak band at 1120-1080 cm$^{-1}$ for AlBF1 characterize vibrations of the $B^{3+}$ bonds in the trigonal and tetrahedral coordination. In NMR spectra on $^{11}$B nuclei one can observe a weak signal at 0 ppm of trigonally coordinated boron and an intensive peak at -19 ppm of tetrahedrally coordinated boron. The absorption band at 1182-1152 cm$^{-1}$ in IR spectra of AlBF2 samples gives evidence for the presence of only tetrahedral coordination of $B^{3+}$.

  As in the structure of oxyfluorides there are $Al^{3+}$ ions in the octahedral coordination and $B^{3+}$ ions in the tetrahedral coordination we have good cause to conclude that these substances have a layered structure similar to that of layered aluminosilicates.

## 3.2. ADSORPTION PROPERTIES

Adsorption isotherms of the hydrothermally synthesized oxyfluorides are similar to those of microporous adsorbents (fig. 1). At $p/p_s > 1.2$ the AlBF1 oxyfluoride does not exhibit any substantial increase of the adsorption of water as it is the case for the adsorption of $H_2O$ on silicalite [5]. This effect is related to the hydrophobic properties of the adsorbent under study. The content of physically sorbed water in the starting substance is very low, which is confirmed by the absence of any no-ticeable deformation vibrations of water in a region of 1600 $cm^{-1}$. It is also known that $F^-$ ions promote hydrophobization of the substances in question. Adsorption of hexane does not occur, therefore the kinetic diameter of micropores is smaller than 4.3 A°. The specific pore volume determined for water vapors is $8.5 \cdot 10^{-3}$ $cm^3/g$.

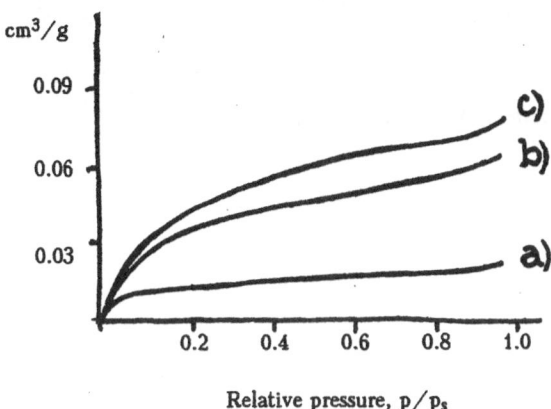

Figure 1. Adsorption isotherms of water vapor for AlBF1 (a), AlBF2 (b), and of hexane for AlBF2 (c).

In the case of AlBF2 the adsorption of water at $p/p_s < 0.2$ is also little, but under higher relative pressures it increases substantially, which is caused by the capillary condensation in the "secondary" pores of the adsorbent. The adsorption sites for water molecules are primarily Ca cations and accessible structural –OH groups. The adsorption capacity for hexane is $6.2 \cdot 10^{-2}$ $cm^3/g$ which is practically equal to that for water namely $6.5 \cdot 10^{-2}$ $cm^3/g$. Comparison of adsorption isotherms gives evidence for an approximately equal contributions of hydrophobic due to the presence of $F^-$ and hydrophilic components to the physical adsorption of molecules. The sizes of micropores are 4.3-5.2 A°. The much smaller adsorptivity of AlBF1 can be attributed to the differences in the composition and, as a consequence, in the crystalline structure of AlBF1 and AlBF2.

## 4. Summary

New microporous aluminoborate oxyfluorides have been synthesized and described whose structural elements are octahedral aluminium, trigonal and tetrahedral boron. The adsorption properties of the materials obtained are affected by fluorine ions.

## References

1. Barrer, R.M. (1982) *Hydrothermal Chemistry of Zeolites*, Academic Press, London.

2. Yu, J., Tu, K., Xu, R. (1994) Synthesis and characterization of a novel microporous boron-aluminium chloride with a cationic framework, *Studies in Surface Science and Catalysis*, **84A**, 315-322.

3. Yu, J., Xu, R., Kan, Q., and Xu, B. (1993) Synthesis and characterization of a novel boron-containing aluminum basic chloride, *J. Materials Chemistry* **3**, 77-82.

4. Nakamoto, K. (1986) *Infrared and Raman Spectra of Inorganic and Coordination Compounds*, Wiley, New York.

5. Shirmer, V. (1985) Properties of micropous adsorbents of the $SiO_2$ type and of highly dealuminized zeolites, in M.M. Dubinin and T.G. Plachenov (eds.), *Adsorbents: Production, Properties, and Applications*, Nauka, Leningrad, pp. 73-79.

# STUDY OF THE POROSITY OF KELEX-IMPREGNATED ORGANIC POLYMERS USING NITROGEN ADSORPTION-DESORPTION AT 77 K AND 129-Xe NMR SPECTROSCOPY.

J.L. BONARDET**, S.ESTEBAN*, AND G. COTE*
*Laboratoire de chimie analytique, ESPCI, 10 rue Vauquelin 75005, Paris, France.
**laboratoire de chimie des surfaces, UPMC case 196, 4 place Jussieu, Tour 55, 75252 Paris cedex 05 France.

## 1. Introduction

The aim of this study is to understand better the physical chemistry of impregnated solids used in hydrometallurgy to extract metallic ions from solutions. We have chosen to interest ourselves in the evolution of the porosity of organic solid polymers impregnated with Kelex 100. This one is a is well known extracting product in liquid-liquid extraction and for which many physicochemical data are available. Moreover, supports loaded with Kelex have an obvious economic interest, since they could be used to extract germanium or gallium ions from electrolytic bath. Such an operation would serve two objectives: recovery of an element of high added-value and also elimination of an inhibitor (Ge) of zinc electrolysis..

## 2. Experimental

### 2.1 NATURE OF THE PRODUCTS.

Two organic polymers were used to prepare the Kelex impregnated phase. Firstly, a hydrophobic copolymer of divinylbenzene and ethylvinylbenzene, (Amberlite XAD 1180), and secondly , a hydrophilic polymethacrylic ester (Amberlite XAD7). The polymers occur in the form of spherical particles of relatively homogeneous dimensions (0.3 <diameter <1 mm)

Kelex (Schering) is a viscous, oil containing about 90% of the chelating agent : 7-(4-ethyl,1-methyl,octyl)-8 hydroxyquinoleine.

Impregnation was performed by the so-called "dry" method: the polymer grains, previously washed and dried, are immersed in the impregnation solution obtained by dissolving Kelex 100 in n-heptane; the mixture is homogenised by stirring, then the

*J. Fraissard (ed.), Physical Adsorption: Experiment, Theory and Applications, 579–586.*
© 1997 Kluwer Academic Publishers.

solvent is slowly eliminated under vacuum at room temperature. The impregnated support is finally left in an oven at 100°C for 12 h.

## 2.2 EXPERIMENTAL TECHNIQUES

Before any adsorption, the samples are outgassed at 333 K for 12 h. The nitrogen adsorption-desorption isotherms at 77K are obtained by volumetric method on a home-made equipment, the accuracy of the measurements being about ± 5%.Xenon is adsorbed at 300 K, the temperature of the NMR probe; spectra are obtained on a Bruker MSL400 (xenon resonance frequency at 110.7 MHz). The chemical shifts are expressed in ppm relative to xenon gas at zero pressure.

## 3. Results and discussion.

### 3.1 NITROGEN ADSORPTION -DESORPTION ISOTHERMS AT 77 K.

Figures 1 and 2 show the nitrogen adsorption-desorption isotherms at 77K for samples XAD 7 and XAD 1180 with different Kelex. loadings. For samples without Kelex or lightly loaded (< 0.6g/g),theses isotherms present a hysteresis loop characteristic of mesopores. When the extent of impregnation is increasing (> 0.6 g/g) this loop disappears and the isotherms go from type IV to type II in the Brunauer classification. Analysis of the t-plots (figures 3 and 4) deduced from these isotherms allows to conclude that there are both micro and mesoporous regions in the clean and lightly loaded samples (<0.2 g/g) On the other hand, when the Kelex content is over 0.2 g/g the micropores are fully filled or blocked by the Kelex. The values of the constant C calculated from the BET isotherm in the range. $0.03 < P/Po < 0.3$, decrease when the degree of impregnation increases. They are very different for the free-Kelex polymers, 237 and 77 for XAD 7 and XAD 1180, respectively, but fall to 35 and 32 when the Kelex loading is over 1 g/g. This result shows that superficial fluid-solid interactions are much stronger for the hydrophilic support and also that Kelex impregnation reduces these interactions.

Figures 5 and 6 show the variation of the BET surface areas of the impregnated polymers (curves 1). As expected, the specific area decreases monotonously when the Kelex loading increases; above a loading of 1.5 g/g the impregnated polymers have a surface area near zero. Since the shape of the graphs related to the evolution of the micropore (obtained from the t-plots, curve 2) and mesopore areas (obtained from the difference: $S_{BET} - S_{micro}$, curve 3) is more instructive. Thus figure 5 for the hydrophilic support, XAD 7 indicates a maximum in the variation of the mesopore area for a Kelex content of 0.1 g/g; a such maximum does not exist for the hydrophobic polymer XAD 1180. We can explain these results in terms of two opposing phenomena: Kelex impregnation of the polymer blocks or fills a part of the micropores and mesopores

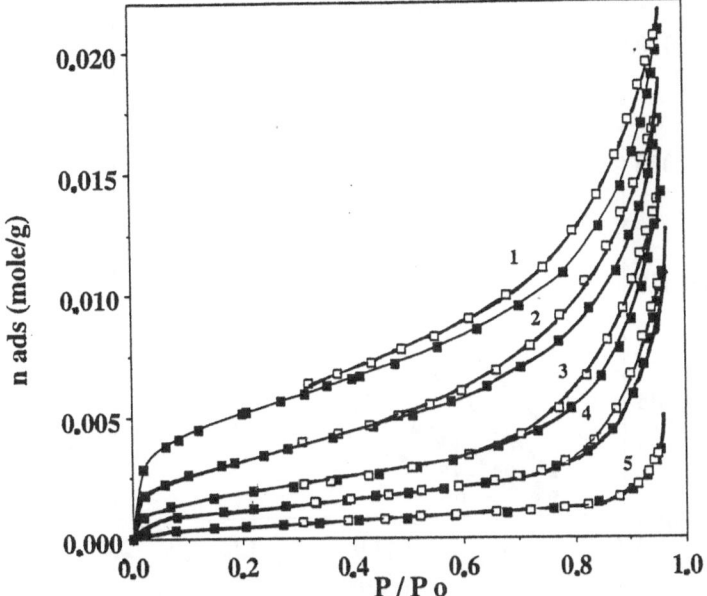

**Figure 1 :**   Ads.(full symbols)-des.(open symbols) isotherms of N2  at 77K
on XAD7 samples :
1: 0.07 g/g;  2: 0.18 g/g; 3: 0.40 g/g; 4: 0.58 g/g; 5: 1.03 g/g.

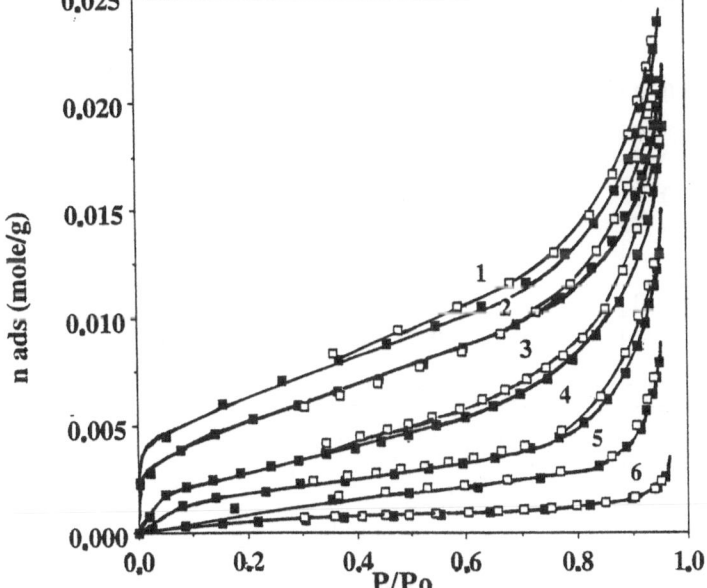

**Figure 2 :**   Ads.(full symbols)-des.(open symbols) isotherms of N2
at 77K on XAD1180 samples:
1:0.00g/g; 2:0.05g/g; 3:0.21g/g; 4:0.40g/g; 5:0.64g/g; 6:1.15g/g

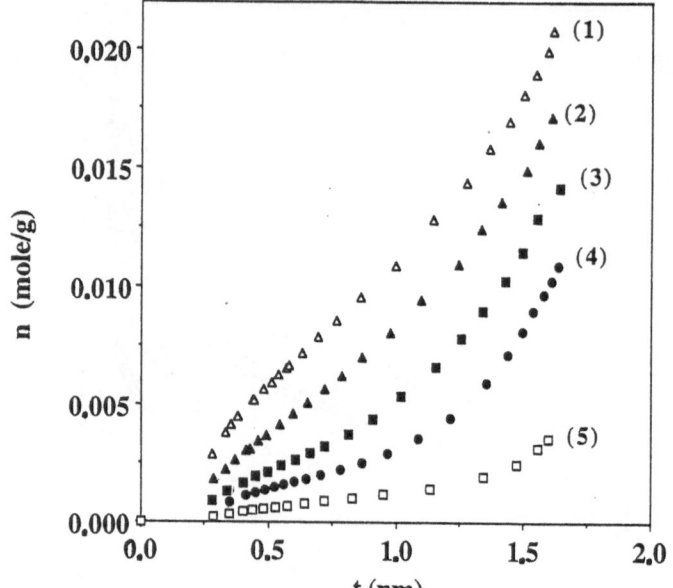

**Figure 3:** t-plots for XAD7 samples:
1:0.07g/g; 2:0.18g/g; 3:0.40g/g; 4:0.58g/g; 5:1.03g/g

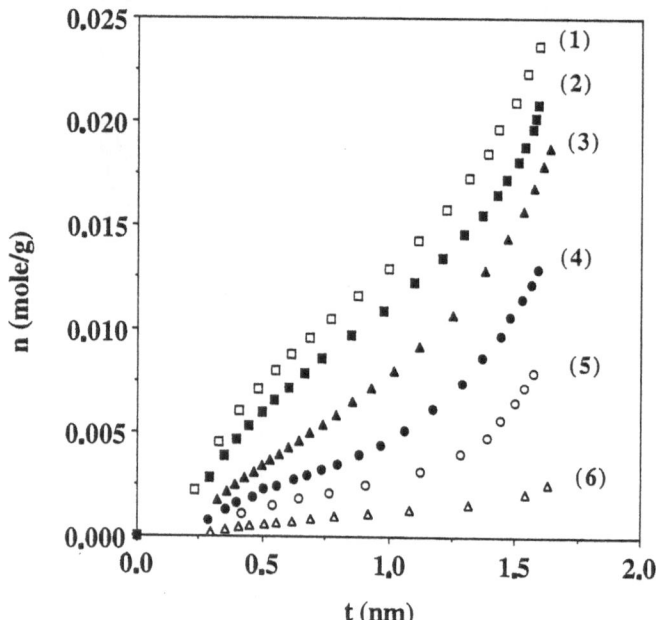

**Figure 4:** t-plots for XAD1180 samples:
1:0.00g/g; 2:0.06g/g; 3:0.21g/g; 4:0.40g/g; 5:0.64g/g; 6:1.15g/g

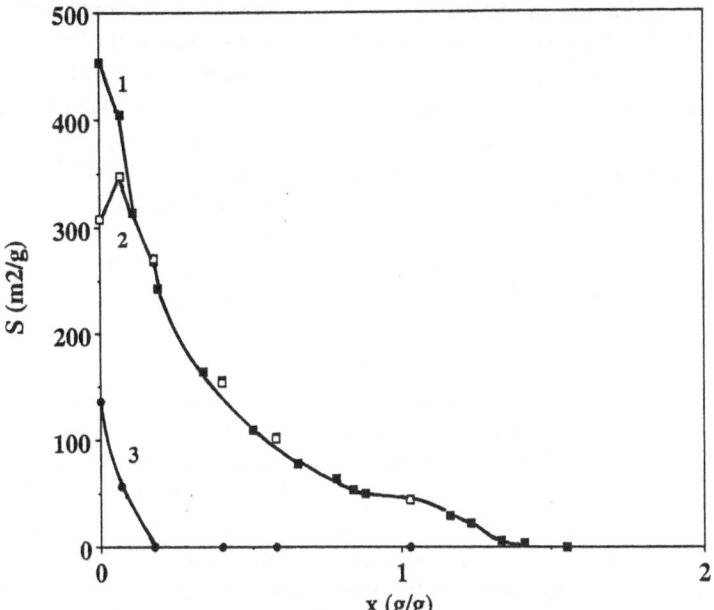

**Figure 5:**     BET surface area (1), mesoporous surface (2) and microporous surface (3) versus kelex loading for XAD7 samples.

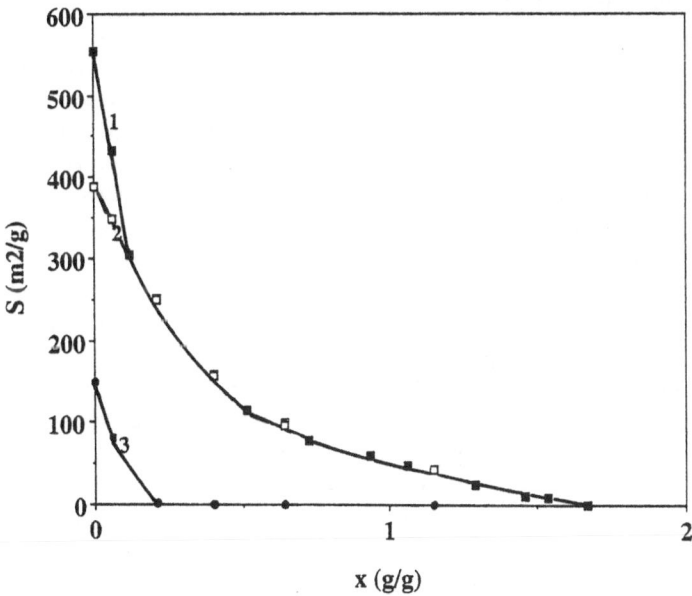

**Figure 6:**     BET surface area (1), mesoporous surface (2),and microporous surface (3) versus kelex loading for XAD1180 samples.

leading to the reduction of the surface area; but this same impregnation causes also swelling of the XAD 7 polymer lattice which is less rigid than that of the polymer XAD 1180, and consequently increases the area. Up to an impregnation of 0.1 g/g swelling is more important than blocking; beyond this, it is the opposite. This interpretation is confirmed by measurements on water adsorption by exposing the polymers to a saturated vapour phase at ambient temperature; this shows that XAD 7 swells by about 70% whereas XAD 1180 is not modified. Analyses of the pore distribution curves obtained by the BJH method (2) confirm this interpretation. The $dV/dr = f(r)$ curves (figure 7 et 8) show that they behave very differently as regards Kelex fixation. In the case of the hydrophobic polymer (XAD 1180), when the amount of impregnated phase on the support is increasing the pore distribution a decrease in the pore volume with however, a certain discrimination as regards the pore size : Kelex 100 appears to be distributed randomly over all the micro and mesopores but, the smaller pores are filled or blocked more quickly and disappear first. For the hydrophilic XAD7 polymer much the same conclusion can be drawn; however, comparing the $dV/dr = f(r)$ curves of the neat solid and the impregnated solids we can observe an increase of the number of mesopores in the range 8-20 nm that reveals expansion of the acrylic lattice upon penetration by Kelex. The internal structure of the grains is seriously disturbed, no doubt because of the flexibility of the macromolecular chains. This swelling of the lattice is thus confirmed.

## 3.2 129-Xe NMR RESULTS.

NMR spectroscopy of adsorbed xenon (129-Xe) is a very useful technique for studying certain physical properties of porous solids such as microporosity, structural defect, loss of crystallinity... Developed at the beginning of the 80s by Fraissard et Ito (3) it was first applied to zeolites, then extended to other porous materials, such as clays, amorphous gels, polymers and coals. This method is exposed in several reviews(4-6).

In our case, we observe that the chemical shift of adsorbed xenon is practically independent of the xenon concentration for a given Kelex content (table 1) and for the two type of polymers. However, the value of the chemical shift depends on the degree of impregnation and decreases when this increases. Beyond a loading of 0.20 g/g it is no longer possible to detect a signal, even at high xenon pressure (> 1 bar). These results show two things: firstly, the resonance signal corresponds to xenon adsorbed in the microporous region of the solid, since it is no longer detected when the microporosity has disappeared (loading> 0.2 g/g) and secondly, this microporosity is open and the adsorbed xenon exchanges rapidly with the xenon gas in the rest of the solid (chemical shift independent of the pressure). Same results were found by Conner et al (7) for compacted silica gels.

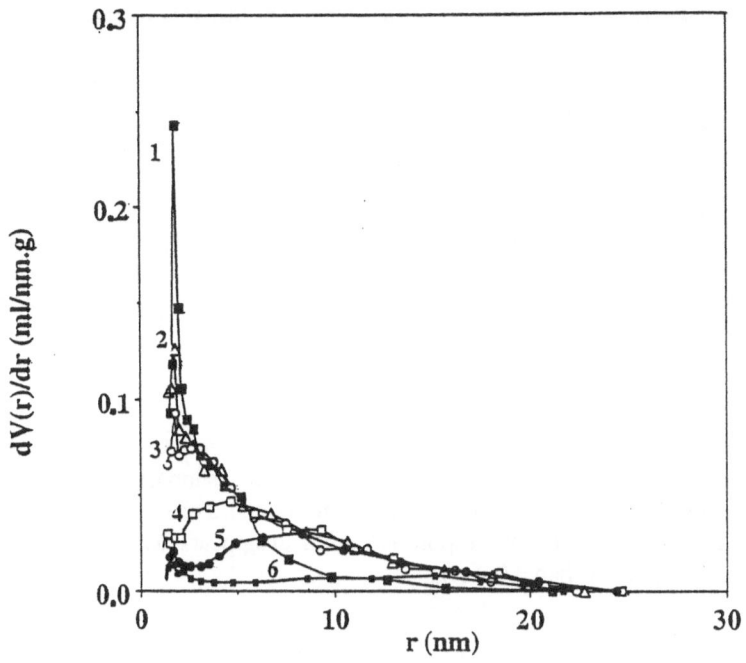

**Figure 7:** pore size distribution for XAD7 samples :
1:0.00 g/g, 2:0.07 g/g, 3:0.18 g/g, 4:040 g/g, 5:0.58 g/g, 6:1.03 g/g

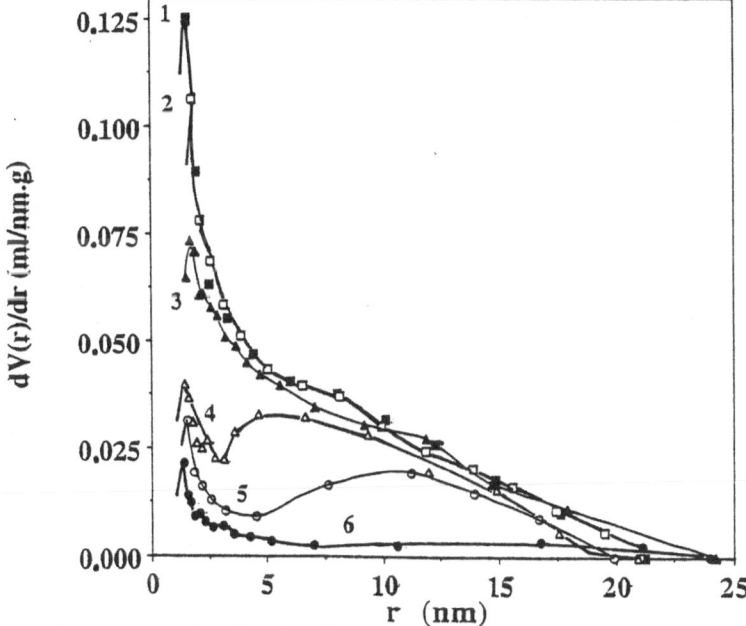

**Figure 8:** pore size distribution for XAD1180 samples:
1:0.00 g/g; 2:0.05 g/g; 3:0.21 g/g; 4:0.40 g/g; 5:0.64 g/g; 6:1.15 g/g

TABLE 1 : *Chemical shift in ppm, of adsorbed xenon versus xenon concentration.*

| [Xe]$10^{20}$atom/g | 1 | 1.5 | 1.7 | 2 | 2.3 | 2.5 | 2.8 | 3.2 |
|---|---|---|---|---|---|---|---|---|
| XAD7 0g/g | | 132ppm | | | 134ppm | | | 135ppm |
| XAD7 0.07g/g | | 119ppm | | | | 116ppm | | |
| XAD7 0.18g/g | | 106ppm | | | | | | |
| XAD1180 0.g/g | 130ppm | 129ppm | | | | | 129ppm | |
| XAD1180 0.05g/g | 125ppm | | 123ppm | | | 125ppm | | |
| XAD1180 0.21g/g | | | | 107ppm | | | | |

## 4. Conclusion

These results indicate that the two organic polymers impregnated with Kelex. 100 behave very differently. Impregnation of the hydrophilic support leads to swelling of the lattice, while the hydrophobic compound is unaffected. For both polymers Kelex is randomly impregnated and the microporous region disappears when the content of impregnated phase is greater than à 0.2 g/g.

## References

1. Warshawsky, A. (1981) *Ion exchanged and solvent extraction*, Marinski J.A. and Marcus Y.(Eds.), M.Dekker publishers, New York.
2. Barrett, E.P., Joyner, L.G.and Halenda. P.P., (1951) *J. Am. Chem. Soc.*, **73**, 373-380.
3. Ito T.and Fraissard.J.,(1982) *J. Chem. Phys.* **76** (11) 5225-5229.
4. Barrie P.J. and Klinowski J., (1992) *Prog.. NMR Spectr.*, **24**, 91-108.
5. Raftery D and. Chemkla B.F. (1994) *NMR Basic Principles and Progress*, **30**, 11-18.
6. Springuel-Huet M.A, Bonardet J.L.and Fraissard J.(1995) *Appl. Magn. Res.*, **8**, 427-456.
7. Conner W.C., Weist E.L., Ito T.and Fraissard J.(1989) *J. Phys. Chem.*, **93**, 4138-4142.

# RELATIONS BETWEEN ADSORBENT MORPHOLOGY AND MECHANISMS OF ADSORPTION IN SOLID-LIQUID SYSTEMS

K. JEŘÁBEK

*Institute of Chemical Process Fundamentals, Academy of Sciences of the Czech Republic, 165 02 Praha 6, Czech Republic*

Polymer adsorbents have a great potential in a number of separations, both on analytical and processes scale. Their properties can be controlled and it offer an opportunity to optimize their morphology according to specific requirements. For such endeavor it is necessary to have a knowledge of relations between adsorbent morphology and its separation properties. Important part of the adsorbents morphology properties is the specific area and pore size distribution. Influences of these factors has been more or less defined for adsorption in solid - gas systems, but the contemporary knowledge of mechanisms of influences of pore size distribution on adsorption in solid liquid systems is much less developed. In this contribution will be presented some examples of different effects of pore dimensions on adsorption in liquid phase.

For investigations of the influence of texture on adsorption are very suitable polymer adsorbents based on the copolymers of styrene and divinylbenzene. With identical surface properties, they are available in a wide selection of the pore dimensions. Recently, we have used a nearly identical series of such polymers in a few separate studies and the comparison of these results makes possible to demonstrate some less common features of the adsorption in solid - liquid systems. The series was composed of the commercial resins Amberlite XAD-2 and XAD-4 (Rohm and Haas, France) and microporous polymers A and B, prepared in our laboratory by hyper-crosslinking of gel (Polymer A) or macroreticular (Polymer B) starting chloromethylated materials using Friedel-Craft reaction [1-3]. In selected cases was included also the starting material for the polymer B and a custom-made polymer adsorbent K5. Texture properties of these adsorbents are shown in Table 1 and Figure 1.

The first study dealt with the separation of furfural from aqueous solutions [3]. Experimental data were correlated using a simple Langmuir-type equation (1),

$$A = k\, Ka\, cfurf./(1 + Ka\, cfurf.) \qquad (1)$$

which makes it possible to separate the shape-sensitive parameter $K_a$ and the scaling factor k (Table 2).

*J. Fraissard (ed.), Physical Adsorption: Experiment, Theory and Applications, 587–590.*

TABLE 1.Data on the dry-state morphology of the polymers A and B before and after the post-crosslinking procedure and the comercial polymer adsorbents Amberlite XAD-2 and XAD-4, determined from the results of nitrogen adsorption measurements.

| | Polymer A (gel-type) | Polymer B (macroreticular) | | Amberlite | |
|---|---|---|---|---|---|
| | Post-crosslinked | Starting | Post-crosslinked | XAD-2 | XAD-4 |
| Spec. surface BET, m$^2$/g | 940 | 139 | 720 | 340 | 804 |
| Volume of micropores (from t-plot), cm$^3$/g | 0.245 | 0.004 | 0.121 | 0.030 | 0.097 |

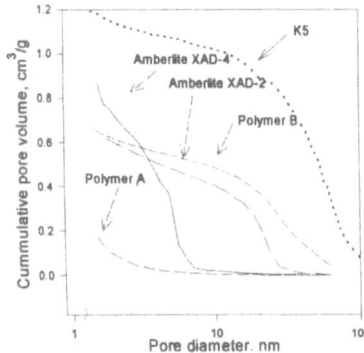

*Figure 1.* Pore size distribution of the polymer adsorbend used. Determined on the base of nitrogen adsorption/desorption isotherms.

TABLE 2.Values of parameters $K_a$ and k of the Eq. 1 for correlation curves on Fig. 2.

| Adsorbent | $K_a$ | k |
|---|---|---|
| Hyper-crosslinked polymer A | 3.44 | 0.48 |
| Starting polymer B | 0.50 | 0.41 |
| Hyper-crosslinked polymer B | 1.85 | 0.39 |
| Amberlite XAD-2 | 0.50 | 0.35 |
| Amberlite XAD-4 | 0.50 | 0.83 |

The sorption capacity of both Amberlites, as expressed by the scaling factor k, is proportional to their BET surface areas (Table 1). It suggests that at these polymers is prevailing the true adsorption of furfural at the polymer surface. Enhanced affinity of the hyper-crosslinked polymers toward the sorbate, as evidenced e.g. by the high values of the shape-characteristic constants $K_a$, is connected with the highly microporous morphology of these materials. Filling of the micropores is known in the gas-solid adsorption process as a factor strongly influencing the starting parts of adsorption isotherms and the existence of this type of interaction should be considered as a possible mechanism in adsorption from solution [4].

The same polymer adsorbents were used also in the study of relation between polymer support morphology and adsorption of di-(2-ethylhexyl)dithiophosphoric acid (DEHTPA) from alcoholic solutions [5]. There was found that DEHTPA in this case does not form an adsorbed layer but the sorption proceeds via a pore filling mechanism. With the increase of the equilibrium concentration in solution were gradually filled all pores with diameter up to about 7 nm. At the same time, the microporosity of the hyper-crosslinked polymers A and B was of no advantage, probably due to the relatively high effective molecular size of the sorbate. After correction for the difference in the molecular size of DEHTPA and nitrogen, the observed adsorption of DEHTPA was quantitatively described using the concept of gradual filling of the pore space with the sorbate only, without a need for corrections for either building of surface layer on the pore walls or modification of the pore structure by swelling due to sorbate absorption (see Figure 2).

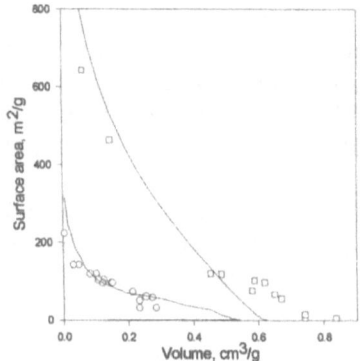

*Figure 2.* Dependence of the residual surface area on the volume of the pore system blocked by adsorption of DEHTPA. O, □ - experimental points for Amberlite XAD-2 and XAD-4, respectively. Solid line computed from the nitrogen pore volume distribution data with a correction for the difference in molecular sizes of DEHTPA and nitrogen (for details see ref.[5])

It seems that the adsorption of DEHTPA proceeds via creation of a separate phase filling the pore volume. The higher the DEHTPA concentration in solution, the wider pores are filled. The nature of the thermodynamic driving force behind the pore filling mechanism of the adsorption of DEHTPA from organic solvents is not obvious. It is probably connected with strong cohesion forces between DEHTPA molecules. However, it is less apparent how the formation of the separate DEHTPA phase is assisted by the pore walls, especially when the pore filling proceeds in pores even one order of magnitude wider than the diameter of the DEHTPA molecule. Elucidation of this problem will require further research.

The third study using the above described set of the polymer adsorbents was an investigation of the adsorption of various pesticides (atrazin, benomyl, chlorotoluron and hexazinon) from water at very low concentrations of these sorbates [6]. Surprisingly, the most effective were not the microporous adsorbents but the extremely wide-porous polymer K5. In the Figure 3 are shown the results for atrazin, for the other pesticides was the overall picture qute similar.

590

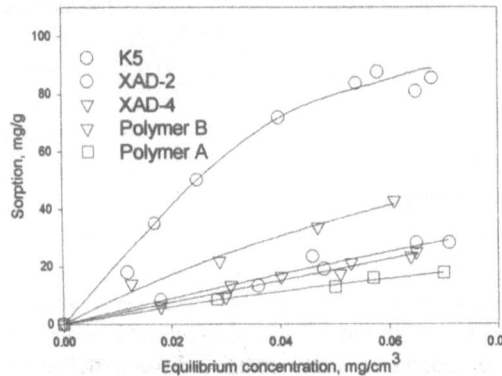

*Figure 3*. Adsorption isotherms of atrazine on various polymer adsorbents.

Currently, the only explanation is that the less efficient adsorption on the other, more narrow-pore adsorbents is a consequence of significant lowering of the sorbate concentration in the pore interior (decisive for the adsorption on the pore walls) due to the steric exclusion phenomena. However, control computations suggest that the explanation of the supremacy of the polymer K5 over the other polymer adsorbents due to this effect requires an effective sorbate size over 3 nm. It could be possible if the pesticides exist in the aqueous solutions not as isolated molecules but as micelles-like clusters. This will have to be proven by a further research.

The purpose of this contribution is not to present definitive explanations of the observed phenomena but to bring attention to some not yet widely known effects in the adsorption in solid-liquid systems and to provoke further development of theory in this domain.

**References**
1.   Davankov, V. A. and Tsyurupa, M. P. (1980) Macronet isoporous styrene polymers: unusual structure and properties, *Angew. Makromol. Chem.* **91**, 127.
2.   Fiestel, L., Schwachula, G., Reute,r H. and Klinkmann, H, (1987) Ger. (East) DD 249,274.
3.   Jeřábek, K., Hanková, L. and Prokop, Z. (1994) Post-crosslinked polymer adsorbents and their properties for separation of furfural from aqueous solutions, *React. Polym.* **23**, 107.
4.   Everet, D. H. (1986) Reporting data on adsorption from solution at the solid/solution interface, *Pure & Appl. Chem.* **58**, 967.
5.   Jeřábek K., Hanková L., Strikovsky A. G. and Warshawsky A. (1996) Solvent impregnated resins: Relation between impregnation process and polymer support morphology: I. di-(2-ethylhexyl)dithiophosphoric acid, *React. Funct. Polym* **28**, 201.
6.   Jeřábek, K., Hanková, L. and Prokop, Z. (1996) Influence of polymer adsorbent morphology on sorption of pesticides from water, *React. Funct. Polym.*, submitted for publication.

# MECHANISM OF INTERACTION OF ORGANIC COMPOUNDS WITH POLYMERIC ADSORBENTS

J. HRADIL

*Institute of Macromolecular Chemistry Academy of Sciences of the Czech Republic, Heyrovsky Sq. 2, 162 06 Praque 6.*

## 1. INTRODUCTION

The sorption and separation media based on inorganic materials and synthetic polymers, respectively, vary in several essential chromatographic properties such as column efficiency and selectivity[1]. The suspension technique is used to produce beads in both cases. Some of autors try to explain the differences by the presence of micropores in polymeric supports[2]. Let analyze the potential reasons for differences mentioned above.

## 2. DISCUSSION

### 2.1. MORPHOLOGY OF POROUS STRUCTURE OF POLYMERIC ADSORBENTS

Polymeric adsorbents are synthesized by two methods: (i) suspension techniques (polymerization or precipitation) and (ii) subsequent crosslinking of copolymers. The polymeric adsorbents are prepared mainly by the first method.

*Regular porous structure* formed by the interstitial volume between spherical particles - nodules. Exception is the bead cellulose, where the globular pores are formed. The spherical particles are formed due to the surface tension between continual phase of water and the discontinual organic phase, which contains monomer and inert diluent. The content

591

*J. Fraissard (ed.), Physical Adsorption: Experiment, Theory and Applications*, 591–597.

of the inert solvent determines the volume of pores. Inorganic adsorbents prepared by the suspension process have also the same geometry of pores. Due to the geometry of pore the model of cylindrical pores is valid only for porosities higher than 60%[3]. In comparison with silica adsorbents, macroporous copolymers are characterized by the bimodal distribution of pores. Such copolymers contain besides macropores (>50nm) and mesopores (from 2 to 50 nm) also an important part of micropores (< 2 nm), as follows also from inversion size exclusion chromatograph measurement[4], fluorescent markers and adsorption measurement. Micropores are formed not only in the solid state of polymer but also by swelling. Also gel-like polymers contain above mentioned three types of pores. In the silica gel supports micropores are syntred by heating to 900°C, as in polymeric adsorbents by heating above $T_g$.

Due to the *different copolymerization parameters* of the monomer and crosslinking agent the regular statistical copolymer cannot be obtained. Copolymerization parameters are defined as a ratio of velocity constant of both monomers. The crosslinking agent reacts usually with greater velocity and therefore the central area of nodules is preferably formed by the crosslinking agent. During further polymerization the nodules are growing and the surface layer are formed by the comonomer with lower degree of crosslinking. As an example can be mentioned copolymerization of styrene with divinylbenzene. The copolymerization parameter of divinylbenzene is larger than one of styrene and therefore at the beginning of polymerization copolymer enriched by divinylbenzene is formed. From that follows that strongly crosslinked nuclea are covered by the shell which is crosslinked to the lower degree. This inhomogenity was experimentally proved by using of fluorescence markers[5] and inversion size exclusion chromatography recently.

## 2.2. POLARITY OF POLYMERIC ADSORBENTS

*Polarity of the polymeric sorbent* can be changed in wide interval from relatively nonpolar styrene - divinylbenzene copolymers toward medium polar ethylene dimethacrylate copolymers to polar acrylonitrile copolymers. The polarity is proportional to the structural substitution parameters as e.g. Hammett substitution constants. As follows from our experience, each of synthetic polymers have a great part of hydrophobic interaction. All biopolymers, e.g. bead cellulose, are more polar than synthetic polymers.

Physical sorption is preferred due to the easy desorption and regeneration of the adsorbent. The commonly known types of interaction[6] adsorbate - adsorbent are formed also by using polymeric sorbents e.g. nonspecific interaction (van der Waals) and specific interaction (Keesom forces - orientation forces, Debye forces - induction forces). The London forces, which are formed due to the dipole moment between the positive charged nucleus and fluctuated electron shell, are proportional to $\alpha_l$ - $\alpha_s$. The Keesom forces are proportional to the interaction between permanent dipoles of the molecules and are equal $\mu_l$ - $\mu_s$. The Debye forces are formed when the dipole of one molecule inducted the dipole in the other. This inductions forces are equal $\mu_l$ - $\alpha_s$. All these forces are a long-range forces. In the one phase are as cohesion forces and on the surface of phase as surface tension.

Free energy of adsorption, on application of aditivity principle, and at the conditions when charge transfer complexes are not formed, is given by the following equation (1).

$$\Delta G_a = \Delta G_d + \Delta G_o + \Delta G_i + \Delta G_h \tag{1}$$

where $\Delta G_d$, $\Delta G_o$, $\Delta G_i$, and $\Delta G_h$ are contributions of dispersion, orientation, and induction forces and of H-bonds.

From this point the free adsorption energy[7] is given by the following equation (2).

$$\Delta G_a = \alpha_s \left[ \alpha_l \, 3 \, I_s \, I_l / 2(I_s + I_l) + \mu_l^2 \right] + \mu_s^2 (2\mu_l^2 / 3kT + \alpha_l) + \Delta G_h \tag{2}$$

where $\alpha_s$ is effective polarizability, $I_s$ ionization potential, and $\mu_s$ the dipole moment of sorbent. Adsorbates are indexed with l.

The free energy of adsorption, calculated from of retention of adsorbates ($V_R$), is given by the folloving equation (3).

$$\Delta G_a = \Delta H_a - T \, \Delta S_a = - RT \, (\ln V_R - 1) \tag{3}$$

The relation between GC quantities and the interaction parameters of solution theories is readily established[8],[9]. In statistical theories of solution thermodynamics, the solute activity is expressed as the sum of two terms, a combinatorinal entropy and a noncombinatorial free energy of mixing. In the Flory-Huggins approximation one has (4)

$$\chi = \ln(273.2R\nu_2/V_g p_1^0 V_1) - [1-V_1/(M_2)_n\nu] - (p_1^0/RT)(B_{11}-V_1) \qquad (4)$$

where $p_1^0$ and $V_1$ are the vapor pressure and molar volume of the solute, $B_{11}$ the second virial coefficient, $M_2$ molecular weight of the stationary phase, and $\nu$ the specific volume of the polymer.

For high molecular weight stationary phases, the second term of eq. becomes equal to -1. More recent polymer solution theories recognize the importance of the free volume dissimilarity of the solute and polymer. This effect, first introduced in theories by Prigodine and coworkers, has important thermodynamic consequences. Flory and collaborators now suggest that noncombinatorial part of the activity coefficient is composed of two terms, an equation of state and a contact interaction contribution. Due to the degree of crosslinking the dispersion forces are growing, but the orientation and induction forces, and also H-bonds are growing too.

During chemisorption, which is not mostly reversible, are formed H-bond, ionic bond, or covalent bond. The energy of two powers is necessary for desorption in comparison with physical adsorption.

## 2.3. SWELLING OF POLYMERIC ADSORBENTS

There is only one difference only between inorganic and carbonaceous adsorbents on the one side and polymeric adsorbents on the other side and it is *swelling of the polymeric matrix* and therefore absorption in the polymeric mass takes part[10]. What is the difference between adsorption and absorption in the polymeric adsorbents?

At low temperature and with adsorbate which didn't swell the polymeric matrix, the adsorption interaction takes part only. It is the two dimensional interaction on the surface, which is stericaly hindered and therefore an optimal configuration of complex is not formed. It is commonly known that the interaction is stronger on edges(contact of two planes) and apexes (peaks), where the interactions may not be exactly two dimensional, but cannot be three dimensional (spacious). Such places are known as active places of the adsorbent.

The polymeric adsorbents, which are prepared by the suspension process, due to the *irregular distribution of the crosslinking agent in the polymers mass*, the low degree of crosslinking agent is on the inner surface of polymer, which can be easily swelled by organic adsorbates with similar chemical structure. Then the absorption mechanism takes part. The polymeric chains are strengthen by swelling, the organic compound is oriented around the chain in the space three dimensional steric position, which is optimal for forming of all types of possible interactions as described above. It must be mentioned that also absorption in the polymeric mass is started by adsorption of the molecule on the polymeric surface. During absorption the pore volume of the copolymer rises. In the macroporous copolymers, where the high content of crosslinking agent is used, the swelling of the polymeric chains is strongly restricted, therefore the changes in volume due to the swelling are low. ($< 5\%$). But it was also observed, that swelling takes place in pores in the form of microgel in such case[11],[12].

The absorption in the polymeric mass of adsorbent was also observed at the temperature above the *glass temperature* $(T_g)$ of the polymer[13], when the morphology of the polymer from oriented crystallites to amorphous phase is changed. It was observed from the retention diagram of organic compounds in wide temperature interval. The discontinuity of the linearity at $T_g$ is formed[14],[15]. Such information gives us the inversion gas chromatography method. But not all of organic compounds cannot swell the investigated polymeric matrix, therefore not all of organic compounds gives us the retention diagram with described discontinuity on the linear dependence.

The changes due to swelling were observed in post-crosslinked adsorbents, but only with compounds which interact with the polymeric matrix (1,2-dichloroethane, ethylalcohol, 2.5 ml/g). The enhanced swelling degree is given by the presence of a part of intramolecular crosslinks and is proportional to the difference between interaction parameters of polymer and solvent ($\delta_s$ - $\delta_l$).

What it is the disadvantage in one case can be an advantage in another. The swelling of polymers is fully exploatated in the superabsorbents. The superabsorbents are copolymers crosslinked to the low degree in which only absorption takes place. Superabsorbents which are working in the water phase are based usually on copolymers of acrylic acid. The superabsorbents which should working in organic phase or/and with enhanced selectivity are under investigation.

## 3. CONCLUSION

The difference between solid phase adsorbents (inorganic and carbonaceous) and polymeric and therefore also in mechanisms of sorption are the following.
- Chemical composition and on composition based polarity and copolymerization parameters
- Irregular (nonstatistic) crosslinking which forms anisotropic microparticles with lower degree of crosslinking in the surface shell.
- Swelling of the polymeric bulk and therefore two mechanisms of interaction are posible:
adsorption and absorption. Swelling rises due to the lower degree of crosslinking and at elevated temperature when transition of crystalline phase to amorphous takes place.
The suspension polymerization technique used to produce polymer beads, the formation of porous structure and the morfology are these factors which contribute substantially to the difference between organic and inorganic separation media. The effect of unequal monomer reactivity rations, swelling of the beads, pore size distribution, and the presence of micropores also takes part.

## 4. REFERENCES

1. Švec, F. (1996) Why is the efficiency and selectivity of Polymeric Stationary Phases for HPLC different from those based on silica, *Chem.Listy* **90**, 103-108.
2. Hosoya, K., Kageyma, Y., Yoshizako, K., Kimata, K., Araki, T. and Tanaka, N. (1995) Uniform-sized polymer-based separation media prepared using vinyl methacrylate as a cross-linking agent. Possible powerful adsorbent for solid phase extraction of halogenated organic solvents in an aqueous environment. *J.Chromatogr.* **711**, 247.
3. Hradil, J. (1976) Contribution to study of submicroscopic structure of macroporous copolymers. Calculation of size of submicroscopic particles and pores from values of specific pore volumes and specific surface area, *Angew.Makromol.Chem.* **66**, 51-66.
4. Hradil, J., Horák, D., Pelzbauer, Z., Votavová, E., Švec, F. and Kálal, J. (1983) Investigation of the porous structure of polymers by chromatographic methods V. Use of gel permeation chromatography in the study of the porous structure of glycidyl methacrylate copolymers, *J.Chromatogr.* **259**, 269-282.

5. Jeřábek, K., Shea, K.J., Sasaki, D.Y. and Stoddard, G.J. (1992) Accessibility of the gel phase in macroporous network polymers: A comparison of the fluorescence probe and inverse size exclusion chromatography techniques. *J. Polym. Sci. Polym. Chem.* **30**, 605-611.

6. Kiselev, A.V. (1967) Problems of the chemistry of surfaces and molecular theory of adsorption, *Zh. Fiz. Khim.* **41**, 2470.

7. Larionov, O.G., Petrenko, V.V., Platonova, N.P., Hradil, J., Švec, F. and Maroušek, V. (1991) Investigation of polymeric adsorbents surfaces by inversion gaz chromatography method. *Zh. Fiz. Khim.* **65**, 1671-1674.

8. Braun, J.M. and Guillet, J.E. (1976) Study of polymers by inverse gas chromatography, *Adv. Polym. Sci.* **21**, 108-145.

9. Gray, D.G. (1979) Gas Chromatographic Measurements of Polymer Structure and Interactions, *Progress in Polymer Science*, **5**, 1-60.

10. Errede, L.A. (1991) Molecular interpretation of sorption in polymers Part I, *Adv. Polym. Sci.* **99**, 1-93.

11. Jeřábek, K., Setínek, K., Hradil, J., and Švec, F. (1987) An investigation of the morphology of glycidyl methacrylate copolymers using inverse size-exclusion chromatography, *Reactive Polym.*, **5**, 151-156.

12. Jeřábek, K. (1985) Determination of pore volume distribution from size exclusion chromatography data. *Anal Chem.* **57**, 1595-1598.

13. Hradil, J. and Švec, F. (1984) Investigation of the surface structure of polymers by chromatographic methods VI. Determination of glass transition temperature of macroporous copolymers by gas chromatography, *J. Chromatogr.* **287**, 67-76

14. Hradil, J. and Švec, F. (1985) Reactive polymers 49. Changes in the porous structure of macroporous copolymers due to succesive effects of solvents and temperature, *Angew. Makromol. Chem.* **130**, 81-90.

15. Hradil, J. and Švec, F. (1985) Reactive polymers 51. The temperature behaviour of macroporous methacrylate sorbents, *Angew. Makromol. Chem.* **135**, 85-97.

# WETTING ON AN ATTRACTIVE SPHERICAL SUBSTRATE

## IOANNIS A. HADJIAGAPIOU
*Solid State Physics Section, Dept. of Physics, University of Athens,*
*Panepistimiopolis, Zografos GR-157 84, Athens, Greece.*

## 1. Theory

The wetting of a spherical substrate, represented by the external potential $V_{ext}(\mathbf{r})$, immersed in a one-component bulk vapour (whose particle's have diameter d, number density $\rho_V$ and critical temperature $T_C$) is studied, as a function of the substrate's radius R and parameter E, using density functional theory. The grand-potential functional for such a system at temperature T is [1],

$$\Omega_V[\rho(\mathbf{r})] = \int_V \left\{ f_h[\rho(\mathbf{r})] + \tfrac{1}{2}\rho(\mathbf{r})\int_V \rho(\mathbf{r}')w(|\mathbf{r}-\mathbf{r}'|)d\mathbf{r}' + \left(V_{ext}(\mathbf{r}) - \mu\right)\rho(\mathbf{r}) \right\}d\mathbf{r} \tag{1}$$

where $\rho(\mathbf{r})$ is the average number density at point $\mathbf{r}$, $\mu$ the bulk chemical potential and V the volume of the system. The repulsive force contribution is treated in the local density approximation, $f_h[\rho(\mathbf{r})]$ is the Helmholtz free energy density of a uniform hard-sphere fluid at density $\rho(\mathbf{r})$, while the long-range attractive forces are treated in mean field approximation, $w(\mathbf{r})$ is the attractive part of the pairwise potential between two fluid molecules. The equilibrium density profile results as a solution to the equation,

$$\mu = V_{ext}(\mathbf{r}) + \mu_h[\rho(\mathbf{r})] + \int_V \rho(\mathbf{r}')w(|\mathbf{r}-\mathbf{r}'|)d\mathbf{r}' \tag{2}$$

where $\mu_h[\rho(\mathbf{r})] = \partial f_h[\rho(\mathbf{r})]/\partial \rho(\mathbf{r})$ is the hard-sphere chemical potential. Following Sullivan and considering the spherical symmetry of the system $\rho(\mathbf{r})=\rho(r)$,

$$w(r) = -\left(\alpha\lambda_{FF}^3/4\pi\right)e^{-\lambda_{FF}r}/\lambda_{FF}r \tag{3}$$

where $\lambda_{FF}$ is the inverse range length, $\alpha = -\int_V w(r)d\mathbf{r}$ and

$$V_{ext}(r) = -E\frac{e^{-\lambda_{WF}r}}{\lambda_{WF}r}\left[\lambda_{WF}R\cosh\lambda_{WF}R - \sinh\lambda_{WF}R\right] \tag{4}$$

For the numerical calculations it was chosen $\lambda_{FF} = \lambda_{WF} \equiv \lambda$ and $\lambda d=1$. All the quantities are transformed to dimensionless units,

599

J. Fraissard (ed.), Physical Adsorption: Experiment, Theory and Applications, 599–603.

$$\mu^* = \beta\mu, \ p^* = \beta d^3 p, \ T^* = T/T_c, \ \gamma^* = \beta d^3\gamma, \ R^* = \lambda R, \ r^* = \lambda r, \ \rho^* = \rho d^3$$

$$\alpha^* = \beta\alpha/d^3 = 11.102/T^*, \ V^*(r) = \beta V(r), \ \Gamma^* = \pi d^2\Gamma/6 \qquad (5)$$

although the asterisks will be suppressed; $\gamma$ is the surface tension, p pressure, $\beta = \left(k_B T\right)^{-1}$, $k_B$ Boltzmann's constant and $\Gamma$ adsorption given by

$$\Gamma = \int_R^\infty \frac{r^2}{R^2}\left[\rho(r) - \rho_v\right] dr . \qquad (6)$$

Substituting the potentials (3, 4) into (2) and differentiating it twice with respect to r, it yields [2,3]

$$\mu_h''(r) = -\frac{2}{r}\mu_h'(r) + \mu_h(r) - \mu - \alpha\rho(r) \qquad (7a)$$

the prime denotes derivative with respect to r. Its solution is uniquely defined if supplemented by proper boundary conditions, which are

$$\mu_h(r) \to \mu_{h,v}, \ \mu_h'(r) \to 0 \qquad \text{as } r \to \infty \qquad (7b)$$

$$\mu_h'(R) = \left[\mu_h(R) - \mu - E\right]\left(\coth R - \frac{1}{R}\right). \qquad (7c)$$

For the calculations the Carnahan-Starling approximation is used,

$$\beta p_h(r) = \rho\left[\left(1 + \eta + \eta^2 - \eta^3\right)/(1-\eta)^3\right]$$

$$\beta\mu_h(r) = \ln(\eta) + \left[\left(8\eta - 9\eta^2 + 3\eta^3\right)/(1-\eta)^3\right] \qquad (8)$$

where $\eta = \pi\rho d^3/6$ the packing fraction.

Another important quantity is the pressure tensor $\mathbf{p(r)}$ which is anisotropic in the interfacial region and the mechanical equilibrium of the fluid under the external potential $V_{ext}(r)$ is $\nabla \cdot \mathbf{p(r)} = -\rho(r)\nabla V_{ext}(r)$. On symmetry grounds, $\mathbf{p(r)}$ depends only on r and is decomposed into the normal $p_N(r)$ and tangential $p_T(r)$ components, so the previous equation can be written

$$p_h'(r) = \frac{2}{r}\left[p_T(r) - p_N(r)\right] - \rho(r)V_{ext}'(r). \qquad (9)$$

Integrating (9) it yields the virial expression for the pressure difference $\Delta p$,

$$\Delta p = \int_R^\infty \left\{\frac{2}{r}\left[p_N(r) - p_T(r)\right] + \rho(r)V_{ext}'(r)\right\} dr , \qquad (10a)$$

and, once $p_T(r)$ is known, the $p_N(r)$-component results

$$p_N(r) = \frac{Q^2}{r^2}p_N(Q) + \frac{1}{r^2}\int_Q^r \left[2u p_T(u) - u^2\rho(u)V_{ext}'(u)\right] du , \qquad (10b)$$

Q is the radial distance where the density $\rho(r)$ attains its bulk value $\rho_V$. The $p_T(r)$-component is minus the grand potential free energy density,

$$p_T(r) = p_h(r) + \tfrac{1}{2}\rho(r)\left[\mu - \mu_h(r) - V_{ext}(r)\right]. \tag{11}$$

The adsorbed-phase is separated from the bulk-phase by a dividing surface where the surface tension acts; as such is the mechanical surface of tension (with radius $R_S$ and tension $\gamma_S$) and the equimolar (with radius $R_e$ and tension $\gamma_e$). Following a similar route as in [4], it is found

$$R_S^2 = \int_R^\infty r^2 \rho(r) V'_{ext}(r)\,dr \Big/ \int_R^\infty \rho(r) V'_{ext}(r)\,dr, \tag{12a}$$

$$R_e^3 = \frac{1}{\eta_V - \eta_L}\left[\int_R^\infty r^3 \frac{d\eta}{dr}\,dr + R^3\left(\eta(R) - \eta_L\right)\right], \tag{12b}$$

$$\gamma_S = \int_R^\infty \left(\frac{R_S}{r}\right)\left[p_N(r) - p_T(r)\right]dr, \qquad \frac{\gamma_e}{\gamma_S} = \frac{2R_e}{3R_S} + \frac{R_S^2}{3R_e^2}. \tag{12c}$$

## 2. Results. Discussion.

The equilibrium density profile results as solution to the boundary value problem (7) for T=0.8. For a specific R, it is obvious from Fig. 1 that as E increases the only possible transition is either from thin to thick film (in the case of partial wetting) or from non-wetting to partial wetting by a thin film; consequently, adsorption $\Gamma$ is negative or positive and increases with E without diverging. The system is either in wetting Class II or III [5].

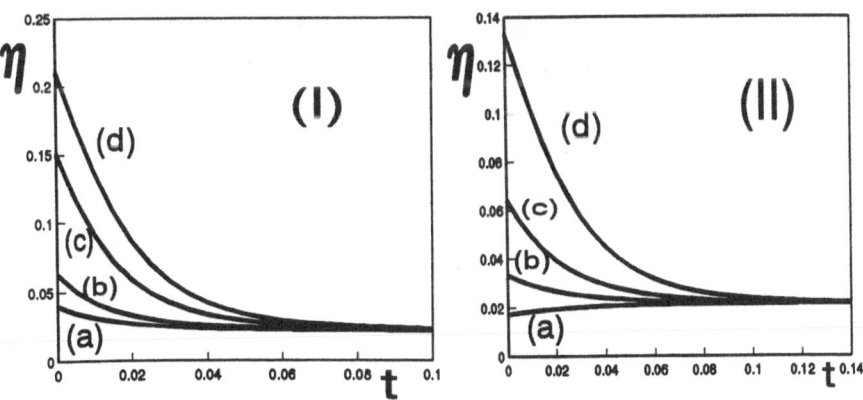

**Figure 1.** Density profiles, labelled by E, for: (I) R=3 (a)3, (b)5, (c)10, (d)13 (II)R=33 (a)0.1, (b)1.5, (c)3, (d)5. t=(r-R)/Q.

Both components of the principal tensor are decreasing in the interfacial region and $p_N(r) > p_T(r)$, interface under tension, in contradistinction to a spherical drop where the interface can also be under compression, $p_N(r) < p_T(r)$, and $p_T(r)$ acquires a deep lobe [2,3]. Both components tend to $p_V$ (the bulk-vapour pressure) as $r \to \infty$.

$R_S$ is, initially, large because, for small E's, $V_{ext}(r)$ is not so strong and the adsorbed phase is more diffused, but on increasing E further $V_{ext}(r)$ becomes stronger and the adsorbed phase is now more localised resulting in a decrease of $R_S$. As the number of adsorbed particles increases with E the radius $R_e$ will also increase to accomodate the larger number of particles (Fig. 2). The associated surface tensions, $\gamma_S$ and $\gamma_e$, on the contrary, do not differ considerably, especially for larger R's (Fig. 3).; this behaviour was also determined in the case of spherical drops [2,3]. Both radii and surface tensions are strongly substrate dependent.

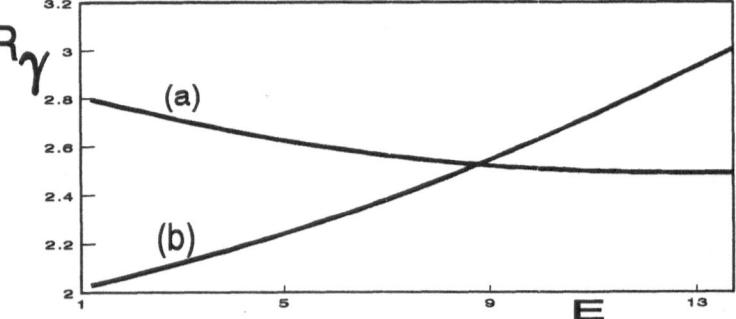

**Figure 2.** The radius of the surface of tension (a)
and the equimolar radius (b), for R=3.

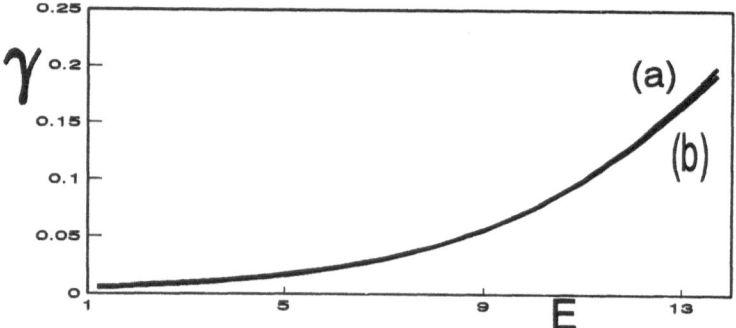

**Figure 3.** The surface tension: equimolar (a)
and mechanical (b), for R=3.

**References.**
[1] Hadjiagapiou, I. and Evans, R. (1985) Adsorption from a binary fluid mixture, *Mol. Phys.* **54**, 383-406.
[2] Hadjiagapiou, I. (1994) Density functional theory for spherical drops, *J. Phys.: Condens. Matter* **6**, 5303-5322.
[3] Hadjiagapiou, I. (1995) Density functional theory for spherical drops: II. Variable-range attractive forces, *J. Phys.: Condens. Matter* **7**, 547-562.
[4] Rowlinson, J. S. and Widom, B. (1982) *Molecular Theory of Capillarity*, Clarendon Press, Oxford, pp. 109-114.
[5] Sullivan, D. E. (1979) Van der Waals model of adsorption, *Phys. Rev. B* **20**, 3991-4000.

# ADSORPTION, DESORPTION AND INDOOR CLIMATE

LIS MARCUSSEN
*Department of Chemical Engineering, Technical University of Denmark, Building 229, DK-2800 Lyngby, Denmark*

ULLA D. KJAER
*Ph.D. student at Department of Chemical Engineering, Technical University of Denmark and Danish Building Research Institute*

AND

PETER A. NIELSEN
*Danish Building Research Institute*
*Postboks 119, DK-2970 Hørsholm, Denmark*

## 1. Introduction

Since the early seventies, there has been an increasing number of complaints from people situated in certain buildings. Typically, the complaints are about symptoms such as fatigue, headache, lack of concentration, mucous membrane irritation in eyes and respiratory passages as well as cutaneous irritation. In addition to human suffering, these symptoms represent an economical burden for the society.
VOCs (Volatile Organic Compounds), present in the indoor air and adsorbed on solid surfaces, are suspected to contribute significantly to the symptoms mentioned above, by respiration of air and dust and by direct contact with the skin. Field studies indicate that the problem is not eliminated by increased ventilation.

## 2. VOC sources

Organic compounds in the indoor air can originate from the outdoor air or from different indoor sources:

*J. Fraissard (ed.), Physical Adsorption: Experiment, Theory and Applications, 605–608.*
© *1997 Kluwer Academic Publishers.*

*Primary emission,* i.e. outgassing from materials in the building (walls, floors, furniture, carpets, repair materials, paint, sealants). The primary emission is a problem which will fade away during the life of the materials.

*Secondary emission* includes emission from indoor surfaces by other mechanisms, and unlike the primary emission, it is a problem which usually does not decrease as time goes by. Secondary emission can be caused by alternating adsorption/desorption processes.

*Emission from activities* in the building (working processes, housekeeping, combustion, hobbies).

## 3. Adsorption/desorption in building materials

Adsorption/desorption processes are expected to contribute to the indoor climate through their influence on the dynamic dispersal and removal of VOCs.

Consequently, it is important, not only to minimize the primary emissions, but also to develop methods for the prediction and control of secondary emissions which are closely related to the sorption properties of building materials and other indoor surfaces.

A research project concerning sorption processes on indoor materials was initiated in 1995; it deals with these theoretical and experimental investigations:

- Lab size experiments for investigation of sorption mechanisms and dynamics for carefully selected building materials and VOCs.
- Full scale experiments for the same materials and VOCs.
- Development of mathematical models for the sorption processes and their influence on VOC concentrations in the indoor air.
- Development of a standard test method for characterization of the sorption properties of indoor materials.

## 4. Adsorption/desorption on dust

Dust can be found almost everywhere. It is a heterogeneous substance consisting of many different organic and inorganic components with different physical and chemical properties. The lighter components are readily respirable, while heavier components tend to agglomerate and serve as an undisturbed habitat for various microorganisms.

The adsorption/desorption studies on dust [1] contribute to a joint venture project between several Danish research institutes (in the fields of building materials, occupational health, medicine, pharmacology, allergology).

1. Pressurized clean air (80% $N_2$, 20% $O_2$)
2. Reduction valve
3. 0.5 $\mu$ filter
4. Needle valve
5. Capillary tube
6. Humidifier
7. Removal of droplets
8. Non-return valve
9. Pollutant evaporator
10. Needle valve
11. Mass flow controller
12. Electronic vacuum balance
13. Mass flow measurement
14. Pump
15. Thermostat
16. Dust sample
17. To IR analysis

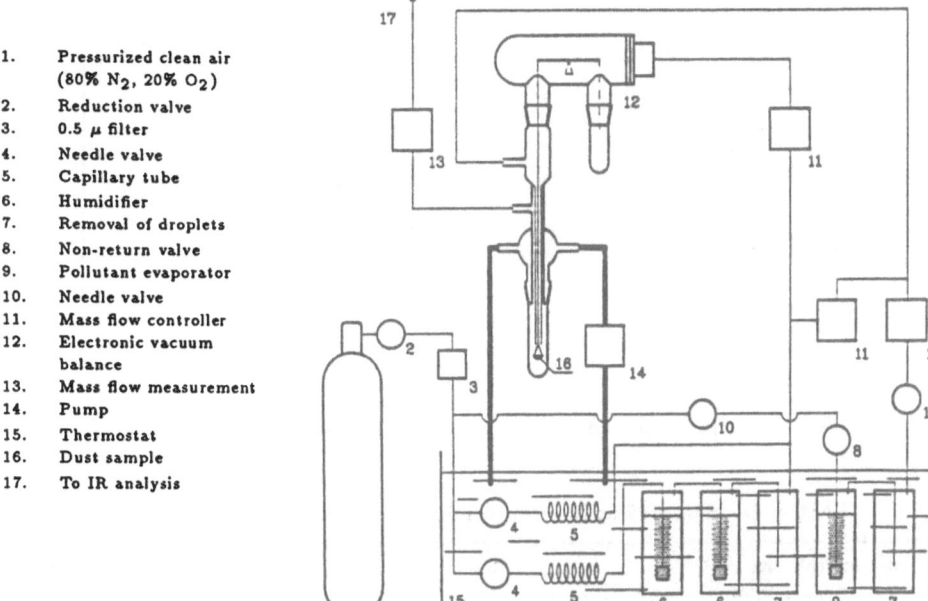

*Figure 1.* Experimental setup for measuring adsorption/desorption on dust.

## 4.1. EXPERIMENTS

The dust samples were collected in different office buildings by a specially constructed cyclone. A microscopic analysis [2] showed that the samples consisted of three major components in varying amounts: human fragments, fibres and particles.

The experimental setup (Figure 1) is based on gravimetric analysis with special reference to the possibility of changing parameters such as temperature, relative humidity, air velocity and pollutant concentration independently of one another.

The experiments consisted of three phases, each lasting 5–7 days: A conditioning phase, where the dust sample was brought to equilibrium with clean, moist air, an adsorption phase where the sample was brought to equilibrium with moist air containing the organic pollutant, and finally a desorption phase where the sample was allowed to give off the adsorbate to clean, moist air of the same composition as in the conditioning phase.

Experimental results for adsorption/desorption (20°C, 105 kPa, air velocity 0.05 m/s) of a gaseous mixture of synthetic air (nitrogen+oxygen), 2-butoxyethanol and water on different dust samples are shown in Table 1 (adsorption equilibria) and in Figure 1 (adsorption kinetics).

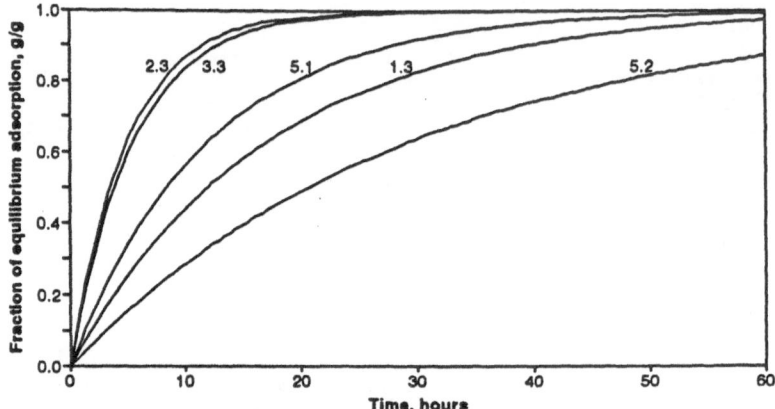

*Figure 2.*  Measured adsorption kinetics for 2-butoxyethanol on dust.

TABLE 1.  Measured adsorption equilibria for 2-butoxyethanol on dust.

| Dust sample no. | 2.3 | 1.3 | 3.3 | 5.1 | 5.2 |
|---|---|---|---|---|---|
| Gas concentration, mg pollutant/$m^3$ | 55.5 | 68.9 | 50.5 | 46.9 | 49.8 |
| Relative humidity, % | 50 | 50 | 47 | 51 | 49 |
| Weight, mg (dust+ads. water at eq.) | 628.22 | 1197.49 | 643.97 | 940.16 | 408.55 |
| $\dfrac{\text{mg pollutant}}{\text{g (dry sample+ads. water)}}$ at eq. | 0.14 | 0.25 | 0.23 | 0.28 | 0.17 |
| Surface area of dry sample [2], $m^2$/g | 0.6 | 0.7 | 0.6 | 0.7 | 0.9 |

## 4.2. DISCUSSION

Since the indoor climate in a building depends on a large number of parameters, it is desirable to simulate mass transfer rates by means of a model which is so simple as possible. Consequently, preliminary theoretical calculations were performed by means of a mathematical model for adsorption/desorption processes in a chamber with air exchange, assuming no interaction between adsorbates as well as linear adsorption/desorption equilibria. The measured desorption from the dust samples was slower than predicted by the model. Better models for adsorption/desorption must be developed and validated for more materials and pollutants.

## 5. References

[1] Kjaer, U.D. (1992) *Adsorption/Desorption on Dust from the Indoor Climate.* M.S. Thesis, Department of Chemical Engineering, Technical University of Denmark and Danish Building Research Institute.
[2] Gyntelberg, F., Suadicani, P., Nielsen, J.W., Skov, P., Valbjoern, O., Nielsen, P.A., Schneider, T., Joergensen, O., Wolkoff, P., Wilkins, C.K., Gravesen, S. and Norn, S. (1994) Dust and the Sick Building Syndrome, *Indoor Air* 4, 223–238.

# THEORETICAL DESCRIPTION OF ION ADSORPTION

# AT THE METAL OXIDE/ELECTROLYTE SOLUTION

# BASED ON THE FOUR LAYER MODEL

R. CHARMAS
*Department of Theoretical Chemistry*
*Maria Curie Skłodowska University*
*Pl. Marii Curie Skłodowskiej 3, 20-031 Lublin, Poland*

## 1. Introduction

Models of surface complexation are the most frequently applied approximation describing ionic adsorption within the double electrical layer formed at the metal oxide/electrolyte solution interface. Of all the models published so far, the most frequently used is so called the Triple Layer Model (TLM) [1], which according to the up-to-date, more realistic outlook was made assuming that the surfaces of real oxides are geometrically deformed and energetically heterogeneous [2-5].

Despite being very successful, TLM requires some modifications. Some basic properties of these adsorption systems still remain unexplained. These fascinating problems are a divergence between the values of PZC (Point of Zero Charge) and IEP (Isoelectric Point). The first papers referred to the cases in which this divergence was significant. With the progress of accuracy in PZC and IEP measurements it was believed, that inequality of PZC and IEP is a fundamental feature of these adsorption systems, being more or less drastic, depending on the adsorption system. This cause as well as others lead to modification of TLM, called the Four Layer Model (FLM).

The idea of the Four Layer Model was introduced to literature by Bowden et al. [6,7] as well as in the papers by Barrow [8,9]. A new layer, (the fourth one as the name indicates, but situated as the second, next to the surface layer "0" where protons are adsorbed), was reserved first for the bivalent metal ions or anions of multiproton oxy-acids. Cations and anions of basic electrolyte were placed still in the same layer as in the Triple Layer Model.

Another view was launched by Bousse et al. [10] who presented in their paper a diagram of the Four Layer Model, in which there are ions of basic electrolyte and the potential determining ions $H^+$. They argued that anions and cations of

*J. Fraissard (ed.), Physical Adsorption: Experiment, Theory and Applications, 609-614.*
© *1997 Kluwer Academic Publishers.*

the basic electrolyte are not located in the same layer (as in TLM), but in two separate ones. A schematic picture of FLM is shown in Figure 1.

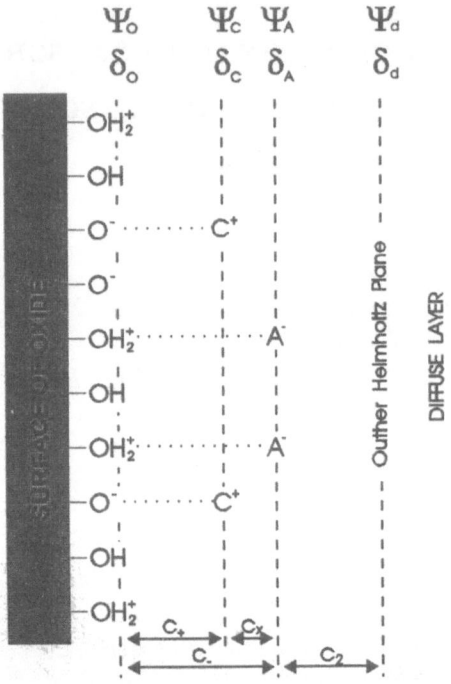

*Figure 1.* Diagrammatic presentation of the Four Layer Model of the metal oxide/electrolyte interface. $\psi_0, \delta_0$ – the surface potential and the surface charge density in the 0-plane; $\psi_C$, $\psi_A$ and $\delta_C$, $\delta_A$ – the potentials and the charges coming from the specifically adsorbed ions (cations $C^+$ and anions $A^-$) of the basic electrolyte in two different planes; $\psi_d, \delta_d$ – the diffuse layer potential and its charge; $c_+$, $c_-$, $c_x$, $c_2$ – the electrical capacitances, constant in the regions between planes.

The first rigorous thermodynamic description starting from this physical model and providing theoretical expressions for all the experimentally measured physico-chemical quantities, has been published recently by us [11]. The aim of presentation is a complete thermodynamic description based on FLM, assuming that the oxide surface is geometrically and energetically homogeneous [11,12].

## 2. Adsorption Model

The potential determining ions $H^+$, and the cations $C^+$ and anions $A^-$ of the basic electrolyte form in the complexation models (TLM and FLM) the following surface complexes: $SOH^0$, $SOH_2^+$, $SO^-C^+$ and $SOH_2^+A^-$, where S is the surface metal atom. The concentrations of these complexes on the surface are denoted by [SOH⁰], [SOH₂⁺], [SO⁻C⁺] and [SOH₂⁺A⁻], respectively. [SO⁻] is the surface concentration of the free sites (unoccupied surface oxygens). Except for protons, which are located in the potential layer $\psi_0$, in FLM the anions are situated within the layer of potential $\psi_A$, whereas the cations are within the layer of potential $\psi_C$.

According to the mass action law, the equations for the intrinsic equilibrium commonly used in the literature constants for the reactions occurring at the oxide/electrolyte interface take in FLM the form presented in Table 1, where $a_C$, $a_A$ and $a_H$ are the activities of cations, anions and protons respectively.

TABLE 1. Surface reactions and equilibrium equations in the FLM.

| Surface reaction | Equilibrium equation |
|---|---|
| $SOH_2^+ \xrightleftharpoons{K_{a1}^{int}} SOH^0 + H^+$ | $K_{a1}^{int} = \frac{(a_H)[SOH^0]}{[SOH_2^+]} \cdot \exp\left\{-\frac{e\psi_0}{kT}\right\}$ |
| $SOH^0 \xrightleftharpoons{K_{a2}^{int}} SO^- + H^+$ | $K_{a2}^{int} = \frac{(a_H)[SO^-]}{[SOH^0]} \cdot \exp\left\{-\frac{e\psi_0}{kT}\right\}$ |
| $SOH^0 + C^+ \xrightleftharpoons{{}^*K_C^{int}} SO^-C^+ + H^+$ | ${}^*K_C^{int} = \frac{[SO^-C^+](a_H)}{[SOH^0](a_C)} \exp\left\{-\frac{e(\psi_0-\psi_C)}{kT}\right\}$ |
| $SOH_2^+A^- \xrightleftharpoons{{}^*K_A^{int}} SOH^0 + H^+ + A^-$ | ${}^*K_A^{int} = \frac{[SOH^0](a_H)(a_A)}{[SOH_2^+A^-]} \exp\left\{-\frac{e(\psi_0-\psi_A)}{kT}\right\}$ |

According to Figure 1 presenting the diagram of FLM, the surface charge density, $\delta_0$, must be proportional to the sum of the concentrations of the following surface complexes:

$$\delta_0 \sim ([SOH_2^+] + [SOH_2^+A^-] - [SO^-] - [SO^-C^+]) \tag{1}$$

and the charges of the specifically adsorbed ions of the basic electrolyte in their planes, are given by:

$$\delta_C \sim ([SO^-C^+]) \qquad \text{and} \qquad \delta_A \sim ([SOH_2^+A^-]) \tag{2}$$

For the whole compact layer, there must be fulfilled the electroneutrality condition:

$$\delta_0 + \delta_C + \delta_A + \delta_d = 0 \tag{3}$$

The value of the diffuse layer charge, $\delta_d$, is given by:

$$\delta_d \sim ([SO^-] - [SOH_2^+]) \tag{4}$$

The total number of the sites capable of forming the surface complexes on the surface, $N_s$, is given by the mass action law, and is equal to:

$$N_s = ([SO^-] + [SOH^0] + [SOH_2^+] + [SOH_2^+A^-] + [SO^-C^+]) \tag{5}$$

The relationship between capacitances takes the following form,

$$\frac{1}{c_t} = \frac{1}{c_+} + \frac{1}{c_x} + \frac{1}{c_2} \tag{6}$$

where $c_t$ is the overall capacitance of the compact part of the double layer.
The relationship between the potentials and the charges within the individual electric layers is the following:

$$\psi_0 - \psi_C = \frac{\delta_0}{c_+} \qquad \psi_C - \psi_A = \frac{-(\delta_A + \delta_d)}{c_x} \qquad \psi_A - \psi_d = \frac{-\delta_d}{c_2} \tag{7}$$

Introducing next the surface coverages $\theta_i$'s by the individual surface complexes $(i = 0, +, C, A)$ and free sites $(i = -)$,

$$\theta_0 = \frac{[SOH^0]}{N_s} \quad \theta_+ = \frac{[SOH_2^+]}{N_s} \quad \theta_C = \frac{[SO^-C^+]}{N_s} \quad \theta_A = \frac{[SOH_2^+A^-]}{N_s} \quad \theta_- = \frac{[SO^-]}{N_s} \tag{8}$$

The equivalent description of the reactions leading to the adsorption of ions onto the free sites $[SO^-]$ are these presented in Table 2.

TABLE 2. Surface reactions describing the adsorption of ions onto free sites, their equilibrium constants and equilibrium equations.

| Surface reaction | Equilibrium constant | Equilibrium equation |
|---|---|---|
| $SO^- + H^+ \longleftrightarrow SOH^0$ | $K_0 = \frac{1}{K_{a2}^{int}}$ | $K_{a2}^{int} \exp\left\{\frac{e\psi_0}{kT}\right\} = \frac{(a_H)\theta_-}{\theta_0}$ |
| $SO^- + 2H^+ \longleftrightarrow SOH_2^+$ | $K_+ = \frac{1}{K_{a1}^{int} \cdot K_{a2}^{int}}$ | $K_{a1}^{int} K_{a2}^{int} \exp\left\{\frac{2e\psi_0}{kT}\right\} = \frac{(a_H)^2\theta_-}{\theta_+}$ |
| $SO^- + C^+ \longleftrightarrow SO^-C^+$ | $K_C = \frac{^*K_C^{int}}{K_{a2}^{int}}$ | $K_{a2}^{int}/{^*K_C^{int}} \exp\left\{\frac{e\psi_C}{kT}\right\} = \frac{(a_C)\theta_-}{\theta_C}$ |
| $SO^- + 2H^+ + A^- \longleftrightarrow SOH_2^+A^-$ | $K_A = \frac{1}{K_{a2}^{int} \cdot {^*K_A^{int}}}$ | $K_{a2}^{int}{^*K_A^{int}} \exp\left\{\frac{e(2\psi_0 - \psi_A)}{kT}\right\} = \frac{(a_H)^2(a_A)\theta_-}{\theta_A}$ |

The set of the nonlinear eqs in Table 2 can be transformed into an equivalent one, having the form of multicomponent Langmuir-like adsorption isotherms of ions $\theta_i$'s $(i = 0, +, A, C)$:

$$\theta_i = \frac{K_i f_i}{1 + \sum_i K_i f_i}, \qquad i = 0, +, A, C \tag{9}$$

where $f_i$'s, $(i = 0, +, A, C)$, are the following functions of activity of protons and salt ions in the equilibrium bulk electrolyte,

$$f_0 = \exp\left\{-\frac{e\psi_0}{kT} - 2.3pH\right\} \qquad f_+ = \exp\left\{-\frac{2e\psi_0}{kT} - 4.6pH\right\} \tag{10ab}$$

$$f_C = a_C \exp\left\{-\frac{e\psi_0}{kT} + \frac{e\delta_0}{kTc_+}\right\} \qquad (10c)$$

$$f_A = a_A \exp\left\{-\frac{e\psi_0}{kT} - \frac{e\delta_0}{kTc_+} - \frac{e\delta^*}{kTc_+} + \frac{e\delta^*}{kTc_-} - 4.6pH\right\} \quad \text{where} \quad \delta^* = \delta_A + \delta_d \qquad (10d)$$

The surface potential function $\psi_0(pH)$ defined in the equations for the equilibrium constants of the surface complexation reactions can be theoretically calculated from the equation, developed by Bousse and co-workers [13], the simplified linear form of which is given by,

$$\psi_0 = \frac{\beta}{\beta+1}\frac{2.3kT}{e}(PZC - pH) \qquad (11)$$

which readily shows the difference between the $\psi_0(pH)$ dependence and the Nerstian one. Only for relatively large values $\beta$, $\frac{\beta}{\beta+1} \cong 1$, the potential change corresponding to one pH unit in this equation becomes that predicted by the Nernst equation i.e. $2.3\frac{kT}{e}$ Volts for a pH unit.

The Rudzinski–Charmas criterion [3,4,11] for the common intersection point (CIP) can be applied to study relations between the intrinsic equilibrium constants of the reactions in Table 1, and the point of zero charge (PZC) for FLM.

Introducing the notation: $PZC = pH_{\delta_0=0,\psi_0=0} = -\log H$, where H is the activity of protons in the bulk solution at PZC, $pK_{ai}^{int} = -\log K_{ai}^{int}$ (i=1,2), $p^*K_i^{int} = -\log^* K_i^{int}$ (i=C,A), the relations reducing the number of the independent equilibrium constants have, for the FLM, the following form [11,12]:

$$PZC = \frac{1}{2}(pK_{a1}^{int} + pK_{a2}^{int} - \log Y^*) \qquad (12a)$$

$$PZC = \frac{1}{2}(p^*K_C^{int} + p^*K_A^{int} + \log X^*) \qquad (12b)$$

where

$$X^* = P^* + a\frac{\partial P^*}{\partial a} \qquad (13a)$$

$$Y^* = a^2\frac{\partial P^*}{\partial a} \cdot \frac{H^2}{K_{a2}^{int}*K_A^{int}} + 1 \qquad (13b)$$

and

$$P^* = \exp\left\{-\frac{eB}{kT}\frac{*K_C^{int}a}{2K_{a2}^{int} + H + 2^*K_C^{int}a}\left(\frac{1}{c_-} - \frac{1}{c_+}\right)\right\} \qquad (14a)$$

$$\frac{\partial P^*}{\partial a} = P^* \ln P^*\left(\frac{1}{a} \cdot \frac{2K_{a2}^{int} + H}{2K_{a2}^{int} + H + 2^*K_C^{int}a} + \frac{d\ln\left(\frac{1}{c_-} - \frac{1}{c_+}\right)}{da}\right) \qquad (14b)$$

and where a is the activity of the cations and anions $(a=a_C=a_A)$ of the 1:1 inert electrolyte considered here.

The procedure of the determination of the equilibrium constants $pK_{a1}^{int}$ and $p^*K_A^{int}$ from eqs 12 for a given pair of $pK_{a2}^{int}$ and $p^*K_C^{int}$ values was discussed in the papers [11,12]. The application of the Rudzinski-Charmas criterion reduces the number of the freely chosen equilibrium constants from four to two.

## 3. Conclusions

As a result some advantages of FLM compared with TLM can be pointed out [11,12]:

- The electrical capacitances are continuous functions of pH, as it should be in case of physical quantities.

- FLM predicts the dependence of the electrical capacitances on salt concentration in the solution, as shown experimentally.

- FLM predicts the difference between PZC and IEP as an essential characteristics of the adsorption at the metal oxide/electrolyte interface.

- The rate of calculations using both models is the same. It takes a few seconds using the computer of PC486 type.

## 4. References

1. Davis,J.A., James,R.O. and Leckie,J.O. (1978) *J. Colloid Inteface Sci.* **63**, 480.
2. Rudziński,W., Charmas,R. and Partyka,S. (1991) *Langmuir* **7**, 354.
3. Rudziński,W., Charmas,R., Partyka,S. and Foissy,A. (1991) *New J. Chem.* **15**, 327.
4. Rudziński,W., Charmas,R., Partyka,S., Thomas,F. and Bottero,J.Y. (1992) *Langmuir* **8**, 1154.
5. Rudziński,W., Charmas,R., Partyka,S. and Bottero,J.Y. (1993) *Langmuir* **9**, 2641.
6. Bowden,J.W., Nagarajah,S., Barrow,N.J., Posner,A.M. and Quirk,J.P. (1980) *Aust. J. Soil Res.* **185**, 49.
7. Bowden,J.W., Posner,A.M. and Quirk,J.P. (1980) in B.K.G.Theng (ed.), *Soil with Variable Charge*, New Zealand Society of Soil Science, Lower Hutt, p.147.
8. Barrow,N.J. (1985) *Adv. in Agronomy* **38**, 183.
9. Barrow,N.J. and Bowden,J.W. (1987) *J. Colloid Interface Sci.* **119**, 236.
10. Bousse,L., de Rooij,N.F. and Bergveld,P. (1991) *Surface Sci.* **135**, 479.
11. Charmas,R., Piasecki,W. and Rudziński,W. (1995) *Langmuir* **11**, 3199.
12. Charmas,R. and Piasecki,W. (1996) *Langmuir* **12**, 5458.
13. Van den Vlekkert,H., Bousse,L. de Rooij,N.F. (1988) *J. Colloid Interface Sci.* **122**, 336.

# INDEX